Animal Agriculture Biotechnology

(Biotechnologies, Policy, Planning, Reviews and Visions)

Animal Agriculture Biotechnology

(Biotechnologies, Policy, Planning, Reviews and Visions)

Author & Editor :

GHULAM MOHYUDDIN WANI

BVSC & AH (Rajasthan), MVSC (Pantnagar)
Ph.D (IVRI), Doctor Medicine Veterinary
(Hannover, Germany) ARS, Fn DAAD / ISSAPU

Patron - in - Chief
Foundation for Advancement of Science & Technology
FAST, Post Box - 461, G.P.O. Srinagar, Kashmir, India

2009

SATISH SERIAL PUBLISHING HOUSE

403, Express Tower, Commercial Complex, Azadpur, Delhi-110033 (India)
Phone : 011-27672852, Fax : 91-11-27672046
E-mail :info@satishserial.com, hkjain1975@yahoo.com
Website : www.satishserial.com

Published by :

SATISH SERIAL PUBLISHING HOUSE
403, Express Tower, Commercial Complex, Azadpur, Delhi-110033 (INDIA)
Phone : 011-27672852 Fax : 91-11-27672046
E-mail : info@satishserial.com, hkjain1975@yahoo.com

ISBN 81-89304-68-2

Typeset at: Typographiya, Delhi # 98 11 48 23 28
Printed at: Salasar Imaging Systems, Delhi-35

Foreword

Biotechnology is the advancement of biological research at molecular level. During last few decades many life saving drugs and diagnostic techniques are credited to biotechnology. I know Prof. G.M. Wani as an eminent and potent scientist whose earlier book on Embryo biotechnology in sheep and goats has acclaimed him a significant position in the field of Biotechnology. His students have been leading biotechnologists in world famous laboratories.

His experience and training in world famous institutes at Hannover, Germany, IVRI, Izatnagar has given him an acumanship which has clear reflections in his book entitled, "Animal Agriculture Biotechnology". The book discusses many techniques and advanced topics like ovarian angionesis, inplantation, foeto-material unions, ultrasonography, laparoscopy, nano technology, stem-cell technology, reproductive manipulations and futuristic visions of livestock biotechnology tools and their use in increasing livestock productivity. Nuclear techniques in use for enhancing animal production and protecting animal health have been reviewed, discussed and envisoned in vivid and clear terms. Pathological aspects of foetal biology and new concepts of invitrofoetus proposed and hypothesized indicated the way forward in animal biotechnology.

The section II agriculture productivity, agro-econ-concerns, agribusiness, agriclinics, agri-entrepreneurships, indigenous-agro-techniques, horticulture development and a vivid new reforms on Extension Education, Food security and e-extension. Who has spent about 2 decades in Extension Education has clearly spelled problems visions, phyto zoo-sanitary and global marketing concerns in this section. On the whole the book is recommended for students, scientists, professionals, NGO, institutions and policy makers not only at state but also at global level. I compliment Prof. Wani for this valuable contribution.

Prof. Anwar Alam
Vice-Chancellor
SKUAST-K, Shalimar-Kashmir

Preface

Under the changing world scenario new global mindset has emerged in the form of fast information and communication, DNA analytic modules, gene mapping with genomic information being reassessed and reconstituted in electronic chips. A new mechanism of ultra sensitive equipments and techniques have emerged to counteract new emerging diseases. A new global system of research has to cope up with these new global challenges. In the words of Prof. Charlottee Krik Baer, director of National Board on Agriculture and Natural Resources, the veterinary Research has to be remandated for protecting public health besides animals health. Thus the wise postulation of Prof. Rudolf Virchow (1821-1902) have come true a century after his death. He postulated that "All Animal Species including Homo Spines are related and that Knowledge gained in one species benefits all the others. This leads to a new concept" "one medicine, one approach". We in bringing this book fulfill his cherished vision with an addition of spiritualistic sciences and its use in future. We add to his phrase all biological life or the holistic life has a common origin (Holly Quran & other Holly Books, the Bible, Gita etc. refers), thus a new vision "One life One Cure". This integrated human-animals-plant and the environment as one entity, therefore, the title "Animal Agriculture Biotechnologies, Policy, Planning, reviews and visions. The whole book has been arranged in two broad sections.

1. Animal Biotechnology

2. Agriculture Scenario & Environment

The first section consists of 28 chapters including last chapter on references cited. The books' initiates with abstract which reflects brief topics and issues discussed in the chapter. Every chapter is a complete entity in itself. One is not bound to read the book in a systemic order. He is free to choose any chapter or even a subheading under a chapter to refresh his knowledge. Every attempt has been made to present latest and futurist view of animal-agriculture biotechnology. The author and editor of this book was the first Animal Biotcchnologist of India to have training and experience of working with Prof. H Geldermann, Willadson of Hannover and Cambridge biotechnology bases. His first discourse with micro manipulation of goat embryos in early 80's has bestowed him a place reflected in to hundreds of websites. His publication on biotechnology, ultrasonography, laparoscopy and fetal biology are references cited with honour everywhere in the global overview of science and technology. The easiest way of finding his contributions is to press his name in www.google.com and the viewers shall come across hundreds of his references and citations. Most of the chapters of this books are either web presentations or presentation of author in National and International Conferences and Seminars. To that extent this book is unique reference and taxt book in India which has been peer reviewed by thousand of scholars around the globe. Their views and suggestions have given it a shine to be one of the unique presentations to date.

The book starts with ovarian angiogenesis discussing its role on ovulation and implantation in mammals. The endocrinological events in reproductive processes have been described in chapter 2. Embryonic and foetal development with morphological and biochemical parameters have been discussed in chapter 3. The techniques of ultrasonography, laparoscopy, have been compiled in chapter 4, chapter 5 is on Reproductive Diseases and Chapter 6 follow with description and use of nano and stem cell technologies. The chapter 7 discusses Reproductive Manipulation Technology and chapter 8 is on ovarian manipulation, and chapter 9 is on concepts of gene targeting, trapping and therapeutic aspects, chapter 10, blends the modern and traditional zoo- phyto -medical practices and therapies for sustenance of productivity and food safety. Earlier chapter 3 basics on foetal morphology and biochemistry understandings shall help to understand clinical / diseases aspects of foetal and uterine interaction discussed in chapter 11 and 12 Reproductive Management, Pregnancy Diagnosis in ruminants have been presented in chapter 13 and 14.

A futuristic pragmatic plan of research on reproduction has been presented as vision 2100 in chapter 15. Wild animal damages to agrarian economy and emerging disease scenario has been reviewed." All global country reports clubbed together in chapter 16.

Nuclear Technique for welfare of protecting life and not destroying it-a nuclear technique for peace has been summed up in chapter 17. Early embryos are the basics of nuclear technique in developmental biology as such chapter 18-22 describes early embryonic development, new concepts and early embryonic pregnancy diagnosis. Reproductive Disease Management, its pathology, reproductive deformities have been discussed in chapter 23 & 25. Research on involution of uterus reviewed by a postgraduate students has been presented as such in chapter 26. The cervical alterations during pregnancy is in chapter 27. Chapter 28 sums up livestock development policy and vision 2050 presented by author in national and international interviews, seminars and workshops, all the cited literature especially in chapters with common titles with foetus embryonic development have been printed as references for more studies.

Section II on agriculture and allied polices has been described briefly in following chapters.

Chapter-29: Agriculture Productivity gains through capacity building, eco-friendly farming and Extension Education Reforms; *Chapter-30* : Success Stories of Agribusiness/Agri-clinics-tool of Agri Entrepreneurship; *Chapter-31* : Eco-Friendly Agriculture Productivity with Indigenous Techniques in Kashmir; *Chapter-32* : Employment Opportunities for Agriculture Graduates; *Chapter-33* : The Horticulture Development in Jammu and Kashmir, Past, Present, Future Research and Outreach Initiatives; *Chapter-34* : Extension Education Reformation: I-New Reformation Models; *Chapter-35* : Extension Education Reformation: II-Technology Use and Knowledge Upgradation; *Chapter-36* : Extension Education Reformation: III-Farmers Perception and Farmers School; *Chapter-37* : Capacity Building and E-Extension for Enhancing Agricultural Productivity in Indian.

Gh. Md. Wani
Author & Editor

Contents

SECTION - II
AGRICULTURE SCENARIO AND ENVIRONMENT

Section - I

ANIMAL BIOTECHNOLOGY

Chapter **1**

Mammalian Ovarian Angiogenesis and Implantation

G.M. Wani

ABSTRACT

Ovarian angiogenesis is name given to increased vascularity of the ovary during the process of ovulation. Vascular endothelial growth factors responsible for these changes have been reviewed. Mammalian embryonic deveolpment in -vitro encounters many problems beyound 8-cell stage. These problems along with new inventions in this era have been discussed.Genome activation factors reported recently have shown that addition of vitamins to the culture media helps in the development of the embryo beyond 8 –cell stage and later too. The role and mechanism of the gamete gene activation has been reviewed. Gap junctions have been researched and may play an important role in future for disease control and restriction of cell to cell infections. The important aspects of biosafety of gametes and implantation processes in mammals have been reviewed. Our own results on slaughter house materials have been referred whereever needed.

1. Ovarian Angiogenesis

Angiogenesis is a process of vascular growth. The ovarian angiogenesis increases new blood vessels in ovary. They supply nutrients and hormones to the follicles. Thus angiogenesis promotes follicular growth and corpusluteum formation (Tamanini & Ambrogi, 2004). The developing follicles have endothelial cells. They form a network in the theca layer. Growth factors mainly produced by granulose cells as VEGE (Vascular endothelial growth factor) and bFgF (basic fibroblast growth factor) act as stimuli for promoting follicular growth and antrum formation (Ziecik, 2004; xiao et al, 2001).

After ovulation, newly formed blood vessels cross basement membrane between theca and granulosa cell layers and undergo a rapid growth to form corpus luteum. The luteal vascular growth varies and angiogenesis and angioregression determine corpusluteum growth and regression. Thus in future the improvement in angiogenic development of ovary may help in reproductive performance (Tamanini and Ambrogi, 2004).

These Hemodynamic changes in ovarian and corpora Luteal walls was studied using colour Doppler ultrasonography. The flow of blood is assessed (Acosta and Miyamoto, 2004). This confirms our earlier studies using Laparoscopy for ovulation (Wani & Sahni, 1988, Wani, 1989 Sharief, 2000).

2. Embryo Block

Embryo blocks action of antibodies and becomes a homograph and not a heterograph Eight cell embryo block was very common block in mammalian embryonic development. Sheep and goat embryos did suffer from this block too in-vitro. An mammalian embryo, makes direct contact with uterine lining only for a short period and thereby thrives on it indirectly through placenta via umbilical cord. It remains protected in an immunologically protected chamber the amnionic sac. Feotal development and biochemical constituents of foetal fluids and morphological development features observed in our studies are reported . (Mufti, 1996, Mufti et al, 2000, Wani, 2004, 2005). Embryo development can be viewed in web images images. Early embryo mesoderm cells called somites differentiate into myoblasts-resulting into muscle and skeletal tissues. Various MRF (muscle regulatory factors) in foetus determine the muscular growth and were professed to be used in future research. Their exploitation would bring muscular lambs at birth. (Wani 2001). Many advanced aspects of myoblast.

Tight junctions prevent ion and water movement between cells and Na/K pumps move into the centre of the blastocyst resulting into osmotic pressure difflections and formation of blastocoel. Plasmogen and its activator results into enzymatic digestion of zona-pellucida during this process. (www.wisc.Edie/ansici-repro/Lec-18 or search Embryo block sheep Metacrawler- Microsoft Internet explorer Figs 9-16).

3. Genome Activation

Activation of embryo is detected by pronuclear formation, cleavage and blastulation. Addition of vitamins during oocyte maturation helps subsequent blastulation and embryo viability (Bormann et al, 2003). Pregnancy associated glycoproteins (PAGS) are antigens synthesized in ruminant trophoblast. They enter maternal blood stream and thus signal early pregnancy (Wani, 2005).

Invitro growth of granulosa cell complexes isolated from cryopreserved ovine ovary has been reported (Newton et al, 1999). These cells have a capacity to produce immunereactive inhibin with FSH, (Compell et all, 1996). The ovine Booroola fecundity gene (Fec B) so widely known has been linked to markers from a region of human chromosome 4q. Thus linking research in ovine or caprine reproduction for improvements or control of reproduction in humans. The VEGF factors (Vascular endothelial growth factors) characterized and expressed in ovine corpus luteum could be of immense "healing touch" value if properly innovated to use in repairs of various body injuries. Ovine FSH has been of value in developing long-term contraceptive vaccine in bonnet (Redmer etal, 1996).

4. Gap-Junction Network

Mammalian oocytes are surrounded by follicular cells. A communication system exists between these oocytes and follicular cells. This system of communication is called Gap -Junction network. This system provides transfer of nutrients and signals between follicular cells and oocytes. The proteins of connexin family form these building blocks. Oogenesis, primodial follicle activation and subsequent ovulation is the result of these and some other proteins factors. They are e-kit receptor, transforming growth factor beta (TGF beta) and bone morphogenic proteins. Progressive morphological changes in the oocytes result into structural rearrangements. These generate energy. The increased energy facilitates oocyte development and ultimately determine its capacity and competence to dominate and ovulate (Fair, 2003).

Pregnancy recognition during day 12-22 in sheep and 15-21 in goats is made on the basis of interferon tau a member of type I interferon. It is the first protein secreted by trophoectoderm of the ruminant conceptus. It blocks the uterine luteolytic mechanism and is responsible for the sustenance of corpus Luteum during early pregnancy days. It acts on the uterine epithelial cells by suppressing their ability for transcription of gene for propogation or activation of oestrogen or oxytocin receptors. It also induces several proteins responsible for embryo implantation.

5. Pre-Implantation Embryo Nutrition

The most important nutrients required by the developing mammalian preimplantation embryo are carbohydrates, amino acids which provide energy. It also prevent cellular stress. Quantification of embryo energy metabolism is thus necessary. An attempt in this direction has been shown in figures. (Jamada, 1998).

The 8-cell embryo block in development is overcome by various growth factors incorporated in culture media. The HB-EGF (heparin-binding epidermal growth factor like growth factor) improves the development of 8-cell embryo to the blastocyst stage. This factor is helpful in preimplantation embryo development. (Tamada et al, 1999).

The method of assessing preimplantation embryo viability by the self-refrencing electophysiological technique is possible. This is an non-invasive measurement. The technique measures the gradients of calcium, potassium, oxygen, and hydrogen peroxide (Gardner, 1998).

Protein, prostaglandins and steroid synthesis in caruncular and intercaruncular endometrium in sheep has been estimated. One day before attachment ie day 15 post coitus few changes exhibited, however, the dynamics of PGF_2 alpha production was effected (Findly et al, 1981).

Comparing invitro and invivo embryonic development in ruminants is possible by use of immunocytochemistry, confocal laser scanning miscroscopy and fluorescence techniques. By combination of these techniques it has been possible to demonstrate that:

1. rRNA gene activation is similar
2. Embryonic nuclear development is similar. However, embryo produced through nuclear transfers are associated with increased apostosis, in the inner cell mass of the blastocysts. Besides embryo produced in-vitro by nuclear transfers or otherwise are associated with mixoploidy. Mixoploidy means embryos with abnormal haploid or polyploidy cells besides normal diploid cells (Cinobell et al, 2003).

Use of anthelmintics like Benacil and tetramisole orally results in toxicity during pregnancy. The embryotoxicity was 100% during 9-18 days post coitus and abnormal lambs delivered by ewes dosed during pregnancy ranged from 34-57% (Grunzhinskas, 1984). Laparoscopy in sheep and goats has been used for ovulation detection, embryo aspiration and diagnosis of pregnancy.

(Wani, 1982 Wani and Buchoo, 1990, 1997, 2000) Embryos recovered through laparoscopy in Boer goats have a recovery of 62% and were economically viable. (Besenfelder et al, 1994).

6. Embryo Biosafety and Disease Risk

Scrapie, an ovine and caprine transmissle spongiform encephalopathy is widely spread among sheep populations in many European Countries. As it is known that susceptibility to scrapie is determined genetically, breeding programmes aiming at providing scrapie- resistant flocks have been established - *(Sipos et al, 2002).*

The pregnant endometrium secreted a number of neutral -to-acidic proteins which constituted, in part, the histotroph. A vitamin A-transport protein, RBP was identified in cultures, conditioned by the endometrium at days 13-21 of pregnancy. The endometrium is the only source of retinol for embryonic tissues. The uterine RBP appears to transport retinol locally toward embryonic tissues. Secretion of RBP by caprine endometrium at days 13, 15 and 21 of pregnancy suggested that retinol played an important role in conceptus development during the periattachment period of early pregnancy *(Lue et al, 2002; Young et al, 2000, 2004).*

Caprine oviduct epithelial cell (COEC) are commonly used in vitro goat embryo production protocols to stimulate early embryonic development. These feeder cells are usually collected from slaughterhouses from unknown serological status animals. In many regions of the world, tissues derived from the resource may be contaminated with CEV and the use of such material in invitro fertilization systems may contribute to transmission of this pathogen to the cultured embryos and dissemination via embryo transfer (ET). The use in invitro fertilization systems of coculture feeder cells that support efficient replication of CAEV to high titers could represent a serious risk for permanent transmission of virus to the cultured embryos and to the surrogate dam involved - *(Lamara et al, 2002).*

7. Early Pregnancy Factors

Pregnancy - associated glycoproteins (PAGs) are antigens synthesized in the superficial layers of the ruminant trophoblast. Initially, they were identified either as proteins differed into the maternal bloodstream (where they have applications in pregnancy diagnosis+PAG or as molecules binding to the LH receptor (PAG2). In this study, double radial immunodiffusion was used to test the ability of antisera raised against different PAG molecules (bovine,

ovine and caprine) to react with placental extracts from non ruminants (rabbit, cat, mouse, pig and wild pig) and ruminants (cow, ewe and goat) Placental extracts from all nonrumiants tested except rabbit reacted with anti bovine. PAG2 (anti- boPAG2). Extracts of ruminant placentas react with different antisera, confirming the expression of various PAG molecules. According to the time at which the placentas were collected (early or middle pregnancy), the reaction differed as regards the thickness, position and number of precipitation lines, suggesting that PAG expression varies as pregnancy progresses. Bos indicus and Bos Taurus placental extracts exhibited different reactions with anti-boPAG2, a single precipitation line in the former case and two lines in the latter. This suggests differential expression of boPAG2related glycoproteins in these two subspecies - (*EL-Amivi, et al, 2001*).

References & Reading Materials

Acosta,T.J. and Miyamoto,A. 2004 Vascular control of ovarian function: ovulation, corpus luteum formation and regression. *Animal Reproduction Science. 82/83: 127-140.*

Argov, N. and Sklan,D. 2004 Expression of MRNA of lipoprotein receptor related protein 8, low density lipoprotein receptor and very low density lipoprotein receptor in bovine ovarian cells during follicular development and corpus luteum formation and regression. Molecular *Reproduction and Development. 68(2): 169-175.*

Bormann,C.L; Ongeri,E.M. and Krisher,R.L.2003 The effect of vitamins during maturation of caprine oocytes on subsequent developmental potential in vitro. *Theriogenology. 59(5/6):1373-1380.*

Battye, K.M; A.J; O-Neill-Co. and Evans, G. 1991 Production of platelet-activating factor by the pre-implantation sheep embryo. *Journal of Reproduction and fertility. 93(2):507-514.*

Buchoo, B.A and Wani, G.M.(1991). Induction of Oestrus in Corriedale ewes during anoestrus. Indian J.Anim sci 60(4):407-408.

Campbell,B.K; Baird,D.T. and Webb.R. 1996 Ovine theca cells are differentially responsive to LH in serum-free culture. *J.Reproduction and Fertility Abstract Series.*

Campbell,B.K; Telfer,E.E; Webb, R. and Baird, D.T. 2000 *Ovarian autographs in sheep as a Molecular and Cellular Endocrinology.*163:131-139.

Campbell, B.K. and Baird, D.T. 2001 Inhibin A is a FsH-responsive marker of granulose cell differentiation which has both autocrine & paracrine actions in sheep. *Journal of Endocrinology. 169: 333-345.*

Campbell,B.K; Webb, R; Dobson, H. and Dobers Ka, C. 2003 *Reproduction in Domestic Ruminant. Symposium on Reproduction in Domestic Ruminants bheld in Crieff, Scotland - U.K. August 2003 PP 103-116.* Reproduction Supplement Series. 61:103-116.

Evans,A.C.2003 Characteristics of ovarian follicle development in domestic animals. *Reprod. Domest. Anim.* 38(4):240-6.

El-Amri, B; Sousa, N.M.De; Mecif,K; Desbuleux,H; Banga-Mboko,H. and Beckers, J.F. 2003 Double radial immunodiffusion as a tool to identify pregnancy-associated glycoproteins in ruminant and nonruminant placentae. *Theriogenology*. 59(5/6):1291-1301.

Fair, T. 2003 Follicular oocyte growth and acquisition of developmental competence. *Animal Reproduction Science*. 78(3/4): 203-216.

Fakowska, Podstawka, M; Wrona, Z. and Krakowski, L. 2002 Biological role of interferon tau. *Medycyna Weterynaryjna*. 58(8):568-571.

Findlay, J.K; Ackland, N; Burton, R.D; Davis, A.J; Walker, F.M.M; Walters, D.E. and Heap, R.B.1981 Protein, prostaglandin and steroid synthesis in caruncular and intercaruncular endometrium of sheep before implantation. *Journal of Reproduction and Fertility*. 62(2):361-377.

Gardner,D.K.1998 Changes in requirements and utilization of nutrients during mammalian preimplantation embryo development and their significance in embryo culture. *Theriogenology*. 49(1):83-102.

Gruzhinskas, V.A. 1984 Effect of the anthelmintics Benacil and tetramisole on prenatal and postnatal development (sheep, swine, cattle, horse). *Veterinariya-Moscow, USSR*. (11): 61-62.

Huskey, R.J. 2001. Myoblast determination flow, www.google.com

Kawate,N; Morita,N; Tsuji,M; Tamada,H; Inaba,T. and Sawada,T.2000 Roles of pulsatile release of LH in the development and maintenance of corpus luteum function in the goat. *Theriogenology*. 54(7):1133-1143.

Kanka, J; Nagai,T. and Piedrahita, J.2003 Gene expression and chromatin structure in the pre-implantation embryo. *Theriogenology*. 59(1):3-19.

Kawate,N; Tsuji,M; Tamada,H; Inaba,T. and Swada,T.2002 Inhibition of luteinizing hormone receptor expression during the development of caprine corpora lutea by administration of gonadotropin-releasing hormone antagonist. *Molecular Reproduction and Development*. 63(4):444-450.

Liu,K.H; Huang, J.C. and Lin, J.H. 2002 Identification of retional binding protein produced by caprine endometrium during periattachment period of early pregnancy. *Asian Australasian Journal of Animal Sciences*. 15(12): 1708-1713.

Lamara,A; Fieni,F; Mselli-Lakhal,L; Tainturier,D. and Chebloune,Y. 2002 Dpithelial cells from goat oviduct are highly permissive for productive infection with caprine arthritis-encephalitis virus (CAEV). *Virus-Research*. 87(1):69-77.

Mufti, A.M; Wani,N.A; Mir,M.M; Khan, M.Z. and Wani,G.M. 1997 Superovulatory response in corriedale ewes during different months of breeding seasons. *Samll Ruminant Research* 25:181-184.

Wani,G.M; Buchoo,B.A. and Wani, N.A. 1997 Use of human gonadotrophin in superovulations of Southdown sheep. *Small Ruminant Research*. 25:93-94.

Wani, G.M. 1996 *Embryo Transfer in sheep and goats. (Lead paper).* Proceedings XIII National Convention of Indian Society for the study of Animal Reproduction, GBPUA&T, Pantnagar.

Wani,G.M. 2001 Enhancement of Chevon production through reproductive Management. *Proc.IV – National Seminar of Indian Society for Sheep and Goat Production and utilization (ISSGPU), Nov. 9-11, 2001,* Bharathiar University, Coimbatore, Tamil Nadu PP 60-68.

Wani,G.M. 2003 Sustainable Livestock Productivity - *Reproductive prospectives - Lead paper B2, National Symposium on* "Strategies for Sustainable Livestock development, 17-19, Oct. 2003, FVSc & AH; SKUAST-K, Alusteng, J&K PP 11-18.

Wani, N.A; Wani,G.M. et al 2003 Use of epididymal spermatozoa from slaughtered sheep for invitro fertilization of invitro-matured oocytes. *Indian. J. Anim. Sci,* 73:143-146.

Wani, G.M and Sahni, K.L.1988 Ovulation detection by Laparoscopy in Goats. *Indian Journal of Animal Sciences* 58 (7):802-804.

Wani, G.M. 1989. Ovulation detection by laparoscopy in Sheep. *Indian Journal of Animal Sciences* 59 (12): 1527-28.

Wani G.M. 1989. Detection of foetal age in sheep and goats by ultrasonic technique. *Indian Journal of Animal Science* 59 (12): 1525-26.

Wani, G.M and Buchoo, BA (1991). Detection of induced ovulation by Laparoscopy in ewe Lambs. *Indian J.Anim.Reprod.*12(1):76-77.

Wani, N.A (1996) MVSc Thesis, SKUAST-K.

Wani, G.M.1995. Early diagnosis of pregnancy in surrogate mothers (Sheep and Goats). *National symposium on Embryo transfer and Genetic Engineering Technologies in Small Ruminants,* May 7-8, CIRG, Makhdoom, Farah, India.

Wani, G.M. 1991. Oestrus activity in goats. *World Review of Animal Production* 26:73-77.

Webb, R; Nicholas, B; Cong,J.G; Campbell, B.K; Garverik, H.A. and Armstrong, D.G. 2003 Mechanisms regulating follicular development and selection of dominant follicle. Reprod. Suppl. 61:71-90.

□□□

Chapter 2

Endocrinological Advances-Vision 2050*

G.M. Wani, Paras Wani & Gazafer Wani

Abstract

Endocrinological advances would need more research on hormone synergism, agonistic and antagonistic actions. Follicular atresia vis-in-vis intrinsic GnRH system would need more experimentation. The deterministic and permissive modes of GnRH action would be applied to reduce hyper stimulation of ovaries. The understanding of dominant follicular dynamics, shall help to harvest Follicular pool in thousands rather than in hundred at present. New biological sources of hormone from fish and other animals should find more clinical trials to popularize cheap and biological rather than chemical hormonal sources.

Sensitive disease diagnostics using RIA, REA, hormone detection kits, ultrasonography, laparoscopy, DNA finger printing, immunological and other psychometric techniques needs consideration. Gap junction and development of conceptus, foetal biological research will need emphasis by 2050. In vitro fertilization and embryo culture shall be replaced with *in-vitro* implantation and in vitro foetal development. Foetomuscular development and foeto-maternal interactions will be more vivid and clear. Reproductive care and disease prevention shall need herbal drug stimulants. Such bio-plant stimulants and medicaments need innovative research attempts. Corpora luteal bio-sensitivity and its use in augmentation of reproduction needs trials. Fertility enhancement needs identification of desired fertility objectives.

The problems and constraints in animal's reproduction have to be addressed by using modern reproductive technique to improve our traditional reproductive management. This all is not possible without basic endocrinological understanding.The advances reproductive techniques need understanding of various endocriological landmarks.

Introduction

The use of artificial insemination (AI), embryo transfer (ET) have helped in elite gene propogation. "Dolley" the super sheep is known and many more genetic clones are in offering. Today our concern is bio-safety of germplasm and the animals products. With high risk of new disease emergence, all researchers are warned to use their materials with caution. The "mad cow disease" and heavy economic losses in U.K. and Europe are eye openers for all of us. The future of gene technology lies in bio-safety and fulfillment of international zoo-sanitation regulations. Disease free zones need more emphasis than producing high yielding livestock from disease infested area. Thus safe and quality animals material (germplasm) and products is the slogan for tomorrow. Zoo-sanitation certifications will demand more trained veterinarians and other para vet staff. This necessities review of human resource development too.

For production and utilization of elite gene pool of animals, we need to know available resources, both biological and material. For example, increasing meat/milk or fibre production from our small ruminants we need to know them better. We must analyse their gene pool both qualitatively and quantitatively as was done in goats (Gall el al., 1997) using DNA techniques. Once we know merits and demerits of our gene pool, could we preserve them or utilize them for further. The methods and postulations in this direction can be reviewed as:

Endocrinological Advances

Ovarian oogenic and endocrine functions are interlocked via Hypothalamo-Hypophysial Gonadal Axis (HHGA). The cyclicity and its behavioral overt and covert phases are intrinsic to these ovarian functions. They have been thus rightly named as " the Zeitgeber or biological clock (**Baird, 1987**). The existence of feedback loops of ovarian steroids with respective gonadotrphin is a biphasic phenomenon, where higher concentrations in blood have inhibitory or suppressive effects, low concentrations have stimulatory effects (Barrac Lough, 1973), Many reproductive mechanisms hang on changes in these ovarian steroids such as labour pains (Csapo, 1981) and ovulatory failures (Snchez-criado et al., 1990).

Interestingly these hormones interact to modify or sometimes antagonize each other. For example P_4 appears to be potent antagonist of E_2 at a combined ratio of 1:1 (Heap and Flint, 1987). The synergism, agonistic and antagonistic actions of these hormone combinations are the basis for all reproduction manipulations viz Control of oestrus, ovulation, conception, gestation, fertility, infertility, contraception and or abortions. Fish and sheep ovarian extracts were tested to find out species sensitivity. It seems it is possible to tap huge fish ovarian extracts for intra and interpolation in mammalian ovary (Najar, 1999). Coming years will see more attempts to use such resources to avoid side effects from chemical and synthetic analogues of hormones.

New Hormonal Roles - Follicular Atresia

GnRH - regulates gonadotrpin release from pituitary, but its continuous and prolonged use in natural form or as its potent agonist induces an inhibitory effect on ovarian steroidogenesis, follicular development and ovulation (Kledzik et al., 1978; Srivastawa et al., 1995). GnRH receptor gene encoding is expressed in the rat ovary (Kaiser et al., 1992). A new concept prevails today which postulates that an intrinisic GnRH system competes with receptor. Thus its biological response exist within the ovary. The follicular atresia is brought by this intrinsic GnRH system (IGS) (Erickson et al., 1994). A new vision of autocrine and paracrine secretion of ovary vis-in-vis GnRH becomes evident. Future modulations in correcting reproductive failure involving hormone therapy needs rational understanding of all endocrine, autorine and related control mechanisms. Ovarectomized ewes when implanted with E_2 and P_4 implants were maintained in an artifical follicular phase. This speaks of an endocrine model compatible to a natural sequencing hormones of desired levels (Krash and evans, 1996).

Intrinisic GnRH System (IGS)

Two theoretical methods called deterministic and permissive models have been put forth to explain role of GnRH (Karsch et al., 1977).

Deterministic Model

Deterministic model envisages that a surge-like increase in GnRH is needed to induce preovulatory LH surge. This model befits animals whose LH release and ovulation is either time bound (Copulation) or signal dependant (cervical signal, rabbit and neural signal Central circadian pacemaker in rata refers).

Permissive Models

Permissive models explain GnRH in relation to ovulatory LH sruge. Here feed back of steroids exert their influence on anterior pituitary gland. This is supported by evidences that LH surge continue in situations where GnRH secretion is apparently abolished. For example in monkeys with hypothalamic lesions, or in women with hypothalamic amenorrhea or congenital GnRH deficiency, normal LH surges occur (Martin et al., 1990).

Hormone Agonists/Antagonists

Hormone antagonist or antihormone are those compounds or elements which prevent an endogenous hormones from exerting its biological effect(s) (Vanlook, 1988), These antihormone need to be safe i.e. free from side effects. The hormone antagonists in principle acts in 3 ways.

i) Inhibition of hormone synthesis

ii) Neutralising hormone action by specific antibodies.

iii) Blocking the receptors of the hormone.

Use of antagonists, agonists have demonstrated their role in ovulation control. Their low histamine-releasing activity (Fraser, 1981) support their wider use which needs further extensive studies. More details are reviewed (Najar, 1999).

Hormones-Biological Sources

The synergism, agonistic or antignostic actions of steroid and other reproductive hormones are the basis of reproductive manipulations like ovulation and fertility control, conception, gestation, fertility, infertility and contraception (Wani, 2002b). Fish and sheep ovarian extracts were tested for interclass sensitivity. It was revealed that intra and interpolation in mammalian ovaries is possible using these materials (Najar, 1999). Future attempts to utilize these cheap and easily available biological materials for ovarian manipulation may reduce the use of high cost chemical and synthetic analogues (Najar, 1999).

The biggest problem in ovarian manipulation faced in the past was due to varied superovulatory response (wani, 1984). The follicular atresia was postulated to reduce this constraint if induced adequately. As our understanding of (IGS) intrinisic GnRH system was poor then, we could not use it to overcome our problems. Now we know follicular atresia is related to this IGS (Erickson et al., 1994) and therefore, our later manipulations incorporated GnRH with comparably better success in ovarian manipulations. A new vision of autocrine and paracrine secretions of ovarian GnRH became the subject of discussion. Estrogen alone can induce ovulation but for CNS (Central Nervious System) priming progesterone combination in desired ratio is essential (Evans et al., 1995). Thus future modeulations in correcting reproductive failures needs a proper understanding of hormonal inter and intra polation with physical, chemical, biological and neuroendocrine secretions. Pheromones and other psycoemotional factors too need an analysis.

In our experiments on infertility (anoestrus) in cows of Kashmir (District Budgam) we were puzzled to see better response with low doses (300 μg) of GnRH than the higher dose of 400 ug (Shah, 1996) . We were unable to explain our results until the recent studies on permissive and deterministic models of GnRH action came to light (Karsch and Evans, 1996,Karsch et al., 1997).

Herbal-Biostimulators

Pheromones, sire introduction effects, sensory reflexus and vast other stimulants increase receptivity as well as fertility in domestic animals. Sprays and herbal cure products will be in great demand to increase or decrease fertility in coming years. The use of stimulants or repellents as sprays and scents may be in need. Herbal cures for various reproductive wastage needs further research. The use of herbal products, fruits, nuts and medicaments in ancient Kashmiri culture after delivery, need to e looked at. The Cichorium spp. Saffron, Cofula anthemodies, Prunella vulgaris, Artemisa and other products in use during post delivery period needs scientific analysis.

The ovarian sensitivity decrease soon after partuition and its resurgence after 14 days in cows, 40 odd days in women and specific periods in other animals, reveal a lot which needs exploration to phase out various sequences of reproduction.

Corpora Luteal Biosensitivity

Corpus luteum growth and development is the first earliest pregnancy signal along with maternal recognition factors discussed (Wani, 1996). The corpus luteum development and receptivity to PGF_2 alpha is now known. The ascending and descending corpora lutea are insensitive to PGF_2 alpha. What makes them insensitive and what resurges their sensitivity need exploration. May be in future, CL insensitivity might have direct effects on survival of the conceptus. Does conceptus make CL more sensitive to PGF_2 alpha? Or in other words, does the gravid CL and non-gravid CL behave similarly or differently to PGF_2 alpha. Many such questions vis-à-vis the ovarian morphology, uterine environment and behavioral patterns need to be studied and co-related to develop a manipulative and curative safe system. In some species CL or pregnancy is necessary till delivery, in others it is needed only till placenta develops. Its complete genesis and biostimulation and bioregulatory roles are obscure and needs further research.

Dominant Follicle Dynamics

The dominant follicle dynamics may be in focus for ovulation control. Dominant follicle or would be bride at each ovulation appears around day 7 of cycle in women. It occurs much earlier on day- 1 to 0 in ruminants (cow, sheep, goats). This time point may be the focus to try future selective interferences in ovulation control. Antagonist use during this phase has shown decreasing levels of estrogens in serum thereby indirectly hinting at dominant follicle alteration, as it is one of chief source of this hormone during this phase (Fraser, 1987). Dominant follicle in its sojourn to freedom and release at ovulation seems to attain some specific qualities. Why? It alone releases itself of bondage's in ovary (ovulation) and rest of its competitors are hindered through the process as atretic follicles. What make this dominant follicle to attain these qualities among a vast pool of follicles needs further study. In coming 50 years one shall possibly understand and interlock a pre-determined messenger system triggering dominance in one or few follicles destined to become the new offspring of that species. This system when decoded shall offer new opportunities in controlling ovulation both for contraception (women) or augmentation (animals). New genepool of dominant follicles from dead high yielding animal ovaries would help to propagate and multiply the elite.

Over stimulatory effect during superovulation could be avoided by use of antagonists as dominant follicles seem to evade suppression, thus follicular selection will prevent excessive stimulation and cystic follicles. Follicular fluid constituents containing occyte awaiting for biochemical signals to resume (meiosis) growth and development to ensure timely and healthy ovulation need more exploration (Lenton et al., 1988). The specific signals triggering occyte release might help in *in vitro* occyte maturation success rate.

Development of Conceptus-*In Vitro* Foetus

The lean meat production needs more studies on development of conceptus. We know that prenatal development undergoes in three stages. The early pregnancy signals skeletal & nervous tissue growth, on priority. During mid-gestation the hind and fore-limbs develop. During late gestation muscular development along with peripheral extremities including hair develop. In order to increase lean meat production, research in needed to investigate relations between hard and soft tissues. Maternal and foetal interactions needs a probe. Some basic studies (Mufti, 1996) indicated prospects of manipulating lean meat production (Wani, 2001, 2002 a) by bio-stimulators during late gestation, which may help, the production of muscular lambs/ kids at birth, besides, increasing birth weight. More details have been described (Wani, 2001, 2002a) Fig.1-5. More studies in future are needed to reveal the exact mechanism involved in critical transition of power control from maternal to embryonic genome. Application of recent techniques of reverse transcription, PCR and others may reveal the details (Schultz et al., 1990).

Slaughter House Follicle Pool

A good number of our endangered species die in wild climatic conditions, which are hard and harsh. Under temperate climatic conditions dead bodies are entrapped in snow for long. It gives a hope to revive their follicular pool. Our studies on resumption of life using follicles from slaughtered sheep and fertilizing them with epididymal semen was an pioneering attempt in this direction (Wani, 1996) (Fig 6-8). The protein content of unfertilized ovum is around 100 mg in rabbit, 27 mg in mouse (Brinister, 1967, 1971). The small genetic material needs extensive analysis to unveil vast informations in its codans and anticodans. Further ovum maturation or resumption of meiosis is initiated during birth or shortly before birth and thus the pool of follicles remains fixed at birth. Therefore, foetal ovary has to be the subject of further manipulations especially in view of its immunological insensitivity (Guraya, 1985).

Gamete Biosafety

Bacterial counts in semen in unprecedented numbers effect semen fertility (Kher, and Dholokia, 1985), The contaminational hazards have increased due to environmental pollution, water and air contamination. Above all the development of resistance of known bateria and other micro-organism to antibiotics (Gupta and Maurya, 1993). We have examined the semen of about 40 bulls maintained at the Frozen semen bank, Rambirbag, Kashmir for bacterial loads in semen and tested antibiotic sensitivity (Koul, 2002).

Presence of bacteria in large numbers in semen, effects its quality and fertility (Kher and Dholokia, 1985). The source of contamination of semen vary from inflammatory foci and lesions on penile apparatus, contaminations on equipments, appliances, diluters, extenders air and laboratory environment.

These contamination hazards have increased with environmental pollution. Many bacteria having developed resistance of routine antibiotics used (Gupta and Maurya 1993) in semen extenders has made inseminations more critical and vulnerable. More studies in this direction are needed to develop a bio-safe and infection free germplasm pool of frozen and fresh semen in future.

Reproductive Physiology

Reproduction is a complex phenomenon, starting with oogenesis as early as 135th day post mating. By 50-100th day post conception (mating) sheep and goat conceptus in-utero shows follicular growth. This follicular growth and development upto primary oocycte stage in some mammals ceases at birth (Wani, 1996). The further growth and maturation of primary occytes is triggered at puberty i.e. at the onset of first sexual activity. The release of ova i.e. ovulation follows. During this proves of folliculogeneis and ovulation, followed by fertilization and development of conceptus, the male and female gametes undergo various morphological and physical changes. They pass through various barriers, reservoirs, under the influence of bio-physical, hormonal,morpho-chemical mechanisms (Wani, 1996, Wani and Buchoo, 1990, Wani, 1984a and b, Wani et al.,1986). More in-depth studies on barrier mechanism vis-à-vis morphological changes during different hormonal milieu is needed.

Foetal Muscular Development

Genetic, hormonal and environmental factors influence foetal growth and development. The proper in-utero maturation and growth of foetus signify proper care and management of gravid nannies. The pre-natal development of foetus is exhibited at birth by birth weight. The head and skeletal tissues grow and mature earlier, followed by fore and hind limbs. The muscular development of foetus occurs during late gestation. Thus, during early pregnancy nervous and skeletal tissues are having priority for nutritive requirements in-utero(Musti, 1996). Maternal nutrition is important for growth and development of meat in off spring in-utero. Muscular growth stimulants may thus help production of quality meat lambs or kinds in future.

Foeto-Maternal Interaction

The role of endometrial stromal cells during pre-implantation stage needs further research. Some studies on ultrastructure of endometrium (Brinsfield et al., 1974) are still not sufficient to answer the questions. What is the role of immune cells and endometrial cells in building of the immunological barrier to protect the foetus. Foetal biology is abarren field of research in the world.

After few initial steps by Dr. Moris Yound in UK and D. Srivastava in India, only a scattered information is available on the subject (Wani and Buchoo, 1990; Mufti, 1996). Much is still to be unveiled by 2050. The placental hormone influence mammary development and milk production of the dam. This needs further research by 2050. Foetal membranes serve as Chief source of nutritive exchanges. The arrangement of cotyledons, caruncles, positioning of single and twin conceptus, indicate complex mechanisms with common roots. (Mufti et al., 2000) These mechanisms of foeto-maternal interactions and relationships emphasize proper foetal number diagnosis and corresponding materno-nutritional management strategies.

Small Ruminant Reproduction

There is a need of ovulation, pregnancy and foetal number detection by methods described (Wani, 1981, 1982; Wani and Sahni, 1980, 1981; Wani and Buchoo, 1991, 1993). These reports demonstrated successfully results ofr ovulation detection, multiple ovulation predictions, early pregnancy and foetal numbers. The induction of multiple ovulation, in-vitro fertilizations, embryo collections, embryo culture, conservation, micromanipulation, splitting, cloning and sexing are some of the reproductive high-tech applications used to increase reproduction rates (Wani 1996; Wani et al., 1990; Wani et al., 2000 a,b; Buchoo et al., 2000). These techniques have but a little application under field conditions in small ruminants especially under extensive management system. However, their use has already revolutionized propagation of elite germplasm and can be used under buck mother farms to produce high quality sires.

The futuristic attempts in this direction should include in vitro post implantation and foetal development experiments. These future manipulations at organogenesis level may help to induce desired quality changes in offspring so that dressing percentages could be increased from present 40-50 per cent to 55-60 per cent or more.

Reproductive Management

The reproductive management strategies to enhance meat production could safely be divided into two kinds:

i. Traditional management: Animal are still managed under zero- input/ migratory /tribal or extensive management system. The application of above described methods will be limited. They may indirectly help in this system by better sire availability.

ii. Modern management: Zero-grazing, stall-fed or semi-intensive, commercial goat farming system may use the above methods more and more to produce high quality germplasm. They may use variety of improved reproductive technique or reported (Wani,1996; Wani et al., 200a; Wani et al., 1998; Mufti et al., 1997; Wani and Sahani, 1981; Wani et al., 1986, 1987, 1988; Wani and Buchoo, 1991; Buchoo and Wani, 1991; Wani, 1989, Wani, 1995; Wani, 1984a and b; Wani and Sahani, 1980a).

Enhancement of Fertility

Desired fertility objectives have been discussed (Wani, 2001). Increased prolificacy, early rebreeding, safe/clean parturition, set-time animal crop, elite gene introduction are some of the desired fertility enhancement goals. The constraints in achieving desired objectives in small ruminant livestock sector are the problems of inbreeding, reproductive losses (Wani, et al., 1980; 1981; 1994), pre-pubertal mating, indiscriminate or illicit mating, venereal diseases, nutritional and other stresses. These above referred factors reduce fertility status of our ruminants especially under transhumane migratory systems.

These constraints mentioned could be overcomed by adopting enhancement strategies (Wani, 2001), which include exchange of sires by tribal, Gujjar, Bakarwals and Chopans to reduce inbreeding. This could be ensured through legislature or forced castrations and introduction of sires from outside the flock. The pregnancy detection methods described (Wani, 1981) and later on used extensive in J&K under farm condition (Wani et al., 1998; Wani, 1989) needs adoption as a routine in each flock for better care of offspring and dam.

Nutritional Bio-Conversion

The nutritional status of dams before mating could be improved by grazing them in pastures having 4-5 cm sward height herbage, yielding 1500kgs and above of dry matter per hectare . Such pasturing shall increase ovulation. Reduced sward height pastures needs supplementation and introduction of nutritional block at least 4-5 weeks before mating. Thus proper feeding 100 days prior to mating and parturition may help in healthy and heavy weight lamb/kid crop.

The pasture nutrient analysis and its supplementation as per requirement may help to avoid losses due to such deficiencies as cobalt, iron, phosphorus etc. which effect herbage and through herbage the animal health. The sire fertility test and use of disease free semen is the foremost requirement towards increased productivity from small ruminants.

The nutrition plays important role in maintaining fertility status of animals. With decrease in land for fodder cultivation, alternate sources and technologies to utilize agricultural waste for animal food is needed. The biotechnological intervention in converting crop residues and other agri-horticulture waste into animal feed needs further research.

Conclusion

Critical gaps identified are poor-zoo-sanitary conditions and sufficient human resources. Both these area need emphasis. Development and application of accurate, sensitive and timely disease diagnosis and surveillance techniques especially in the field of animal reproduction should invite more research proposals in future. This will ensure biosafe livestocks and germplasm. Gene propogation of elite germplasm should ensure increase in per animal production. This would means more production from less livestock, thereby relieving soil, water and environmental pressures.

Endocrinological advances should aims at replacing chemical hormones with biological hormones, which could be cost-effective and without side effects. The use of fish, sheep and other ovarian extracts need further research. Follicular atresia, intrinsic GnRH system, hormone agonists and antagonists should attract more funds and research attempts. The use of herbal stimulators and medicaments in caring and prevention of reproductive diseases and incapacities needs immediate attention. More innovative approaches using these indigenously available materials need to be prioritized.

Copora lutea biosensitivity, development of conceptus, follicular dynamics, gamete safety procedures, foeto-maternal interaction, stromal cell studies would be future areas of frontline researches in animal reproduction.

A policy in view of majority 85% of our livestock production systems being small holder enterprises. Research and extension establishments and NGO in animal reproduction needs reorganization to serve as a protective armour around small producers and farm women.

Blending of advanced technology available as embryo transfer, AI, blastomer splitting, DNA micro-injections, DNA finger prints, gene mapping, DNA hybridization, ovulation control, in vitro fertilization, foetal monitoring, ultrasonography with traditional modes of livestock production system needs further innovative attempts. This should aim to further fortify our cost-efficient livestock productivity.

Hitherto untapped areas of wild and zoo animal conservation, protection and propagation, pet animal care, wildlife retrieval and referral clinics are the need of the hour. Nutritional requirement during foetal life, prenatal and neonatal life needs more emphasis along with care and management of dams during mating, pregnancy and parturition. Increased prolificacy, early rebreeding, set-time parturitions are some of the desired fertility enhancements goals. The problems of inbreeding, reproductive losses, pre-pubertal and illicit mating, nutritional stresses could be overcome by adopting fertility enhancement procedures like sire exchanges, pregnancy detection, better nutrition before mating, during pregnancy and after parturition.

Extension and information technology missions for animal reproduction are to be intensified especially in rural India. Scientist Industry interaction, technology interaction under rural sectors and linkages between research, extension, NGO and farmers needs priority by 2050.

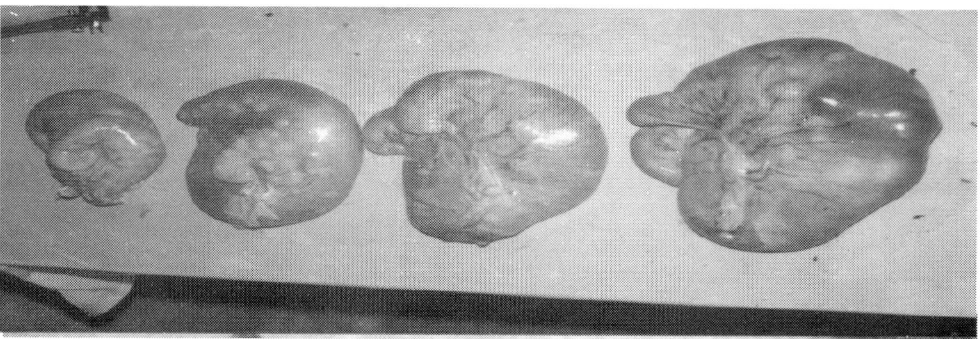

Fig. 1 : Gravid uterus during different stages or gestation (Classified arbitrarity as desc page 48)

Fig. 1 :

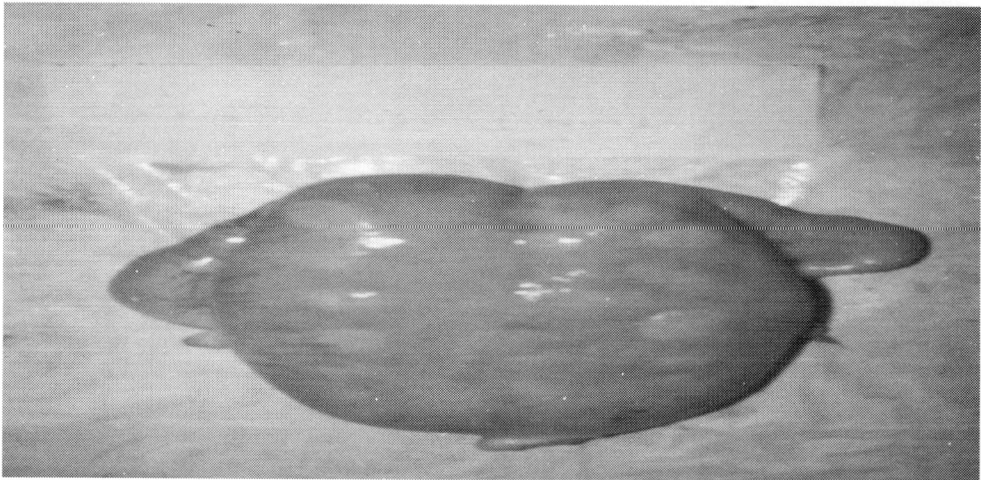

Fig. 2 : Gross appearance of twin pregnancy

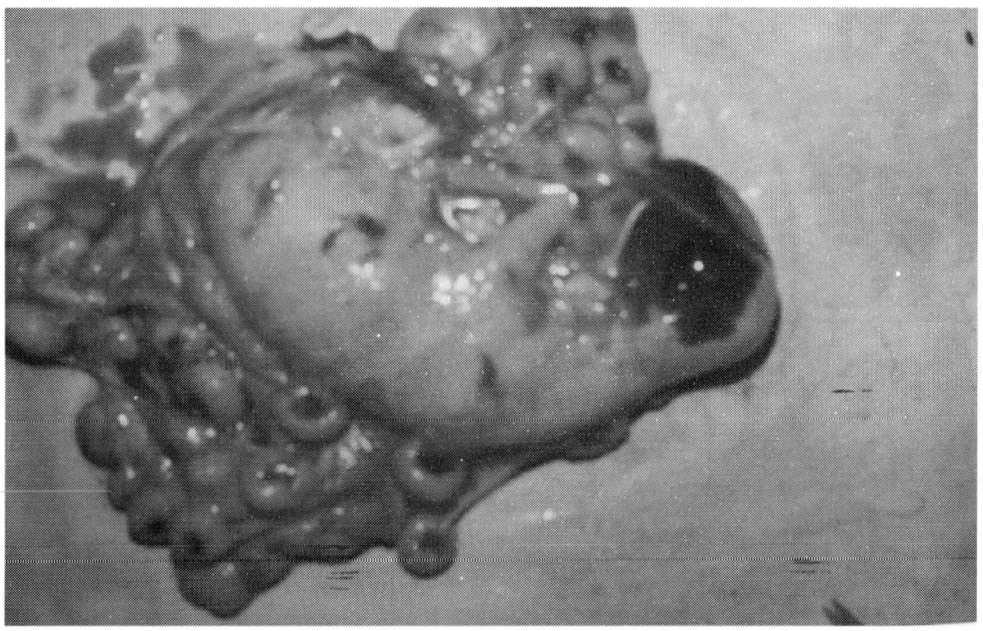

Fig. 3 : Intact ambiotic Sac in single ovine foetus

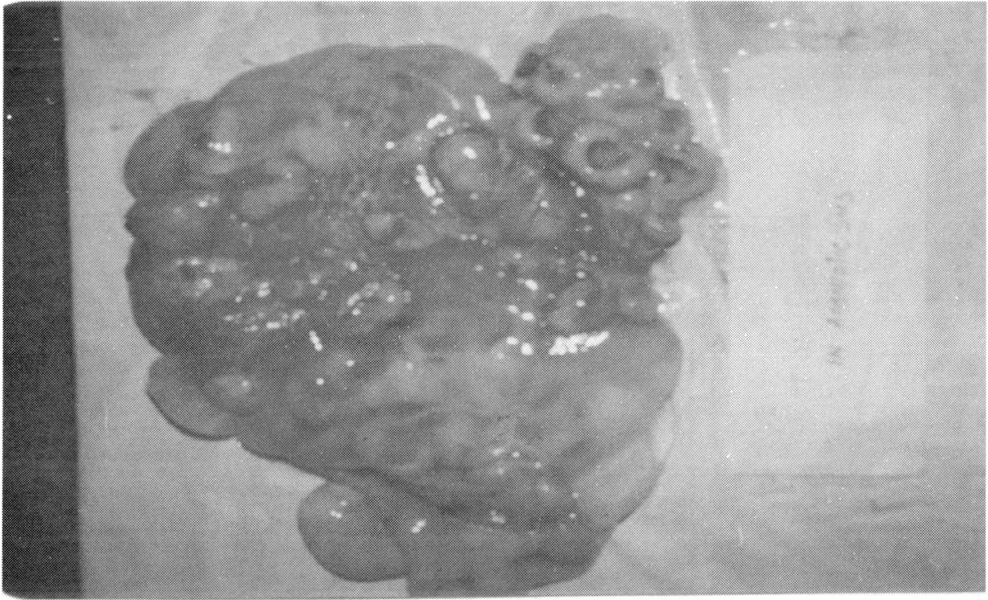

Fig. 4 : Arrangement of cotylredons in twin pregnancy.

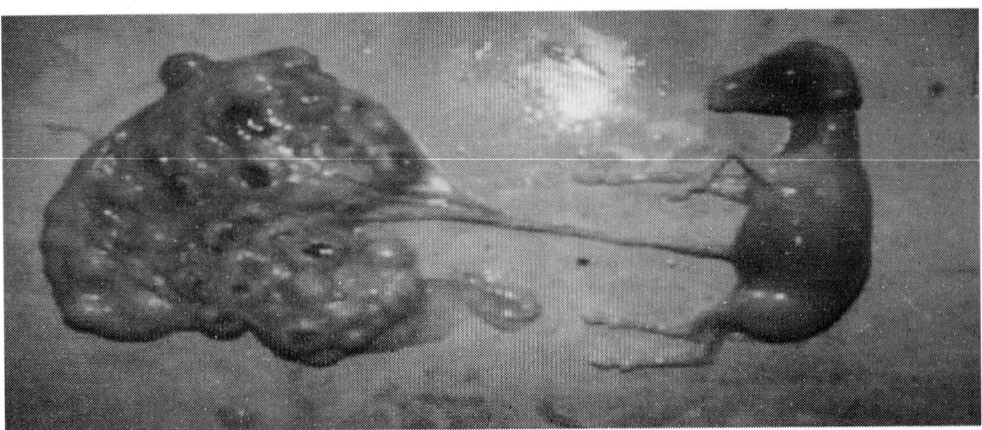

Fig. 5 : Foetus at stage 1 (30-60 days) of gestation with intact cotyledons

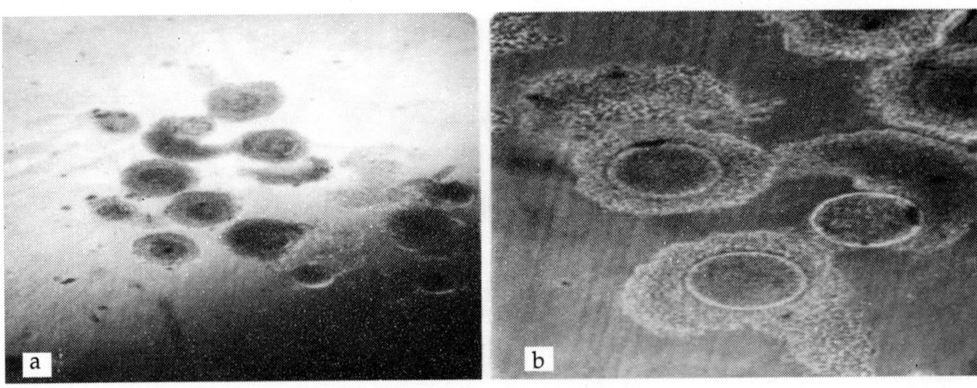

Fig. 6 : a. A pool of good and fair occytes before insubation.
b. Expansion of cumulus cells after incubation for 24 h.

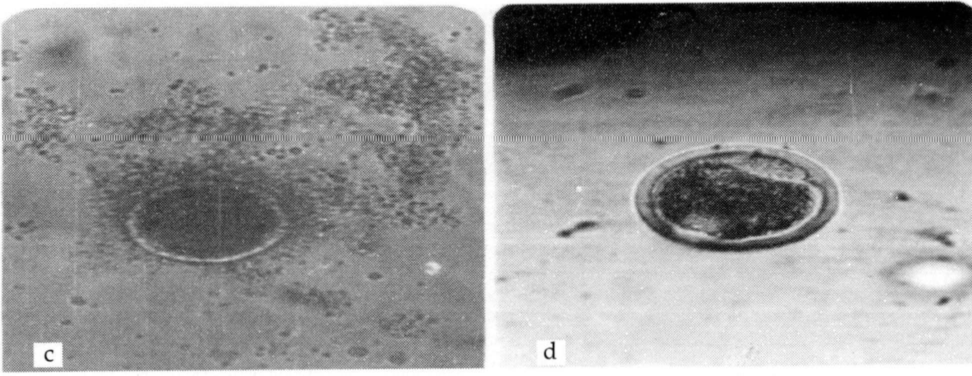

Fig. 7 : c. & d. Extrusion of first polar body in an occytoc after incubation for 24 h. The occytes has been compressred between a slide and cover slip.

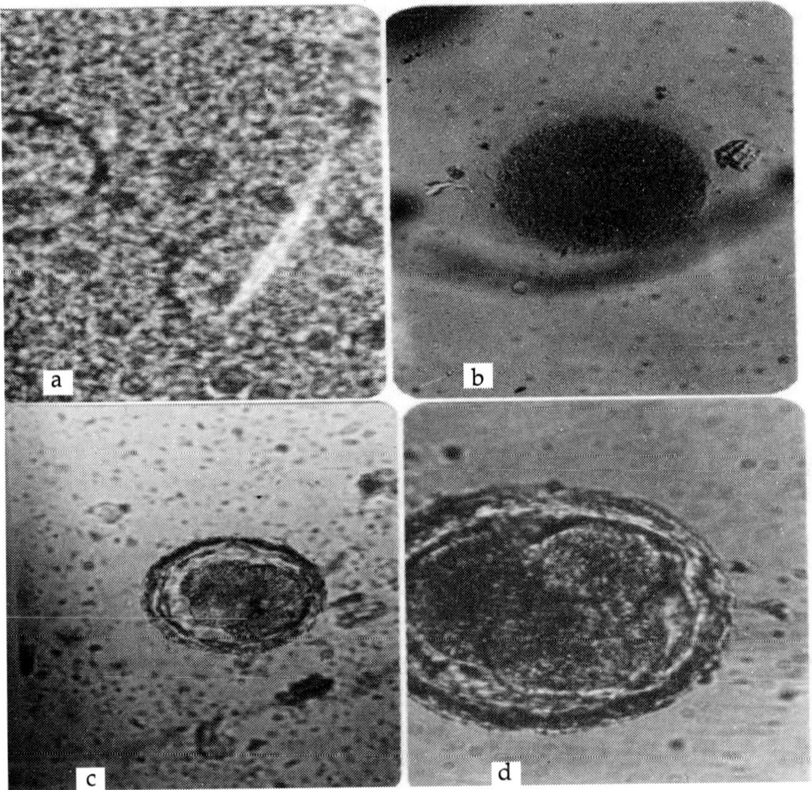

Fig. 8 : a & b. Formation of male and female pronuclei in *in vitro* matured ancytes 18 h after insemination.
c & d. Multicellular embryos obtained after IVm-IVF procedures.

References

Barraclough, Charles A. 1973. Sex steroid regulation of reproductive neuroendocrine processes.In :Handbook of Physiology, Section 7-Endocrinology Vol.II-Female Reproductive system, Part ! American Physiological Society, Washington, printed in USA by Waverly press, Inc., Baltimore, Maryland-21202.

Boshier, D.P. Fariclough, R.J. and Holloway, H.1987. *J. Reproduction Fertility, 79:569-573*

Brinsfield, T.H., Fisher, M and Hawak, H.W.1974. *J.Anim.Sci.39:565-574.*

Buchoo, B.A. and Wani, G.M.1991.Induction of oestrus in Corriedale ewes during anoestrus. *Indian Journal of Animal Science, 60(4) : 407-408.*

Buchoo, B.A.,Wani,N.A.,Khan,M.Z and Wani, G.M.2000.Laparoscopic detection of Ovulation and pregnancy in different sheep breeds. *Indian Journal of Animal Sciences 70(2):61-62.*

Csapo, A.I. (1981). Force of labor in :Principles and practice of Obstetrics and Perinatology. Vol.2 (L.Iffy and MA Kaminetzky, eds.), pp.761-799,*John Wiley and Sons, New York.*

Erickson, G.E.,Lid, Sadrkhanloo R.; Liu X-J, Shimasaki,S. and Ling, N. (1994). Extrapituitary actions of gonadotropin releasing hormone in stitulation of insulin-like growth factor-binding protein and atresia. *Endocrinology 135:1365-1372.*

Evans, N.P., Dahl, G.E., Maugar, D. and Karsch, F.J.1995. Estradiol induces both qualitative and quantitative changes in the patterns of GnRH secretion during the pre-surge period in the ewes. *Endocrinology 136: 16-3-1609.*

Fraser, H.M. (1981). Effect of oestrogen on gonadotrophin release in stumptailed monkeys (Macaca arctoides) treated chronically with an agonist analogue of LHRH. *Journal of Endocrinology 91:525-530.*

Fraser, H.M. (1987). LHRH antagonists and female reproductive function. In: LHRH and its analogs: Contraceptive and Therapeutic Applications. Part 2 (BM Vickery and JJ Nestor Jr.eds.) pp.227-244,M.T.P. press, Lancester.

Gall, C., Gelderman, H. and Wani, G.M. 1997. Analysis of genetic resources of goat breeds in Kashmir. Final Report of Research Co-operation between SKUAST and University of Hohenheim, Germany.

Gupta, R.S. and Maurya, S.N. 1993.Microbial flora of bovine semen and their antibiotic sensitivity pattern. *Indian Journal of Animal Reproduction, 14(3):107-108.*

Guraya, S.S. (1985). Biology of ovarian follicles in mammals, *Springer-Verlag (Publishers),Berlin.*

Heape, R.B., Wang, M.W.,Whyte, A. and Taussing, M.J. 1990. In:Preoceedings of the international Conference on Frontiers in reproductive Physiology, AIIMS, New delhi,8-10 November,1990 L17.

Kaiser, H.B.; Zhao, D.; Cardona, G.R, and Chin, W.W.(1992). Isolation and characterization of cDNAs encoding the rat pituitary gonadotropin-releasing hormone receptor. *Biochemical and Biophysical Research Communications 189:1645-1652,*

Karsch,Fred J.; Bowen, Jennifer M.; Caraty, Alain; Evans, Neil P, and Moenter, Suzanne M.(1997) Gonadotropin-Releasing Hormone Requirement for Oculation. Ovulation. *Biology of Reproduction 56:303-309.*

Karsch, Fred, J. and Evans, Neil P.(1996) Feedback actions of estradiol on GnRH secretion during the follicular phase of the estrous cycle. *Acta Neurobiol. Exp. 56:715-725.*

Kher, H.N and Dholakia, P.M. 1985. Bacteriological studies of bovine semen. *Indian Journal of Animal Reproduction, 5(1):78-84.*

Kledzik, G.S.; Cusan, L.; Auclair, C.; Kelly, P.A. and Labrie F. (1978). Inhibitory effects of luteinizing hormone releasing hormone agonist on rat ovarian LH and follicle stimulating hormone receptor levels during pregnancy. *Fertility and sterility*, 29: 560-564.

Koul, N.A.2002 Bacterial flora of bovine semen and their antibiotic sensitivity pattern. M.V.Sc thesis, SK. University of Agricultural Sciences & Technologyof Kashmir, Srinagar-191 121.

Lenton, E.A.; King, H.; Thomas, E.J.; Smith, S.K. MeLachlan, R.I.; MacNeil, S. and Cooke, I.D. (1988). The endocrine environment of the human oocyte. *J.Reprod.Fert.82:827-841.*

Martin, K.; Santoro, N.; Hall J.; Filicori, M.; Wurman, M. and Crowelery, W.F. Jr. (1990). Management of oculatory pulsatile gonadotropin-releasing hormone. *J. Clin Endorcinol and Metab. 17: 1081A-1081G.*

Mufti, A.M.1996.Studies on morphological characters and some metabolic constituents at different stages of development of ovine foetus. M.V.Sc. thesis, Sher-e-Kashmir University of Agricultural Sciences and Technology, Shalimar, Srinagar.

Mufti, A.M., Wani, N.A.; Wani, G.M., Mir, M.M. and Khan, M.Z,1997 Superovulatory response in sheep during different months of breeding season.*Small Ruminent Research,* 25:181-184.

Najar, A.M.(1999). Effects of fish ovarian extracts on mammalian ovary, Ph.D. Thesis of University of Kashmir, Srinagar-190006.

Sanchez-Criado, J.E.; Bellido, C.; Galiot, F.; Lopez, F.J. and Gaytan, F. (1990). A Possible dual mechanism of the anovulatory action of antiprogesterone RU 486 in the rate. *Biology of Reproduction 42:877-886.*

Schultz, G.A., hahnel, Ann., Watson, A.J and Hojan, A. 1990. In:Proceedings of the international Conference on Frontiers in Reproductive Physiology AIIMS, New delhi, 8-10 November, 1990. L20.

Shah, N.A. 1996. Effect of gonadotrophin releasing hormone (GnRH) on anoestrous crossbred cows under temperate climatic conditions. MVSc Thesis submitted to Sher-e-Kashmir University of Agricultural Sciences & Technology, Srinagar.

Srivastawa, R.K.; Krishna, A. and Sridaran R. (1995). Ovarian development, steroidogeness and gonadotrophin secretion in response of long-term treatment with a gonadotrophin-releasing hormone agonist in the rat. *Journal of endocrinology 146:349-357.*

Van Look, P.F.A. (1988). Hormone fertility reguation at the crossroads. A short review of the potential of hormone antagonists . In Proceedings of a symposium on the Use of Hormone Antagonists for Fertility Regulation organized by the Indian Society for the Study of Reproduction and Fertility and the World Health Organisation's Special Programme of Research, Development and Research Training in Human Reproduction on November 4, 1988 at Mumbai, India. Edited by Chander P.Puri and Paul, F.A.Van Look, Printed by Good Print, Dr.Ambekar Road, Parel Mumbai-400012 (India).

Wani, G.M. 1998.Ultrasonic pregnancy diagnosis in Sheep and goats. *World Review of Animal production, 17:43-48.*

Wani, G.M. 1982. Laparoscopy in farm animals. *World review of Animal Production, 18: 7-13.*

Wani, G.M. 1984a. Vet suche Zur Embryo Production and Kultur bei Ziege. Dr. Med.Vet Dissertation, Tieraertzlichen Hochschule, Hannover, Germany.

Wani, G.M. 1984b. Investigations on ovarian activity, ovulation and pregnancy diagnosis in sheep and goats. Ph.D. Thesis, IVRI, Rohikhand University, Indian Veterinary Research Institute, Izatnagar, U.P.

Wani, G.M. 1986. Embryo transfer in goats. 6th National Congress on Animal Reproduction, Guwhati 28-30th November, 1986.

Wani, G.M. 1989.Detection of foetal age in sheep and goats by ultrasonic technique. *Indian Journal of Animal Science, 59(12):525-30.*

Wani, G.M. 1989. Detection of foetal age in sheep and goats by ultransonic technique. *Indian Journal of Animal Science, 59(12):1525-26.*

Wani, G.M. 1989. Ovulation detection by laparoscopy in goats. *Indian Journal of Animal Science. 59(12):1527-28.*

Wani, G.M. 1995. Early diagnosis of pregnancy in surrogate mothers. National symposium on embryo transfer and genetic engineering technologies in small ruminants. May 7-8, 1995, Makhdoo, Farrah. India.

Wani, G.M. 1996. Embryo biotechnology in sheep and goats. Published by Valley Book House, Srinagar- 190006.

Wani, G.M. 2001. Enhancement of chevon production through reproductive management. In: Proceedings of IV National Seminar of Indian Society for Sheep and Goat Production & Utilization (ISSGPU), held at Bharathiar University, Coimbatore from November 9-11, 2001.pp.60-68.

Wani, G.M. 2002. Goat Husbandry/multipurpose management. Encyclopedia of Dairy Sciences. Edited by Roginishi, Hubert, Enquay, Jhon W. and Fox patrict, Academic Press, London, NW 17B4MS192.

Wani, G.M. and Buchoo, B.A.1990. laparoscopic view of ovine conceptus. *Indian Journal of Animal Science, 60(5):564-568.*

Wani G.M. and Buchoo, B.A. 1991. detection of induced ovulation by laparoscopy in ewe lambs. *Indian Journal of Animal reproduction, 12(1):76-77.*

Wani, G.M. and Sahni, K.L. 1980. Ultrasonic pregnancy diagnosis in sheep and goats. 2nd *All Indian Symposium on Animal Reproduction, Banglore, pp. 1-5.*

Wani, G.m. gelderman, H. and Hahn, J. 1986. *Brunstsychronization bei Ziege. Zuchthygiene, 21: 1-2.*

Wani, G.M., Gelderman, H. and Hahn, J. 1990. Superovulations during early luteal phase in goats. *World Review of Animal Production, 29(2):41-43.*

Wani, G.M., Risam, K.S. and Nowshahri, M.A. 1988. Effect of oestrus synchronization on lambing in corriedale ewes. *Indian Journal of Animal Science, 58(7):800-801.*

Wani, G.M. 1989.Plasma progesterone concentrations during oestrus cycle in goats. *Indian Journal of Animal Reproduction, 19(1):44-46.*

Wani, G.M. Sadat Rehman and Shah Masood 2000. Higher Education and Human resource Development: Knowledge Perspectives in Current Millennium. Paper presented at the Invernational Conference on Higher Education for Human development, organized by Association of Indian Universities at New Delhi on February6 22-24, 2000.pp.239-247.

Wani, G.M., Sinha, N.K. and Khan, B.U. 1987. Oestrus synchronization with progestagens in Muzzaffarnagri ewes. *Indian journal of Animal Science. 57(12)*

Wani, G.M., Sinha, N.K. and Sahni, K.L, 1980. Haematic type of foetal mummification Muzaffernagri ewes. *Indian Veterinary Medical Journal, 4:46-47.*

Wani, G.M, Sinha, N.K. and Sahni, K.L. 1980. Note on gestational oestrus in the Muzzafernagri ewes. *Indian Journal of Animal Science, 50(11):1009.*

Wani, G.M and Sahni, K.L. 1980. An ultrasonic technique for detecting pregnancy in sheep and goats. *Veterinary Research Bulletin, 3:35-37.*

Wani, G.M., Sinha, N.k. and Sahni, K.l. 1980. Note on seasonality of breeding behaviour in Jamunapari goats, *Indian Journal of Animal Science, 50:1153-55.*

Wani, G.M., Sinha, N.k. and Sahni, K.L. 1981. Incidence of abortionsand stillbirths in Muzzaffarnagri and crossbred ewes. *Indian Veterinary Journal, 58:558-561.*

Wani, G.M., Tripathi, S.S. and Saxana, V.B. 1980. Sialoproteins in the cervicaql mucus of repeat breeding cows during oestrus. *Indian Veterinary Medical Journal, 4:49-53.*

Wani, N.A., Wani, G.M., Mir, M.M. and Beg, M.A. 1998. Ultrasonic pregnancy diagnosisin Gaddi goats. *Small Ruminant Research, 29: 239-240.*

Wani, G.m. (2002b). Advances in animal Reproduction in: Proceedings of NationalSeminar on "*Recent Research Trends in Life Sciences*", (August 26-28, 2002) organized by PG Department of Envirornmental Science and Centre of Research for Development, The University of Kashmir, Srinagar- 190006.

❏❏❏

Chapter 3

Morphological & Biochemical Changes in Genital Organs During Embryonic and Foetal Development

G.M. Wani & A. Khatoon

Abstract

The blood of mother and foetus never mixes freely in ovine or caprine foetus. The fertilized zygote develops into a morula which decends and reaches uterus in 3-4 days. The presence of Corpus Luteum Haemorrhagicum (CLH) and its subsequent growth ensures progesterone secretion.

Endometrial receptivity was morphologically documented in this study. The solid tooth shaped projections of non-pregnant uterus change into mucin coated cup-shaped caruncles. Concomitant to these changes blastocysts elongates. Within elongated blastocysts cotyledons develop which fix into caruncles to form the placenta.

The unique feature presented in this paper is recognition of non-self (foetus) by self (endometrium).We present here the following observations"

1. The caruncles in non-pregnant uterine horns consist of solid projection arranged in rows. They are red and upright ville-like. No Corpus Luteum Haemorrhagicum (CLH) was present.
2. In the presence of CLH these caruncles change in form, colour and structure. They are cup-shaped , mucin coated, whitish in colour and give an impression as if a whitish cover or membrane overwrap them.

3. In the meantime blastocyst elongates and covers the whole uterine horn. Within the blastocysts or now we can call it an elongated ballon shaped foetus floating over hypertrophied caruncles. Within this developing foetal sac small round, red shaped ball develop which are called cotyledons. These cotyledons further fix themselves into cup-shaped caruncles. The formation of placentomes begins both within foetal membrances and endometruim cover over cotyledons and caruncles respectively, before union. Thus an indirect attachment of cotyledons and caruncles occurs which is separated by different membranous coverings described.

The mechanism of such attachment and a system of manifold protective covers prevent direct contact of foetus and maternal tissues. The minute details of attachments are supported by histological nuclear unions too. To the best of our knowledge, it is the first finding of the kind being published in literature.

Gross Appearance of Genitalia

The vaginal and cervical lengths in different stages of pregnancy varied from stage to stage but the differences were not significant in case of vagina while, the cervix showed significant increase in length with advanced pregnancy. The reason might be due to the loosening of the muscle fibres, age, breed, plane of nutrition, parity etc. Many authors like Singh *et al.*, 1974 in goat, Pareek *et al.*, 1981in camel, Hafez (1987) in non pregnant adult sheep and Khatri *et al.*, 2004 in buffalo reported the vagina and cervical lengths in non pregnant animal. But they did not compare these lengths with early pregnancy. That is why we could not compare our observations with their findings. However, the vaginal and cervical lengths in non pregnant animal in present study were 6.50 ± 0.23 and 5.30 ± 0.24 respectively; these values were lower compared to pregnant animal of present study.

The lengths from cervix to left and right utero-tubal junction, cervix to inter-cornuate ligament and body of uterus were increasing significantly with the increase in gestational age and the reason might be due to the loosening of muscle fibres and connective tissue fibres, especially elastic fibres. Singh *et al.*, 1974 in goat, Pareek *et al.*, 1981 in camel, Hafez (1987) in non pregnant adult sheep, Shalini and Sharma in goat and Khatri *et al.*, 2004 in buffalo reported the length of uterine horns and body in non pregnant genital organs. The dimensions observed in the present study could not be compared with their studies. However, we have divided the early pregnancy from 14 to 57 days in five different groups and made comparison amongst them. This type of systematic groupings lacks in experimental design cited by other authors. Our results were more systematic and specific as far as the different stages of pregnancy are concerned, while the other authors reported it only for the non pregnant animal. To that extent our findings in this regards are pioneering. However, in present study we have recorded the lengths from cervix to left and right utero-tubal junction, cervix to inter-cornual ligament and body of uterus in non pregnant animals were 19.84 ± 1.16, 20.00 ± 1.17, 10.86 ± 0.60, and 1.31± 0.09 respectively, which was less compared to pregnant animal. These non-pregnant values were almost supported by above authors findings.

Size of Ovaries

The size of ovaries varied from stage to stage during early pregnancy. The mean dimensional changes resulted due to growth and presence of corpus luteum. In the present study the ovarian length, breadth and thickness also changed amongst the groups and varied between left and right side of the ovaries. This stage wise variation could be due to presence of corpus luteum either in left or right ovary or presence of pregnancies on one side of the horns in that particular group. However, our present observation indicated that the right ovary was more active than the left, more corpus luteum and pregnancies were established on the right side of the horns than the left. Wani (1984) also indicated more ovulation on right ovary in sheep and goat than the left. The present study is also supported by the findings of Mufti (1995). The lengths, breadth and thickness of the left and right ovary in non pregnant animal at present study were 1.27 ± 0.04, 0.89 ± 0.04, 0.60 ± 0.03 and 1.34 ± 0.05, 0.97 ± 0.03, 0.69 ± 0.03 respectively. These findings of present study in non pregnant animal were supported by Singh *et al.*, 1974 and Shalini and Sharma, 2004. These values were lower in comparison with pregnant animals of present study.

Changes of Length of Allantoic and Amniotic Cavity and Diameter of Amniotic Cavity

The length of amniotic and allantoic cavity and diameter of amniotic cavity showed significant increasing trend during early pregnancy, reason might be due to the continuous increase of foetal size, fluid and its associated cavities. No literature is available to compare the present study, so in this regards our work was pioneer and we have observed systematically from 14 to 57 days of gestation.

Changes in the Caruncle

The average number of unattached caruncles decreased significantly during early pregnancy whereas, the number of attached caruncles increased non-significantly as the pregnancy advanced. This is natural and could be explained in developing foeto-maternal union during different stages of gestation. In early pregnancy in GpA_1 (Upto 39 days of gestation) none of the caruncle and cotyledons were attached. We could not observe gross caruncular-cotyledonary attachment. The number of unattached caruncle in non pregnant uterine horns in the present study both in left and right horns were 43.00 ± 1.63 and 43.00 ± 1.54 respectively. The numbers of caruncles collectively were almost similar to those reported by Hafez (1987). The exact number of attached and unattached ratio as found in this study could not be compared due to lack of reports in the literature.

Changes in the Foetal Fluid Volume

The volume of amniotic and allantoic fluid increased significantly in different groups during early pregnancy. The possible explanation might be as the pregnancy advanced fluid accumulation increased continuously to the concerned cavities to adjust the growing foetus and also secretory activity of the amniotic and allantoic fluid and foetal waste increased which indicated the metabolic activity of the foetus. The present study was supported by Malan and Curson (1937), Cloete (1939), Bongso *et al.*, 1979 and Goldstein *et al.*, 1980, Amle *et al.*, 1992 and Mufti *et al.*, 2000. Amle *et al.*, 1992, postulated that variations of foetal fluid might be due to the contributions of the foetal and maternal compartments and the control of fluid compartments regulated by foetal endocrine system and foetal kidneys.

Changes in the Foetal Fluid Colour and Consistency

The changes of foetal fluid colour and consistency during the different stages of pregnancy were similar to the reports of earlier author, Bongso *et al.*, 1979, Raj Sundrum *et al.*, 1990, Arthur, 1957 and Amle *et al.*, 1992. The possible reason of changes of colour and consistency during different stages of pregnancy may be due to the entry of various metabolic constituents and foetal wastes to the foetal fluids (Raj Sundrum *et al.*, 1990 and Amle *et al.*, 1992).

Changes of pH in the Foetal Fluid and Maternal Serum

The pH of foetal fluids and maternan serum were alkaline in different groups during early pregnancy. The alkaline nature of foetal fluids and maternal serum could be due to the chemical constituents of the fluids. No suitable literature is available to compare the present study.

Glucose

Glucose in allantoic fluid showed decclining trend but the differences were non-significant. In case of serum glucose level decreased upto GpA_3 but again showed increasing trend in further groups. But in amniotic fluid it showed significant decreasing trend. Increasing or decreasing trend of glucose levels in foetal fluids and maternal serum were fluctuating in between groups of early pregnancy, reason of frequent changes during early pregnancy might be changing metabolic demand of the growing foetus. Till date nobody has done work in this way as present work has done to compare it. But the increased of glucose concentration in maternal blood serum is in agreement with those of Baetz *et al.*, 1976; Lathura *et al.*, 1987; McCrabb *et al.*, 1991 and Mufti (1995).

Total Protein

The total protein in amniotic, allantoic and maternal serum increased as the pregnancy advanced was in agreement with Mufti, 1995, Kancko *et al.*, 1970, McDonald, 1980, Lathura *et al.*, 1987, Bazar and Sharp, 1991. The allantoic fluid has more protein than amniotic but lowers than that of dam serum (Table: 4B.4). Similar observations were observed by Lathura *et al.*, 1987, Mufti (1995). The increased protein concentration in case of serum and also the increasing trend in amniotic and allantoic fluid may be due to increase demands of foetus with growing age and also the results of metabolic activity of the foetus. McDonald, 1980 reported that an increase protein concentration in maternal serum indicative a positive association with pregnancy establishment and fluxuations revealed its demand and utilization for the physiological events that took place during fertilization. Bazar and Sharp (1991), suggested that a protein produced by sheep conceptus between 12-21 days of pregnancy called ovine trophoblastic protein –1 (OTP- 1) that enhanced protein concentration.

Cholesterol

The levels of serum cholesterol were higher than amniotic and allantoic fluids in this study. Mufti, 1995 have shown decreasing trends of cholesterol in serum. Almost similar results were observed in the present study also. Similarly, the cholesterol content of amniotic and allantoic fluid were almost constant but showed slight increasing trend. Changes of cholesterol in amniotic fluid could not be compared with those of Mufti (1995) because he did not estimate the cholesterol from amniotic fluid. The increased of cholesterol levels in foetal fluids during pregnancy has been postulated to be due to the increase amount of its need for progesterone. Cholesterol is the precursor for progesterone (Hafez *et al.*, 1958). To that extend the decrease of cholesterol in blood serum and almost constant in foetal fluids could be explained.

Creatinine, Urea and Urea Nitrogen

The mean creatinine, urea and urea nitrogen concentration during the pregnancy in dam serum and foetal fluids has been analysed. There has been increasing trends of creatinine in all the three fluids. Similar increases were in urea and also in urea nitrogen.

Higher levels of creatinine in allantoic fluid were observed by Hafez and Kamal, 1955; Alexander, 1958 and Lupke, 1967. In direct evidence of increase of creatinine in pregnancy was reported by Dukes, 1995 by establishing increase of creatinine in uterine muscle during pregnancy. The increased levels of creatinine in pregnant animal serum as well as in foetal fluids have been explained by Stainer, 1965. He has reported that lack of exchange of solutes from allantoic fluid may be the cause of increase. Baetz *et al.*, 1976 reported that creatinine and uric acid concentration increased in serum and foetal fluids during pregnancy. Increased level of urea and urea nitrogen during pregnancy reported by Alexander (1958), Wales *et al.* (1973), Mellor and Slatter (1971 and 1972), Soliman (1975) and Chauhan *et al.* (1975). The observations of present study were similar to those earlier reported by Hafez and Kamal, 1955; McDonald, 1980 and Lathura *et al.*, 1987. Some of the studies have also been reported by Mufti *et al.*, 2000.

Changes of Foetal Fluids Contents During Pregnancy

Sodium and Potassium

Variations in the macro and micro minerals of foetal fluids and maternal serum have been presented. Sodium concentration in amniotic and allantoic fluids and maternal or dam serum had shown an almost consistent values during the entire pregnancy. Concentration of potassium was different than those of sodium. In present study amniotic fluid contained more sodium and allantoic fluid contained more potassium supported by Hafez, 1987. There was decrease of potassium in maternal serum whereas, potassium concentration increased in both the foetal fluids and reversed trend shown by sodium in all the fluids. Similar trends have been observed by Rajsundram, 1990; Chauhan *et al.*, 1992; Mellor and Slater (1971 and 1972) and Wales and Murdoch, 1973 and Mufti, 1995. The concentration varied between the different stages of pregnancy. Their variations indicated active metabolic activities of foetus and its healthy kidney functioning. Sodium and potassium are known to be the chief constituents of electrolytes and ionic exchange system, thereby there decrease and increase in the foetal fluids and dam serum for indicative of life within the foetus.

Copper, Zinc, Manganese and Iron

The contents of copper, zinc, iron, manganese have been reported in serum of ovine during early pregnancy, however no study seen have been made for determination of these constituents during early pregnancy as was done in this studies. Our observation, indicate that the copper values in amniotic, allantoic and in serum fluxuate during early pregnancy, were almost constant in serum and decreasing trend in amniotic and allantoic fluid Table: 4B.5. Zinc concentration in amniotic and allantoic fluids increased almost significantly between groups where as their trends fluxuating in GpA_3 and GpA_4 where as zinc in maternal serum showed decreasing trend in serum but also were fluxuating in GpA_3 and GpA_4. Manganese were fluxuating in different groups of gestation but almost have shown decreasing trend both in foetal fluids and maternal serum. The mean iron concentration have shown increasing trend both in amniotic and allantoic fluid but decreasing trend in maternal serum. Our results were partly supported by those of Baetz *et al.*, 1976, Dhami *et al.*, 2003 and Xu- HuiZhong *et al.*, 2004 but they did not estimate systematically as we have done in our study.

Changes of Enzymes in Foetal Fluids and Maternal Serum

Enzymes of foetal fluids and maternal serum were analysed and described. GOT/AST and GPT/ALAT concentration in foetal fluids were fluctuating in between groups but in maternal serum showed decreasing trend and alkaline phosphatase were increasing both in foetal fluids and maternal serum. Changes of enzymes during early pregnancy revealed the metabolic activity of the conceptus. Our present study can partly be compared with the earlier author Manish *et al.*, 2004, Antunovic and Steiner, 2004, Pourouchottamane *et al.*, 2005, but they have estimated only from serum and not even systematically.

Hormonal and their Receptor Changes

The Estrogen and progesterone concentration increased significantly from gestation 14 to 57 days (From GpA_1 to GpA_5), and narrated under results 7.23 to 7.26. Rodrigues *et al.*, 1998, reported that serum levels of progesterone increased during gestation decrease abruptly only by two days before preterm. Hafez and Hafez, 2000 and Schular, 2002 suggested that both estrogen and progesterone is a placental hormone and ewe is a placenta depended animal

(Hafez, 1987 and Hafez and Hafez, 2000). Hewitt *et al.*, 2002 illustrated that estrogen was essential for epithelial proliferation and embryo implantation and was dispensable for stromal desidualization. The estrogen and progesterone concentration in present study in non pregnant animal was 451.25 pg ± 9.53, and 1.2ng ± 0.05 respectively, where estrogen was much higher and progesterone was much lower in non pregnant animals compared to pregnant animal. Estrogen and progesterone receptor showed increasing trend during early pregnancy but their differences between the groups were very less and they were almost constant in all the groups. Both transitory attachment on day 18 and gross caruncular cotyledonary attachment on day 40th of gestation were supported by progesterone receptor. The observation of present study supported by earlier author Hafez and Hafez, 2000, they have postulated that decreasing concentration of progesterone receptor reduce MUC-1 production and open a receptive state for conceptus attachment. Our study was supported by earlier reports given by Thomas and fuller (1995). They have postulated that increased level of progesterone suppress both estrogen and progesterone receptor. Hewitt and Korach, 2000 postulated that the progesterone attenuates the proliferative effect of estrogen in the uterus, where as estrogen also induces the progesterone receptor. The present results with respective to estrogen and progesterone in serum and those of estrogen and progesterone with reported by Hafez and Hafez, 2000, Cengiz *et al.*, 2000, Kurita *et al.*, 2001 and Roth *et al.*, 2004. Differentiation and caruncular growth are known to be regulated by estrogen and progesterone (Schular, 2002). The endometrial receptivity changes both in the endometrium and placentation as seen morphologically with due evidences from histopathology could be properly explained in view of our studies on estrogen, progesterone and their receptor concentration estimated throughout the gestation in this study. Many questions regarding the role of ovine and caprine ovarian steroids in initiating endometrial receptivity as observed in this study could be easily explained in view of the review reported by Huct - Hudson *et al.*, 1989, Wani, 1995, Wani, 1996, Wani, 2003, Wani, 2005, Wani, 2006 . Proliferation of uterine cells is influenced by E_2 and P_4 receptor . Our study corroborated and confirmed the studies of Lee and Dee Mayo, 2004. The ability of uterine caruncular epithelium to through inter digital projection which fit into the evaginated projections of the cotyledones have been confirmed by ER and PR receptor studies in our work. This to a broader extend explains the concept of implantation as put forth by Lee and Dee Mayo, 2004. The details of estrogen, progesterone concentration and their receptors are given (Khatoon, 2007). The estrogen and progesterone receptor concentration in present study in non pregnant animals were 0.55 ng ± 0.01, and 2.19 ng ± 0.08 respectively.

Gross and Histological Genital Changes in Pregnancy

Gross changes in non pregnant genitalia were externally symmetrical horns and internally the caruncles were pointed and histologically, the size and shape of the caruncle and the endometrial glands in the intercaruncular region were smaller and caruncular epithelium was smooth and regular. Few blood vessels were present beneath the caruncle. These changes were supported by Wani *et al.*, 2006. As per pregnancy concerned we have studied from 15 days upto 135 days of gestation (Range: 14 to 140 days). Caruncular size and vascular proliferation were increasing continuously as the gestation advanced. At 15 days the horns were slightly asymmetrical but the base of the horn (Part of the horn nearer to the body of the uterus) bulged dorsally, the reason might be due to the presence of conceptus at the base of the uterine horn. The blastocyst was not completely elongated upto the tip of the horns and also the cavities were not completely filled up with foetal fluid. At 15 days, histologically, the size of the caruncles were larger in comparison with non pregnant genitalia. Glandular proliferation started in the inter-caruncular region but caruncular epithelial lining was regular and smooth just like non pregnant sample. Vascular proliferation observed at the base of the caruncle. The foetal membrane did not show any specific changes except, the presence of some basophilic cells. Asymmetry of the horns was observed from 18 days onwards and was increasing as the gestation period advanced. At 18 days of gestation the changes in the caruncle (histologically) were almost same with 15 days of gestation except, the increasing size of caruncle and vascular proliferation. Important observation histological at 18 days of gestation was appearance of binucleate cells (binucleate cells observed constantly in the fetal membrane from 18 days onwards) and villi (papillae) like extension in the fetal membrane and endometrial glandular opening in the inter-carucarular region, which was an indication of transitory attachment supported by (Wani, 2008), Hafez and Hafez (2000). Dellmann and Eurell (1998) postulated that histotrophe, secreted by uterine glands nourished the embryo before implantation. The postulation of Dellmann and Eurell (1998) also supported our observation. This attachment was also supported by progesterone receptor, where progesterone receptor decreased slightly (Hafez and Hafez, 2000). Upto 20 day's corpus luteum possessed neck and was bulged out from the ovarian surface but from 23 days onwards it started embedding within the ovary and was present upto the term. From 15 days upto 23 days of gestation grossly, the caruncles were rounded and swollen type that might be the effect of progesterone (Schular, 2002) but at 25 days of gestation a circular depression was observed at the centre of the caruncle and gradually it became flattened. Continuously presence of depression at the centre of the caruncle was observed from 40 days and it was increasing as the gestation advanced and gradually they became cup shaped. Important point to be noted that as the age advanced the size of the caruncles were continuously increasing both gross and histologically (Schular, 2002). At 15 days the blastocyst was not elongated upto the tip of the horns and embryo was just like a red clot, it could only be recognized after microscopic observation. At this very stage the amniotic cavity was not distinct and fluid accumulation to the allantoic cavity was very less. But from 18 days onwards there was continuous accumulation of foetal fluids. From 20 days onwards the blastocyst was completely elongated upto the tip of the horns but was not completely filled with fluids. Emerging blood vessels from the hind gut of foetus was grossly recognized from 20 days onwards. Two vessels emerged from the hind gut of foetus but actually it was not two, were total of four vessels recognized after 25 days of gestation and two of each entered either of the horns. As the gestation period advanced the vessels were becoming thick and prominent and also the ramifications from the main vessels were increasing. From 25 days onwards the amniotic cavity could be recognized grossly. At 36 days of gestation, blastocyst was completely filled with fluid and whole blastocyst ramified with blood vessels and it was almost reddish in colour. From this stage head, eye bud and tail could be recognized grossly and this might be the beginning of foetal stage (Hafez and Jainndeen, 1993 and Hafez and Hafez, 2000). At 39 days single layered reddish granules were present mostly in the membrane of the left side were nothing but the cotyledons that appeared for the first time. Cotyledones were arranged in single layered which was reddish in colour and granular. Very important changes were observed from 40 days of gestation onwards. At 40 days we have observed gross caruncular-cotyledonary attachment that was supported by histological changes and progesterone receptor concentration (Hafez and Hafez, 2000), which was the main motto of our research. At this stage histologically, a typical change was observed in the caruncular epithelial lining. There were some projections rather to say dentations on the caruncular epithelial lining. Gradually, from 40 days of gestation onwards the lengths of these projections were increasing. Within these projections the trophoblastic villi were invading or infiltrating gradually and developed syncytium (Hafez and Hafez, 2000 and Gordon, 2004). In our studies we have observed the formation of syncytia from 42 days onwards. Hafez and Hafez, 2000 and Gordon, 2004 mentioned about syncytial formation but they did not cite the actual time of syncytial formation which might be cleared from our observation. From 18th days of gestation upto 40th day's changes in the fetal membrane were presence of binucleate cells, villi like extension and increasing vascular proliferation, but from 40 days onwards other changes were continuing except villi like extension i.e. the height of villi like extension reduced. From 40 days upto 47 days of gestation the caruncular-cotyledonary attachment was very fragile and they got separated by holding the uterine horn without imposing any pressure or by putting very slight pressure. But from 48 days onwards the attachment could not be separated by holding the uterine horn, it was becoming strong gradually. Depression at the centre of the caruncles was observed from 40 days onwards and gradually it was becoming deep and rounded. The cotyledonary invasion increased gradually as if the placentomes were engulfing the cotyledons. From 80 days onwards the placentomes completely attached the cotyledons, leaving no space at the centre of the placentome. The placentome became completely rounded, this gross change was also supported by histological observation of our study. Histologically, the foetal and maternal parts were completely intermixed and a capsule was present around the placentome. From this stage onwards it was difficult to separate the fetal and maternal parts and needed very strong pressure. Our observation partly supported the postulation of Gordon (2004). He postulated that the attachment process in ruminant was superficial and non invasive, actually as per our observation the attachment was non invasive at the early stage only but it was becoming strong from 48 days onwards and cotyledons were completely attached by the placentomes from 80 days onwards. As per our observation attachment started microscopically from 18 days of gestation (embryonic stage), it might be started from 17 days as we did not have 17 days of pregnant sample (Roberts, 1971), and completed grossly by establishing caruncular-cotyledonary attachment at around 40 days (Foetal stage) of gestation. Our observation also supported the report of Roberts (1971). He reported that ovine embryonic attachment begins at about 17 to 20 days and is completed by about 28 to 30 days (Roberts, 1971). Our observation also supported by Bazer and Roberts (1983), Hafez (1987), Wani (1996), Gray *et al.* (2002) and Wani *et al.* (2006) directly or indirectly.

Histology of Cervix

We have studied histology of cervix in non-pregnant and different stages of pregnancy i.e. early, mid and late pregnancy. In case of non-pregnant animals' cervix contained short mucosal folds. The lining epithelium was simple low columnar type. Numerous tiny mucous glands were present in the lamina propria-submucosa. This study was supported by Banks (1986) and Hafez (1987).

In early pregnancy, the tunica mucosa became more folded with primary, secondary and tertiary folds. As the gestational period progressed the folds became elongated with more branching that occupied most part of the lumen.

The lamina epithelialis mucosa was lined by simple columnar with goblet cells during early pregnancy. As the gestation period advanced height and loosening of the lining epithelial cells increased. The secretory activity of the mucosa increased with higher gestational age and formed cervical seal in the cervical lumen (Banks, 1986).

The lamina propria-submucosa consisted of loose connective tissue with dense irregular arrangement (Banks, 1986). It also contained numerous glands and blood vessels in early gestation period. As the pregnancy advanced the glandular and vascular proliferation occurred. The glands were mucous in type and became larger with higher gestational age due to increase secretory nature.

The muscle layers of tunica muscularis were arranged as inner circular and outer longitudinal pattern. Loosening of connective tissue fibres occurred with higher gestation. The tunica muscularis was rich in collagen and elastic fibres (Banks, 1986 and Hafez, 1987) as these fibres have a high tensile strength.

Tunica serosa was rich in loose connective tissue fibres and blood vessels. Vacularization increased with higher gestational ages.

Foetal Attachment and Placentation

Ovine embryonic attachments are reported to begin at about 17-20 days of post conception and have been reported to be completed by about 28-30th day of gestation (Roberts, 1971). In our study the actual attachment has been reported microscopically when examined it could detect villi like projection in the foetal membrane. The actual morphological union macroscopically could not be established before day 40th of gestation in the present study. As the above authors announcing or describing the attachment around day 17-20 in sheep did not specify whether they have observed it macroscopically or microsopically. The narration of the above authors pointed to microscopic observation which could not be corroborated in their studies. However, on the basis of microvilli projections between foetal and maternal tissues one could safely confirm the transitory attachment begins between 18-20 days and was visible macroscopically around day 40th of gestation. To that extend our observations have given a new concept of foeto-maternal attachment around day 40th. However, transitory microscopic connection has been established at 18th days of gestation. We have seen hatched elongated blastocyst with a red clot was nothing but embryo (recognized under microscope) at 15 days of gestation and initial development of the placenta. As the time advanced there were development of amniotic and allantoic cavity and was increasing accumulation of fluids. Gradually there were increasing vascular proliferation in the fetal membrane and whole fetal membrane ramified with blood vessels was also completely filled with foetal fluids and head, eye bud and tail could be recognized grossly, this might be the beginning of foetal stage. At 39 days of gestation some single layered reddish granules were present in the fetal membrane were nothing but the cotyledons, appeared for the first time. Cotyledones were arranged in single layered which was reddish in colour and granular. Both this process i.e. formation of placenta and organogenesis of the fetal parts occured within unattached (before morphological attachment) elongated blastocyst. On the basis of our observation, one could safely deduce that the placentation begins with the thickening of the blood drop. The embryonic mass is enveloped within a small cavity which appeared like a round mass and gradually converted to a fluid filled cavity and ramification and networking of blood vessels occured around the conceptus. The conceptus was fully formed within the elongated blastocyst. We have recorded sequential images of beginning of placentation and formation of cotyledons. The whole process begins when the conceptus still wrapped up within the elongated blastocyst. The placental membranes differentiated within elongated blastocyst before physical or gross attachment. The vascularisation of conceptus begins during the earliest stage. Caruncles also undergone tremendous changes and the size were increasing as the gestation advanced (Schular, 2002). It consisted of angiogenic (increased blood supply); mucification (formation of a mucous coat) and finally solid projected caruncles were differentiated into cup shaped structures. The morphological differentiation of caruncles during different stages of pregnancy is shown.

In brief, our studies have revealed that during non pregnant condition the caruncles were projected type. Changes in the caruncles started from the beginning of pregnancy and caruncles were swollen and rounded type during the early pregnancy that might be the early effect of progesterone, gradually developed depression at the centre and ultimately became cup shaped. Histologically, proliferated endometrial glands were observed in the intercaruncular region from 15 days of gestation onwards and from 18 days onwards glandular openings were present in the intercaruncular epithelial lining. The lining cells of the gland were simple cuboidal, indicated secreting something and might be histotrophe (Dellmann and Eurell, 1998). On this vary stage a specific phenomenon observed through histological observation were binucleate cells and villi (papillae) like extension in the fetal membrane, indicated the development of transitory attachment by projecting villi (papillae) like extension to the opening of endometrial gland presented in the intercaruncular region.

Details changes of caruncles, intercaruncles, fetal membranes, conceptus both grossly and histologically described. There are very little informations available on caruncular development, morphological status of conceptus in literature. However, our earlier observations have been confirmed by the details of present study (Wani, 2006).

In our preliminary studies we had not incorporated histological, endocrinological and other morphological details, which have been under taken in these studies. Although all the observations reported by the above authors have been partially confirmed, like nobody reported regarding the actual day of suncytial appearance in ovine and many others. The present study have come up with more elaborate evidences which clearly showed that the process of attachment starts almost during the 3rd week of pregnancy, where as the fully formed attachment consisting union of cotyledons and caruncles observed at 40th day of gestation.

In our observations we have found that the elongated blastocyst contains the fully differentiated conceptus, wrapped up in a finely coated membrane which gradually gave rise to amniotic sac. The only connection between the rest of the foetal membrane and growing conceptus was through a tiny blood vessel which eventually transformed into umbilical chord. This blood vessels coming out of the conceptus spread into a network throughout the blastocyst. Subsequently, with the development of the pregnancy the interviewing blood vessels meeting at various ends developed into a ball of blood mass. As per our observations each blood vessel terminated into a small blood drop. Thus, initially the cotyledonary mass looks like scattered red dots which eventually joint together and formed a loose mass of blood vessels. With compaction they became solid red round masses which were later on shaped themselves as cotyledons. These cotyledons during their formation may ramify and send projections across the blastocyst borders. As per our histological evidences we have noticed such projections even during early stages (From 18 days of gestation). Macroscopically, we were unable to find such projections or indication of such attachment until day 40th of gestation. Morphologically, from 40 days we had evidences of caruncular-cotyledonary attachments which have been described. We claimed these results to be reported for the first time. Literature reviewed did not indicate any such reports till date. However, the reports of King *et al.* (1981), Wani (1996), Hafez (2000), Gray *et al.* (2002), Gordon (2004) and Wani *et al.* (2006) indicated certain biochemical markers which supported histological and morphological observations found in this study. The observations on the appearances of changes in epithelium, appearance of binucleate cells and the glandular development as observed in this study was in confirmatory with the findings of Bhavsar *et al.* (1974), King *et al.* (1981), Michel and Stief (1985), Skidmore *et al.* (1997), Hafez (2000), Paul *et al.* (2003) and Rodriguez *et al.* (2005). We do differ with the observations with some of the above authors; they have based their conclusion on scanty observation. Our attachment process as detailed above has been hinted by Roberts (1971) and Hafez (2000) in their observation, where in Roberts (1971) stated that ovine embryonic attachment begins at about 17 to 20 days and is completed by about 28th to 30th days of gestation and Hafez and Hafez (2000) stated that placental attachment in ruminants was characterized by the appearance of binucleate cells arising from uninucleate cells of the trophoblast, first appeared on day 17 and are present throughout the gestation. These cells migrated and fused with the underlying uterine surface epithelium to form multinucleate cells or syncytium and conceptus attachment in ruminant species involves both caruncular and inter-caruncular areas of the endometrium. A transitory attachment first occured as the bovine and ovine conceptus developed finger like villi (Papillae) that projected into the lumen of the uterine glands. These papillae provided a temporary anchor and absorptive structure for the conceptus as more complete attachment progressed. Our studies were in confirmation of this earlier speculation. To that extend the evidences put forth in this study fortified the theoretical model proposed by Roberts (1971) and Hafez and Hafez (2000).

Placentation observed by earlier authors was described as superficial and non-invasive type, Gordon (2004). Our observation in this study did not fully support the views of the above author. He postulated that the attachment process in ruminant was superficial and non invasive, actually as per our observation the attachment was non invasive at the early stage only but it was becoming strong from 48 days onwards and cotyledone completely invaded the caruncle from 80 days onwards and after 80 days of gestation it was very difficult to separate the foetal cotyledon and maternal caruncle, it needed strong pressure to separate and also sometime the foetal and maternal parts torned out. However, the presence of mononucleate cells and binucleate cells in the foetal membrane observed at 15th day and 18th day of gestation, gross/morphologic fusion of foeto-maternal cells observed at 40th days of gestation and formation of syncytium was observed in 42 days of gestation respectively in the present study. In present study binucleate cells were observed from 18 days of gestation but syncytial formation occurred only after gross or morphological attachment of foetal and maternal tissue. At the beginning the binucleate cells were within the foetal membrane but from 36 days onwards we have observed free binucleate cells in the foetal membrane that indicated the migration of binucleate cells to the maternal tissue and formed syncytium from 42 days and observed continuously as pregnancy progressed. Our studies were in complete agreement with Gordon (2004) so far as the release of binucleate cells through secretory granules is concerned. Gordon has postulated that the binucleate cells secrete pregnancy associated glycoproteins and fascilitates the transfer of complex molecules across the placental barrier into the maternal circulation. He has also postulated that as attachment is initiated, the trophoblast binucleate cells migrate and fuse with maternal endometrial cells and formed syncytium. Hafez and Hafez (2000) postulated that placental attachment in ruminants is characterized by the appearance of binucleate cells arising from uninucleate cells of the trophoblast, first appears on day 17 and are present throughout the gestation. These cells migrate and fuse with the underlying uterine surface epithelium to form multinucleate cells or syncytium, may be involved in immunologic protection of the ceonceptus and the transfer of placental lactogen. The reports of Gordon (2004) and Hafez and Hafez (2000) in case of binucleate and syncytial formation were complete agreement with our study.

Summary and Conclusion

The whole study consisted of collecting eighty six samples of gravid and non-gravid uteri from the different local slaughter houses. The blood samples were also collected from each of the animals. In the present study the gravid uteri of sheep ranging from 14 to 140 days was divided into five groups e.g. Group A (GpA), Range: 14 – 57 days and Mean ± S.E : 41.40 ± 2.17, Group B (GpB), Range: 58 -74 and Mean ± S.E : 67.87 ± 1.38, Group C (GpC), Range: 75 – 93 and Mean ± S.E : 89.20 ± 2.33, Group D (GpD), Range: 94 – 120 and Mean ± S.E : 110.60 ± 3.14 and Group E (GpE), Range:121 – 140 and Mean ± S.E : 130.0 ± 2.79. Again, GpA is subdivided into five groups viz. Group A_1 (GpA_1), Range: 14 – 39 days and Mean ± S.E: 27.5 ± 2.70, Group A_2 (GpA_2), Range: 40-44 days and Mean ± S.E: 42.16± 0.75, Group A_3 (GpA_3), Range: 45-48 days and Mean ± S.E: 46.8 ± 0.58, Group A_4 (GpA_4), Range: 49-52 days and Mean ± S.E: 51.4 ± 0.51 and Group A_5 (GpA_5), Range: 53-57 days and Mean ± S.E: 55.75 ± 0.63. Foetal age determination in between 15 to 39 days was difficult; no formula could be used directly due to some drawback. Therefore, we have innovated a new graph/scale known as *"Khatun-Wani graph, 2007"*.

The blood samples were collected by jugular venupuncture, decanted serum and were analysed for various biochemical constituents, minerals, enzymes and hormones. The foetal fluids were also collected to estimate above all the constituents except, hormones to ascertain systemic changes with increasing gestational age. The tissue samples were collected from uterine horn kept in the sterile plastic vial containing phosphate buffer saline (PBS) and stored at deep freeze (-70°c) for estrogen and progesterone receptors estimation and remaining parts of the genital organ kept at 10 % formalin for histological examination.

Pregnancy is an intimate association between mother and foetus. Foetus is given a special status of "Homograph" in spite of its foreign or heterographic structure. This protective immunological cover to foetus is hidden in the formation and exact attachment of the foeto-maternal placenta viz. cotyledons and caruncles union. This main objective was the preamble of this investigation, with following objectives:-

1. Comparisons of female genitalia, biometric changes during pregnancy and non pregnancy.
2. Correlate external and internal morphology of the uterine caruncles during pregnancy with non pregnancy.
3. The histological changes during the attachment of cotyledons with caruncles and their mode of union.
4. To study conceptus attachment and changes occurring in dams oestrogen and progesterone, estrogen and progesterone receptor and other biochemical constituents ranging from glucose, protein and enzymes to trace elements. The chronological observations concomitant with the above objectives can be briefly presented as :

At 15 days of gestation, caruncles were swollen and mucinous. Around day 18th, an elongated worm like blastocyst with small comma or curved embryonic mass was recorded. Around day 29th microscopically, somites, head, eyebud, tailbud and heart with emerging blood vessels were observed. Around day 36th of gestation elongated blastocyst with ramification and networking of the blood vessels was seen. At day 39th of pregnancy, cotyledons were seen as small reddish granules as if blood vessels ends joining together gave rise to these round red balls i.e. cotyledones. The first attachment or formation of placentome was observed around day 40th of the gestation. Since this time line an intimate association between foetal cotyledons and uterine caruncles gave rise to prominent placentomes, in uterine tissue which could easily seen grossly.

Various changes in the uterine caruncles and intercaruncular area observed have been described. Trophoblastic villi were observed around day 18th and projections or dentations in the caruncles were discerned around day 40th of gestation. Syncytial formation was observed around day 42nd of post conception. From day 80th onward complete intermixing of cotyledons and caruncles was observed.

The changes in allantoic and amniotic cavities along with other parts of foetal and maternal components have been recorded and analyzed. The data revealed significant changes between pregnancy stages. The main characteristics of foetal fluids were analyzed. The pH of amniotic fluid and maternal serum was alkaline but in allantoic fluid became acidic towards term. The concentration of glucose in whole gestation period differed in GpA (4.33 ± 0.42) , GpD (3.41 ± 0.21), GpE (2.92 ± 0.15) in amniotic fluid, In GpA (3.57 ± 0.34), GpE (2.3 ± 0.43) in allantoic fluid and GpA (56.85 ± 2.22), GpC (59.8 ± 0.88), GpD (63.42 ± 1.75), GpE (65.22 ± 1.90) in serum (Whole gestation period), again in case of early pregnancy differed in GpA_1 (6.87 ± 0.43), GpA_2 (3.52 ± 0.22), GpA_3 (3.82 ± 0.90), GpA_4 (3.11 ± 0.64), GpA_5 (3.09 ± 1.20) in amniotic fluid, GpA_1 (68.53 ± 2.46), GpA_2 (52.51 ± 2.66), GpA_3 (41.92 ± 0.91), GpA_4 (54.04 ± 2.15) in serum.

The concentrations of total protein in whole gestation period differed in GpA (0.40 ± 0.06) and GpE (0.84 ± 0.26) in amniotic fluid), in GpA (0.48 ± 0.06), GpD (0.93 ± 0.09), GpE (1.42 ± 0.20) in allantoic fluid, in GpA (6.94 ± 0.82), GpB (11.10 ± 0.62), GpC (13.0 ± 0.75), GpD (13.11 ± 0.97) and GpE (16.92 ± 0.69) in serum whereas, the similar parameter for the early pregnancy differed in GpA_1 (0.20 ± 0.03) and GpA_5 (0.60 ± 0.15) in amniotic fluid, GpA_1 (0.25 ± 0.11), GpA_3 (0.50 ± 0.10), GpA_4 (0.68 ± 0.12), GpA_5 (0.51 ± 0.12) in allantoic fluid; GpA_1 (2.96 ± 0.15), GpA_3 (10.24 ± 0.43), GpA_4 (11.85 ± 0.55), GpA_5 (12.51 ± 0.69) in serum.

The concentrations of cholesterol in whole gestation period differed in GpA (3.13 ± 0.19) and GpB (3.13 ± 0.08) in allantoic fluid, GpA (107.62 ± 3.44), GpB (82.0 ± 5.2), GpC (65.80 ± 2.40), GpD (63.80 ± 1.91), GpE (57.75 ± 1.31) in serum. Similar parameter for the early pregnancy differed in GpA_1 (3.96 ± 0.17) and GpA_3 (5.68 ± 0.37) in amniotic fluid, GpA_1 (2.45 ± 0.15), GpA_3 (5.06 ± 0.22), GpA_4 (3.18 ± 0.13), GpA_5 (2.98 ± 0.06) in allantoic fluid and GpA_1 (121.06 ± 8.41), GpA_2 (107.67 ± 1.58), GpA_3 (99.18 ± 3.18), GpA_4 (98.20 ± 3.02), GpA_5 (96.25 ± 5.48) in serum. The concentrations of creatinine in whole gestation period differed in GpA (1.52 ± 0.26), GpB (5.34 ± 0.35), GpC (7.34 ± 1.19), GpD (9.30 ± 0.65) and GpE (13.67 ± 0.75) in amniotic fluid; GpA (2.21 ± 0.37), GpB (6.09 ± 0.43), GpC (11.36 ± 0.88), GpD (15.01 ± 0.10) and GpE (16.77 ± 0.31) in allantoic fluid and GpA (2.23 ± 0.36), GpB (6.25 ± 0.26), GpC (11.64 ± 1.26), GpD (16.54 ± 0.58) and GpE (23.88 ± 1.14) in serum. Similar parameter for the early pregnancy differed in GpA_1 (0.16 ± 0.04), GpA_2 (0.96 ± 0.21), GpA_3 (1.61 ± 0.47), GpA_4 (2.28 ± 0.57) and GpA_5 (3.69 ± 0.11) in amniotic fluid, GpA_1 (0.33 ± 0.08), GpA_2 (1.01 ± 0.21), GpA_3 (2.12 ± 0.29), GpA_4 (3.85 ± 0.38) and GpA_5 (5.39 ± 0.23) in allantoic fluid and GpA_1 (0.54 ±

0.07), GpA_2 (0.86 ± 0.05), GpA_3 (2.51 ± 0.32), GpA_4 (4.08 ± 0.34) and GpA_5 (5.5 ± 0.22) in serum.

The concentrations of urea in whole gestation period differed in GpA (14.01 ± 0.72), GpB (20.16 ± 0.66), GpC (23.32 ± 1.20), GpD (34.30 ± 1.41) and GpE (38.10 ± 0.47) in amniotic fluid; GpA (16.60 ± 1.02), GpB (28.93 ± 1.26), GpC (37.63 ± 3.48), GpD (52.08 ± 2.96) and GpE (72.36 ± 1.08) in allantoic fluid and GpA (61.93 ± 2.46), GpB (88.19 ± 3.42), GpC (96.95 ± 0.39), GpD (107.70 ± 4.96) and GpE (114.71 ± 1.68) in serum. Similar parameter for the early pregnancy differed in GpA_1 (9.92 ± 0.84), GpA_2 (12.90 ± 0.35), GpA_3 (14.91 ± 0.86), GpA_4 (16.92 ± 0.74) and GpA_5 (18.56 ± 1.10) in amniotic fluid, GpA_1 (11.15 ± 0.91), GpA_2 (14.46 ± 0.58), GpA_3 (17.22 ± 0.47), GpA_4 (18.07 ± 0.82) and GpA_5 (26.22 ± 0.53) in allantoic fluid and GpA_1 (49.20 ± 3.00), GpA_2 (61.06 ± 0.10), GpA_3 (67.21 ± 2.44), GpA_4 (73.49 ± 1.39) and GpA_5 (78.00 ± 1.34) in serum.

The concentrations of urea nitrogen in whole gestation period differed in GpA (6.54 ± 0.34), GpB (9.41 ± 0.31), GpC (10.89 ± 0.56), GpD (16.02 ± 0.66) and GpE (17.79 ± 0.22) in amniotic fluid; GpA (7.75 ± 0.48), GpB (13.51 ± 0.59), GpC (17.57 ± 1.62), GpD (24.32 ± 1.38) and GpE (33.79 ± 0.50) in allantoic fluid and GpA (28.92 ± 1.15), GpB (41.19 ± 1.60), GpC (45.28 ± 0.18), GpD (50.30 ± 2.32) and GpE (53.57 ± 0.79) in serum. Similar parameter for the early pregnancy differed in GpA_1 (4.63 ± 0.39), GpA_2 (6.02 ± 0.16), GpA_3 (6.96 ± 0.40), GpA_4 (7.90 ± 0.35) and GpA_5 (8.67 ± 0.51) in amniotic fluid, GpA_1 (5.20 ± 0.43), GpA_2 (6.75 ± 0.27), GpA_3 (8.05 ± 0.22), GpA_4 (8.44 ± 0.39) and GpA_5 (12.24 ± 12.24) in allantoic fluid and GpA_1 (22.98 ± 1.40), GpA_2 (28.51 ± 0.46), GpA_3 (31.39 ± 1.14), GpA_4 (34.32 ± 0.65) and GpA_5 (36.43 ± 0.63) in serum.

The concentrations of sodium in whole gestation period differed in GpA (96.06 ± 3.14) and GpE (90.48 ± 0.86) in amniotic fluid; GpA (88.07±1.19), GpC (78.8 ± 3.15), GpD (75.20 ± 1.46) and GpE (72.0 ± 1.58) in allantoic fluid and GpA (120.36 ± 3.36), GpD (136.97 ± 4.40) and GpE (138.70 ± 3.45) in serum. Similar parameter for the early pregnancy differed in GpA_1 (94.50 ± 1.22), GpA_2 (88.50 ± 0.76), GpA_3 (87.20 ± 1.50), GpA_4 (81.60 ± 2.00) and GpA_5 (80.50 ± 2.87) in allantoic fluid, GpA_1 (113.79 ± 4.48), and GpA_5 (124.98 ± 1.81) in serum.

The concentrations of potassium in whole gestation period differed in GpA (5.23 ± 0.14), GpC (6.92 ± 0.70), GpD (8.13 ± 0.44) and GpE (7.69 ± 0.37) in amniotic fluid; GpA (8.97 ± 0.36), GpB (12.25 ± 0.31), GpC (14.39 ± 0.47), GpD (16.80 ± 0.73) and GpE (20.86 ± 1.33) in allantoic fluid and GpA (7.33 ±

0.14), GpB (6.77 ± 0.12), GpC (6.40 ± 0.11), GpD (6.22 ± 0.12) and GpE (6.18 ± 0.11) in serum. Similar parameter for the early pregnancy differed in GpA_1 (4.99 ± 0.22), and GpA_5 (5.38 ± 0.10) in amniotic fluid, GpA_1 (6.73 ± 0.26), GpA_2 (8.96 ± 0.24), GpA_3 (9.21 ± 0.39), GpA_4 (11.09 ± 0.22) and GpA_5 (11.60 ± 0.07) in allantoic fluid and GpA_1 (7.83 ± 0.15), GpA_4 (6.65 ± 0.16) and GpA_5 (6.11 ± 0.36) in serum.

The concentrations of copper in whole gestation period differed in GpA (2.54 ± 0.22), GpB (2.03 ± 0.03), GpC (2.05 ± 0.10), GpD (1.98 ± 0.01) and GpE (1.88 ± 0.05) in amniotic fluid; GpC (3.12 ± 0.46), GpD (2.20 ± 0.10) and GpE (2.31 ± 0.07) in allantoic fluid and GpA (3.18 ± 0.24), GpB (2.54 ± 0.24), GpC (2.34 ± 0.07), GpD (2.09 ± 0.09) and GpE (2.20 ± 0.08) in serum. Similar parameter for the early pregnancy differed in GpA_1 (2.74 ± 0.83), GpA_2 (3.20 ± 0.36) in amniotic fluid, GpA_1 (2.13 ± 0.16), GpA_2 (2.78 ± 0.38), GpA_3 (2.75 ± 0.32) and GpA_4 (2.45 ± 0.26) in allantoic fluid and GpA_1 (2.95 ± 0.48), GpA_3 (2.47 ± 0.14), GpA_4 (4.61 ± 0.60) and GpA_5 (2.95 ± 0.55) in serum.

The concentrations of zinc in whole gestation period differed in GpA (5.84 ± 0.39), GpD (8.35 ± 0.30) and GpE (8.18 ± 0.23) in amniotic fluid; GpA (5.83 ± 0.33), GpB (8.88 ± 0.94), GpD (10.90 ± 1.05) and GpE (12.13 ± 0.13) in allantoic fluid and GpA (6.42 ± 0.49) and GpE (4.93 ± 0.15) in serum. Similar parameter for the early pregnancy differed in GpA_2 (6.78 ± 1.00) and GpA_3 (4.70 ± 0.57) in amniotic fluid, GpA_1 (4.89 ± 0.63) and GpA_5 (8.10 ± 0.06) in allantoic fluid and GpA_1 (6.52 ± 1.16) and GpA_5 (4.10 ± 0.33) in serum.

The concentrations of manganese in whole gestation period differed in GpA (3.37 ± 0.15), GpB (2.75 ± 0.09), GpC (3.06 ± 0.04) and GpE (2.90 ± 0.07) in amniotic fluid; GpA (3.17 ± 0.10) and GpB (2.81 ± 0.10) in allantoic fluid and GpA (3.58 ± 0.17), GpB (3.18 ± 0.06), GpC (3.20 ± 0.08), GpD (2.61 ± 0.16) and GpE (2.50 ± 0.03) in serum. Similar parameter for the early pregnancy differed in GpA_1 (4.19 ± 0.51), GpA_2 (3.15 ± 0.15), GpA_3 (3.26 ± 0.20), GpA_4 (3.05 ± 0.09) and GpA_5 (3.00 ± 0.00) in amniotic fluid, GpA_1 (3.47 ± 0.29) and GpA_5 (2.50 ± 0.14) in allantoic fluid and GpA_1 (4.34 ± 0.41), GpA_2 (3.34 ± 0.16), GpA_3 (3.45 ± 0.11), GpA_4 (3.19 ± 0.18) and GpA_5 (2.86 ± 0.21) in serum.

The concentrations of iron in whole gestation period differed in GpA (8.61 ± 0.34), GpB (9.50 ± 0.37), GpC (9.43 ± 0.27), GpD (10.17 ± 0.48) and GpE (10.16 ± 0.44) in amniotic fluid; GpA (5.82 ± 0.15), GpB (6.89 ± 0.28), GpC (7.76 ± 0.23), GpD (8.22 ± 0.21) and GpE (9.43 ± 0.54) in allantoic fluid and GpA (13.05 ± 0.28), GpB (12.43 ± 0.35), GpC (12.34 ± 0.39), GpD (10.00 ± 0.45) and GpE (8.62 ± 0.49) in serum. Similar parameter for the early pregnancy differed in GpA_1 (6.68 ± 0.11), GpA_2 (8.00 ± 0.52), GpA_3 (9.66 ± 0.42), GpA_4 (10.31 ± 0.08) and GpA_5 (10.94 ± 0.71) in amniotic fluid, GpA_1

(5.31 ± 0.25), GpA_2 (5.58 ± 0.36), Gp A_3 (6.04 ± 0.12), GpA_4 (6.18 ± 0.19) and GpA_5 (6.73 ± 0.28) in allantoic fluid and GpA_1 (14.37 ± 0.53), GpA_2 (12.50 ± 0.56), GpA_3 (12.58 ± 0.24), GpA_4 (12.40 ± 0.40) and GpA_5 (12.00 ± 0.41) in serum.

The concentrations of Glutamate Oxalo Transaminase (GOT) in whole gestation period differed in GpA (15.26 ± 0.45), GpB (13.18 ± 1.07) and GpC (13.40 ± 0.40) in amniotic fluid; GpA (13.48 ± 0.42) and GpC (12.43 ± 0.50) in allantoic fluid and GpA (104.10 ± 4.03), GpB (94.38 ± 3.05), GpC (86.60 ± 3.79), GpD (83.60 ± 3.83) and GpE (78.50 ± 1.32) in serum. Similar parameter for the early pregnancy differed in GpA_1 (122.90 ± 8.76), GpA_2 (88.83 ± 2.39), GpA_3 (92.80 ± 4.10), GpA_4 (106.00 ± 4.74) and GpA_5 (91.75 ± 2.63) in serum.

The concentrations of Glutamate Pyruvate Transaminase (GPT) in whole gestation period differed in GpA (7.56 ± 0.24), GpB (6.36 ± 0.39) and GpE (6.83 ± 0.33) in amniotic fluid; GpA (6.34 ± 0.24) and GpB (5.49 ± 0.37) in allantoic fluid and GpA (34.63 ± 1.19), GpB (29.75 ± 0.97), GpC (28.20 ± 0.80), GpD (26.00 ± 1.30) and GpE (24.00 ± 2.04) in serum. Similar parameter for the early pregnancy differed in GpA_1 (7.45 ± 0.24), GpA_2 (6.38 ± 0.41) and GpA_3 (8.56 ± 0.70) in amniotic fluid, GpA_1 (6.65 ± 0.24), GpA_2 (5.57 ± 0.30) and GpA_5 (5.28 ± 0.45) in allantoic fluid and GpA_1 (37.00 ± 3.19), GpA_2 (31.50 ± 1.78) and GpA_3 (32.20 ± 0.66) in serum.

The concentrations of alkaline phosphatase in whole gestation period differed in GpA (30.73 ± 1.51), GpB (35.75 ± 1.63), GpC (40.20 ± 1.50), GpD (41.20 ± 1.59) and GpE (43.75 ± 1.31) in amniotic fluid; GpA (27.33 ± 1.42), GpB (31.88 ± 1.44), GpC (33.60 ± 1.60), GpD (32.40 ± 2.40) and GpE (36.50 ± 1.26) in allantoic fluid and GpA (60.70 ± 3.43), GpB (71.88 ± 2.96), GpC (76.80 ± 4.99), GpD (78.80 ± 2.75) and GpE (92.50 ± 4.99) in serum. Similar parameter for the early pregnancy differed in GpA_1 (20.90 ± 1.85), GpA_2 (33.50 ± 1.15), GpA_3 (35.40 ± 1.72), GpA_4 (37.60 ± 0.51) and GpA_5 (36.75 ± 2.36) in amniotic fluid, GpA_1 (19.10 ± 1.86), GpA_2 (28.67 ± 1.69), GpA_3 (32.00 ± 2.61), GpA_4 (34.00 ± 2.10) and GpA_5 (31.75 ± 1.55) in allantoic fluid and GpA_1 (37.50 ± 2.62), GpA_2 (66.67 ± 2.40), GpA_3 (70.80 ± 3.65), GpA_4 (81.60 ± 2.46) and GpA_5 (71.00 ± 5.26) in serum.

The concentrations of progesterone and estrogen in whole gestation period differed in GpA (5.32 ± 0.33), GpB (8.26 ± 0.11), GpC (7.70 ± 0.23), GpD (7.08 ± 0.35), GpE (7.00 ± 0.32) and in GpA (138.50 ± 11.42), GpB (250.00 ± 0.00), GpC (230.00 ± 12.25), GpD (210.00 ± 18.71), GpE (210.00 ± 10.00) respectively in serum.

The concentrations of progesterone and estrogen receptors in whole gestation period differed in GpA (3.81 ± 0.04), GpB (4.33 ± 0.06), GpC (4.38 ± 0.04), GpD (4.09 ± 0.09) and in GpA (0.37 ± 0.01), GpB (0.41 ± 0.01), GpC (0.41 ± 0.01) respectively in uterine tissue.

The concentrations of progesterone and estrogen during early pregnancy differed in GpA_1 (3.26 ± 0.32), GpA_2 (5.19 ± 0.08), GpA_3 (6.00 ± 0.14), GpA_4 (7.00 ± 0.21), GpA_5 (7.89 ± 0.08) and in GpA_1 (74.00 ± 8.59), GpA_2 (113.33 ± 6.67), GpA_3 (160.00 ± 8.37), GpA_4 (198.00 ± 5.00), GpA_5 (236.25 ± 2.39) respectively in serum.

The concentrations of progesterone and estrogen receptors during early pregnancy differed in GpA_1 (3.74 ± 0.05), GpA_3 (3.82 ± 0.04), GpA_4 (3.98 ± 0.02), GpA_5 (4.06 ± 0.02) and in GpA_2 (0.36 ± 0.01), GpA_4 (0.39 ± 0.01), GpA_5 (0.37 ± 0.01) respectively in uterine tissue.

Status of progesterone (5.19 ± 0.08) and estrogen (113.33 ± 6.67) in dam serum and progesterone receptor (3.62 ± 0.09) and estrogen receptor (0.36 ± 0.01) in uterine tissue were there during attachment.

On the basis of above results we deduce that the caruncular-cotyledonary attachment occurred at 40 days of gestation but microscopically transitory attachment occurred at 18 days of gestation by the trophoblastic villi. The morphological changes leading to intimate association viz. formation of cotyledons, changes in caruncles, microvilli projections and final union as placentomes is completed with 18-40th day of post conception. As a result of our above study we dare to claim first inventions as follows:

1. Preparation of a graph for predicting the age of conceptus on the basis of whole length, crown-anus length which we name as "Khatun-Wani graph, 2007".
2. We describe the possible mechanism of formation and union of cotyledons and caruncles and fostering a new idea that elongated blastocyst upto day 36th could recovered from uterus and cultured invitro, thereby preparing the first step for invitro foetus investigation in sheep.

References

Alexander, D. P., Nixon, D. A., Widdas, W. F. and Wohlzogen, F. X. 1955. Gestational variation in the composition of foetal fluids and foetal urine in sheep. *Journal of Physiology*, 140: 1.

Alexander, D. P., Nixon, D. A., Widdas, W. F. and Wohlzogen, F. X. 1958. Renal function in sheep foetus. *Journal of Physiology*, 140: 14-22.

Aminoor-Rahman, A. N. M., Snibson, K. J., Lee-Chee-Seong and Meeusen, E. N. T. 2004. Effects of implantation and early pregnancy on the expression of cytokines and vascular surface molecules in the Sheep endometrium. *Journal of Reproductive Immunology*, 64 (1/2): 45-58.

Amle, M. B., Chinckar, S. R., Hukari, V. B. and Deopurkar,V. L. 1992. Studies on Foetal Fluids of Buffaloes. *Indian Journal of Animal Reproduction*, 13 (2): 165- 167.

Antunovic, Z., Speranda and M., Steiner, Z. 2004. The influence of age and reproductive status on the blood indicators of ewes. *Archiv-fur-Tierzucht*, 47(3): 265-273.

Arthur, G. H. (1956).In: Vety. Reproduction and Obstetrics, 5th edition, p. 43.

Arthur, G.H. (1957). Some notes on quantities of foetal fluids in ruminants. *British Vety. Journal*, 113: 17.

Bacha J. William, and Jr.Linda M. Wood, 1990. In: Colour Atlas of Vety. Histology. Lea & Febiger, Philadelphia, London. pp 207.

Baetz, A. L., Hubbert, W. T. and Graham, C. K. 1976. Changes of biochemical constituents in bovine foetal fluids with gestational age. *American Journal of Veterinary Research*, 37(9): 1047- 1052.

Banks, W.J. 1986. Applied Veterinary Histology, Hosy Year book, Philadelphia, p. 503.

Bazer, F. W. and Roberts, R. M. 1983. Biochemical aspects of conceptus - endometrial interections. *J. Exp. Zool.*, 228: 373.

Bazer, F.W. and Sharp, D.C. 1991. In: Reproduction in farm animals by Hafez, 6th edition, p. 83.

Bhavsar, B. K., Pattabiraman, S. R. and Venkataswami, V. 1974. Endometrial changes during early pregnancy of Bubalus bubalis. *Indian Veterinary Journal*, 51(3): 175-176.

Bongso, T. A., Edirisinghe, R. and Athuraliya, D. 1979. Studies on foetal fluids in goat. *Indian veterinary journal*, 56(7): 562- 569.

Bongso, T.A and Basrur, P.K.1976. Foetal fluids in cattle. *Canadian Journal*, 17(2): 38-41.

Cengiz, F., Yaman, K. and Galip, N. 2000. Studies on progesterone, estrogen and calcium values in cows before and after calving. *Veteriner-Fakultesi-Dergisi-Uludag-Universitesi*, 19(3): 37-43.

Chauhan, R. A. S. and Tiwari, S. 1996. Variation in the transaminase and alkaline phosphatase activity of caprine foetal fluids with gestational age. *Indian Veterinary Journal*, 73(10): 1045- 1047.

Chauhan, R.A.S., Tiwari, S. and Datta, I. C. 1992. Changes in the volume and biochemical constituents of caprine foetal fluids with gestational age. *Indian Journal of Animal Reproduction*, 13(1): 72-74.

Cloëte, J. H. L. 1939. Onderstepoort. *J. Vet.Sci.Anim.Indust.*, 13: 418.

Dellmann, H. Dieter and Eurell Joann. 1998. In: Textbook of Veterinary Histology,Fifth Edition, Lippincott Williams & Wilkins

Dey, S. K. 2004. Focus on implantation reproduction, 128: 655-656.

Dhami, A. J., Patel, P. M., Lakum, P. D., Ramani, V. P. and Pande, M. B. 2003. Micronutrient profile of blood plasma in relation to age reproductive status of Holstein Friesian cattle. *Indian-Journal-of-Animal-Nutrition,* 20(2): 206-211.

Dukes, H. H. 1955. Protein Metabolism. Physiology of Domestic Animals. 7th Edition, pp 562.

Eulenberger, K. H., Neubauer, K. and Elze, K. 1977. Histological studies of the endometrium and chorioepithelium of sows during early pregnancy. *Monatshefte- fur-Veterinarmedizin,* 32(9):333-336.

Fakowska - Podstawka, M., Wrona, Z. and Krakowski, L. 2002. Biological role of interferontau. *Medycyna-Weterynaryjna,* 58(8): 568-571.

Garland, H. O. 1983. Cited in "The effects of pregnancy and sodium intake on water and electrolyte balance in sheep. *British Veterinary Journal,* 144: 147.

Goldstein, M. H., Bazer, F. W. and Barron, D. H. 1980. Characterisation of changes in volume, osmolarity and electrolyte composition of porcine foetal fluids during gestation. *Biology of Reproduction,* 22(5): 1168 - 1180.

Gordon Ian, 2004. In: Reproductive technologies in farm animals. CABI publishing.

Gray, G. A., Barghardt, R. C., Johnson, G. A., Bazer, F. W. and Spencer, T. E. 2002. Evidence that absence of endometrial gland secretions in uterine gland knockout ewes compromises conceptus survival and elongation. *Reproduction,* 124: 289-300.

Green, W.W. and Winters, L. M. 1945. Tech. Bull. Minn. Agric. Exp., Sta No. 169.

Guillomot, M., Flechon, J. E. and Wintenberger Torres, S. 1982. Cytochemical studies of uterine and trophoblastic surface coats during blastocyst attachment in the ewe. *Journal of Reproduction and Fertility.* 65(1): 1-8.

Hafez, E. S. E. 1987. In: Reproduction in farm animals, 5th edition, Lea & Febiger, Verghese Company

Hafez, E. S. E. and Hafez, B. 2000. Reproduction in farm Animals. Seventh Edition, Lippincott Williams & Wilkins, A Wolters Kluwer Company, pp. 126-139.

Hafez, E. S. E. and Jainudeen, M. R. 1993. Gestation, Prenatal physiology and parturition. In: Reproduction in farm animals. 6th edition, pp. 225-226.

Hafez, E. S. E. and Kamal, M. A. M. 1955. Physiochemical investigation on placental fluids of buffaloes. *Indian Journal of Vety. Science,* 25(3): 39-45.

Hafez, E. S. E., Magharabi, M. M. and Anwar, A. 1958. Some factors effecting blood cholesterol level in buffaloes. *Indian Journal of Vety. Science,* 28(3): 133-138.

Henricks, D. M and Harris, R. B. Jr. 1978. Cytoplasmic estrogen receptors and estrogen concentration in bovine uterine endometrium. *Endocrinology*, 103(1):176-185.

Hewitt, S. C. and Korach, K. S. 2000. Progesterone action and responses in the alpha ERKO mouse. *Steroids*, 65(10/ 11): 551- 557.

Hewitt, S. C., Goulding, E. H., Eddy, E. M. and Korach, K. S. 2002. Studies using the estrogen receptor alpha knockout uterus demonstrate that implantation but not decidualisation associated signaling is estrogen dependent. *Biology of Reproduction*, 67(4): 1268-1277.

Huct-Hudson, M. Y., Andres, G. K. and Dey, S. K. 1989. Cell type specific localization of C-MYC-Protein in mouse uterus-modupition by steroid hormones and analysis of the pre implantation period. *Endocrionology*, 125: 1683-1690.

Hugget, A. St. G. and Widdas, W. F. 1951. The relationship between mammalian foetal weight and conception age. *Journal of Physiology*, 114: 306- 317.

Hugh Y. Yee, Edward S. Jenest, and Fred R. Bowels, 1971. Modified manual or automated o- toluidin system for determining glucose in serum, with an improved aquous reagent. *Clin Chem.*, 17(2):103-107.

Ismail, Z. B., Qureshi, T., Levy, M., Khamas, W. and Nour, A. 2003. Preliminary report on some physiological parameters in pregnant female camels (Camelus dromedarius) in North America. *Journal-of-Camel-Practice-and-Research*, 10(2): 125-129.

Joshi, C. L., Nanda, B. S. and Saigal, R. P. 1976. Histomorphological and histochemical studies on the female genitalia of ageing goat, glands in cervix uteri. *Anatomischer-Anzeiger*, 139(1/2): 193-199.

Joubert, D. M. 1956. A study of prenatal growth and development of components of caprine conceptus. *Journal of Agricultural Science*, 47: 382- 428.

Kancko, J. J. and Corneleous, C. E. 1970. Clinical biochemistry of domestic animals. Volume I, 2nd edition, pp. 313-394.

Khatri Pershotam, Hamayun Khan and Rind, M. M. 2004. Biometrical studies of reproductive organs of non pregnant buffaloes slaughtered in Hyderabad. *Journal of Animal and Veterinary Advances*, 3(8): 500-502.

Kind, P. R. .M. and Kings, E. J. 1954. *J. Clin. Pathol.*, 7: 322.

King, G. J., Atkinson, B. A and Robertson, H. A. 1981. Development of the intercaruncular areas during early gestation and establishment of the placenta. *Journal of Reproduction and Fertility*, 61(2): 469-474.

Kleflin, A., Zobundzija, M., Martincic, T. and Sladic, Z. 1988. Concentration changes of potassium and sodium ions in the pigs foetal fluids during the first half of gravidity. *Veterinarski - Archiv*, 58 (5): 225- 232.

Kolmer, J. A., Spanbling, E. H. and Robinson, H. W. 1951. Approved Laboratory Techniques.Appleton Century Crofts, New york.

Kurita, T., Lee - Kijun, Saunders, P. T. K., Cooke, P. S., Taylor, J. A., Dahiya, R. and Cunha, G. R. 2001. Regulation of progesterone receptors and decidualisation in uterine stroma of the estrogen receptor- alpha knockout mouse. *Biology of Reproduction,* 64(1): 272-283.

Lathura, R. A., Gupta, R. C., Khar, S. K. and Baru, P. 1987. Variation in metabolic constituents of caprine foetal fluids during different stages of gestation. *Indian Journal of Animal Reproduction,* 8(2): 133-137.

Lee, K. Y. and De Mayo, F. J. 2004. Animal models of implantation. *Reproduction.* 128: 679-695.

Luna, L. G. 1968. Manual of Histological Staining methods of the Armed Forces Institute of pathology, 3rd Edn. The Blakistan Division, McGraw-Hill Book company, New Work

Lupke, H., Mulling, M. and Sohn, E. 1967. Maternal foetal relationship in cattle, content to total nitrogen, urea, uric acid and creatinine in foetal fluids. *Vety. Bulletin,* 38: 2047.

Malan, A. P. and Curson, H. H. 1937. Onderstepoort. *J. Vet. Sci. Anim. Indust.,* 8: 417.

Manish - Mahawar, Jain, A. K. and Puneet – Kumar, 2004. Clinically important enzymes in Jamunapari and Marwari goats. *Indian-Journal-of-Animal-Sciences,* 74(9): 924-925.

McCrabb, G. J., Egan, A. R. and Hosking, B. J. 1991. Maternal under nutrition during mid pregnancy in sheep. *British Journal of Nutrition,* 65(2): 157-168.

McDonald, L.E. 1980. *Veterinary Endocrinology and Reproduction,* 3rd edition, p. 61, pp. 388-389.

Mellor, D. J. and Slatter, J. S. 1971. Daily changes in foetal fluids during last three months of pregnancy in ewes. *Journal of Physiology,* 217: 573-604.

Mellor, D. J. and Slatter, J. S. 1972. Daily changes in foetal urine and relationship with foetal fluids and maternal plasma during last three months of pregnancy in ewes. *Journal of Physiology,* 227: 503-525.

Michel, G. and Stief, E. 1985. Placentation and foetal development in sheep. *Monatshefte-Fur-Veterinarmedizin,* 40(10): 344- 347.

Mufti, A. M. 1995. Thesis on Morphological characters and some metabolic constituents at different stages of development of ovine foetus. SKUAST-K, Shalimar.

Mufti, A. M., Wani, G. M., Wani, N. A., Buchoo, B. A. and Khan, M. Z. 2000. Prenatal development of ovine fetus. *Small-Ruminant-Research,* 38(1): 87-89.

Natelson, Scatt, M. L. and Beffa, C. 1951. *Americ. J. Chem, Path.* 21: 275.

Noden M. Drew and Lahunta De Alexander, 1985. In: Embryology of Domestic Animals. Williams & Wilkins.

Owen, J. A., Iggo, B., Scand rett F. J., Stewart, C. P. 1954. *Biochem. J,* 58: 426.

Pareek, P. K., Joshi, C. K. and Vyas, K. K. 1981. Note on the biometry of genital organs of female camel. *Indian Journal of Animal Sciences.* 51(10): 1010.

Paul, S., Ghosh, R. K., Das, P., Mahata, T. K. and Thakur, A. M. 2003. Studies on gross anatomy and histology of the non pregnant and pregnant (upto 90 days) uterus in the non descript cow. *Indian Journal of Animal Health,* 42(2): 178- 181.

Pourouchottamane, R., Chatterjee, A., Sheikh, I. U., Sarkar, M. and Gogoi, D. 2005. Blood biochemical constituents of female yaks in different physiological status. *Indian-Veterinary-Journal,* 82(10): 1108-1109.

Raj Sundrum, R. C., Kathireson, D. and Pattabiraman, S. R. 1990. Changes in biochemical constituents of foetal fluids with increase in gestation period in buffaloes. *Indian Journal of Animal Reproduction,* 11(1): 24-28.

Ramadhar Singh, 1988. Ageing and age. Essential of Animal Production and Management. First Edition. Kalyani publisher, pp. 210- 227.

Ramana Rao, K.V. 2006. In: General studies for Agricultural Research services and other Competitive Examinations. 4th Revised and Enlarged Edition, Jain Brothers, New Delhi, pp. 35-36.

Reinhold, J.G. 1953. In: standard methods of clinical chemistry (Reiner M.ed), New York and London, Academic press, 1: 88.

Reitman, S. and Frankel, S. 1957. A colorimetric method for the determination of serum glutamic oxaloacetic and glutamic pyruvic transaminases. *Am. J. Clin.Pathol.,* 28: 56- 63.

Richardson, C. 1980. In: Vety. Reproduction and Obstetrics (Editor, Noakes David E, Parkinson Timothy J, England Gary C W) Eighth Edition, p. 68.

Roberts J. Stephen, 1971. Veterinary Obstetrics and Genital Diseases, Second Edition (Indian Edition), CBS Publishers and Distributiors, pp. 36-47.

Rodrigues, P. A., Vicenta, W. R. R., Cipolla-Neto, J. and Oliveira, C. A. 1998. Evaluation of the progesterone hormone profile in ewes at early pregnancy, pre-term and early lactation. *Arquivo-Brasileiro-de-Medicina-Veterinaria-e-Zootecnia,* 50(2): 133-139.

Rodriguez-Marquez, J. M., Hernandez-Vasquez, A., Hindalgo-Delgado, G. 2005. Ovine endometrial lining modification during the implantation in natural and induced multiple pregnancies. *Revista-cientifica, Facultad -de- Ciencias-Veterinarians, Universidad-del-Zulia,* 15(4): 326-333.

Roth, T. L., Bateman, H. L., Kroll, J. L., Steinetz, B. G. and Reinhart, P. R. 2004. Endocrine and ultrasonographic characterization of a successful pregnancy in a sumatran rhinoceros (*Dicerorhinus sumatrensis*) supplemented with a synthetic progestin. *Zoo-Biology,* 23(3): 219-238.

Rowe, A. J., Wulff, C. and Fraser, H. M. 2003. Localization of MRNA for vascular endothelial growth factor (VEGF), angiopoitins and their receptors during peri-implantation period and early pregnancy in marmosets (*Callitrix jacchus*). *Reproduction,* 126: 227-238.

Sastry, N. S. R., Thomas, C. K. and Singh, R. A. 1982. Dentition and aging of animals. Farm Management and Poultry Production. Second revised Edition. Vikas Publishing House Pvt. Ltd., pp. 43-50.

Schular, 2002. In: Reproductive Technology in Farm Animals. CABI publishing, pp. 237-254.

Shalini and Sharma, D. N. 2004. Biometrical study of internal genital organs of gaddi goats. *Indian Journal of Animal Reproduction*, 25(2): 148-150.

Shameha-Intan, A. R., Zuki, A. B., Nornajiban, K. and Zamri-Saad, M. 2005. Microscopic evaluation of uterine mucosa in pregnant and non pregnant rats. *Journal of Animal and Veterinary Advances*, 4(6): 590- 594.

Singh Yashwant, Sharma, D. N. and Dhingral, L. D. 1979. Morphogensis of the testis in goat. *Indian J. Anim. Sci.*, 11: 925-931.

Singh, S. K, Bhattacharya, A. R and Luktuke, S. N. 1974. Studies on biometry of genital organs of female goat. *Indian Veterinary Journal*, 51(2): 81- 85.

Skidmore, J. A., Wooding, F. B. P. and Allen, W. R. 1997. Placentation during the first 60 days of gestation in the dromedary camel (*camelus dromedarious*). *Journal of Camel Practice and Research*, 4(2): 199-202.

Snedecor, G. W. and Cochran, W. G. 1980.Statistical Methods. 6th Indian Edition. Oxford and IBH Publishing Company, New Delhi

Soliman, M. K. 1975. Studies on physiological chemistry of allantoic and amniotic fluids of buffaloes at various stages of gestation. *Indian Vety. Journal*, 52: 106-112.

Stainer, M. W. 1965. Transfer of radioactive water, urea and glycine between maternal and foetal body fluids in rabbit. *Journal of Physiology*, 178: 127-149.

Surgeev, L., Kleeman and Walker, 1990. Prediction of ovine foetal age using real time ultrasound imaging. Australian Society, *Animal Production*, 18: 547.

Thailender, G., Eriksson, H., Edquist, L. E., Rodriguez-Martinez, H. 1990. Variations in oestradiol and progesterone receptor levels in the porcine myometrium and cervix during pregnancy and parturition. *Journal of Veterinary Medicine*, Series- A, 37(5): 329 -338.

Thomas E. Spencer and Fuller W. Bazer, 1995. Temporal and spatial alterations in uterine estrogen receptor and progesterone receptor gene expression during the estrous cycle and early pregnancy in ewe. *Biology of Reproduction*. 53: 1527-1543.

Wales, R. G. and Murdoch, R. N. 1973. Changes in the composition of sheep foetal fluids during early pregnancy. *Journal of Reproduction and Fertility*. 33(2): 197- 205.

Wallace, L. R. 1948. *Journal of Agricultural Science*, 38: 93, 243.

Wani, G. M and Buchoo, B. A. 1990. Laparoscopic view of ovine conceptus. *Indian Journal of Animal Science*, 60(5): 564 – 568.

Wani, G. M. 1995. Early diagnosis of pregnancy in surrogate mothers (Sheep and Goat). In National Symposium on Embryo Transfer and Genetic Engineering Technologies in Small Ruminants, May 7-8, CIRG, Makhdoom, Farah, India.

Wani, G. M. 1996. Embryo Biotechnology in Sheep and Goats. Publisher Valley Book House, University Road, Hazratbal, Srinagar, Kashmir.

Wani, G. M. 2000. Effect of ocyte harvesting techniques on invitro maturation and invitro fertilization in sheep. *Small Ruminant Research*, 36(1): 63-67.

Wani, G. M. 2002. Multipurpose Management/ Goat Husbandry. Encyclopedia, Dairy sciences, Editor Roginski, H, Academic Press, An imprint of Elsevier Science, London, Boston, Amsterdam, New York, Oxford paris, San Diego, San Francisco, Singapore, Sydney, Tokyo, II: 1259-1270.

Wani, G. M. 2003. Biotechnological advances in goats production. In *National Work-Shope on "Recent Advances in Goat Rearing"*. MPKKV, Agric.Univ. Rahuri, 29-30 December.

Wani, G. M. 2005. Advances in ovine and Caprine Reproduction, www.word-mart.com. Research Paper, 9: 1-25.

Wani, G. M. 2006. Biotechnological Advances in Goat Production. www.word-mart.com. Research paper, p. 10.

Wani, G. M., Khatun, A., Wani, G., Goswami, P. and Khan, M.Z. 2006. Preimplatation endometrial and foetal structural changes in sheep. Paper presented, In *International Symposium on "Emerging Trends in Genomic and Proteomic Sciences"* Mumbai, India, 15-18 october.

Wani, G.M.1984. Investigation on ovarian activity, ovulation and pregnancy diagnosis in sheep and goats. Ph.D.Dissertation. Rohilakhand University, Bareily, Izatnagar.

Xu-HuiZhong, Liu-Dan, Zhong-YuanQiang and Yang-FengXing, 2004. Mineral element contents in the blood of cattle and their variations. *Journal-of-Southwest-Agricultural-University*, 26(3): 236-239.

Zietzschmann, O., Krolling, Oh. and Vitmus, 1955. In: Vety. Obstetrics and Genital Diseases (S.J. Roberts eds), 2nd Indian Edition, p. 31.

Zlatkis, A., Zak, B. and Boyle, G. J. 1953. *J. Lab. Clin. Med.*, 41: 486.

Zogno, M. A., Miglino, M. A. and Oliveira, M. F. 2004. Biochemical analysis of fetal fluids and cytology of the amniotic fluid of female moco (*Kerodon upestris*). *Brazilian-Journal-of-Veterinary-Research-and-Animal-Science*, 41(4): 228-235.

□□□

Chapter 4

Ultrasonography and Laparoscopy in Farm Animals

G.M. Wani & Paras Wani

Abstract

The use of laparoscopy for ovulation detection in Small Ruminants (sheep, goats) has revealed that ovulation occurs earlier (12-24 hrs) in ewes than in goats (24-42 hrs) after onset of heat. It has been used to study development of conceptus in ewes and was used to predict ovarian response in superovulated ewes and ewe-lambs with 73-95% accuracy.

Use of both A-mode and B-mode ultrasonography in small ruminants has been reviewed. The present day, real time ultrasonic imaging has made it possible to detect pregnancy in sheep and goats as early as day 20. The accuracy of B-mode (linear array and sector scanner systems) is higher than A-mode equipments. (doppler or amplitude modes). The B-mode realtime imaging systems allow non-invasive visualization of the internal reproductive organs and their conceptus. Its present and future use in the study of maternal recognition of pregnancy , transuterine migration, Foeto-maternal interaction, Preimplantation embryo signals, early embryonic mortality and diagnosis of reproductive disorders and failures is reviewed and discussed.

Laparoscopy

Since the adoption of laparoscopy in sheep and goats (Phillipo et al 1971, Wani, 1982), a number of investigations have been carried in these species through laparoscopy. We shall present a review of our studies here under following headings.

Ovulation Detection in Muzaffarnagri Sheep

The appearance of corpus luteum (CL) was observed at laparoscopy after every 6 hourly intervals from the beginning till the end of heat (0-48 hrs). The laparoscopy was thereafter conducted at 3 days interval from 3rd to 28th day post mating. The ovulation was observed to occur between 12024 hrs after the onset of heat. The corpus lateum started regressing around day 9th and regresses between 12-15 day of the cycle. (Wani and Sahni, 1988).

Ovulation Detection in Jamunapari Goats

Similar procedures as in sheep were adopted in goats. About 87% examined in this study ovulated within 24-36 hrs after the onset of heat. All ovulated between 24-42hrs. The ovulation rate was 1.55 CL. (Corpus luteum) started regressing around the day 12th in goats and regressed between 15th-18th day of the cycle (Wani, 1989).

Development of Conceptus in Corriedale Ewes

The uterine distention (development of conceptus) was examined through laparoscopy. First group was laparoscopied between 28-35 days post mating. Group 2 ewes were examined around 58± 0.6 days post-mating and Group 3 ewes were examined on 93 ± 0.8 days post-mating. The laparoscopic observations are shown in Tab-1. For confirmation of these observations laparotomy was conducted immediately after laparoscopy (Wani and Bachoo, 1999.

Introduction

The infertility and ovulation diagnosis is difficult in small ruminants and primates.sheep and goats serve as model animals for the primates and human diagnostics .a review of the techniques used by the author have been described.

1. Ovulation Detection in Muzaffarnagri Sheep

The appearance of corpus luteum (CL) was observed at laparoscopy after every 6 hourly intervals from the beginning till the end of heat (0-48 hrs). The laparoscopy was thereafter conducted at 3 days interval from 3rd to 28th day post mating. The ovulation was observed to occur between 12024 hrs after the onset of heat. The corpus lateum started regressing around day 9th and regresses between 12-15 day of the cycle. (Wani and Sahni, 1988).

2. Ovulation Detection in Jamunapari Goats

Similar procedures as in sheep were adopted in goats. About 87% examined in this study ovulated within 24-36 hrs after the onset of heat. All ovulated between 24-42hrs. The ovulation rate was 1.55 CL. (Corpus luteum) started regressing around the day 12th in goats and regressed between 15th-18th day of the cycle (Wani, 1989).

3. Development of Conceptus in Corriedale Ewes

The uterine distention (development of conceptus) was examined through laparoscopy. First group was laparoscopied between 28-35 days post mating. Group 2 ewes were examined around 58± 0.6 days post-mating and Group 3 ewes were examined on 93 ± 0.8 days post-mating. The laparoscopic observations are shown in Tab-1. For confirmation of these observations laparotomy was conducted immediately after laparoscopy (Wani and Bachoo, 1990).

Table 1 : Development of Conceptus in Sheep

Group -I Laparoscopy Laparotomy Signs Uterine horns, thin fluid filled distended, transparent appearance. Asymmetry of gravid horns. Uniform distention, uterine Blood vessels hypertrophied. (Diagnosis Early Pregnancy - 1st phase 0-7 wks).

Group-II. Pregnant uterine horn distended, Gravid horn distended like Like a balloon. Fallopian tubes looked as if Enlarged . Gravid horn thick. Ballons are joined together Small marks as budding Distended uterine horn Cotyledons. Difficult to bring out of Incision. (Diagnosis Mid Pregnancy - 2nd phase 8-13 wks).

Group-III Pregnant horn like a big ballon, Pregnant uterine horns large Foetus recognized, cotyledons much distended, Identified as small rounded dark. Buds. Full foetus felt. (Dianosis Late Pregnancy - 3rd phase 14- term).

Ref. Wani and Bachoo, 1990

4. Prediction of Ovarian Response in Superovulated Ewes

Eleven superovulated corridale ewes were laparoscopied on 5th or 6th day post-mating. The number of corpora lutea and ovulated follicles were counted. Accuracy % was calculated by comparing observations at laparscopy with those actually observed at laparotomy. The overall accuracy in predicting ovulating rate and unovulated was 73.7% and 80.4% respectively. (Wani and Bachoo, 1990).

(ii) In Superovulated Ewe Lambs

Twenty two crossbred ewe lambs were superovulated as per Wani et al (1989). Those in heat were laparoscopied 5-6 day post -mating . The number of corpora lutea and un-ovulated follicles were counted at laparoscopy and compared with actual observations at laparotomy. The overall occuracy % of predicting unovulated follicles and ovulations varied from 75-95% respectively (Wani et al, 1990).

The use of laparoscopy in future may include aspiration of follicular fluid, estimation of foetal numbers, flushing of superovulated donors, diagnosis of reproductive disorders intrauterine insemination, laparoscopic embryo transfer and study of foeto-maternal relationship.

Ultrasonography

Ultrasonography consists of two system: The A-mode ultrasounds: consists of amplitufe mode, sonar or echo systems. For two decades now, these systems have been in use of pregnancy diagnosis in small ruminants. (Wani, 1981, Wani and Sahni, 1981, Watt et al, 1984).Now real-time B-mode ultrasonography is available. If used correctly, it can help not only to detect pregnancy as early as 20-25 days post-mating, but real time imaging systems can help in the diagnosis of reproductive disorders both in male and female small ruminants. It can also help, in the monitoring of embryo development, transfer implantation, foeto-maternal relationship and foetal growth and placentation. Let us review the use of ultrasonography under following three heads.

a. Past use of ultrasounds
b. Present use of ultrasonography and
c. Future prospects of ultrasonography.

(a) Past Use of Ultrasounds

The ultrasound equipments mostly used in the past were A-mode equipments. They were mostly of two types:

a) Doppler System

The doppler equipments worked on the principle of altered ultrasound frequency. It mainly employed detection of foetal pulse, movements and uterine artery circulation as a means of detecting pregnancy (Wani, 1981, Wani and Sahni, 1980, Wani and Sahni, 1981). The accuracy of pregnancy detection after 50 days post-mating ranged between 75-90% in ewes and she-goats.

The equipment was used to predict foetal age too. This was done by counting foetal heart heats.

b) Amplitude Mode

Another type of early ultrasonic equipments were based on the detection of the difference in a coustic impedance between the contents in the gravid uteres and the abdominal viscera. These equipments were Amplitude depthscope, or amplitude-depth ultrasonic analyzer. They were mostly used by Australian workers with 90-95% accuracy after 2 months of pregnancy (Lindhi, 1966, 1969).

II. Present use of Ultrasonography

With the availability of real-time ultrasound systems, the past A mode systems are almost replaced. These B Mode real-time systems allow non-invasive visualization of the internal anatomy of the reproductive organs and their conceptus (Buck-rell, 1988). These systems have been described as most advanced technology of reproductive research after RIA (Ginther, 1986). The advantages/non-pregnancy tests in small ruminants, can provide informations on foetal numbers, foetal viability and diagnosis of reproductive tract diseases.

Equipment

The equipments marketed for use in animals have been mostly modeled for Mare. Two systems are in use for veterinary purposes.

(i) Linear array system

(ii) Sector Scanner

However, now some systems have the capability to provide both linear and sector scanning systems. Sector systems have some advantage over linear array, system, such as,

i. They require less skin surface contact which reduces the scan time required per ewe.

ii. They have superior resolution power especially in its electronic model

The cost of these systems vary between 10,000 to 15,000 US dollars. Sectors scanners are constlier than linear array models. Image quality comparison are suggested before purchase (Buckrell. 1988).

Transducers

Transducers are available both for abdominal and rectal use. The head frequency of a transducer is important for getting better images. Higher the transducer head frequency, the better will be the resolution power. With higher and better resolution, image quality shall improve. With higher and better resolution, image quality shall improve. At present three frequency transducers are available 7,5,5 and 3 MHz. For rectal use the 7.5 MHz transducer gives better results, for visceral use 5 MHz transducers may be used. For deeper penetration (e.g.flank) the lower frequency transducers 3 MHz are used.

Advantages and Limitations

The major benefit of using present day real-time systems is that we get image copy on a polarid film, slides, video or papers. This record serves as teaching aid or for self learning. The image can be frozen and tissues can be identified, measured before recording the image.

The another advantage of real-time systems is that the image can be freezed or we can have a hard copy of the image the chances of error in making diagnosis is minimized. The accuracy in diagnosis using real-time systems may be as high as 90% even during early gestation; however it depends on the choice of region and position while scanning i.e. rectal, flank standing, inguinal, clipped unclipped belleys etc.(Logue et al, 1987). As the accuracy and efficiency of using real-time ultrasound may be influenced by many factors. It is advisable to provide following information while publishing date on these systems:

a) System used- "Linear or Sector"
b) Transducer frequency.
c) Stage of gestation
d) Region of scanning
e) Duration of scan
f) Fasting
g) Breed and parity of the ewe or she-goat
h) Operator
i) Early embryonic death.

False positive diagnosis may result from an early diagnosis of pregnancy in ewes between day 20-25 post-mating. This may be the result of subsequent embryonic death or abortion (Fowler and Wilkins, 1984).

III. Future Prospects of Ultrasonography

The efficiency of the real-time systems is expected to be increased, both in terms of resolution and real-time elaborative imaging system. It is hoped that use of ultrasonography shall help in increasing our understanding, on feeto-placental, foeto-maternal, embryo- endometrial interactions, besides monitoring embryo-foetal development. Some of its present days used could be improved, such as:

a. Maternal Recognition of Pregnancy

The study of pre-implantation embryos, estimations of endometrial secretions and real-time imaging of the events involved in implantation may help us to understand the factors involved in the maternal recognition of pregnancy. Some studies using ultrasonography have suggested that maternal recognition of pregnancy takes place around 14-17 days in ewe and does, respectively (Riera, 1984). Ultrasonography along with assays for detection of antiproges terone antibodies in the uterine epithelium could help us to understand the signals responsible for sustained progesterone secretion (C.L. growth) during early pregnancy. The understand shall increase on the embryoendometril interactions. This can be possible by biopsy studies on morphological charges in endometruim, modulated and monitored by use of ultrasonography.

b) Trans Uterine Migration

trans uterine migration of embryos has been found to occur in 8% ewes with single ovulation, in 87.%% ewes with multiple ovulations. The assessment of blastocyst effects on the endothelium, especially to assess the role of embryonic stages involved in transition of maternal to embryonic control may be possible by combined use of ultrasonography with histochemistry and recent methods of reverse transcription of RNA into DNA (Schultz,1990).

c) Foetal-Maternal Interaction

Ultrasonography has been used to understand the mechanism of epithelial-chorial placentation in ewes. In ewes 60-80 concave carundes were differentiated. In doe this number was more 160-180. Evidence of caruncular development could be undertaken in future, using real-time, freezed mode imaging systems. The study of fluid dynamics in relation to expanding vehicle may help cur understanding of the foeto-maternal relationship in future.

d) Detection of Early Pregnancy and Embryonic Motility

Using a 5 MHz rectal probe pregnancy was detected as early as day 20 post mating. On screen such pregnancies produce an image resembling small pockets of fluid ventral and anterior to bladder. Embryonic death in an ewe too was captured on avideo attached to ultrasonography (Buckreli, 1980). Between 40-50 days post mating accuracy % has been near 100% with these B-mode systems.

References

Phillippo, M. Swapp, G.h.Robinson , J.J and Gill, G.C. (1971). The diagnosis of pregnancy and estimation of foetal numbers in sheep by laparoscopy. *Journal Reproduction Fertility*. 27: 129-132.

Wani G.M. (1982). Laparoscopy in farm animals. *World Rev. Anim. Prod.* 18, 7-13

Wani G.M.., Sahni K.L., (1988) Ovulation detection by laproscopy in sheep. *Indian Jour.Animal Sci*, 58 (7):802-804

Wani, G.M. (1989) Plasma progesterone concentrations during oestrus cycle in goats. *Indian Journal of Animal Reproduction*, 10 (1):44-46

Wani G.M. (1990) Post-partum breeding in goat-a review *Livestock Advisor* 15(1):39-42

Wani G.M. (1990) Superovulations during early luteal phase in goats. *World Review of Animal Production*, 29 92) 41-43

Wani G.M. (1981) Augumentation of fertility in goats : *Oestrus synchronization Annual report.*, Cent.Ins.Res.Goats.Makhdoom, India.

Wani G.M..and Buchoo ,B.A.1990, Laparoscopic view of ovine conceptus,*Indian Journal of Animal Science*.60(5): 564-568.

Wani, G.M; Sinha, N.K. and Sahni, K.L. 1981. Incidence of abortions and stillbirths in Muzaffernagri and crossbred ewes. *Indian Veterinary Journal*. 58: 558 561.

Wani, G.M. and Sahni, K.L 1980 Pregnancy diagnosis in Sheep and Goats. *Indian Farming* 5:25-27.

Buckrell, B.C. (1988). Application of ultrasonography (USG) in reproduction of sheep and goats. *Theriogenology*. 31:115-125.

Lindhi, J.L.(1966) A pregnancy tester for ewes. *Farm Journal October* (cited)

Lindhi, J.L. (1969) Pregnancy diagnosis in dairy goats using ultrasonic Doppler instruments, *J.Dairy Sci*.52:4.

□□□

Chapter 5

Small Ruminant Reproductive Diseases

G.M. Wani

Abstract

Reproduction in important for the survival of all be it humans or animals. No life without wife although a common solagan for human is true for all life on this Universe, be it insect or Elephant. The holy "Quran" says God created every living entity on earth in pairs. Thus reproductive management in general and reproductive health management is essential for conservat5ion, propagation and sustenance of life on earth. Sheep and goats have an early domestication history. Domestic sheep (Ovis aries), order Artiodactyla and family Bovidae have been derived from ovis musimon, a species distributed across Mediterranean islands. Their domestication began some 10,000 years ago. Longevity differs among breeds but maximum life span is 19-20 years. Keisler (1999) in Encyclopedia of Reproduction has given broadly reviewed global reproductive parameters of sheep and goats. The reader is also referred to review on sheep and goat production in general and goat management system in Encyclopedia of Dairy sciences, published by Academic Press. U.K in 2003 (Wani, 2003). Immunological aspects of sheep and goat health in general have been reviewed and depicted as html on http:/medicihne.ucsd.edu/cpa/sheep.html.2003.

Sheep and goats harbor Chlamydia psittaci and weak and dead foeti are born or aborted. Various diseases affecting reproductive health or sheep & goats have been reviewed (Smith et al, 1972). Various physiological data on sheep has been used to study human prenatal development. Many studies of value to control and cure reproductive losses in human were studied on sheep models. Like regulation of amniotic fluid in sheep suggested that (VEGF) vascular endothelial growth factor increases intramembranous abortion and micro vesicular transport. Similarly, technetium crossed ovine placenta more rapidly than smaller sodium ions. (Daneshmand et al 2003; Gilbert et al, 1996). These and other studies have revolutionized the health care of human and animal especially related to reproductive disease diagnosis and management.

The above short description and brief review envisages the importance of the topic under discussion. In order to reduce reproductive losses, disease, nutrition and breeding management schedules are to be advocated. It may need summing up information and methodology of clinical examination of reproductive organs, a checklist of physical signs of diseases associated with genital system. Besides clinical examination of goats their abnormalities needed to be discussed and imagined. For proper strategies of reproductive health management we need information on:

Incidence of reproductive disease and disorders. Diagnostic features of reproductive disorders. Management tools for prevention and cure of these diseases. This paper intends to review the data available on the topic and discussed it in the light of our own limited observations and profess a reproductive health management schedule from lamb/ kid born to lamb-kid born from her. Seed to seed plant or offspring to offspring cycle.

Common Reproductive Diseases

In Nov, 2006, a book detailing common reproductive diseases and disorders has been published by Saunders, edited by youngquist and Walter on ovine and caprine theriogenology. The common diseases listed are abortions, dystocia, listeriosis, mastitis, pregnancy toxemia, vaginal prolapse, ring womb, uterine prolapse, vibriosis (compylo bacteriosis), white muscle disease, toxoplasmosis, cache valley virus, Chlamydia, leptospirosis, Q fever, rift valley fever, salmonellosis, metritis, peritonitis, blue tongue, Fuzz lamb syndrome (Border lamb disease), foetal mummification, foetal macerations, cystic ovaries, persistant corpus luteum, trichomoniasis. We have also observed and reported Haematic foetal mummification in an ewe at CIRG, Makhdoom, first report of its kind in India (Wani et al, 1980) and a case of schitoschoma reflexus in an ewe at Srinagar. Sheep Breeding Farm, SKUAST (Wani et al 2000). Besides a number of reports on reproductive diseases of sheep & goat from CIRG, CSWRI and other SAU have been reported and presented in various health related seminars or ISSGAPU in 1981, 1982, 1987, 1991, 2001, 2002, 2003, 2004, 2005, 2006 and published.

Incidence of Reproductive Diseases

A. (Europe) Brucellosis in Sheep & Goats has less occurrence than cattle in Europe (Cyprus). Its Incidence is 0.0005% to 0.013% only. The engootic abortions, Q fever incidence in affected flocks were over 40%. The incidence of Hydatidosis in sheep was 0.79% but in goats it was neglible 0.09%, so one could profess goat meat safer than sheep, so far as hydatidosis as zoonotic disease is concerned. Blue tongue, toxoplasmosis incidence vary from 5-11% in sheep & Goats (Polydorou 1983). Chlamydia in sheep 11% and goat 9% is less than cattle (24%). Campylobacteriosis has an incidence of 1.2% in both sheep & Goats. Coxiella burnetic infections are 3 & 6% in sheep & Goats respectively. Toxoplasmosis is higher in sheep 23% than goats 7%, may be because of their clean and selective feeding habits. Listeria monocytogenes infection was very low in sheep (0.7%) than goats (2.2%) or cattle (4.6%). Brucellosis has host preference for cattle but sheep and goats too can harbour the infection (Luchsinger & Anderson, 1979).

B. (America) Blue Tongue was first reported from USA in 1948 in taxas sheep. Now virus has been detected in its 18 states. Initially its morbidity was 50-70% and mortality 20-50%. Subsequently with proper reproductive health measures its incidence has its incidence has been greatly reduced to minimal 1-2%. Cattle are the virus reservoirs. Culicoides is insect vector. Virus lias been isolated from sheep kids (Melophagus ovinus) and cattle lice (Haematopinus curysternus Denny). Dairy goats or Angora goats do not habour nutural Blue tongue. Vaccines have been produced to reduce its incidences. (Hourrigan and Klingsporn, 1975). Bluetongue can spread through winds as reported in Cyprus (Pitzolis, 1987).

C. Global: Blue tongue occurance is traced to Europe, 125 years ago. It spread to all most all countries globally, through distribution and sale of European animals to Africa, Asia & Other countries. Blue tongue is spread by a biting gnats or midges, genus culicoides. (Walton, 2004). Caprine arthritis-encephalitis is seen often in new born kids but not before birth. The disease symptoms are similar to Visna in sheep, where Leuksencephalomyelitis occurs. (Adams & Crawford, 1980). DNA technology using 2 PCR tests for diagnosis of Paratuberculosis in goats was reported from Germany (Schroeder et al 2001). Flow cytometery using monoclonal antibody detector was used for diagnosis of paratuberculosis in Czech goats. (Faldyna et al, 1999). Toxoplasmosis in goats was reported from Malaysis. The incidence was 35% in affected herds incidence was higher 50% in small farms, Selanger and lowest (20%) in Melaka & Negri Nembilan farms. Incidence was more in Malay farmer flocks 40% than Chinese run farms 20%. No infection was detected in large commercial farms. (Dorry et al, 1993). A steering group on evolving national Livestock policy perspective in their report of 1996, describe, Blue tongue, Brucellosis, Chlamydia, Maedi-visna disease, (Sheep), mastitis affecting small ruminants, besides other diseases like John's disease goat pox, sheep pox, enteroloxaemia, foot and mouth disease, PPR, Rinder pest & pneumonia. On the basis of above facts a considerable economic loss is inflicted on national and international livestock economic through reproductive diseases. Therefore, it is very important to draw reproductive health management schedules to reduce the loss. In this regard we shall devote our attention in next few pages under following different heads:

1. Examination Methodology

Unless our vets, NGO, scientists and those working with sheep & goats are conversant with methods of detecting deformities it is difficult to identify and cure than. Thus examination of external and internal genital organs shall be discussed and demonstrated through slides and figures.

2. Investigative Technology

The tehnique used for clinical diagnosis of reproductive diseases and deformities shall be shown in slides under presentation and shall be discussed through figures and charts.

3. Projcctive Strategies

Reproductive health checklist shall be discussed to help farmers, NGO & others to operate efficient reproductive health management.

4. Examination Methodology

i Male External Genitalia

Examination is performed under standing conditions. Visual and palpation of scrotum, spermatic cord may be made. Size, texture may hint towards abnormalities prolapse of intestine, presence of hernia or inguinal ring or ruptures may be looked for. Testes should be symmetrical, mobile, having an consistency like a ripe tomato. The total scrotal diameter in an adult Ram or back should be around 33 cms. Enlarged, painful or Hard testicles indicate disease and thus need further laboratory investigation. Such Rams or bucks should be immediately isolated and not used for breeding till clearance from diagnostic examination.

Testicular atrophy & hernia may be indicated on examination physically or through ultrasonography. Infections epididimytis can be detected as enlarged epididymitis similarly asmetrical spermatic cords are on disease indication scars on scrotum detect casectomized ram and bucks. Examination of penis and prepuce in necessary. The method of its examination is shown in figs. Inflammation or ulceration may be seen. Many disease like Fizzle rot need to be scored oat by seminal and blood examinations. (Jackson and cockroft, 2002).

ii. Female Genital Tract

The examination of udder, vulva, vagina and cervix becomes necessary. The method of examining udder is shown in Figs. Mastitis, gangrene, prolapse or vaginal secretion may reveal various disease conditions. Small necrotic plaques on vaginal wall be examined through head light and speculum. Locomotor problems of limbs and feet, lamness, nevous movements may invent many discuses like border disease tremors in young lambs, paresis due to spinal cord abscess in 3 month old lambs, circling, head tilt conditions due to coenuriasis in 3 month to 2 years age sheep and goat. Adult sheep nervous conditions include cervical subluxation (Paresis) as in listeriosis, brain abscess and scrapid (ataxia, pruritus, loss of weight). Metabolic disease include recumbancy, dullness and head pressing in pregnancy toxaemia, weakness, recombency and Haccid paralysis in hypocalcaemes, hyperasthesis, staggers in hypomognesaemia and kangaroo gait (ataxia).

Similarly intersexes in goats could be diagnosed by external examination of female genitalia, as shown in fig. Similarly pseudopregnancy and other genital tract defects or deformities could be detected via ultrasonography or hormonal, biochemical examination too. Radiograph or detection of estrone sulphate and progesterone in blood, milk may indicate pregnancy (wani, 1989). Similarly examination of vaginal discharge, cervix could help us to detect retained placenta, necrotic vaginitis. Acute or septic metritis. endometritis, chronic vaginitis, vaginal cysts or other deformities. Various shapes and sizes of these reproductive disorders, abnormalities will be as slides 1-29. Various lamb mortality in exotic sheep has been documented with managemental too for reduction (Suden et al, 1990).

Investigative Technology

A. Use of ultrasonography, laparoscopy and radiation tools have been used to explore reproductive diseases and disorders. Since their first use in India around 1978-82 (Wani, 1981, 1982) many groups have now emerged. Even our pioneering work at Makhdoom has been rated in pioneering, first of its kind by British & American Journal reviews notably theriogenology (1988) and 2002 website www.googli.com 21 years of ultrasound use. Now its use has been popularized and almost all veterinary college of India use it for research purpose. It is now time in 11 induct few diagnostic summer or winter school on it, so as to train atleast 10-20 veterinarian in each state for its use under farm conditions. Their use even in pre-slaughter animals may save economic resources and loss of lambs. We used many investigative technologies for small ruminants. Bachoo & Wani, 1991; Sudan et al, 1990; wani, 1981, 82; Wani et al, 1998; Wani & Bachoo, 1990. Wani & Bachoo, 1993, 1991; Wani, 1989.

B. Radio Immunoassays : This technology use have predicted pregnancy in goats, sheep by blood, milk examinations. Its use for diagnosis of foetal abnormalities needs to be explored. In collaboration with ultra-sonography, hormone, estrogen/progesterone values may help to diagnose foetal deformities, ovarian disorders or even placental defects. However, this technology has to be remodeled and a area of clinical diagnostics in reproduction is on the walls in future. More landmarks as normal values during early, mid or late pregnancy images have to be corborated. Foetal dimensions, placements and growth factor study should be multiplied. The lone experiments at vety. College, Srinagar (Mufti et al, 2000) he fortified in all veterinary colleges as was done in ultrasonography. Various details of the internal structures of genitalia have been depicted by wani et al, 2006 a,b,c, 2007. For further use and standardization of foeto-maternal interactions various articles of the author are available free of cost on www.wanigm.net or www.word-mart.com. Even clicking WANI.G.M on www.Googli.com search engine shall make available all these paper to the interested readers. (Wani, 1989; Sharief 1996).

Prospective Strategies

i. Even USA, national Academy has given green light for intensive study of small ruminant research on reproduction, which may serve as fertility control models in human too. Thus more funds and emphasis in research on basic, strategic and applied aspects is to be given in India too.

ii. An examination and investigative schedule for breeding rams needs to be formulated into a regulation. An organization as per Indian standard Institution needs to be established, which shall take all measures to safeguard our stock from emerging diseases through imports under W'FO mid others. Massive breeding ram investigations need to be advocated. Similarly, female before breeding or at culling need to be singled out.

iii. Common recommendations on the topic.

a. Sheep and goat enterprise needs encouragement and adequate funds for establishing reproductive disease diagnostic centers at state/regional levels immediately. The emphasis on zoo-sanitation arising under WTO & import/ export restrictions needs these technical Improvements. These labs could be located in veterinary colleges for survey and identification of these diseases. Use of advance techniques needs priority.

b. Clinical aspects of Veterinary institutes including IVRI, NDRI, CSWRI or CIRG are poorly developed. ICAR should not forget that these institutions have public health importance. Their care and facilities should be of referral institutes like AIIMSW or other national or state Medical college. Even American NAS have now laid many emphasis on renewal of clinical diagnostic in small ruminant reproduction.

c. Feeding mechanization, using cheap and easily available field staffs for sheep and goat wholesome and balanced feeds needs to be tested and used.

d. A proper identification of the sheep and goat resources needs to be made using DNA finger printing and selection of our best breed semen, embryo needs propagation. It is prelude to seed HIM whole sheep & Goat population free of all diseases especially sexually transmitted diseases. A proper diagnostic checks are necessary for all imports or exports of semen or Embryos.

e. The excellent facilities at CSWRI & CIRG need to be utilized as referral clinical diagnostic Ihi5 | itials as in America, Europe and especially Germany. A sheep and goat clinic needs to be established in each SAU to serve as referral clinical at state level white as CIRG & CSWRI* should act as National referral institutes for Goat and sheep respectively at par with IVRI which looks after large animals.

f. The global importance of sheep and goat meat being comparatively free of major zoonotic diseases threats needs moe emphasis on health, reproduction, physiology, nutrition, meat technology and extension of these technologies to poorer of the poor in this field. It is high time to re-organize these two species-specific institutes at par with IVRI & NDRI as deemed University as they are already serving as national institutes for sheep and goats.

References

Adams, D.S and Crawford, T.B, 1980. CAE (Caprine arthritis-encephalalitis)- a viral syndrome in goats. *Inht. Goat & Sheep Research* 1 (2): 168-172.

Bachoo, B.A and Wani, G.M, 1991. Induction of oestrus in corriedale ewes during anoestrus. *Indian J of Anim sci* 61(4): 407-408.

Charke, G.L and Osburn, B.I, 1975. Border disease like syndrome found in California lambs. Proc. 18th Annual meeting of American Association of Vety. Lab. Diagnostics 1975: 303-325.

Danesmand, S.S; Cheung, C.Y and Bruce, A.A 2003. Regulation of amniotic fluid volume by intramembranous abortion in sheep. Role of passive permeability of vascular endothelial growth factor. *Amer. J. Obstet. Gynecol.* 1988: 786-793.

Dorny, P; Gasman, C; Sani, R and Vercruysee, J. 1993. Toxoplasmosis in goats - a sero-epidemiological study in pennensular Malaysia. *Annals of Tropical Medicine and Parasitology*. 87(4): 407-410.

Faldyna, M; Toman, M and Pavlik, I. 1999. Leutocyte counts and Lymphocyte subpopulalzons in peripheral blood of pygmy goats from herd infected with Mycobacyterium avium subspecies pseudotuberculosis veterinarrie- Medicina. 44(9): 259-262.

Gilbert, W.M; Newman, P.S; Eby-wilkens, E and Brace, R.A 2002. Technetium to 99 m rapidly crosses the ovine placenta and intramembranous pathway. *Amer. Obstet. Gynecol.* 175: 1557-1562.

Haurrigan, J.L and Klingsparn, A. L, 1975. Epizootology of Blue tongue- the situation in USA. *Australian Vet. Journal.* 1975; 51: 203-208.

Jackson, P.G. G and Cockroft, P.D, 2002. Clinical examination of farm animals. Backwell Science Publishing Co, Australia. Website - www.blackwell- science.com.

Keisler, D:H, 1999. Sheep and goats in Encyclopedia of Reproduction. E. Knobil and J.D Neill NLS. Vol 4, pp 479-492. Academic Press, San Diego, 1999.

Muttoo, F.A; Buchoo, B.A and Wani, G.M. 1989. Lambing process in ewes. *Int. J. Anim. Sci.* 4: IK.S-I85.

Pitzolis, G. 1987. A review of Blue tongue in Cyprus with evidence of windborne spread vectors. Dept. of vet. Services, Ministry of Agriculture and Natural Resources, NICOSIA, Cyprus.

Polydorou, K. 1983. Cyprus Annual report of department of vet. Services. 1983-84. pp 224.

Lushsinger, D.W and Anderson, R.K. 1979. Longitudinal studies of naturally acquired brucella tiltorlion infection in sheep. *American. J. Vet. Res.* 40(9) : 1307-1312.

Srhroeder, C; Seelinger, F; Gaede,W; Westermeier and Granter, M. 2001. Tierarzt liche-Praxis - Ausgabe- G-Grosstier/Nutztiere. 29(1): 19-26.

Smith, IIA; Jones, T.C and Hunt, R.D veterinary pathology. Lea & Febiger Philadelphia, 1972.

Hutitk-M, N.A; Risam,K.S; Mattoo, F.A and Wani, G.M, 1990. Lamb mortality in exotic sheep. *Indian Vet. J.* 67:51-54.

□□□

Chapter 6

Nano - Stem Cell Technology Use in Mammalian Reproduction

G.M. Wani, Paras Wani & Gazanfer Wani

ABSTRACT

Early embryonic development needs detailed investigations if we have to reduce embryonic mortality. Embryonic mortality details have been reviewed ***(wani, 2005).*** The reproductive abnormalities have been discussed and a sequential photographic presentation of embryonic mortality have recently been published ***(Khatoon et al 2007).*** The critical period when maximum embryonic losses go unnoticed is the period before attachment. Losses before actual union of foeto-maternal tissues can be missed and calculations of embryonic mortality estimates remain unaccounted for. Some recent studies of ***(Wani 2006, Khatoon et al 2006)*** have paved a new innovative model for investigation on foeto-maternal development and its attachment. This paper describes the detailed embryonic development, blastocyst elongation, formation of cotyledons within elongated blastocyst, their ramifications and networking of vasculature and formation of foetal placental unit. Conversely, a detailed photographic evidence of uterine receptivity, formation of a multicoated immunological barrier- the beginning of foeto- maternal- barrier mechanism along with caruncle formation, uterine milk formation and finally union of cotyledous with caruncles has been photographically documented.

Introduction

Second and 3rd generation animal biotechnologies as reviewed (Wani, 2005) are multiple ovulation, embryo-transfer, preservation, cloning, micromanipulation splitting of embryos and nuclear transfer. The 4th Generation reproductive technologies have advance beyond our imagination and now we do have transgenic goats giving milk rich in insulin. (Wani 2007). The stem cell technology, embryonic cell culture, pronuclear micro-injection along with sperm-mediated, transgenesis have been in use and are precisely reviewed and discussed (Wani 2005, 2007).

The application of this advancement for diagnostic, therapeutic and economic uses especially its transfer to human medicine or for field application in animal industry needs further research and evaluation.

Another fact which can be of importance is the number of cells per embryos. During first day these may be two cells but after 8 days the cell number increases to 120 or more. It takes 32 hours for the oocyst after fertilization to reach the two cell embryo. Another 13 hours are needed to develop it into a four cell embryo. In another 15 hours 8 cell stages is reached. It is from here onwards that it takes 30 hours for a embryo to be converted into a Morula. This has been one of the most critical periods of embryonic development in ovine. A thematic diagram of various stages of follical maturation of oocytes development and further development has been reported (*Wani 2005a*) figures from web site also show that the protein content of 16 day old embryo increased by 7500 times. Morphologically the embryo size varies from spherical, ovid to elongate after day 13th past coitus. New protein synthesis per unit of embryo is maximum during 8-13th day old embryos. Similarly phosphorylation per unit of embryo is also ten times more in 8 day old embryos than the later stages. Glucose and lactose production increased up to day 16th and the maximum metabolic weight is up to 14th day past coitus. For increased productive synthesis, aminoacid requirements are made, through production of allanic and glutonic acids by the embryos. The matobolic products get rid of amines. The signal transduction from maternal to foetal controls have been discussed (Moris *et al.* 2001).

Dominant Follicle Dynamics

The new animal, products, medico-Veterinary packages, profitable genetic (trait) improved DNA Chips, more resistance giving gene make ups shall have a new role in future. For the next coming super advanced technology driven century, we must be prepared to use the 4th generation technologies for animal Welfare and Protective mechanisms. We thought and used

Laparoscopy and ultrasonography 30 year ago (Wani *et al,* 1979, Wani 1981,1982), which has now become a "buzz" word and further sophisticated and modernized by my students across the world (Aggarwal, 2006, Wani, 2006). Our concept of DNA finger printing of cashmere goats in 1990's (Khatlani *et al,* 1995) has now crossed the barriers of technology advancement. Today, in these pages, we shall be presenting our own results on :

1. Embryonic development, formation of cotyledons, blactocyst development ramification of vascular system, net working of blood vessels to form initial placenta, their union and attachment processes. We shall describe morphological physiological, endocrinology, histochemical and receptor analysis to support our observations. This is 5th generation reproductive technology base for development of in-vitro-foeto models.

   For 6th–10th generation reproductive scientists we shall present theoretical hypothesis a vision -2050 in the form of.

2. Use of stem-cell, germ-cell and Nano-technologies for production of disease free, highly prolific reproductive gametes, their preservation and conservation. The use of these and other molecular and new-generation DNA-related bio-diagnostics shall also be discussed.

Uterine Morphological Changes

The early pregnancy uterine morphological land marks were described only externally per rectum in bovine (***Zamznis, 1976***) some of its significant features have been summarized in a booklet (***Wani, 1986***). It describes easy palpable signs of early pregnancy in bovine and compares its observations with equines, sheep and goats. Some of the earliest indications of pregnancy by physical examination were:

i) Presence of corpus luteum of early pregnancy 15 day post-coitus

ii) Asymmetry of uterine horns 30-35 days post coitus

iii) Fluctuation around 40-49 days in cattle. Thereafter the advanced pregnancy signs like foetal membrance feel and slips. Palpation of placentomes and full foetal or known foetal membrane slip are felt.

Various developments of foetal signs have been described (Wani, 1986). The month wise morphological assessment of foetus in sheep has been reported (Mufti, 1995, Mufti *et al,* 1997). Various reviews reports and books now describe events in development of foetus (Wani, 2005, 2006, 2007).

Development of Conceptus-*In Vitro* Foetus

Pregnancy is an intimate association between mother and foetus. The protective covering on maternal caruncles and over wrap of foetal cotyledons with Blastocyst membranes, transformed into foetal membranes give the "Homograph" status to foetus. The mechanism of ruminant uteri has been discussed (Gray *et al.* 2002). Its more recent findings and attachments are the topic of one of our students.

We have tried to univel many obscure mechanism on blastocyst formation, plancentation and development of cotyledons. The protective mucin coverings over uterine caruncles and their union with cotyledons has been snaped.

The internal changes have been correlated with biochemical, morphological and histochemical and histoanatomic feature of the gravid and non-gravid uteri.

Foeto-Maternal Interaction

The estrogen, progesterone in foetal and maternal blood have been estimated along with estrogen progesterone receptor concentrations in the pregnancy and non pregnant states.

Endometrial receptivity has been discussed (Huct-Hudson, et al 1989, Wani 1996, Wani, 2003, 205, 2006). Proliferation of uterine cells is influenced by E2 and P4. These hormones have their receptors in uteri called as E2 (ER), P4 (PR) receptors (Lee and Demyo 2004) we have also investigated the same and are being analysed. Our studies indicate certain changes which project a new mechanism of foeto-maternal unions (Khatoon and Wani, 2007).

Various growth factors described (Wani 2006) may be helpful in future for invitro-foetal growth.

Reproductive Management

The reproductive management strategies to enhance meat production could safely be divided into two kinds:

i. *Traditional management:* Animal are still managed under zero- input/ migratory /tribal or extensive management system. The application of above described methods will be limited. They may indirectly help in this system by better sire availability.

ii. *Modern management:* Zero-grazing, stall-fed or semi-intensive, commercial goat farming system may use the above methods more and more to produce high quality germplasm. They may use variety of improved reproductive technique or reported (Wani,1996; Wani et al., 200a; Wani et al., 1998; Mufti et al., 1997; Wani and Sahani, 1981; Wani et al., 1986, 1987, 1988; Wani and Buchoo, 1991; Buchoo and Wani, 1991; Wani, 1989, Wani, 1995; Wani, 1984a and b; Wani and Sahani, 1980a).

Enhancement of Fertility

Desired fertility objectives have been discussed (Wani, 2001). Increased prolificacy, early rebreeding, safe/clean parturition, set-time animal crop, elite gene introduction are some of the desired fertility enhancement goals. The constraints in achieving desired objectives in small ruminant livestock sector are the problems of inbreeding, reproductive losses (Wani, et al., 1980; 1981; 1994), pre-pubertal mating, indiscriminate or illicit mating, venereal diseases, nutritional and other stresses. These above referred factors reduce fertility status of our ruminants especially under transhumane migratory systems.

These constraints mentioned could overcome by adopting enhancement strategies (Wani, 2001), which include exchange of sires by tribal, Gujjar, Bakarwals and Chopans to reduce inbreeding. This could be ensured through legislature or forced castrations and introduction of sires from outside the flock. The pregnancy detection methods described (Wani, 1981) and later on used extensive in J&K under farm condition (Wani et al., 1998; Wani, 1989) needs adoption as a routine in each flock for better care of offspring and dam.

Corpora Luteal Biosensitivity

Corpus luteum growth and development is the first earliest pregnancy signal along with maternal recognition factors discussed (Wani, 1996). The corpus luteum development and receptivity to PGF_2 alpha is now known. The ascending and descending corpora lutea are insensitive to PGF_2 alpha. What makes them insensitive and what resurges their sensitivity need exploration. May be in future, CL insensitivity might have direct effects on survival of the conceptus. Does conceptus make CL more sensitive to PGF_2 alpha? Or in other words, does the gravid CL and non-gravid CL behave similarly or differently to PGF_2 alpha. Many such questions vis-a-vis the ovarian morphology, uterine environment and behavioral patterns need to be studied and co-related to develop a manipulative and curative safe system. In some species CL of pregnancy is necessary till delivery, in others it is needed only till placenta develops. Its complete genesis and biostimulation and bioregulatory roles are obscure and needs further research.

Dominant Follicle Dynamics

The dominant follicle dynamics may be in focus for ovulation control. Dominant follicle or would be bride at each ovulation appears around day 7 of cycle in women. It occurs much earlier on day- 1 to 0 in ruminants (cow, sheep,goats). This time point may be the focus to try future selective interferences in ovulation control. Antagonist use during this phase has shown decreasing levels of estrogens in serum thereby indirectly hinting at dominant follicle alteration, as it is one of chief source of this hormone during this phase (Fraser, 1987). Dominant follicle in its sojourn to freedom and release at ovulation seems to attain some specific qualities. Why? It alone releases itself from bondages in ovary (ovulation) and rest of its competitors are hindered through the process as atretic follicles. What make this dominant follicle to attain these qualities among a vast pool of follicles needs further study. In coming 50 years one shall possibly understand and interlock a pre-determined messenger system triggering dominance in one or few follicles destined to become the new offspring of that species. This system when decoded shall offer new opportunities in controlling ovulation both for contraception (women) or augmentation (animals). New genepool of dominant follicles from dead high yielding animal ovaries would help to propagate and multiply the elite.

Over stimulatory effect during superovulation could be avoided by use of antagonists, as dominant follicles seem to evade suppression, thus follicular selection will prevent excessive stimulation and cystic follicles. Follicular fluid constituents containing occyte awaiting for biochemical signals to resume (meiosis) growth and development to ensure timely and healthy ovulation need more exploration (Lenton et al., 1988). The specific signals triggering occyte release might help in *in vitro* occyte maturation success rate.

Development of Conceptus-*In Vitro* Foetus

The lean meat production needs more studies on development of conceptus. The prenatal development undergoes in three stages. The early pregnancy signals, skeletal & nervous tissue growth, on priority. During mid-gestation the hind and fore-limbs develop. During late gestation muscular development along with peripheral extremities including hair develop. In order to increase lean meat production, research in needed to investigate relations between hard and soft tissues. Maternal and foetal interactions needs a probe. Some basic studies (Mufti, 1996) indicated prospects of manipulating lean meat production (Wani, 2001, 2002 a) by bio-stimulators during late gestation, which may help, the production of muscular lambs/kids at birth, besides, increasing birth weight. More details have been described (Wani, 2001, 2002a). More studies in future are needed to reveal the exact mechanism involved in critical transition of power control from maternal to embryonic genome. Application of recent techniques of reverse transcription, PCR and others may reveal the details (Schultz et al., 1990).

Slaughter House Follicle Pool

A good number of our endangered species die where climatic conditions are hard and harsh. Under temperate climatic conditions dead bodies are entrapped in snow for long. It gives a hope to revive their follicular pool. Our studies on resumption of life using follicles from slaughtered sheep and fertilizing them with epididymal semen was an pioneering attempt in this direction (Wani, 1996). The protein content of unfertilized ovum is around 100 mg in rabbit, 27 mg in mouse (Brinister, 1967, 1971). The small genetic material needs extensive analysis to unveil vast informations in its codans and anticodans. Further ovum maturation or resumption of meiosis is initiated during birth or shortly before birth and thus the pool of follicles remains fixed at birth. Therefore, foetal ovary has to be the subject of further manipulations especially in view of its immunological insensitivity (Guraya, 1985).

Gamete Biosafety

Bacterial counts in semen in unprecedented numbers effect semen fertility (Kher, and Dholokia, 1985), The contaminational hazards have increased due to environmental pollution, water and air contamination. Above all the development of resistance of known bateria and other micro-organism to antibiotics (Gupta and Maurya, 1993). We have examined the semen of about 40 bulls maintained at the Frozen semen bank, Rambirbag, Kashmir for bacterial loads in semen and tested antibiotic sensitivity (Koul, 2002).

Presence of bacteria in large numbers in semen, effects its quality and fertility (Kher and Dholokia, 1985). The source of contamination of semen vary from inflammatory foci and lesions on penile apparatus, contaminations on equipments, appliances, diluters, extenders, air and laboratory environment. These contamination hazards have increased with environmental pollution. Many bacteria having developed resistance to routine antibiotics used (Gupta and Maurya 1993) in semen extenders has made inseminations more critical and vulnerable. More studies in this direction are needed to develop a bio-safe and infection free germplasm pool of frozen and fresh semen in future.

Foeto-Maternal Interaction

The role of endometrial stromal cells during pre-implantation stage needs further research. Some studies on ultrastructure of endometrium (Brinsfield et al., 1974) are still not sufficient to answer the questions. What is the role of immune cells and endometrial cells in building of the immunological barrier to protect the foetus. Foetal biology is abarren field of research in the world. After few initial steps by Dr. Moris Young in UK and D. Srivastava in India, only scattered information is available on the subject (Wani and Buchoo, 1990; Mufti, 1996). Much is still to be unveiled by 2050. The placental hormone influence mammary development and milk production of the dam. Foetal membranes serve as Chief source of nutritive exchanges. The arrangement of cotyledons, caruncles, positioning of single and twin conceptus, indicate complex mechanisms with common roots. (Mufti *et al.*, 2000). These mechanisms of foeto-maternal interactions and relationships emphasize proper foetal number diagnosis and corresponding materno-nutritional management strategies.

Nanotechnology

In the proceeding pages we have seen the role of biotechnology in animal reproduction. The coming generations will harvest the gains in the form of more micron-level-accurate technology being replaced with nanogram-level-invisible technology- the nanotechnology. Now even converging technologies are shaped resulting from merger of Nano- science and Nanotechnology with biomedicine, bio-sciences, biotechnology, genetic engineering and information technology.

The cognitive sciences like Neuro- sciences, computing, communication combined with nanotechnology may be so mystic and deterministic that science shows will appear as "*Magic shows*". This new technology wave may define our spiritualistic wavelengths. Many unsolved or resolved facts on birth and genetic make up shall send such messages as was known of Sagic saints, peers and Fakirs.

The birth of a technology needs patience of public and policy makers. In the past "*Michael Farady*" - the inventor of electric generator, was asked by the then Chancellor of Exchequer, as to what use is the electricity you are generating. He argues but no argument convinced the accountant. Finally, he promised him that the day will come when you can tax it. Today a remote destination on this earth demands 24 hr electricity. They never argue its price, but its availability as savoir to life be it medical, veterinary or any other field. Thus like Farady I profess and advocate experiments on use of nanotechnology in animal reproduction today. It may sound strange, but yes in my life I shall fore see a role of nanotechnology in Animal reproduction.

What is Nanotechnology

Technology related to features of nanometer scale. Thus we will unveil gametes in minute details; sperm head colliding zona-pellucida shall be filmed. The radial blast of cortical reaction shall be known. All chemical messengers shall be unveiled at nano-level.

The shape of gamete, Zygote, embryo and foetal development at nanolevel shall bring forth new additions to our knowledge. The use of such information through use of computer languages, both hard and soft shall make this knowledge dispensable and useable at remote-corners of the Universe. Combing cognitive, mythical and mysterious happening within womb could be registered and translated via nano-ultra-sonography.

Thus futuristic view of nanotechnology will be nanographic follow of information and has been described (Mordini, E. 2007).

Nanotechnology may allow us to send tiny nano microprocessors into the animal womb to send us signals which we wish to know for prevention of diseases. The stem-cell-germ-cell technology combine with Nanotechnology shall bring new hope and future for the propapagation of Gene-technology-ETT- A-I Combine. It may sound like sending a wave of spirit to fertilize an "egg" - thus proving the God Word in *Quran* of creating the *Chirst* through mystic spirit. This and other convergence of known reproductive technologies are challenge to us in future.

Nanaoparticels developed can be agents which would use blood and carry durgs for curing disease at molecular levels. Thus a new mechanism of targeting infective agents of smaller dimension could be a possibility.

Stem-Cell-Technology

Stem- cells are undifferentiated cells. The embryo at morula stages have undifferentiated blastomeres. These cells can transform themselves into 200 or more cell type, which could be used to repair or regenerate new desired cells. This advanced cell research may help diseased people like those suffering from cancer, Alzheimer's disease syndrome, Parkinsons disease and even paralysis.

The stem cells are derived from 4-5 day old embryos or fertilized cells. The stem-cell-technology has a new role to play in Animal reproduction.

Firstly, the stem-cell source could be pooled out of slaughter-house oocytes or from the vast pool of embryos hatching out in many animal species.

Secondly, the cell source could be obtained from foetal culture and if new-grown-immunoprotectant cell of foetus could help or provide germ-cell specific to any particular organ, the repair of organs could be a new revolution.

References

A. Khatoon, G.M. Wani, A.R. Choudhury, M.Z. Khan, R.Islam and B.A.Pandit,(2007) Incidence of reproductive abnormalities in small ruminants. *The Indian Journal of Animal reproduction*: 28(1):53-56; June 2007 www.buzzle.com.

Buchoo, B.A. and Wani, G.M.1991.Induction of oestrus in Corriedale ewes during anoestrus. *Indian Journal of Animal Science, 60(4) : 407-408.*

Buchoo, B.A.,Wani,N.A.,Khan,M.Z and Wani, G.M.2000.Laparoscopic detection of Ovulation and pregnancy in different sheep breeds. *Indian Journal of Animal Sciences 70(2):61-62.*

Karsch, Fred, J. and Evans, Neil P.(1996) Feedback actions of estradiol on GnRH secretion during the follicular phase of the estrous cycle. *Acta Neurobiol. Exp. 56:715-725.*

Karsch,Fred J.; Bowen, Jennifer M.; Caraty, Alain; Evans, Neil P, and Moenter, Suzanne M.(1997) Gonadotropin-Releasing Hormone Requirement for Oculation. Ovulation. *Biology of Reproduction 56:303-309.*

Kher, H.N and Dholakia, P.M. 1985. Bacteriological studies of bovine semen. *Indian Journal of Animal Reproduction, 5(1):78-84.*

Kledzik, G.S.; Cusan, L.; Auclair, C.; Kelly, P.A. and Labrie F. (1978). Inhibitory effects of luteinizing hormone releasing hormone agonist on rat ovarian LH and follicle stimulating hormone receptor levels during pregnancy. Fertility and sterility, 29:560-564.

Koul, N.A.2002 Bacterial flora of bovine semen and their antibiotic sensitivity pattern. M.V.Sc thesis, SK. University of Agricultural Sciences & Technologyof Kashmir, Srinagar-191 121.

Lee,K.Y and De-Mayo, F.J; 2004, Animal Models of Implantation Reproduction. 128, 679-695.

Lenton, E.A.; King, H.; Thomas, E.J.; Smith, S.K. MeLachlan, R.I.; MacNeil, S. and Cooke, I.D. (1988). The endocrine environment of the human oocyte. *J.Reprod.Fert.82: 827-841.*

Martin, K.; Santoro, N.; Hall J.; Filicori, M.; Wurman, M. and Crowelery, W.F. Jr. (1990). Management of oculatory pulsatile gonadotropin-releasing hormone. *J. Clin Endorcinol and Metab. 17: 1081A-1081G.*

Mordini, E, 2007. Coverging technologies- the next challenge. Conference on converging technologies. University of Montreatl (Quebec) Canada. 22-23 Oct, 2004 pp 1-8.http:11 agora.qc.ca/collogue/gga.nsf/conferences/converging-technologies.

Mufti, A.M., Wani, N.A.; Wani, G.M., Mir, M.M. and Khan, M.Z,1997 Superovulatory response in sheep during different months of breeding season. *Small Ruminent Research*, 25:181-184.

Mufti, A.M.1996.Studies on morphological characters and some metabolic constituents at different stages of development of ovine foetus. M.V.Sc. thesis, Sher-e-kashmir University of Agricultural Sciences and Technology, Shalimar, Srinagar.

Najar, A.M.(1999). Effects of fish ovarian extracts on mammalian ovary, Ph.D. Thesis of University of Kashmir, Srinagar-190006.

Wani G.M. and Buchoo, B.A. 1991. detection of induced ovulation by laparoscopy in ewe lambs. *Indian Journal of Animal reproduction, 12(1):76-77.*

Wani G.M. Goswami P; Khatoon A; Wani P, Wani G. 2006 Mammalian Foeto-maternal attachment- a new invention www.buzzle.com http://www.buzzle com./edit orials/9-5-2006-107646 asp.

Wani G.M., A. Khutun, P.Goswami; G.Wani and M.Z. Khan (2006[a]) Uterine changes under normal and abnormal corpus luteum in sheep and goats. Lead paper presented in XII Annual ISSAR Conference, held at MHOW, M.P (India), November 10-12, 2006.

Wani G.M., A.Khutun, P. Goswami and M.Z.Khan (2006_b) Ovarian and foetal abnormalities in small ruminants . Pesented in National seminar on Innovations and Recent advances in reproduction for augmenting small ruminant production held at CSWRI, Avikanagar (India), December 28-30, 2006

Wani G.M., (2007) Recent Trends in Policy Initiatives & Technological intervention Goat production as livelihood resource, presented at Sri Venkateswara Veterinary University, Tirupati w.e.from 20-22, June,2007

Wani G.M.,Khatoon A., 2006 ,Early embryonic development in sheep and goats www.buzzle.com http:/www.buzzle.com/edit orials/8-7-2007-104739 asp.

Wani, G.M and Sahni, K.L. 1980. An ultrasonic technique for detecting pregnancy in sheep and goats. *Veterinary Research Bulletin, 3:35-37.*

Wani, G.M, Sinha, N.K. and Sahni, K.L. 1980. Note on gestational oestrus in the Muzzafernagri ewes. *Indian Journal of Animal Science, 50(11):1009.*

Wani, G.M. (2002b). Advances in animal Reproduction in: Proceedings of National Seminar on "*Recent Research Trends in Life Sciences*", (August 26-28, 2002) organized by PG Department of Environmental Science and Centre of Research for Development, The University of Kashmir, Srinagar- 190006.

Wani, G.M. 1982. Laparoscopy in farm animals. *World review of Animal Production, 18: 7-13.*

Wani, G.M. 1984a. Vet suche Zur Embryo Production and Kultur bei Ziege. Dr. Med.Vet Dissertation, Tieraertzlichen Hochschule, Hannover, Germany.

Wani, G.M. 1984b. Investigations on ovarian activity, ovulation and pregnancy diagnosis in sheep and goats. Ph.D. Thesis, IVRI, Rohikhand University, Indian Veterinary Research Institute, Izatnagar, U.P.

Wani, G.M. 1986. Embryo transfer in goats. 6th National Congress on Animal Reproduction, Guwhati 28-30th November, 1986.

Wani, G.M. 1989. Detection of foetal age in sheep and goats by ultransonic technique. *Indian Journal of Animal Science, 59(12):1525-26.*

Wani, G.M. 1989. Ovulation detection by laparoscopy in goats. *Indian Journal of Animal Science. 59(12):1527-28.*

Wani, G.M. 1989.Detection of foetal age in sheep and goats by ultrasonic technique. *Indian Journal of Animal Science, 59(12):525-30.*

Wani, G.M. 1995. Early diagnosis of pregnancy in surrogate mothers. National symposium on embryo transfer and genetic engineering technologies in small ruminants. May 7-8, 1995, Makhdoom, Farrah. India.

Wani, G.M. 1996. Embryo biotechnology in sheep and goats. *Published by Valley Book House, Srinagar- 190006.*

Wani, G.M. 1998.Ultrasonic pregnancy diagnosis in Sheep and goats. *World Review of Animal production, 17:43-48.*

Wani, G.M. 2001. Enhancement of chevon production through reproductive management. In: Proceedings of IV National Seminar of Indian Society for Sheep and Goat Production & Utilization (ISSGPU), held at Bharathiar University, Coimbatore from November 9-11, 2001.pp.60-68.

Wani, G.M. 2002. Goat Husbandry/multipurpose management. Encyclopedia of Dairy Sciences. Edited by Roginishi, Hubert, Enquay, Jhon W. and Fox patrict, Academic Press, London, NW 17B4MS192.

Wani, G.M. and Buchoo, B.A.1990. laparoscopic view of ovine conceptus. *Indian Journal of Animal Science, 60(5):564-568.*

Wani, G.M. and Sahni, K.L. 1980. Ultrasonic pregnancy diagnosis in sheep and goats. 2nd *All Indian Symposium on Animal Reproduction, Banglore, pp. 1-5.*

Wani, G.M., Sinha, N.K. and Khan, B.U. 1987. Oestrus synchronization with progestagens in Muzzaffarnagri ewes. *Indian journal of Animal Science. 57(12)*

Wani, G.M., Sinha, N.K. and Sahni, K.L, 1980. Haematic type of foetal mummification Muzaffernagri ewes. *Indian Veterinary Medical Journal, 4:46-47.*

Wani, G.M., Sinha, N.k. and Sahni, K.I. 1980. Note on seasonality of breeding behaviour in Jamunapari goats, *Indian Journal of Animal Science, 50:1153-55.*

Wani, G.M., Sinha, N.k. and Sahni, K.L. 1981. Incidence of abortionsand stillbirths in Muzzaffarnagri and crossbred ewes. *Indian Veterinary Journal, 58:558-561.*

Wani, G.M., Tripathi, S.S. and Saxana, V.B. 1980. Sialoproteins in the cervicaql mucus of repeat breeding cows during oestrus. *Indian Veterinary Medical Journal, 4:49-53.*

Wani, N.A., Wani, G.M., Mir, M.M. and Beg, M.A. 1998. Ultrasonic pregnancy diagnosis in Gaddi goats. *Small Ruminant Research, 29: 239-240.*

❑❑❑

Chapter 7

Reproductive Technology Applications

G.M. Wani, Paras Wani, Talat Zahir & Gazanfer Wani

Abstract

Reproductive health management in Ruminants is a buzz word called Total Quality Management (TQM). Goals to achieve reproductive efficiency means 12-13 months calving interval, low repeat breeding and early rebreeding after parturition. Less than 2% abortion & less than 10% retained placenta and metritis in herds may ensure less than 18% culling rates.

In a developed country like USA reproductive efficiency has decreased. A total conception of 51% in 1955 has reduced to 38%. Does it speak we have to make use of modern technologies of ETT (Embryo transfer technology). The old conventional sleeve, rectal palpation, CL evaluation, reproductive disease diagnostic, has to be replaced with ultrasonography, laparoscopy, DNA guided diagnostic and improved fertility assessment technologies. Various modern technological impact on improving reproductive management have been discussed.

Uterine diagnostics for reducing (COD) Cystic ovarian disease, anoestrus, reproductive losses and infertility needs emphasis. The reproductive techniques consist of cryo-preservation, sperm sexing, spermatogenesis, invitrofertilization, nuclear transfers and introduction of disease resistance and animal welfare too. The reproductive difficulties in artificially produced calves, and stress to their mothers are averse to animal welfare. Various methods and their effects have been reviewed.

A new role of foeto maternal effects of hybridization of breeds has come to surface. Various such events have been expressed. New reproductive concepts such as Nitric oxide as a signaling molecule, Endothelial, nitric oxide synthase and its effects on reproductive ability have been explained. New knowledge models, like development of ovine concepts, foeto maternal interaction, extra cellular hormone action, cytokines, relay signals and their conversion to monomeric and dimeric forms have been postulated.

Hypothalmic astrocyte concept model with its biological control system has been detailed. Major technological innovations by the author in last 2 decades have been shown as pictorial monographs. A new concept of Dairy economic Vission 2050 for India has been forecasted.

A global policy prospect of Livestock industry for India has been attempted to glow Livestock marketing competition. A possible mega role for small farmers enterprise with mixed farming modules is presented. New generated technology transfer, research needs and gaps have been identified for young researchers.

Introduction

1. Total Quality Management

(TQM) is the current buzzword in the production of quality food animal products for human consumption. The TQM concept implies that appropriate attention is given to all facets of dairy herds management in an effort to satisfy the consumer and their needs. Maintaining good records and using computer driven technology is recommended. Goals that are realistic and made possible by adopting a reproductive health programme are:

i. Average calving interval between 12 to 13 months
ii. Breeding efficiency below 1.5 services per conception and less than 30 percent returns after 60 days.
iii. Postpartum interval to first standing heat between 30 to 40 days.
iv. Postpartum interval to first breeding no greater than 55 to 70 days.
v. Repeat breeders of 8 to 10 percent or less,
vi. Abnormal anestrus after 60 days postpartum reduced to 2 to 15 percent.
vii. Abortions of 1 to 2 percent or less.
viii. Retained placentas of 5 to 10 percent or less.
ix. Metritis of 5 to 10 percent or less.
x. Culling percentage for reproductive failure below 18 percent.

2. Reproductive Efficiency

The famous reproductive scientist and clinician Zemjanis, compared the reproductive efficiency attained in U.S.A during 1955 with that of herds maintained in 1985. The first service conception had decreased from 53.4% to 38.7 in 1985 and the corresponding figures for total conception rate were reduced from 50.7% to 37.7%. The percentage of open cows at P.D. examination increased from 22 to 27. Repeat breeding was around 15% calving interval was 13.2 - 13.4 months. The prenatal deaths increased from 3.4 to 9%. Thus decline of reproductive efficiency in 1985 or 2007 over past 3-4 decades was observed. Is it that technologies of 1955 are not able to deliver efficient reproduction in 2007 ?. During the period a huge new set of technologies came into existence. Is that we need to incorporate new technology refinement in herd management especially reproductive ability ?. Should we go beyond traditional rectal palpation or sleeves and visits to new monitoring and evaluation through ultrasonic or other new technology?. A variety of application of new technologies are needed for heat detection. CL evaluation, reproductive disease diagnosis, semen evaluation, pregnancy diagnosis, infertility and other reproductive losses. This is where we need "Reproductive health Management through new techniques". (Zem janis, 1987).

3. Uterine Diagnostics

The Uterine Condition can be disgnosed by Rectal palpation, vaginal speculum examination, uterine swab culture, uterine biopsy, ultrasonic imaging of uterus and ovaries, and progesterone testing. Abnormal postpartum cows are identified based on uterine horn size and texture in addition to vulvar discharge. Cows identified as abnormal are treated, with intrauterine infusion of antibiotics, antiseptics, chemical irritants, prostaglandin injection, or a combination of these agents. Most cows (>80%) should ovulate by 30 days postcalving. Cystic ovarian disease (COD) may occur in 5 to 10% of cows postpartum and delay first ovulation and first breeding. Anestrous cows will have small inactive ovaries with little change over sequential examinations. Follicles <15 mm in diameter may be found on examination, but fail to ovulate. Diagnosis of pregnancy at 35 days has enhanced value compared to 45 days or later. Pregnancy exam at 35 days reduces days between breeding and the percentage of cows over 48 days between breeding. (Ferguson, 1995).

4. Reproductive Biotechniques

Based on the progress in scientific knowledge of endocrinology, reproductive physio- biology and embryology during the last fifty years, new bio-techniques have been developed and introduced into animal breeding and husbandry. Among them are oestrussynchronisation/ ovulation induction, artificial insemination, multiple ovulation induction, embryo transfer (MOET), in vitro embryo production (IVP) and cloning by nuclear transfer. The aims of these reproductive technologies were initially to speed up the genetic improvements of farm animals by the increase of offspring of selected males and females, and the reduction of the generation intervals. The techniques of cloning by nuclear transfer applied for experimental purposes, have the prospect of more practical implementation in near future. The uniformity of herds for an early management or for the multiplication of transgenic animals after gene-targeting shall be its outcome. Within the farm animal biotechnologies, animal welfare should receive special attention. Transfer of embryos into dairy cows may lead to more dystocia. Intrauterine insemination via laparoscopy or laparotomy with minimal anaesthesia by breeder organizations operates stress in animals.

Both multiple ovulation induction and embryo transfer are generally accepted technically. However the transfer of embryos after multiple ovulation increase in embryonic death, larger calves with longer gestation times, and more dystocia (van wagtendo et al 2000). Some of the biotechnologies used in modern animal reproduction management can be summarized and named as :

5. Reproductive Techniques

Reproductive techniques have been reviewed, wani, 2004 and are briefly reproduced as :

i. Ist Generation Reproductive Technologies

a. Artificial insemination technology
b. Cryo-preservation Techniques.
c. Sperm sexing
d. Opportune time insemination
e. Spermatogenesis.

ii. 2nd Generation Reproductive Technologies

a. Multiple Ovulation Embryo Transfer (MOET).
b. Embryo Development Blocks.
c. Embryo Flushing & evaluation.
d. Embryo preservation and culture.

iii. 3rd Generation Reproductive Technologies

a. Invitro fertilization (IVF)

b. Clonning Techniques.

iv. Future Generation Reproductive Technologies

a. Stem cell technology

b. Embryonic cell lines.

c. Pro-nuclear microinjections.

d. Nuclear Transfers.

e. Viral Mediated Transgenesis.

v. Reproduction Biotechnology Application

a. New animal products.

b. Profitable production traits.

c. Disease resistance

d. Animal welfare and protection.

Successful somatic cloned animal production has been reported in various domesticated species, including cattle; however it is associated with a high rate of pregnancy failure. The low cloning yield could possibly arise from either an abnormal and/or poorly developed placenta. In comparison to control cows, fewer placentomes were found in somatic cell nuclearrecipient (NT) cows at day 60 of gestation, suggesting a retardation of fetal /placental growth in these animals. NT cows not only had fewer numbers of chorionic villi but also had poorly developed caruncles. Macroscopic examination revealed a typical development of the placentomes in terms of shape and size. Histological disruption of chorionic villi and caruncular septum was found in NT cows. Of particular interest was that the expression of genes, as well as proteins in the placentome, was disparate between NT and artificially inseminated cows, especially placental lactogen (PL) and pregnancy associated glycoprotein (PAG). In contrast, prolactin-related protein (PRP-I) signals were comparable across cows, including NT cows carrying immotile fetuses. The expression of extracellular matrix degrading molecule, heparanase (HPA) in NT cows was divergent from that of control cows. Microarray data suggest that gene expression was disorientated in early stages of implantation in NT cows, but this was eliminated with progression of gestation. These findings strongly support a delay in trophoblast development during early stages of placentation in NT cows, and suggest that placental specific proteins, including

PLs, PAGs, and HPA are key indicators for the aberration of gestation and placental function in cows. (Kruip and Van Re, 2008) .There is a large body of evidence demonstrating that, in comparison with in vivo programme controls, the size and weight of IVP calves is higher (30% over 50 kg), the gestation period is longer, the % dystocia and the incidence of caesarean sections is much high. The % abortions and perinatal death are also higher (Behboodi et al, 1995; Kruip and Den Daas 1995; Wagtendon-de Leeuw et al. 1998; 2000). In general the calves are less active (Reinders et al. 1995). In addition the % of hydroallantois and congenital malformation including abnormal limbs and spinal cords, is increased in IVP calves and lambs. Taking together these problems are defined as the large offspring syndrome (LOS) (Young 1998). (Farin & Farin; 1995) and Sinclair et al. (1997) found a differential growth of organs (liver, heart , kidneys and adrenal gland) after IVP. Postnatally too some IVP calves anomalies were observed. (deRoos et al.2000). MOET, including synchronization and induction of oestrus and AI, as well as IVP,NT are sometimes with serious consequences. We suggest (potential) risks of biotechnologies for farm animal welfare should be systematically assessed. The type of research should be multidisciplinary and should make use of appropriate scientifically valid experimental designs and protocols. Results obtained accordingly in developing and using the safest biotechnological methods and procedures. Technological progress which is ethically justified and beneficial for society in general the scientific and agricultural community i,e made use of for increased productivity.

6. New Concepts

Nitric Oxide as a Signaling Molecule

Many years scientists in reproductive research talked of endothelium-derived relaxing Factor (EDRF). This was thought to be a signaling molecule. The discovery that EDRF was in fact nitric oxide (NO). Over 60,000 papers have been published in last 10 years on NO. Nobel prize of 1998 was on this discovery. Nitric oxide is responsible for neurotransmission, immune defence, cell death, (apoptosis) and cell motility. It has a half life of few seconds. Enzyme producing it are (NOS) nitric oxide synthases (Phil, 2007). Nitric Oxide being a small molecule, diffuses rapidly across cell membrane. Its diffusion distances can be several hundred microns and its biological effects are mediated through a range of targets as haem groups, cysteine residues, iron or zinc clusters. Nitric oxide (NO) is now recognized as an important intracellular and inter cellular messenger. It relaxes uterine tissues, by inhibiting uterine contractility. Thus, by reducing uterine movement it helps in uterine quiescence. This is habitual of progesterones and antignostic to PGF2 alpha. NO has also a role in ovulation by interplaying ovarian steriodogenesis and prostaglandins. Progesterone synthesis is reported to have increased in NO dose dependent rats (Yuan et al, 1999) Nitric oxide decreases oestradiol secretion in rats. It is established that NO plays a role in endothelium dependent vascular relaxation, phagocytic cell cytotoxicity, cenbtral nervous system neurotransmission and uterine relaxation (Moha et al, 2001). Granulosa cells and luteal cells in the rat ovary synthesize nitric oxide. Preovulatory follicles in their theca and stromal cell layers have enzyme synthesis. Endothelial nitric oxide synthase (NOSIII) and inducible nitric oxide synthase (NOS II) are among these enzymes. Nitric Oxide is a fast acting signal which has the capacity to block entrance of blood components into follicular fluid. New concept of NOS system operating in ovulation and follicular rupture is now postulated. NOS acts via vascular dilation and ovulatory leukocyte distribution. The reorganization of follicular rupture and formation of corpora lutea requires nitric oxide. Thus, NO is having a role in ovarian physiology and steriodogenesis. A neuropathway may be evinced in future. There is evidence of hypothalmic NOS - containing neurons, that regulate LH. Moreover, NOS containing (NERVE) fibers have also been found in rats. Does NO act as a local modulator of steriodogenesis? as many of its actions are mediated by Iron-containing enzymes, like guanylate cyclase and cyclo oxygene. The upregulation of progesterone and down regulation of oestradiol production in ovaries is now ascribed to NOS - system. NO may act by binding to the iron-sulphur moiety of the enzymes involved in the process of steriodiogenesis and effecting their production. (Yuan et al, 1999; Motta et al. 2001).

7. New Knowledge Models

a. Development of Ovine Conceptus

- Protective covering on caruncles
- Protective mucin covering.
- 18th day transitory attachment.
- 39th day cotyledons appear.
- 40th day projections from elongated Blastocysts into caruncular Ėpithelia.
- 42th day syncytial formation.
- 48th day attachment fixed as placentomes.
- Non-invasive attachment
- Invasive attachment post 40th day.

- 80th Day of Gestation – placentomes, - structured appearance of recognizable individual cotyledons & caruncles disappear
- Intra caruncular and caruncular cotyledonary attachment is supported by evidence from:
  - Histological change
  - Progesterone / Estrogen receptors
  - Biochemical & Morphological evidences.

    (Khatoon, 2007 – M.V.Sc thesis).

b. Foeto Maternal Interaction Transitory Attachment

- Binucleate cells (Gestation day) 18
- Proliferation of Endometrial cells
- E2/P4 receptors
- Growth factors.
- Progesterone receptors increase with gestation progress
- Estrogen E2 receptors almost constant during pregnancy
- Increased E/P ratio.
- Fluctuations in mineral, glucose, enzyme and proteins
- Vaginal, cervical and uterine epithelial changes observed as pregnancy progressed.

C. Extracellular Hormones Actions (Concept by Canadian Scientists)

Growth factor or Cytokines ...... Transcription of genes through specific receptor ..Intracellular signaling molecules (Groner, 2002).

Conversion Monomeric to Dimeric form

Modus: Tyrosine phosphorylation

D. Hypothalmic Astrocytes Employ

*..... Transforming Growth Factor (TGF) Directly to regulate GnRH ..... controls ... Sexual maturation ... & ....Adult reproduction functions.*This is what we name now as: Astrocyte – GnRH – Neuron signaling Pathway

Techniques used : Situ hybridization immunohistochemistry ... ... Motion communication processes.

Bouret et al, 2004 France

E. Biological Control

Multiple signals .... Receptivity or un-receptivity (Ringo et al, 1996 – U.S.A) Hypothalmo -pituitary – gonodotrophic – axis .... Mid gestation develops in sheep - late gestation ... GnRH neuron centers develop in pregnancy or during foetal life ... Only Maturational changes occur at Puberty – so as to form a GnRH neuronal system.

F. Hypothalmo

– pictuitory – somatorophic axis begin in fetal life Midgestation .... But Growth hormone R.H neuronal system does not develop until 10 weeks after birth. (Polkowska, 1995) Poland.

8. "Major Technological Advances by Author in Last 10 Years"

i. An attempt using ovaries and sperms from slaughtered sheep for invitro -fertilization and *in-vitro* maturation techniques were used to produce morula. These techniques innovated and published have aroused world wide recognitions, as awards. This work has been cited by Histcite-Index Lan wilmut founder of Dolly, the Sheep. (Wani & Wani, 2000, 2003). In this era of biosafety and international Zoo and phyto sanitation, an attempt was made to limit bacterial and microbial loads in semen. Invitro bacterial sensitivity and bacterial load of semen was reported, (Seh *et al.* 2000) (Koul 2002) which stimulated a new era of biosafety of gametes and higher fertility rates in cattle.

ii. The hitherto untapped research area of pre-natal development of ovine foetus was attempted (Mufti *et al.* 2000, Wani and Buchoo, 1990). The investigation aroused much interest and citations. Very few studies on this area are available in world. This study is of great value for human and animal health. It has been further investigated during last few years. For the first time in world we have presented the mammalian foeto maternal union in its true form. This is an indirect association. A new concept of foetal growth, differentiation, foetal cotyledons within the elongated blastocysts has aroused the hope for preparation of a *in-vitro* Foetal–Development Model. (Wani, *et al.* 2006a, 2006b). These new concepts were presented in International conferences on Genomic and Proteomic concepts. The models were highly appreciated for use in various drug–gene–disease control interactions. (Wani *et al.* 2006c; d; e). The binuclear and hybrid nuclear cells have been identified to specify syncytia and related mechanisms of foeto-maternal attachments. These concepts shall revolutionize disease resistance, control and prevention.

iii. We tried fish ovarian extract (waste material) to replace costly synthetic hormones. This is ready for patatening. Laparoscopy for ovarian prediction along with ultrasonic use for pregnancy diagnosis in Sheep & goats was innovated and perfected. (Wani et al, 1998; Mufti et al, 1998).The use of hcG for improving lambing rates (Wani et al 2000) and use of laparoscopy for ovulation detection are new replications of our earlier investigations (Buchoo et al 2000).World famous Elsevier / Academic Press invited me to write a paper for Encyclopedia of Dairy Sciences Vol II PP 1259-1270. The only Asian to be invited to contribute on Goat Management Systems is in itself significant achievement summed up by the world famous Prof. John R.Campbell, the president Emeritus of Oklahama State University as "World renowned disciplinary expert, Wise and excellent author". Our work on Economics of Pashmina goats and their Genetic characterization has been applauded by world famous Scientists and has helped Pashmina Industry in Ladakh to produce fine pashmina.iv. Our recent prenatal and research on foeto maternal interactions is one of its unique in the world. Various gynaecological land marks, placental and immunological barriers, caruncle, cotyledon development and their association is of unique nature and can be reviewed.

9. Dairy Economics

Dairy Farming in particular and Livestock enterprises in General suffer from two problems:

1. Higher production and maintenance cost.
2. Low market prices of livestock products.

Global milk pricing varies with Dairy Farm Structures. Dairy Farm Structures are either small 2-5 animal farms as ours or large more than 50 or 100. The countries like USA, EU, Canada, Isreal, Hungry, USA, South America and Oceania have large farm structures. Except Oceania none produce milk at world market price of 15-22 dollar a quintal. Thereby, signifying large farm structures are not always economical. Our small holding size of dairy structure is more competitive than West. This advantage needs to be harvested under WTO-Liberalization Global System. Should this need strengthening of small farmer's co-operative? Do we needs implementation of small farmer friendly policies. The sale of 1 litre of milk in India is less than a litre of water. This needs serious consideration by our planning commission.

a. Higher Production Costs

The costs of milk production are high in Scandinavian Countries and Spain. The yesterday giants are todays loosers. The highest production price is also in counties like Hungary, Czech Republic, Isreal, Bangladesh, Thailand and parts of China. Their cost of production is more than global average cost of production viz 15-22 dollars/quintal. Those who produce at world market prices of 15-22 dollars include Estonia, Poland, Chile, Brazil, India, Vietnam and Oceania. Thus, we have tough competition with these nations to gain export markets. The Switzerland, Norway and Canada produce milk at much higher price than EU and USA, their markets in middle east are our export pastures.

b. Lower Milk Prices

The low milk prices than world market averages are in Argentina and Pakistan. However, they have distribution advantage rather than production axis, as farmers demorcating forums are non-existant. As such they may not pose any competition with us. We have export advantage in Asia as our other neighbour China too produces milk at higher costs. We must prepare ourselves as "Future milk Exporters in Asia". This needs a preparation for zoo-sanitation and export oriented shift in policy decision.

c. North-West-Export-Zone

The northern-west states have self sufficiency in milk. We have to make this North-West-Zone comprising of J&K, H.P., Haryana, Punjab, Rajasthan and Gujarat as our potential milk export-zone. We will have to induce mission modes for its animal disease-free-status, so that we meet international-zoo-Sanitary standards. Poor zoo-sanitation is our major block to exports of milk. Outreach to farmers on these lines is suggested.

10. Global Meat Industry

The next thrust area under transfer of technology has to make India "major meat exporting" country. The global meat industry is slowly moving in the hands of Brasallians and Chinese farmers. Thanks to our vegetarian majority diets we too are on run. Against, daily needs of 25g/person/day we produce only half of it on per capita basis, but our advantage with goat meat abundance can differentiate our export potential from the rest. Even global super powers USA have more demand for Cheron than what is produced. A policy change and frame work has to be drafted. Our product technology packaging and consumer preferences have to be webbed through transfer of technology to export concerns in particular and in general to home users. The focus has to be centered around hygienic meat production, ecological preservation, processing and product technology. This in principle means "reshaping livestock range and forage management" through effective communication and transfer of technology across its terminal and potential users and beneficiaries.

11. Generated Technology Transfer

We have to lead in Embryo transfer technology in Asia as has been done by us in dissemination of AI technology, in collaboration with Swedish expertise. Many international experiences are needed in this and other Transfer of Technology missions. We need bio-technological links to be developed with industry and user departments. Licensing in service mode is desired in technology transfer. We can do it on Cornell University pattern. Strengthening pre-clinical, clinical and basic diagnostic skills to glow animal science Institutes as a word referral clinical centers. This would need emphasis on technician training facilities for physiology, Reproduction, AI, Biochemistry neurosciences, biology and like. Linking, research and extension, faculty, students and farmers under KVK pattern are needed for quick animal science technology transfer. The referral TOT in animal sciences is ready to be switched on in the following area in next few years.

- Rapid diagnostic tools and techniques.
- Tannin degradation technology and low- cost feed packaging.
- High security disease prevalence checks. May be its sharing of emerging disease pattern with public health experts give us better dividends.
- Biotechniques for germplasm propagation and improvements.

12. In short we need Emphasis on

- Global-food- system-chain-factors(GFSCF).
- Research on competitive comparative health and disease (CCDH).
- Internationalize Veterinary and animal science education.
- Out reach for global farming through TOT.(Transfer of Technology).

13. Future Research Needs or Gaps

- Avoiding emerging diseases threats.
- Refined disease diagnostics.
- Drug - disease - targeting.

International Zoo-Sanitation regulations and product safety measures to fit us in Global - food - chain - operative systems.

14. Future Research Targets

i. Targeted gene delivery.
ii. Neuroregulatory mechanisms.
iii. Innate immune system.
iv. Protein interactions - transcriptional conformational mechanism.
v. Gene - drug - designing. Molecule targeting under Genome.
vi. Natural receptivity associated proteins for immune protection imaging and mapping.

References

Wani S.M; P.Goswarrv, A. Khatun; P.Wani and G.Wani, 2006. Mammalian foeto-material attachment- c new invention www.buzzle.com/editorials/9.5.2006.107646 asp.

Wani G.M.; Khatun, A; P. Goswami; S. Mir; Darzu, M.M and M.Z. Khan 2006, Early Embryonic development in sheep and Goats www.buzzle.com/editorials/8.7.2006 104739 asp-29k.

Wani, N.A and Wani, G.M. 2003. Use of epididymal spermatozoa from slaughtered sheep for in vitro fertilization of in vitro matured oocytes. *Ind. J.Anim. Sci.73:143-146.*

Wani, N.A, Wani, *G.M.* et al. 2000. Effect of oocyte harvesting techniques on in vitro maturation and in vitro fertilization in sheep. *Small Ruminant Research 36: 63-67.*

Wani, N.A; Buchoo, B.A; Khan, M.Z. and Wani, G.M. (2000). Use of human chorionic gonadotrophin for implanting lambing rates in different breeds of sheep. *Indian J. Animal Reproduction, 21(2) : 45.*

Sen, B.A; Wani, S.A; Khan, M.Z. and Wani, G.M. (2000). In vitro sensitivity pattern of bad eric isolated from uterine secretions of repeat breeding cows. *Indian J. Anim. Reprod 21(1): 61-62.*

Mufti, A.M; Wani, G.M; Wani,N.A and Khan,M.Z. (2000). Prenatal development of ovine foetus. *Small Ruminant Research. 38: 87-89*

Wani, N.A; Wani, G.M; Mir, M. M and Beg, M.A. 1998. Ultrasonic pregnancy diagnosis in Gaddi Goats. *Small Ruminant Research. 29 : 239-240.*

Mufti, A.M; Wani, G.M; Mir, M.M. and Khan, M.Z. (1997). Superovulatory response in sheep during different months of Breeding season. *Small Ruminant Research. 25: 181-184.*

Wani, G.M. and Buchoo, B. A. (1990). Laparoscopic view of ovine conceptus. *Indian Journal of Animal Science. 60(5): 564-568.*

Koul. NA. Bacterial flora of bovine semen and their antibiotic sensitivity pattern. MVSc thesis 2002, SKUAST-K.

Worn, G.M; Buchoo, B.A. and Wani, N.A. (1997). Use of human Gonadotrophin in supei'ovulations of South down sheep. *Small Ruminant Research 25:93-94.*

Wani, et al, 2006c. Pre-Implantation endometrial and foetal structural changes in sheep. Paper presented at International Symposium on *"Emerging Trends in Genomic and Proteomic Sciences" 15-18* Oct. 2006, Mumbai, India.

Wani, et al, 2006d, Uterine changes under normal and abnormal corpus luteum in sheep and Goats. Lead paper XXII Annual convention & National Symposium on *"Innovative Technologies for Fertility Enhancement in Livestock"*. 10-12 Nov. 2006.

Wani, G.M., et al.2006e. Foeto-maternal attachments in Small Ruminants. *Small Ruminant Research* (In press).

Jeffrey F. Keown et.al. 2006. *Dairy Management for Optimum Production and Reproductive. Performance,* University of Nebraska, Lincoln Extension, Institute of Agriculture and Natural Resources.

Phil Dash website, 2007. Reproductive and cardio vascular disease research group. Nitric oxide. www.sgul.ac.uk/depts/immunology 1-dash/no 11/13/2007.

Yuan-Lin-Dong, Pandu, R.R; Li Fang and Vallampalli, C. 1999. Nitric oxide reverses prostaglandin induced inhibition of ovarian progesterone secretion in rats. *Human Reproduction*. 14 (1) : 27-32.

Moptta, A.B; Estevez, A, Tognetti T; Gimeno, MAI and Franchi, A.M, 2001. Dual effects of nitric oxide in functional and regressing rat corpus luteum. *Mol. Hum. Reprod.* 7(1): 43-47.

James D.Ferguson, 1995. Reproductive Health Programmes. University of Pennsylvania, School of Veterinary Medicine, Centre for Animal Health and fteproductivity, Kennet Square, PA, U.S.A.

Kruip and Van Re.2008. Biotechnology of Reproduction and Farm Animals Welfare. Institute for Animal Science and Health BV, Netherlands.

Zimjanis, 1987. Bovine Theriogenology - Yesterday, today and tomorrow. Proceedings of the Society of Theriogenology Meetings. Sept. 16-18, 1987 Austin, TX.pp. 332-339.

□□□

Chapter 8

Ovarian Manipulation and Ovulation

G.M. Wani

Abstract

Ovarian function and early ambryonic development has been discussed , Our own results have been discussed in detail.A ll these results have benn published in national and international journals of repute by the author during his research carreer of past 35 years. The Ovulation process and mechanism as observed by the author on sheep and goats through laraotomy and laparoscopy is reported. The recent advances in ovulation process too have been reviewed.

Laparoscopy

Since the adoption of laparoscopy in sheep and goats (Phillipo et al 1971, Wani, 1982), a number of investigations have been carried in these species through laparoscopy. We shall present a review of our studies here under following headings.

1. Ovulation Detection in Muzaffarnagri Sheep

The appearance of corpus luteum (CL) was observed at laparoscopy after every 6 hourly intervals from the beginning till the end of heat (0-48 hrs). The laparoscopy was thereafter conducted at 3 days interval from 3rd to 28th day post mating. The ovulation was observed to occur between 12024 hrs after the onset of heat. The corpus lateum started regressing around day 9th and regresses between 12-15 day of the cycle. (Wani and Sahni, 1988).

2. Ovulation Detection in Jamunapari Goats

Similar procedures as in sheep were adopted in goats. About 87% examined in this study ovulated within 24-36 hrs after the onset of heat. All ovulated between 24-42hrs. The ovulation rate was 1.55 CL. (Corpus luteum) started regressing around the day 12th in goats and regressed between 15th-18th day of the cycle (Wani, 1989).

3. Development of Conceptus in Corriedale Ewes

The uterine distention (development of conceptus) was examined through laparoscopy. First group was laparoscopied between 28-35 days post mating. Group 2 ewes were examined around 58± 0.6 days post-mating and Group 3 ewes were examined on 93 ± 0.8 days post-mating. The laparoscopic observations are shown in Tab-1. For confirmation of these observations laparotomy was conducted immediately after laparoscopy (Wani and Bachoo, 1990).

Table 1 : Development of Conceptus in Sheep.

Group - I	Laparoscopy	Laparotomy
	Uterine horns, thin fluid filled Transparent appearance. Uniform distention, uterine Blood vessels hypertrophied	Uterine horns distended Asymmetry of gravid horns.
Group - II	Pregnant uterine horn distended Like a balloon. Fallopian tubes Enlarged . Gravid horn thick. Small marks as budding Cotyledons.	Gravid horn distended like a ballon. Looked as if few Ballons are joined together Distended uterine horn Difficult to bring out of Incision.
Group - III	Pregnant horn like a big ballon, Foetus recognized, cotyledons Identified as small rounded dark Buds.	Pregnant uterine horns very much distended whole gravid horn unable to be examined. Full foetus felt.

....Ref. Wani and Bachoo, 1990

4. Prediction of Ovarian Response in Superovulated Ewes

Eleven superovulated corridale ewes were laparoscopied on 5th or 6th day post-mating. The number of corpora lutea and ovulated follicles were counted. Accuracy % was calculated by comparing observations at laparscopy with those actually observed at laparotomy. The overall accuracy in predicting ovulating rate and unovulated was 73.7% and 80.4% respectiveiy. (Wani and Bachoo, 1990).

In Superovulated Ewe Lambs

Twenty two crossbred ewe lambs were superovulated as per *Wani et al (1989)*. Those in heat were laparoscopied 5-6 day post –mating. The number of corpora lutea and un-ovulated follicles were counted at laparoscopy and compared with actual observations at laparotomy. The overall occuracy % of predicting unovulated follicles and ovulations varied from 75-95% respectively (Wani *et al,* 1990).

The use of laparoscopy in future may include aspiration of follicular fluid, estimation of foetal numbers, flushing of superovulated donors, diagnosis of reproductive disorders intrauterine insemination, laparoscopic embryo transfer and study of foeto-maternal relationship.

Ultrasonography

Ultrasonography consists of two system: The A-mode ultrasounds: consists of amplitufe mode, sonar or echo systems. For two decades now, these systems have been in use of pregnancy diagnosis in small ruminants. (*Wani, 1981,* Wani & Sahni, 1981, Watt *et al,* 1984). Now real-time B-mode ultrasonography is available. If used correctly, it can help not only to detect pregnancy as early as 20-25 days post-mating, but real time imaging systems can help in the diagnosis of reproductive disorders both in male and female small ruminants. It can also help, in the monitoring of embryo development, transfer implantation, foeto-maternal relationship and foetal growth and placentation. Let us review the use of ultrasonography under following three heads.

a. Past use of ultrasounds

b. Present use of ultrasonography and

c. Future prospects of ultrasonography.

(a) Past Use of Ultrasounds

The ultrasound equipments mostly used in the past were A-mode equipments. They were mostly of two types:

Doppler System

The doppler equipments worked on the principle of altered ultrasound frequency. It mainly employed detection of foetal pulse, movements and uterine artery circulation as a means of detecting pregnancy (Wani, 1981, Wani and Sahni, 1980, Wani and Sahni, 1981). The accuracy of pregnancy detection after 50 days post-mating ranged between 75-90% in ewes and she-goats. The equipment was used to predict foetal age too. This was done by counting foetal heart heats.

Amplitude Mode

Another type of early ultrasonic equipments were based on the detection of the difference in a coustic impedance between the contents in the gravid uteres and the abdominal viscera. These equipments were Amplitude depthscope, or amplitude-depth ultrasonic analyzer. They were mostly used by Australian workers with 90-95% accuracy after 2 months of pregnancy (Lindhi, 1966, 1969).

Present Use of Ultrasonography

With the availability of real-time ultrasound systems, the past A mode systems are almost replaced. These B Mode real-time systems allow non-invasive visualization of the internal anatomy of the reproductive organs and their conceptus (Buck-rell, 1988). These systems have been described as most advanced technology of reproductive research after RIA (Ginther, 1986). The advantages/non-pregnancy tests in small ruminants, can provide informations on foetal numbers, foetal viability and diagnosis of reproductive tract diseases.

Equipment

The equipments marketed for use in animals have been mostly modeled for Mare. Two systems are in use for veterinary purposes.

(i) Linear array system

(ii) Sector Scanner

However, now some systems have the capability to provide both linear and sector scanning systems. Sector systems have some advantage over linear array, system, such as,

i. They require less skin surface contact which reduces the scan time required per ewe.

ii. They have superior resolution power especially in its electronic models.

The cost of these systems vary between 10,000 to 15,000 US dollars. Sectors scanners are constlier than linear array models. Image quality comparison are suggested before purchase (*Buckrell. 1988*).

Transducers

Transducers are available both for abdominal and rectal use. The head frequency of a transducer is important for getting better images. Higher the transducer head frequency, the better will be the resolution power. With higher and better resolution, image quality shall improve. With higher and better resolution, image quality shall improve. At present three frequency transducers are available 7,5,5 and 3 MHz. For rectal use the 7.5 MHz transducer gives better results, for visceral use 5 MHz transducers may be used. For deeper penetration (e.g.flank) the lower frequency transducers 3 MHz are used.

Advantages and Limitations

The major benefit of using present day real-time systems is that we get image copy on a polarid film, slides, video or papers. This record serves as teaching aid or for self learning. The image can be frozen and tissues can be identified, measured before recording the image.

The another advantage of real-time systems is that the image can be freezed or we can have a hard copy of the image the chances of error in making diagnosis is minimized. The accuracy in diagnosis using real-time systems may be as high as 90% even during early gestation; however it depends on the choice of region and position while scanning i.e. rectal, flank standing, inguinal, clipped unclipped belleys etc. (Logue *et al,* 1987). As the accuracy and efficiency of using real-time ultrasound may be influenced by many factors. It is advisable to provide following information while publishing date on these systems:

a) System used- "Linear or Sector"
b) Transducer frequency
c) Stage of gestation
d) Region of scanning
e) Duration of scan
f) Fasting
g) Breed and parity of the ewe or she-goat
h) Operator
i) Early embryonic death

False positive diagnosis may result from an early diagnosis of pregnancy in ewes between day 20-25 post-mating. This may be the result of subsequent embryonic death or abortion (Fowler and Wilkins, 1984).

c. Future Prospects of Ultrasonography

The efficiency of the real-time systems is expected to be increased, both in terms of resolution and real-time elaborative imaging system. It is hoped that use of ultrasonography shall help in increasing our understanding, on feeto-placental, foeto-maternal, embryo-endometrial interactions, besides monitoring embryo-foetal development. Some of its present days used could be improved, such as:

Maternal Recognition of Pregnancy

The study of pre-implantation embryos, estimations of endometrial secretions and real-time imaging of the events involved in implantation may help us to understand the factors involved in the maternal recognition of pregnancy. Some studies using ultrasonography have suggested that maternal recognition of pregnancy takes place around 14-17 days in ewe and does, respectively (Riera, 1984). Ultrasonography along with assays for detection of antiproges terone antibodies in the uterine epithelium could help us to understand the signals responsible for sustained progesterone secretion (C.L. growth) during early pregnancy. The understand shall increase on the embryoendometril interactions. This can be possible by biopsy studies on morphological charges in endometruim, modulated and monitored by use of ultrasonography.

Trans Uterine Migration

Trans uterine migration of embryos has been found to occur in 8% ewes with single ovulation, in 87.%% ewes with multiple ovulations. The assessment of blastocyst effects on the endothelium, especially to assess the role of embryonic stages involved in transition of maternal to embryonic control may be possible by combined use of ultrasonography with histochemistry and recent methods of reverse transcription of RNA into DNA (Schultz, 1990).

Foetal-Maternal Interaction

Ultrasonography has been used to understand the mechanism of epithelial-chorial placentation in ewes. In ewes 60-80 concave carundes were differentiated . In doe this number was more 160-180. Evidence of caruncular development could be undertaken in future, using real-time , freezed mode imaging systems. The study of fluid dynamics in relation to expanding vehicle may help cur understanding of the foeto-maternal relationship in future.

Detection of Early Pregnancy and Embryonic Motility

Using a 5 MHz rectal probe pregnancy was detected as early as day 20 post mating. On screen such pregnancies produce an image resembling small pockets of fluid ventral and anterior to bladder. Embryonic death in an ewe too was captured on avideo attached to ultrasonography (*Buckreli, 1980*). Between 40-50 days post mating accuracy % has been near 100% with these B-mode systems.

References

Phillippo, M. Swapp, G.h.Robinson , J.J and Gill, G.C. (1971). The diagnosis of pregnancy and estimation of foetal numbers in sheep by laparoscopy. *Journal Reproduction Fertility. 27: 129-132.*

Wani G.M. (1982). Laparoscopy in farm animals. *World rev.anim.Prod. 18, 7-13*

Wani G.M.., Sahni K.L., (1988) Ovulation detection by laproscopy in sheep. *Indian Jour.Animal Sci, 58 (7):802-804*

Wani, G.M. (1989) Plasma progesterone concentrations during oestrus cycle in goats. *Indian Journal of Animal Reproduction*, 10 (1*):44-46*

Wani G.M. (1990) Post-partum breeding in goat-a review livestock advisor 15(1):39-42

Wani G.M. (1990) Superovulations during early luteal phase in goats. *World Review of Animal Production, 29 92) 41-43*

Wani G.M. (1981) Augumentation of fertility in goats : Oestrus synchronization. *Annual report.,* Cent.Ins.res.Goats.Makhdoom, India.

Wani G.M..and Buchoo ,B.A.1990, *Laparoscopic view of ovine conceptus, Indian Journal of Animal Science.* 60(5): *564-568.*

Wani, G.M; Sinha, N.K. and Sahni, K.L. 1981. Incidence of abortions and stillbirths in Muzaffernagri and crossbred ewes. *Indian Veterinary Journal. 58: 558-561.*

Wani, G.M. and Sahni, K.L 1980 Pregnancy diagnosis in Sheep and Goats. *Indian Farming 5:25-27.*

Buckrell, B.C. (1988). Application of ultrasonography (USG) in reproduction of sheep and goats. *Theriogenology. 31:115-125.*

Lindhi, J.L.(1966) A pregnancy tester for ewes. *Farm Journal* October (cited)

Lindhi, J.L. (1969) Pregnancy diagnosis in dairy goats using ultrasonic Doppler instruments. *J. Dairy Sci. 52:4.*

□□□

Chapter 9

Stem Cell Technological Applications

G.M. Wani, P. Wani & G. Wani

ABSTRACT

Un-differentiates cells are called blank cells. Grown in cell culture for few generations, they remain un-differentiated. How to know they are un-differentiated has been discussed in this paper. The embryonic stem cells have been used to recover many diseases by direct differentiation. Various mechanisms and methods described. The planning and policy situation of Stem Cell Technology in its country of origin has been reviewed. The pituitary hypoplasia diagnosed to be a stem cell defect. Foreign gene expression in embryos, transgenic and animal modeling for disease cure has been postulated, on the basis of research results reported. Various embryo manipulation, transfer of gene, nuclear injection, gene mapping, isolation and analysis of gene in future have been reported and discussed. Embryonic Stem Cell Therapeutics, Gene targeting, cell death, cell differentiation, programming molecules research has been reviewed. Mutation defects in mice have been related to stem cell defects. A 2 step process of gene targeting mechanism has been developed which help a lot in gene therapy and animal genetic manipulations. Gene trap strategies and mechanism of its use for human and animal good has been briefly hinted at for future researchers to initiate new research mechanism.

1. Introduction

The concerns shadowing our livestock improvements are poor zoo-sanitation. Our huge product humps needs export market. Our poor disease safety procedures need to be sharpened. Our milk products at present are imported by Bangladesh and Sri- Lanka alone leaving a huge surplus in our Gowdowns. International zoo sanitary conditions need products from disease-free animals. This is possible by preventing animal diseases. Therefore a new thrust on immunological studies like stem cell, primodial Germ Cell, fetal and cord blood cells shall be our endeavor to discuss in these pages.

2. Stem Cell Technology

Stem cells are blanks cells, which can be cultured and give rise to any kind of cells and tissues. They are pluoripotent cells, which proliferate and can be differentiated into different cells types. Various other cells have this capacity which is briefly explained (Puri 2003). They have a capacity to revitalize diseased tissue and can replace organ transplantation. Stem cell technology, promise safety against life threatening diseases and organ replacements. This is possible by creating identified cell lines which will give rise to tissue and organs for replacing diseased parts, thus we see a futuristic replacement of blood bank with stem cell bank.

Stem cell line availability may help cloning. The evidence of such miracle was born in February, 1997 named "Dolly the Sheep." Dolly came into existence by fusion of an adult mammary cell nucleus of a pregnant Finn Dorset White Sheep with enucleated Oocyte of Scottish black Face Ewe. The embryo so developed invitro was reared to birth in the uterus of a black faced ewe. Since then cloning of pets and animals is in practice and even human clones are the fantary of HBO film displays. This has brought home many realities of spiritualism like birth of Christ and above all the birth of Isaq. A hope to infertile couple to have babies of their own origin. Here genetic make up is not borrowed but is ones own. Unlike in vitro fertilization the need of sperm cell too is warded off. So a baby to a virgin is a reality. It has many benefits, risks and applications too. A brief is as:

Advantages

1. Organ replacement, no illegal sale of organs and exploitations.
2. Drug sensitivity test- animal friendly- animal welfare approach.
3. Genomic research possible- many unanswered questions shall be replied.

Risks

Oncogenic potential and ethical conflicts.

Applications

Animal with higher yield could be produced.

Gene mapped animals may be possibility with simple means.

Embryonic development solutions could be found.

1. Aging research may reveal many secrets of longevity. Models for diseases investigations with cells and tissue can save animals and humans. Drug sensitivity tests shall be simpler. Gene targeting will be possible.

Cell Death Warded Off

The Somatic cells have DNA polymerase activity which is unable to replicate the end of a linear molecule, so it is unable to make up for the ends of eukaryotic chromosomes, which shorten at ends so the aging effect. This is why chromosome becomes progressively instable and thus leads to cell death. This formative defect is awarded off in stem cell. These cells are toti potent and hence have no chromosomal abrasion possibilities, nicely passaged germ cells without having clubbing into Embryoid.

Embryonic Stem Cell

Foetal tissues- a source of embryonic stem cell can too differentiate into various cell types through three germ layers.

Embryonic stem cells (ESC) can be had from inner cell mass of pre implantation embryo and cultured in mouse embryonic feeder cells. Embryonic development after fertilization has been detailed (Wani, 1996). After fertilization within 30 hrs zygote divides and becomes a morula in 3-4 days post coitus. A blastocyst is seen within 5-6 days, p.c. The 150μ blastocyst (1/ 7 th of a mm) has already differentiated into outer trophoblast (70 cells and inner cell mass, cluster of 30 cells). They are multipotent and rise to germ layers ectoderm, mesoderm and endoderm. Maintenance of ICM cell lines, culture feeder layers under undifferentiated state in now possible by Leukemia Inhibitory Factors (LIF) addition to growth culture.

The research has entered an advance phase and we have many tests which can differentiate stem cell from un-differentiated cells.

Differentiation

ICM cell proliferate and undergo differentiation. The activity of lineage specific genes is the evidence of differentiation. Lineage commitment is infact the beginning of the differentiation. Thus totipotencey is infact multilineage expression now. The trancriptosome activity is thus a signal. Its minimal activity maintains the cell in open state. The levels are low but in detectable levels, so its standardization could help to maintain them?

Programming Molecules

The Demethylases- reprogramming molecules present in blastocyst cytoplasm may be a positive factor for maintenance of open transcriptosome activity. Similarly, heterochromatin modeling and regulation could be negative mechanism. The inadequate knowledge of regulatory gene today hamper our understanding but tomorrow we could know switch off and on of these regulators & various interfering mechanism like methylation effects and SIRNA (Small Interfering RNA). In future, we dream for Genome shuffling i.e. shape of the genome profile. The gene profile like Oct. 4 embryos & BMP4 etc. has already been identified. More identification is on; solutions are expected, through research.

A broad review of about 100 research publications in presented in these pages, under various heads.

Stem -Cell –Technology

Stem- cells are undifferentiated cells. The embryo at morulla stages has undifferentiated blastomers. These cells can transform themselves into 200 or more cell type, which could be used to repair or regenerate new desired cells. This advanced cell research may help diseased people like those suffering from cancer, Alzheimer's disease syndrome, Parkinson's disease and even paralysis.

The stem cells are derived from 4-5 days old embryos or fertilized cells. The stem-cell- technology has a new role to play in animal reproduction.

Firstly, the stem cell source could be pooled out of slaughter house oocytes or from the vast pool of embryos hatching out in many animal species.

Secondly, the cell source could be obtained from foetal culture and if new-grown--immunoprotectant cell of foetus could help or provide germ-cell specific to any particular organ, the repair of organs could be a new revolution.

Embryoid Bodies.

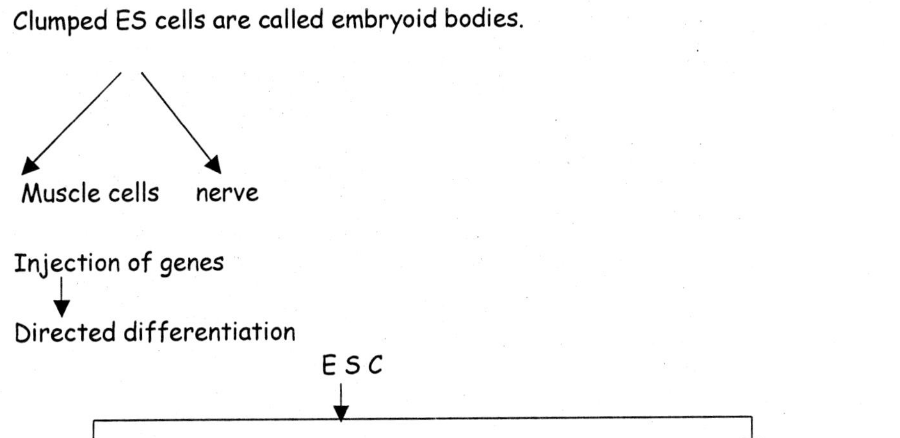

U.K Cabinet Decisions in 1990

Biotechnology in general and stem-cell-nanotechnology in particular is still not in its applied stage even in U.K. A balanced information system, inter-research council is suggested. Cabinet office (1990) U.K Govt. has started supporting embryonal stem cell technology. Manpower shortage in Embryo manipulation in cattle, protein engineering, animal trans genesis, stem cell biology is felt. Thus these fields are future emerging technologies to enhance livestock and agri-productivity. More and more researchers shall be attracted to this area, as more jobs securities are evident in industry related to stem cell technologies.

Mutation Defects Investigated

A stem cell defect in mice has been related to mutations that cause pituitary dwarfism. The cause lies in the gene encoding the transcription of growth hormone factor 1. (GHF 1 or pit 1). The dwarf mice were found to be deficient in growth hormone (GH), prolactin synthesis. Thus they exhibit pituitary hypolasia -a- stem cell defect. Tolipotent embryonic stem cells can be identified by the methods described (Reddy *et. al.* 1992). Foreign gene expression in embryos had been described (Kondoh e*t. al.*,1991)This has resulted in success of transgenesis of mice,fowls and fishes.

Thrust Areas of Research

Agricultural and food research council of UK has stressed some of the scientific thrust areas like molecular biology, biochemistry, intracellular signaling stem cell biology, global climatic change and agro-based food production & safety. The transgenic animal modeling to develop disease targeting as reviewed (Wani, 2007) has been the main focus of human disease cure models in centre of genomic research university of Edinburgh. The focus has been on use of transgenic in the fields of toxicology, cardiovascular diseases, pulmonary diseases, inflammatory and immunological diseases. Many neuropsychiatries, reproductive, endocrinological, embryological disorders could be evaluated and studied through these animal transgenic models. High priority in funding in suggested for initiating these areas. (Lathe & Mullins, 1993, Wani, 2008).

Animal Genetic Uses

Animal breeding has entered an era of embryo manipulation, transfer of genes, nuclear injections, gene mapping, isolation and analysis of genes for future genetic gains. (Bul field *et. al.,* 1994). This would have tremendous impact on animal related development. The detailed procedures have been described (Houdebine.1992). Insertional mutagenesis and mouse development have been reported (Kuelin *et. al.,* (1992). Various mice were generated with disrupted TAPS gene using embryonic stem cell Technology (Kaer-L-Van *et al,* 1992). Ruminant microbial degradation using stem cells were used to improve ruman feed degradability. (Goto *et. al.,* 1993). Embryonic stem cells were used in embryonic stem therapeutics to generate various modes of Trypanosoma cruzi, parasitaemia, CD4 or CD8 molecules differed in interferon mRNA contents. This mechanism when moduled into theraphy modes may yield resistance, against Trypanosama cruzi or brucei (Rottenberg *et. al.,* 1993). Bone healing was achieved through Bone morphogenetic proteins (BMPS). These are differentiative factors whose principal function is to induce transformation of undifferentiated mesenchymal cells into chondroblasts and osteoblasts in dose-dependent manner. (Kirker *et. al.,* 1995). Germ- line chimeras have been produced in mice. (Uchida et al 1995).

Gene Targeting-2 Step Process

Gene targeting strategy was developed. This was used to develop many mice strains. The so called double replacement gene targeting is a 2-step process in which a region of the gene of interest is first replaced with a selectable marker. An inactivated allele so formed is retargeted with another vector to yield or reconstruct the inactivated allele. (Moore *et.al.,* 1995). Pierre-Robin

Syndrome in human signaling skeletal and facial abnormalities is of great importance. Similar abnormalities reminiscent of this disease was generated using activin receptor type II. This Act Rc II was found to signal pituitary gonadotrophins and reproductive performance was defective. All Act Rc II deficient mice did not show the skeletal or facial abnormalities, but showed pituitary release or FSH suppression defects. (Matzuk et al, 1995).

Gene Therapy

Embryonic Stem Cell Technology uses have been discussed in theraphy of human diseases. (Ramirez & Bradley, 1994). Inhibin deficient mice are prone to gonadal cancers. (Mutzuk *et. al.*, 1994). Livestock transgenesis and potentiality of stem-Cell Technology use is reviewed (Seemark, 1994). Gene transgetting in reproductive studies has been reported as earlier as 1995. (Camper et al, 1995). Experiment on laboratory Animal, has been concern of animal welfare associations. Transgenic technology can reduce animal use in future. (Gordon et al, 1997). Procine Chimeras have been produced using primodial germ cell derived colonies. (Piedrahita *et. al.,* 1998). Introducing mutations using gene targeting technique in mice has been reported. (Torres & Kulin, 1997). A detailed review covering transgenic animal modeling has been reviewed. The technologies like DNA microinjection, embryonic stem cell technology, traits affecting domestic animal. Productivity, animal bioreactors have been discussed (Pinkert et al, 1996).

Directed Differentiation

New concept on FSH, mechanism action using embryonic stem cell technology has been studied. FSH is needed for follicle maturation in females. Its role in male fertility, as was not established (Kumar *et. al.*, 1997). Stem cell biology has revealed that stem cells have the capacity to renew their self. Some stem cells can been directed to choose a selected pathway. It means they can differentiate into particular cells only. One can use growth factors cell lineage identification mechanism is needed to achieve a break through. Various questions are still not clear like regeneration in response to injury. The system recognition and regeneration manipulative mechanisms are being searched.(Morrison at al, 1997).Embryonic stem cell lines has been established in many specie besides mice like cattle, rats etc (stranzinger,1996).

Gene-Trap Strategies

Various reviews on reproductive physiological research and future prospects have been reported. (Foote, 1996) more recent reviews have been published (Wani, 2007, 2008). New techniques, describing gene trap strategies or embryonic nervous systems have been in use since last one decade in Japan (Shiraieh *et al.*, 1996). Yamada *et al.* (1994) described gene targeting approaches. His review describes methods of using embryonic stem cells for gene targeting growth and differentiation studies. Hair development is now possible using keratinocyte- growth factors isolates. (Muridae *et al.*, 1996). Using stem cell factor melanoma in gray horses was investigated. No such association with coat colour was established (Rieder *et al.*, 1999). Embryonic sex was determined in pre-implantation embryos, using green fluorescent protein (GFP) gene on X chromosome. (Little 1998).

References

Nair, M.D, 2001. Legislation affecting Science and Technology. *J. Intellectual Property.* 6:5, 398-403.

Zhang/XueMing *et al.*, 2001. Effect of several factors on mouse spermatogonial stem cells. *J Vety. Sci.*, 21:5,497-500.

Wobus, A.M *et al.*, 2000. Embryonic stem cell technology. J.Life Science, Italy 2000-214.

Huelsken, J., *et al.*, 2001, Beta Catenin controls hair follicle morphogenesis and stem cell. J. Cambridge, 4: 533-545.

Gustafsson, E., *et al.*, 2001. Directed expression of re recombinase in endothelial cells of Embryoid bodies and transgenic mice. *J. Cell Sci.* 114: 4, 671-576.

Vasquez, K.M *et al.*, 2000. Specific mutation induced. *J. Sci.Washington,* 290:5491, 530-533.

Reijo, R.A., *et al.*, 2000. DAZ family proteins exist throughout male germ cell. *J. Biology Repd.* 63:5, 1490-1496.

Ledermann, B., 2000. Embryonic stem cells and gene targeting. *J. Exp. Physiology,* 85:6, 603-613.

Telfer, J.C., *et al.*, 2001. Expression and function of stem cell. *J. Dev. Biology,* 229:2, 363-382.

McLarn, A., 2000. Cloning pathways to a pluripotent future. *J. Sci. Washington,* 288:5472, 1775-1780

Schwenk, F., *et al.*, 2003. Hybrid embryonic stem cell. *J. Mole. & Cellular Bio.* 23:11, 3982-2989.

Cho-ChungHee, Jung, Ha, HS, *et al.*, 2003. Protamine 2 deficiency leads to sperm DNA damage. *J. Bio. Repd.* 69: 211-217.

DU-Lixin, *et al.*, 2003. The investigation of cell culture conditions to maintain chicken embryonic stem cells as totipotent cells. *J. Animal Sci,* 16:8, 1102-1107.

Zhou-ZgenMing, *et al.*, 2001. Isolation and culture of ES-like cell. *J. Agri. Univ.* 25: 3, 73.

Hasthorpe, S. 2003. Clonogenic culture of normal spermatogonial. *J. Bio.Repd.* 68:4, 1354-1360.

Pant, H.C, Agarwal, S.K., 2002. Increasing reproductive efficiency and productivity of cattle and buffaloes. *J.Izatnagar,* 14-23.

Arney, K.L. et al 2002. Genomic imprinting. *J. Animal dev.* 233-264, 277-280.

Rintu, Banerjee, et al, 2002. Developments in Biotechnology. *J. Bio.* 1:1, 9-16.

Humpherys, D, et al 2002. Abnormal gene expression in cloned mice. *J. National Acad.Sci,* 99:20, 12889-12894.

Takeda, M., et al, 2001. Cockroach midgut peptides. *J. Bio.Animal* 37:6,343-347.

Nakajima, K., et al, 2002. Control of cell division. J.Supp. S265-S276.

Bruce, D.M. 2002. Stem cells, embryos and cloning. *J. Molecular Bio.* 319:4, 917-925.

Brinster, C.J.2003. Restoration of fertility of germ cell. *Bio. Repd.* 69:2, 412-420.

Park, S.M et al 2004. Generation of embryonic stem cell. *J. Animal Sci.* 17(12): 1641-1646.

Rocha, A., et al 2004. Application of inducible. *J. Biotechnology,* 6(2) :118-127.

Paz, A.H. et al., 2003. Mouse embryonic stem cells. *J. Scientiae Veterinariae,* 31(1): 51-54.

Yang-Weifeng,et al 2004. Premature corpus luteum failure in superovulated Boer goats. *J. An.Bio. Bulletin,* 9(1): 449-454.

Jing, XiaoQi., et al 2004. Nucleaolar changes in nuclear transfer embryos. *J. Ani. Bio. Bulletin,* 9 (1): 491-495.

Lei, AnMin, et al 2004. Effect of different media. *J. Ani. Biot.Bulletin,* 9(1) 669-670.

Gao, ZhiMin, Yang, ChunRong et al., 2004. Application of embryo transfer in Boer goats. *J. Ani . Biot. Bulletin,* 9(1): 467-474.

Dong-WuZi et al 2004. Studies on male germ-line stem cell meiosis. *J. Ani. Biot. Bulletin,* 9(1): 828-834.

Yan, XingRong., et al 2004. Effects of mouse oocyte age on spindle location. *J. Ani. Biot. Bulletin,* 9(1): 678-682.

Ge, XiuGuo, et al 2004. Migration of the mouse primordial germ cell. *J. Ani. Biot.Bulletin,* 9(1): 694-701.

Sang-RunZi., et al., 2004. Research on the effect factors of isolation. *J. Ani. Biot. Bulletin,* 9(1):733-739.

Tang-WeiFeng, Dou, ZgongYing et al 2004. Comparison of methods for diffentiating mouse embryonic stem. *J. Ani. Biot. Bulletin,* 9(1): 757-764.

XU-XiaoMing et al., 2004. Isolation and culture of embryonic stem cells. *Ani. Biot. Bulletin,* 9(1):843.

XU, XiaoMing et al,2004. Isolation and culture bovine embryonic germ. *J. Ani. Biot. Bulletin,* 9(1):844.

HU, JunHe, et al., 2004. Profess of porcine somatic cloning. *J. Ani. Biot.-Bulletin,* 9(1): 479-486.

Liu, Yuxiao., et al 2004. The recent progress of murine primordial germ cells. *J. Ani. Biot. Bulletin,* 9(1): 523-529.

Lei-AnMin, et al., 2004. Relationships between the morphology and quality of embryos cloned. *J. Ani. Biot. Bulletin.* 9(1):671-672.

Lei-AnMin, et al., 2004. Interspecific and intraspecific comparisons of the developmental ability of cloned embryos. *J. Ani.Biot. Bulletin,* 9(1):677.

Wng, Han et al, 2004. Sex determination in embryos with polymerase chain reaction. *J. Ani. Biot. Bulletin,* 9(1): 555-559.

Admas, D.J., et al 2004. Mutagenic insertion and chromosome Engineering Resource. *J. Naure, Gen.* 36(8): 867-871.

Tam, P.P,L., et al 2003. Mouse embryonic chimeras. *J. Dev. Cambridge,* 130 (25): 6155-6163

Reddy, G.V. , et al 2005. Stem cell homeostasis and growth dynamics can be uncoupled in the Arabidopsis shoot apex. *J. Sci. Washington.* 310(5748): 663-667.

Baska, K.M., et al 2005. Ubiquitination and its consequences for mammalian. *J. Repd.System,* 83-114.

Dobrinski, I. 2005. Germ cell transplantation and testis tissue xenografting in domestic animals. *J. Ani. Repd. Sci.* 89(1/4):137-145.

Laowtammathron,C., et al., 2005. Factors affecting cryosurvival ofnuclear transferred bovine and swamp. *J. Theriogenology,* 64(5): 1185-1196.

Misra, A.K., et al., 2005. Embryo transfer technology (ETT) in cattle and buffalo. *J. Ani. Sci.* 75(7): 842-857.

Misra, A.K., et al 2005. Application of reproductive biotechnology. *J. Ani. Sci.* 75(7): 893-904.

An, Jing et al., 2003. Cultivation and identification of cheicken embryonic stem cells. *J. Acta, Zoologica, Sinica,* 49(5):698-703.

Morgan, H.D. et al, 2005. Epigenetic reprogramming in mamals. *J. Molecular Gene,* 14 (Review 1): R47-R58.

Li-LianJun, et al., 2005. Factors affecting renewal and differentiation of spermatogonial stem cells. *J. Yunnan Agri. Univ.,* 20(1):77-82.

Oatley, J.M., et al., 2005. Changes in spermatogenesis and endocrine function in the ram testis due to irradiation. *J. Ani. Sci.,* 83(3):604-612.

Deng, Yun, et al. 2005.Molecular cloning of MSRG-11. *J. Subuca,* 37(3): 159-166.

Li-JiXia, et al., 2004. Research Progress of embryonic stem cell differentiation induced in vitro. *J. Vety. Sci. & Tech.* 34(6): 40-43.

□□□

Chapter 10

Sustainable Livestock Productivity Reproductive Prospective

G.M. Wani & Gazanfer Wani

Abstract

Indian livestock production is economical due to our low-cost management systems. This is our strength, which empowered us to have the first rank in global milk production. Due to our poor zoo-sanitary conditions, we are unable to export our surplus milk. Thus our first requirement is sensitive, accurate and timely reproductive and other disease surveillance, diagnosis and control.. Against per capita per day requirement of 25 gms meat, we at national level have an availability of 9 gms only. The enhancement of meat production should be our 2nd priority area by 2050. These strengths and weaknesses of our livestock sector have been reviewed. Our thrust, however, is limited to reproductive management. We have to intensify bio-safety measures for production of disease free germplasm (semen, embryo) and other livestock products. Gene propagation and its utilization has to be specific to our needs and should have disease control checks . This will need huge human resources. We would need 32,000 more vets. Our rate of entry in veterinary colleges. We will need more postgraduate vets. This will need reorganization of the veterinary education system. Endocrinological advances would need research on hormone synergism, agonistic and antagonistic actions. Follicular atresia vis-a-vis Intrinsic GnRH system would need more experimentation. The deterministic and permissive modes of GNRH action would be applied to reduce hyper stimulation of ovaries. The understanding of dominant follicular dynamics, shall help to harvest follicle pool in thousands rather than in hundred at present. New biological sources of hormone from fish and other animals should find more clinical trials to popularize cheap and biological rather than chemical hormonal source. Sensitive disease diagnostic using RIA, REA hormone detection kits, ultrasonography, laparoscopy, DNA finger printing, immunological and other psychometric, techniques needs consideration. Gap junction and development of conceptus, foetal biological research will need emphasis by 2050, In vitro fertilization and embryo culture shall be replaced with in vitro implantation and in vitro foetal development. Foetomuscular development and foeto-maternal interactions will be more vivid and clear. Reproductive care and disease prevention shall be need herbal drug stimulants. Such bio-plant stimulants and medicaments need innovative research attempts. Corpora luteal biosensitivity and its use in augmentation of reproduction needs trails. Fertility enhancement needs identification of desired fertility objectives. The problems and constraints in animals reproduction have to be addressed by using modern reproductive technique to improve our traditional reproductive management.

Introduction

India has approximately one billion animals for a billion humans. We need less animals to sustain our present productivity if proper gene propagation and animal reproduction methods used. Our per animal production is much less. Inspite of this weakness our per unit cost of production of livestock products is not only less but is economical. This is why, our livestock farming exists without any subsidy and has great potential to be sustainable. Thus, competitive economic livestock production systems are our strengths. Details of economics of goat management systems have been reviewed (*Wani, 2002*). Similar evaluation in other livestock species viz. Cattle, buffaloes and sheep do reveal positive trends. The improvements in our traditional and reproductive management of buffaloes have won us the first rank in world milk production. Our grains are however over shadowed by export restrictions to developed and developing countries except Bangladesh and Sri Lanka. Our poor zoo-sanitary conditions are responsible for it. Thus our first requirement in future is to provide sensitive, accurate and timely disease surveillance and diagnostics, so as to make our milk or meat surplus areas as disease-fre-zones. Per capita meat requirement as per ICMR recommendations is 25 gms per day. We produce at present only 9 gm per person per day. This critical gap abridgement needs more emphasis on enhancing meat production. These two priorities should be the main objectives of this review.

Zoo-Sanitation and Bio-Safety

With production of 78 million tones of milk in 1999-2000, India ranks first in global milk production. It has Rs. 113 million worth milk products in export godowns. There are few buyers. Except Sri Lanka and Bangladesh none dares to import our cheap milk and meat productions, for fear of disease transmission. Our poor zoo-sanitary condition is our weakness. Our products fail to complete products from other countries, as their products appear to be safer. The lobbing by rich, media hippie world puts our livestock products at disadvantage even with our own Asian neighbours, whose bio-safety measures are more accurate and sensitive than us.

The use of artificial insemination (AI), embryo transfer(ET) have helped in elite gene propagation. "Dolley" the super sheep is known and many more genetic clones are in offering. Today our concern is bio-safety of germplasm and the animal products. With high risk of new disease emergence all researchers are warned to use their materials with caution. The "mad cow disease" and heavy economic losses in U.K. and Europe are eye openers for all of us. The future of gene technology lies in bio-safety and fulfillment of international zoo-sanitation regulations. Disease free zones need more emphasis than producing high yielding livestock from disease infested areas. Thus safe and quality animal material (germplasm) and products is the slogan for tomorrow. Zoo sanitation certifications will demand more trained veterinarians.

Gene Propagation and Utilization

For production and utilization of elite gene pool of animals, we need to know our own resources, both biological and material. For example, increasing meat/milk or fibre production from our small ruminants we need to know them better. We must analyse their gene pool both qualitatively and quantitatively as was done in goats (*Gall et al., 1997*) use DNA techniques. Once we know merits and demerits of our gene pool, could we preserve them or utilize them for further up-gradation. A small attempt of buck exchange in Ladakh and better use from identified tribes helped in the fine pashmina production, which awaits suitable markets. Mongolia crossed their pashmina goats with Russian "Don goats". The quantity of pashmina increased but the quality reduced. We are fortunate that due to our criticism, J&K avoided introduction of Don goats in Ladakh, otherwise by now we would have lost the only quality "fine pashmina" of Kashmir (*Wani et al., 195; Gall et al., 1997*).

Endocrinological Advances

Ovarian oogenic and endocrine functions are interlocked via Hypothalamo-Hypophysial-Gonadal Axis (HHGA). The cyclicity and its behavioural overt and covert phase are instrinsic to these ovarian functions.

The have been thus rightly named as the Zeitgeber or biological clock (*Baird, 1987*). The existence of feedback loops of ovarian steroids with respective gonadotrphin is biphasic phenomenon, where higher concentrations in blood have inhibitory or suppressive effects, low considerations have stimulatory effects (*Barac Lough, 1973*). Many reproductive mechanisms hang on changes in these ovarian steroids such as labour pains (*Csapo, 1981*), ovulatory failures (*Sancheq-Crado et al., 1990*). Interestingly these hormones interact to modify or sometimes antagonize each other. For example P_4 appears to be potent antagonist of E_2 at a combined ration of 1:1 (*Heap and Flint, 1987*). The synergism ,agonistic and antagonistic actions of these hormones combinations are the basis for all reproduction manipulations viz. Control of oestrus, ovulation, conception, gestation, fertility, infertility, contraception and or abortions. Fish and sheep ovarian extracts were tested to find out interclass sensitivity. It seems it is possible to tap huge fish ovarian extracts for intra and interpolation in mammalian ovary *(Najar, 1999)*. Coming years will see more attempts to use such resources to avoid side effects from chemical and synthetic analogues of hormones.

Follicular Atresia

GnRH- regulates gonadotrpin release from pituitary, but its continuous and prolonged use in natural from or as its potent agonist induces an inhibitory effect on ovarian steroidogenesis , follicular development and ovulation (*Kledzik et al., 1978; Srivastawa et al., 1995)* GnRH receptor gene encoding is expressed in the rat ovary (*Kaiser et al., 1992*). A new concept prevails today which postulates that an intrinsic GnRH system competes with receptor. Thus its biological response exist within the ovary. The follicular atresia is brought by this intrinsic GnRH system (IGS) (*Erickson et al., 1994*) A new vision of autocrine and paracrine secretion of ovary vis-in-vis GnRH becomes evident.Future modulations in correcting reproductive failure involving hormone thereby needs rational understanding of all endocrine, autorine and related control mechanisms. Ovarectomized ewes when implanted with E_2 and P_4 implants were maintained in an artificial follicular phase. This speaks of an endocrine model compatible to a natural sequencing hormones of desired levels (*Karsh and Evans*, 1996).

Instrinsic GnRH system (IGS)

Two theoretical models called deterministic and permissive models have been put forth to explain of GnRH (*Karsch et al., 1977).*

Deterministic Model

Deterministic model envisages that a surge-like increase in GnRH is needed to induce preovulatory LH surge. This model befits animals whose LH release and ovulation is either time bound (Copulation) or signal dependant (cervical signal, rabbit and neural signal Central circadian pacemaker in rats refers).

Permissive Models

Permissive models explain GnRH in relation to ovulatory LH surge. Here feed back of steroids exert their influence on anterior pituitary gland. This is supported by evidences that LH surge continue in situations where GnRH secretion is apparently abolished. For example in monkeys with hypothalamic lesions, or in women with hypothalamic amenorrhea or congenital GnRH deficiency, normal LH surges (*Martin et al., 1990).*

Hormone Agonists/Antagonists

Hormone antagonist or anti-hormone are those compounds or elements which prevent an endogenous hormones from exerting its biological efft(s) (*Vanlook, 1988*). These anti-hormone need to be safe i.e. free from side effects. The hormone antagonists in principal acts in 3 ways.

i) Inhibition of hormone synthesis

ii) Neutralising hormone action by specific antibodies.

iii) Blocking the receptors of the hormone/Use of antagonists, agonists have demonstrated their role in ovulation control.

Their cow histamine-releasing activity (*Fraser, 1981*) support their wider use which needs further extensive studies. More details are reviewed (*Najar, 1999*).

Reproductive Failures –GnRH Mechanism

The biggest problem in ovarian manipulation faced in the past was due to varied superovulatory response (*Wani, 1984*). The follicular atresia was postulated to reduce this constrain if induced adequately. As our atresia was postulated to reduce this constraint if induced adequately. As our understanding of (IGS) intrinsic GnRH system was poor them, we could not use it to overcome our problems. Now we know follicular atresia is related to this IGS (*Erickson et al., 1994*) and therefore, our later manipulations incorporated GnRH became the subject of discussion. Estrogen alone can induce ovulation but for CNS (Central Nervous System) priming progesterone combination in desired ratio is essential *(Evans et al., 1995).* Thus future modulations in correcting reproductive failures needs a proper understanding of hormonal inter and intra polation with physical , chemical, biological and neuroendocrine secretion. Pheromones and other phycoemotional factors too need an analysis.

In our experiments on infertility (anoestrus) in cows of Kashmir (District Budgam) we were puzzled to see better response with low doses (300 μg) of GnRH than the higher dose of 400 μg (*Shah, 1996*). We are unable to explain our results until the recent studies on permissive and deterministic models of GnRH action came to light (*Karsch and Evans, 1996, Karsch et al., 1997).*

Herbal - Biostimulators

Pheromones, sires introduction effects, sensory reflexus and vast other stimulants increase receptivity as well as fertility in domestic animals. Sprays and herbal cure products will be in great demand to increase or decrease fertility in coming years. The use of stimulants or repellents as sprays and sprays and scents may be in need. Herbal cures for various reproductive wastage needs further research. The use of herbal products, fruits, nuts and medicaments in ancient Kashmiri culture after delivery, need to be looked at. The *Cichorium* spp, saffron, *Cofula anthemoides, Prunella vulgaris,* Artemisa and other products in use during post delivery period needs scientific analysis.

The ovarian sensitivity decrease soon after parturition and its resurgence after 14 days in cows, 40 odd days in women and specific periods in other animals, reveal a lot which needs exploration to phase out various sequences of reproduction.

Corpora Luteal Biosensitivity

Corpus luteum growth and development is the earliest pregnancy signal along with maternal recognition factors discussed (*Wani, 1996*). The corpus luteum development and receptivity to PGF_2 alpha is now known. The ascending and descending corpora lutea are insensitive to PGF_2 alpha. What makes them insensitive and what resurges their sensitivity need exploration. May be in future, CL insensitivity might have direct effects on survival of the conceptus. Does conceptus make CL more sensitive to PGF_2alpha? Or in other words, does the gravid CL and non-gravid CL behave similarly or differently toPGF_2 alpha. Many such questions vis-a-vis the ovarian morphology, uterine environment and behavioural patterns need to be studied and co-related to develop a manipulative and curative safe system. In some species CL of pregnancy in necessary till delivery, in others it is needed only till placenta develops. Its complete genesis and biostimulation and bioregulatory role are obscure and needs further research.

Dominant Follicle Dynamics

The dominant follicle dynamics may be in focus for ovulation control. Dominant follicle or would be bride at each ovulation appears around day 7 of cycle in women. It occurs much earlier on day - 1 to 0 in ruminants (*cow, sheep, goats*). This time point may be the focus to try future selective interferences in ovulation control. Antagonist use during this phase has shown decreasing levels of oestrogens in serum thereby indirectly hinting at dominant follicle alteration, as it is one of chief source of this hormone during this phase (*Fraser, 1987*). Dominant follicle in its sojourn to freedom and release at ovulation seems to attain some specific qualities. Why? It alone releases itself of bondage's in ovary (Ovulation) and rest of its competitors are hindered through the process as atretic follicles. What makes this dominant follicle to attain these qualities among a vast pool of follicles needs further study. In coming 50 years one shall possibly understand and interlock a pre-determined messenger system triggering dominance in one or few follicles destined to become the new offspring of that species. This system when decoded shall offer new opportunities in controlling ovulation both for contraception (women) or augmentation (animals). New genepool of dominant follicles from dead high yielding animal ovaries would help to propagate and multiply the elite.

Development of Conceptus – *In Vitro* Foetus

The lean meat production needs more studies on development of conceptus. We know that prenatal development undergoes in three stages. During early pregnancy skeletal & nervous tissues grow, on priority. During mid-gestation the hind and fore-limbs develop. During late gestation muscular development along with peripheral extremities including hair develop. In order to increase lean meat production, research in needed to investigate relationship between hard and soft tissues. Maternal and foetal interactions needs a probe. Some basic studies (Mufti, 1996) indicated prospects of manipulating lean meat production (*Wani, 2001, 2002 a*) by bio-stimulators during late gestation, which may help, the production of muscular lambs/kids at birth, besides, increasing birth weights. More details have been described (*Wani, 2001, 2002a*) (Fig. 1-5). More studies in future are needed to reveal the exact mechanism involved in critical transition of power control from maternal to embryonic genome. Application of recent techniques of reverse transcription, PCR and other may reveal the details (*Schultz et al., 1990).*

Slaughter House Follicle Pool

A good number of our endangered species die in wild , where climatic conditions are hard and harsh. Under temperate climatic conditions dead bodies are entrapped in snow for long. It gives a hope to revive their follicular pool. Our studies on resumption of life using follicles from slaughtered sheep and fertilizing them with epididymal semen was an pioneering attempt in this direction (*Wani, 1996*) (Fig 6-8). The protein content of unfertilized ovum is around 100 mg in rabbit, 27 mg in mouse (*Brinster, 1967, 1971*). This small genetic material needs extensive analysis to unveil vast information in its condans and anticodnas. Further ovum maturation or resumption of merosis is initiated during birth or shortly before birth and thus the pool of follicles remains fixed at birth. Therefore, foetal ovary has to be the subject of further manipulations especially in view of its immunological insensitivity (*Guraya, 1985*).

Gamete Biosafety

Bacterial counts in semen in unprecedented numbers effect semen fertility (Kher and Dholokia, 1985), The contaminational hazards have increased due to environmental pollution, water and air contamination. Above all the development of resistance of known bateria and other micro-organism to antibiotics (*Gupta and Maurya, 1993*). We have examined the semen of about 40 bulls maintained at the Frozen semen bank, Rambirbag, Kashmir for bacterial loads in semen and tested antibiotic sensitivity (*Koul, 2002*).

Presence of bacteria in large numbers in semen, effects its quality and fertility (*Kher and Dhololkia, 1985*). The source of contamination of semen vary from inflammatory foci and lesions on penile apparatus, contaminations on equipments, appliances, diluters, extenders air and laboratory environment. These contamination hazards have increased with environmental pollution. Many bacteria having developed resistance of routine antibiotics used (*Gupta and Maurya, 1993*) in semen extenders has made inseminations more critical and vulnerable. More studies in this direction are needed to develop a bio-safe and infection free germplasm pool of frozen and fresh in future.

Foetal Muscular Development

Genetics, hormonal and environmental factors influence foetal growth and development. The proper in-utero maturation and growth of foetus signify proper care and management of gravid nannies. The pre-natal development of foetus is exhibited at birth weight. The head and skeletal tissues grow and mature earlier, followed by fore and hind limbs. The muscular development of foetus occurs during late gestation. Thus, during early pregnancy nervous and skeletal tissues are having priority for nutritive requirements in-utero (*Mufti, 1966*). Maternal nutrition is important for growth and development of meat in off spring in-utero. Muscular growth stimulants may thus help production of quality meat lambs or kinds in future.

Foeto-Maternal Interaction

The role of endometrial stromal cells during pre-implantation stage needs further research. Some studies on ultrastructure of endometrium (*Brinsfield et al., 1974*) are still not sufficient to answer the questions. What is the role of immune cells and endometrial cells in building of the immunological barrier to protect the foetus. Foetal biology is a barren field of research in the world. After few initial steps by Dr. Moris Young in UK and Dr. Srivastava in India, only a scattered information is available on the subject (*Wani and Buchoo, 1990; Mufti, 1960*). Much is still to be unveiled. The placental hormone influence mammary development and milk production of the dam. This needs further research . Foetal membranes serve as chief source of nutritive exchange (Fig. 9). The arrangement of cotyledons, caruncles, positioning of single (Fig. 10) and twin (Fig. 11) conceptus, indicate complex mechanisms with common roots. These mechanisms of foeto-maternal interactions and relationships emphasize proper foetal number diagnosis and corresponding materno-nutritional management strategies.

Small Ruminant Reproduction

There is a need of ovulation, pregnancy and foetal number detection by methods described (*Wani, 1981, 1982; Wani and Sahni, 1980,1981; Wani and Buchoo, 1991, 1993*). These reports demonstrated successful results ovulation detection, multiple ovulation predictions, early pregnancy and foetal numbers. The induction of multiple ovulation, in-vitro fertilisations, embryo collections, embryo collections, embryo culture, conservation, micromanipulation, splitting, cloning and sexing are some of the reproductive high-tech applications used to increase reproduction rates (*Wani, 1996; Wani et al., 1990; Wani et al.,2000 a,b; Buchoo et al ., 2000*). These techniques have put a little application under field conditions in small ruminants especially under extensive management system. However, their use has already revolutionized propagation of elite germplasm and can be used under buck mother farms to produce quality sires.

The futuristic attempts in this direction should include in vitro post implantation and foetal development experiments. These future manipulations at organogenesis level may help to induce desired quality changes in offspring so that dressing percentages could be increased from present 40-50 percent to 55-60 per cent or more.

Reproductive Management

The reproductive management strategies to enhance meat production could safely be divided into two kinds:

i. Traditional management : Animals are still managed under zero-input/ migratory / tribal or extensive management system. Here the application of above described methods will be limited. The may indirectly help in this system by better sire availability.

ii. Modern management : Zero-grazing, stall-fed or semi-intensive, commercial small ruminant farming system may use the above methods more and more to produce high quality germplasm. They use variety of improved reproductive techniques described or reported (Wani, 1966; Wani *et al* ., 2000a; Wani *et al* .,1998; Mufti *et al.*, 1997; Wani and Sahni, 1981; Wani *et al.*, 1986, 1987, 1988; Wani and Buchoo, 1991; Buchoo and Wani,1991; Wani, 1989; Wani,1995; Wani, 1984a and b; Wani and Sahni, 1980a).

Enhancement of Fertility

Desired fertility objectives have been discussed (*Wani,2001*). Increased prolificacy, early rebreeding, safe/clean parturition, set-time animal crop, elite gene introduction are some of the desired fertility enhancement goals. The constraints in achieving desired objectives in small ruminant livestock sector are the problems of inbreeding, reproductive losses (*Wani, et al., 1980; 1994*), pre-pubertal mating, indiscriminate or illicit mating, venereal diseases, nutritional and other stresses. These above referred factors reduce fertility status of our ruminants especially under trans-humane migratory systems.

These constraints mentioned could be overcomed by adopting enhancement strategies (*Wani, 2001*), which include exchange of sires by tribal, Gujjar, Bakarwals and Chopans to reduce inbreeding. This could be ensured through legislature or forced castrations and introduction of sires from outside the flock. The pregnancy detection methods described (*Wani, 1981*) and lateron used extensively in J&K under farm condition (*Wani et al., 1998; Wani, 1989)* needs adoption as a routine in each flock for better care of offspring and dam.

Nutritional Bio-Conversion

The nutritional status of dams before mating could be improved by grazing them in pastures having 4-5 cm sward height herbage, yielding 1500 kgs and above of dry matter per hectare. Such pasturing shall increase ovulation. Reduced sward height pastures needs supplementation and introduction of nutritional block atleast 4-5 weeks before mating. Thus proper feeding 100 days prior to mating and parturition may help in healthy and heavy weight lamb/kid crop.

The pasture nutrient analysis and its supplementation as per requirement may help to avoid losses due to such deficiencies as cobalt, iron, phosphorus etc. which effect herbage and through herbage the animal health. The sire fertility test and use of disease free semen is the foremost requirement towards increased productivity from small ruminants.

The nutrition plays important role in maintaining fertility status of animals. With decrease in land for fodder cultivation, alternate sources and technologies to utilize agricultural waste for animal food is needed. The biotechnological intrervention in converting crop residues and other agri-horticulture waste into animal feed needs further research.

Summary

Critical gaps identified are poor-zoo-sanitary conditions and insufficient human resources. Both these area need emphasis. Development and application of accurate, sensitive and timely disease diagnosis and surveillance techniques especially in the field of animal reproduction. The aim be biosafe livestock products and germplasm. Gene propagation of elite germplasm of elite germplasm should ensure increase in per animal production. This would means more production from less livestock, thereby relieving soil, water and environmental pressures.

Endocrinological advances should aims at replacing chemical hormones with biological hornmones, which cold be cost-effective and without side effects. The use of fish, sheep and other ovarian extracts need further research. Follicular atresia, intrinsic GnRH system, hormone agonists and antagonists should attract more funds and research attempts. The use of herbal stimulators and medicaments in caring and prevention of reproductive diseases and incapacities needs immediate attention. More innovative approaches using these indigenously available materials need to be prioritized.

Copora lutea biosensitivity, development of conceptus, follicular dynamics, gamete safety procedures, foeto-maternal interaction, stromal cell studies would be future areas of frontline researches in animal reproduction.

A policy in view of majority 85% of our livestock production system being small holder enterprises. Research and extension establishments and NGO in animal reproduction needs reorganization to serve as a protective armour around small producers and farm women.

Blending of advanced technology available as embryo transfer, AI, blastomer splitting, DNA micro-injections, DNA finger prints, gene mapping, DNA hybridization, ovulation control, in vitro fertilization, foetal monitoring, ultrasonogrphy with traditional modes of livestock production system needs further innovative attempts. This should aim to further fortify our cost-efficient livestock productivity.

Hitherto untapped areas of wild and zoo animal conservation, protection and propagation, pet animal care, wildlife retrieval and referral clinics are the need of the hour. Nutritional requirement during foetal life, prenatal and neonatal life needs more emphasis along with care and management of dams during mating, pregnancy and parturition. Increased prolificacy, early rebreeding, set-time parturitions are some of the desired fertility enhancements goals. The problems of inbreeding, reproductive losses, pre-;pubertal and illicit mating, nutritional stresses could be overcome and adopting fertility enhancement procedures like sire exchanges, pregnancy detection, better nutrition before mating, during pregnancy and after parturition.

Extension and information technology missions for animal reproduction are to be intensified especially in rural India. Scientist-industry interaction, technology integration under rural sectors and linkages between research, extension, NGO and farmers need priority.

References

Barraclough, Charles A. 1973. Sex steroid regulation of reproductive neuroendocrine processes. In : *Handbook of Physiology, Section 7-Endocrinology Vol.II-Female.*

Reproductive system, Part ! American Physiological Society, Washington, printed in USA by Waverly Press, Inc., Baltimore, Maryland-21202.

Boshier, D.P. Fariclough, R.J. and Holloway, H.1987. *J. Reproduction Fertility,* 79:569-573

Brinsfield, T.H., Fisher, M and Hawak, H.W.1974. *J.Anim.Sci.* 39:565-574.

Buchoo, B.A. and Wani, G.M.1991.Induction of oestrus in Corriedale ewes during anoestrus. *Indian Journal of Animal Science,* 60(4) : 407-408.

Buchoo, B.A.,Wani,N.A.,Khan,M.Z and Wani, G.M.2000.Laparoscopic detection of Ovulation and pregnancy in different sheep breeds. *Indian Journal of Animal Sciences* 70(2):61-62.

Csapo, A.I. (1981). Force of labor in :Principles and Practice of Obstetrics and Perinatology. Vol.2 (L.Iffy and MA Kaminetzky, eds.), pp.761-799,John Wiley and Sons, New York.

Erickson, G.E.,Lid, Sadrkhanloo R.; Liu X-J, Shimasaki,S. and Ling, N. (1994). Extrapituitary actions of gonadotropin releasing hormone in stitulation of insulin-like growth factor-binding protein and atresia. *Endocrinology* 135:1365-1372.

Evans, N.P., Dahl, G.E., Maugar, D. and Karsch, F.J.1995. Estradiol induces both qualitative and quantitative changes in the patterns of GnRH secretion during the pre-surge period in the ewes. *Endocrinology* 136: 16-3-1609.

Fraser, H.M. (1981). Effect of oestrogen on gonadotrophin release in stumptailed monkeys (Macaca arctoides) treated chronically with an agonist analogue of LHRH. *Journal of Endocrinology* 91:525-530.

Fraser, H.M. (1987). LHRH antagonists and female reproductive function. In: LHRH and its analogs: Contraceptive and Therapeutic Applications. Part 2 (BM Vickery and JJ Nestor Jr.eds.) pp.227-244,M.T.P. press, Lancester.

Gall, C., Gelderman, H. and Wani, G.M. 1997. Analysis of genetic resources of goat breeds in Kashmir. Final Report of Research Co-operation between SKUAST and University of Hohenheim, Germany.

Gupta, R.S. and Maurya, S.N. 1993.Microbial flora of bovine semen and their antibiotic sensitivity pattern. *Indian Journal of Animal Reproduction*, 14(3):107-108. Guraya, S.S. (1985). Biology of ovarian follicles in mammals, Springer-Verlag (Publishers),Berlin.

Heape, R.B., Wang, M.W.,Whyte, A. and Taussing, M.J. 1990. In:Preoceedings of the international Conference on Frontiers in reproductive Physiology, AIIMS, New Delhi,8-10 November,1990 L17.

Kaiser, H.B.; Zhao, D.; Cardona, G.R, and Chin, W.W.(1992). Isolation and characterization of cDNAs encoding the rat pituitary gonadotropin-releasing hormone receptor. Biochemical and Biophysical Research Communications 189:1645-1652. Karsch, Fred J.; Bowen, Jennifer M.; Caraty, Alain; Evans, Neil P, and Moenter, Suzanne M.(1997) Gonadotropin-Releasing Hormone Requirement for Oculation. Ovulation. *Biology of Reproduction* 56:303-309.

Karsch, Fred, J. and Evans, Neil P.(1996) Feedback actions of estradiol on GnRH secretion during the follicular phase of the estrous cycle. *Acta Neurobiol. Exp.* 56:715-725.

Kher, H.N and Dholakia, P.M. 1985. Bacteriological studies of bovine semen. *Indian Journal of Animal Reproduction*, 5(1):78-84.

Kledzik, G.S.; Cusan, L.; Auclair, C.; Kelly, P.A. and Labrie F. (1978). Inhibitory effects of luteinizing hormone releasing hormone agonist on rat ovarian LH and follicle stimulating hormone receptor levels during pregnancy. *Fertility and sterility*, 29:560-564.

Koul, N.A.2002 Bacterial flora of bovine semen and their antibiotic sensitivity pattern. M.V.Sc thesis, SK. University of Agricultural Sciences & Technologyof Kashmir, Srinagar-191 121.

Lenton, E.A.; King, H.; Thomas, E.J.; Smith, S.K. MeLachlan, R.I.; MacNeil, S. and Cooke, I.D. (1988). The endocrine environment of the human oocyte. *J.Reprod.Fert.*82:827-841.

Martin, K.; Santoro, N.; Hall J.; Filicori, M.; Wurman, M. and Crowelery, W.F. Jr. (1990). Management of oculatory pulsatile gonadotropin-releasing hormone. *J. Clin Endorcinol and Metab.* 17: 1081A-1081G.

Mufti, A.M.1996.Studies on morphological characters and some metabolic constituents at different stages of development of ovine foetus. M.V.Sc. thesis, Sher-e-kashmir University of Agricultural Sciences and Technology, Shalimar, Srinagar.

Mufti, A.M., Wani, N.A.; Wani, G.M., Mir, M.M. and Khan, M.Z,1997 Superovulatory response in sheep during different months of breeding season.*Small Ruminent Research,* 25:181-184.

Najar, A.M.(1999). Effects of fish ovarian extracts on mammalian ovary, Ph.D. Thesis of University of Kashmir, Srinagar-190006.

Sanchez-Criado, J.E.; Bellido, C.; Galiot, F.; Lopez, F.J. and Gaytan, F. (1990). A Possible dual mechanism of the anovulatory action of antiprogesterone RU 486 in the rate. *Biology of Reproduction* 42:877-886.

Schultz, G.A., hahnel, Ann., Watson, A.J and Hojan, A. 1990. In:Proceedings of the international Conference on Frontiers in Reproductive Physiology AIIMS, New Delhi, 8-10 November, 1990. L20.

Shah, N.A. 1996. Effect of gonadotrophin releasing hormone (GnRH) on anoestrous crossbred cows under temperate climatic conditions. MVSc Thesis submitted to Sher-e-Kashmir University of Agricultural Sciences & technology, Srinagar.

Srivastawa, R.K.; Krishna, A. and Sridaran R. (1995). Ovarian development, steroidogeness and gonadotrphin secretion in response of long-term treatment with a gonadotrophin-releasing hormone agonist in the rat. *Journal of Endocrinology* 146:349-357.

Van Look, P.F.A. (1988). Hormone fertility reguation at the crossroads. A short review of the potential of hormone antagonists . In Proceedings of a symposium on the Use of Hormone Antagonists for Fertility Regulation organized by the Indian Society for the Study of Reproduction and Fertility and the World Health Organisation's Special Programme of Research, Development and Research Training in Human

Reproduction on November 4, 1988 at Mumbai, India. Edited by Chander P.Puri and Paul, F.A.Van Look, Printed by Good Print, Dr.Ambekar Road, Parel Mumbai-400012 (India).

Wani, G.M. 1998.Ultrasonic pregnancy diagnosis in Sheep and goats. *World Review of Animal production,* 17:43-48.

Wani, G.M. 1982. Laparoscopy in farm animals. *World review of Animal Production,* 18: 7-13.

Wani, G.M. 1984a. Vet suche Zur Embryo Production and Kultur bei Ziege. Dr. Med.Vet Dissertation, Tieraertzlichen Hochschule, Hannover, Germany.

Wani, G.M. 1984b. Investigations on ovarian activity, ovulation and pregnancy diagnosis in sheep and goats. Ph.D. Thesis, IVRI, Rohikhand University, Indian Veterinary Research Institute, Izatnagar, U.P.

Wani, G.M. 1986. Embryo transfer in goats. 6th National Congress on Animal Reproduction, Guwhati 28-30th November, 1986.

Wani, G.M. 1989.Detection of foetal age in sheep and goats by ultrasonic technique. *Indian Journal of Animal Science,* 59(12):525-30.

Wani, G.M. 1989. Detection of foetal age in sheep and goats by ultransonic technique. *Indian Journal of Animal Science,* 59(12):1525-26.

Wani, G.M. 1989. Ovulation detection by laparoscopy in goats. *Indian Journal of Animal Science.* 59(12):1527-28.

Wani, G.M. 1995. Early diagnosis of pregnancy in surrogate mothers. National symposium on embryo transfer and genetic engineering technologies in small ruminants. May 7-8, 1995, Makhdoo, Farrah. India.

Wani, G.M. 1996. Embryo biotechnology in sheep and goats. Published by Valley Book House, Srinagar- 190006.

Wani, G.M. 2001. Enhancement of chevon production through reproductive management. In: Proceedings of IV National Seminar of Indian Society for Sheep and Goat Production & Utilization (ISSGPU), held at Bharathiar University, Coimbatore from November 9-11, 2001.pp.60-68.

Wani, G.M. 2002. Goat Husbandry/multipurpose management. Encyclopedia of Dairy Sciences. Edited by Roginishi, Hubert, Enquay, Jhon W. and Fox patrict, Academic Press, London, NW 17B4MS192.

Wani, G.M. and Buchoo, B.A.1990. laparoscopic view of ovine conceptus. *Indian Journal of Animal Science,* 60(5):564-568.

Wani G.M. and Buchoo, B.A. 1991. detection of induced ovulation by laparoscopy in ewe lambs. *Indian Journal of Animal reproduction,* 12(1):76-77.

Wani, G.M. and Sahni, K.L. 1980. Ultrasonic pregnancy diagnosis in sheep and goats. 2nd All Indian Symposium on Animal Reproduction, Banglore, pp. 1-5.

Wani, G.m. gelderman, H. and Hahn, J. 1986. Brunstsychronization bei Ziege. *Zuchthygiene,* 21: 1-2.

Wani, G.M., Gelderman, H. and Hahn, J. 1990. Superovulations during early luteal phase in goats. *World Review of Animal Production,* 29(2):41-43.

Wani, G.M., Risam, K.S. and Nowshahri, M.A. 1988. Effect of oestrus synchronization on lambing in corriedale ewes. *Indian Journal of Animal Science,* 58(7):800-801.

Wani, G.M. 1989.Plasma progesterone concentrations during oestrus cycle in goats. *Indian Journal of Animal Reproduction,* 19(1):44-46.

Wani, G.M. Sadat Rehman and Shah Masood 2000. Higher Education and Human resource Development: Knowledge Perspectives in Current Millennium. Paper presented at the Invernational Conference on Higher Education for Human development, organized by Association of Indian Universities at New Delhi on February6 22-24, 2000.pp.239-247.

Wani, G.M., Sinha, N.K. and Khan, B.U. 1987. Oestrus synchronization with progestagens in Muzzaffarnagri ewes. *Indian journal of Animal Science.* 57(12)

Wani, G.M., Sinha, N.K. and Sahni, K.L, 1980. Haematic type of foetal mummification Muzaffernagri ewes. *Indian Veterinary Medical Journal,* 4:46-47.

Wani, G.M, Sinha, N.K. and Sahni, K.L. 1980. Note on gestational oestrus in the Muzzafernagri ewes. *Indian Journal of Animal Science,* 50(11):1009.

Wani, G.M and Sahni, K.L. 1980. An ultrasonic technique for detecting pregnancy in sheep and goats. *Veterinary Research Bulletin,* 3:35-37.

Wani, G.M., Sinha, N.k. and Sahni, K.l. 1980. Note on seasonality of breeding behaviour in Jamunapari goats, *Indian Journal of Animal Science,* 50:1153-55.

Wani, G.M., Sinha, N.k. and Sahni, K.L. 1981. Incidence of abortionsand stillbirths in Muzzaffarnagri and crossbred ewes. *Indian Veterinary Journal,* 58:558-561.

Wani, G.M., Tripathi, S.S. and Saxana, V.B. 1980. Sialoproteins in the cervicaql mucus of repeat breeding cows during oestrus. *Indian Veterinary Medical Journal,* 4:49-53.

Wani, N.A., Wani, G.M., Mir, M.M. and Beg, M.A. 1998. Ultrasonic pregnancy diagnosisin Gaddi goats. *Small Ruminant Research,* 29: 239-240.

Wani, G.M. (2002b). Advances in animal Reproduction in: Proceedings of NationalSeminar on "Recent Research Trends in Life Sciences" , (August 26-28, 2002) organized by PG Department of Envirornmental Science and Centre of Research for Development, The University of Kashmir, Srinagar- 190006.

□□□

Chapter 11

Clinical Aspects of Embryonic and Fetal Development in Small Ruminants

G.M. Wani & A. Khatun

Abstract

The white revolution has been the result of propagation of better animal germplasm. Meat and animal bye products from small Ruminants are our main export items. We earn a sizable foreign exchange from hides, fur, wool, pashmina and other animal byproducts. Chevon and mutton consumption in world is increasing. The use of 117 kg of large ruminant meat in USA per person is yet adding 1.7 kgs of chevon too. Higher reproductive rates of Small Ruminants is a mean to boost small ruminants products. Various technology advancements like biotechnology, stem cell and nano cell technologies along with future technologies have been reviewed by us. (Wani, 1996, Wani, 2001, Khatoon et al 2007, Wani *et al*, 2006, 2007 a and Wani 2007, 2008).

This paper presents our own results on clinical aspects of embryonic and foetal development in Small Ruminants. We shall be presenting a review and our results with discussions on:

i) Estrogen-progesterone milieu in non-pregnant and pregnant animals during different (pregnancy).

ii) The new concepts of Estrogen: Progesterone receptors shall be compared in non-pregnant and pregnant animals.

iii) These hormonal values shall be co-related with morphological and physiological development of caruncles, blastocyst, foetus, with histopathological and actual visual appearances. This shall help clinico-reproductive diagnostics in small ruminants.

Introduction

Embryonic losses in sheep and goats are very high. About (22-44%) embryos are lost during early pregnancy, maximum losses occur during pre-implantation period (Wani, 2006). Various biochemical, nutritional and genetic causes have been discussed (Lee and De Mayo, 2004 and Wani, 2005).

Ovine pre-implantation blastocyst undergoes elongation for upto 30 days and feeds on uterine milk through histotrophs (Wani *et al.*, 2006). Uterine epithelium undergoes various changes. The morphological nature of these changes is not recorded. The changes in caruncles and endometrium are the beginning of immunological protective cover which prevents rejection of conceptus (Wani, 1996). Therefore, this study was planned to find out morphological, physiologic and histological changes. They were viewed with biochemical changes too.

Review

Uterine Physiology

The uterine endometrium of pregnant females is dominated by progesterone which stimulates development of glandular epithelium that becomes highly secretory and produces histotrophe which is essential for conceptus development.

Placental production of progesterone is known to increase markedly between 70th and 100th days of gestation in the ewe after the placenta has reached its maximum size (Gordon, 2004). The corpus luteum of pregnancy is essential throughout pregnancy in all farm animals except, ewe and mare as placenta produces sufficient progesterone to maintain pregnancy in these species.

Endometrial oestrogen and plasma estrogen levels were also inversely correlated with plasma P4 levels. Thus, P4 levels influences ER levels, the hormone may make the tissue less responsive to ovarian oestrogen at such times as the luteal phase of the cycle and during pregnancy (Henricks and Harris, 1978). Therefore, the concentrations of Estrogen Progesterone receptors have an importance for pregnancy.

Estrogen/Progesterone Receptors

The concentration of oestradiol cytosolic rcceptors (ERc) and oestradiol nuclear receptors (ERn) in non placental and placental myometrium was low in early and mid pregnancy, increasing in late pregnancy and was highest at parturition. The levels of progesterone cytosolic receptor (PRc) in non placental regions were low in early pregnancy, increased in late pregnancy but declined during parturition whereas progesterone cytosolic receptor (PRc) in placental regions and progesterone nuclear receptor (PRn) in placental and non placental regions showed reverse character.

Thomas and Fuller (1995) suggested that uterine estrogen receptor (ER) and progesterone receptor (PR) gene expression is regulated in a tissue and cell.

In ovine uterus, estrogen increases both estrogen receptor (ER) and progesterone receptor (PR) expression, where as progesterone (P) inhibits estrogen receptor (ER) and the expression of its own receptor. Further they have recorded that alterations in tissue and cell specific expression of uterine estrogen receptor (ER) and progesterone receptor (PR) not only depend on circulating steroid hormone concentrations but also depend on locally produced, paracrine acting factors.

Serum progesterone rapidly rises following formation of corpus luteum. Timing of implantation may be regulated by length of time the uterine endometrium is exposed to progesterone stimulation. The conceptus can not attach to the uterine epithelial surface until the large glycoprotein called MUC-1 has been removed. Over 8-10 days of exposure to progesterone, nuclear progesterone receptors in the uterine epithelium down-regulate leading to a loss of the direct effect of progesterone on this cell type. Since epithelial synthesis of MUC-1 is stimulated by progesterone, loss of progesterone receptor from the uterine epithelium would reduce MUC-1 production and open a receptive state for conceptus attachment (Hafez and Hafez, 2000).

Endometrial Receptivity

Endometrial receptivity is stimulated by ovarian steroids. The role of ovine and caprine ovarian steroids in initiating endometrial receptivity has been reviewed by various workers (Huct-Hudson *et al.*, 1989, Wani, 1995, Wani 1996, Wani, 2003, Wani 2005 and Wani, 2006). Proliferation of uterine cells is influenced by estrogen (E2) and progesterone (P4). These hormones have their cognate nuclear receptors in ovine/ caprine uteri called E2 receptor (ER) and P4 receptor (PR). (Lee and De Mayo, 2004). This intrinsic ability of uterine epithelium has been reviewed (Dey, 2004).

IFN-tau is believed to act by down-regulating oestrogen receptor thus preventing the oxytocin receptor and regulate the progesterone secretion (Gordon, 2004).

Serum progesterone concentration remained at luteal levels 1.5 ± 0.5 ng/ml for the first two months of pregnancy and then gradually increased in pregnant Sumatran rhinoceros. However, progesterone decreased almost to luteal levels during the 5th months after that it increased again and eventually reached peak concentration (13.3 ± 1.9 ng/ml) shortly before parturition (Roth *et al.,* 2004).

Histological Changes in the Female Genital Organ

Extensive erosion of pseudostratified columnar epithelium of the uterus occurred at the time of invasion of binucleate cells of the trophoblast (Bhavsar *et al.,* 1974).

On day 18th the presence of blastocyst reduced the height of the uterine epithelium, on day 20th the trophoblast was starting to contact with maternal epithelium; while on day 21st and 30th attachment became more complex. King *et al.* (1981)

This review suggests that biochemical changes occur in the composition or distribution of the trophoblastic cell coat during the process of blastocyst attachment in the ewe (Guillomot *et al.,* 1982).

Upregulation of progesterone receptors (PR) in endometrial stroma was mediated by estrogen signal which induces stromal cells to differentiate into decidual cells (Kurita *et al.,* 2001).

Our Own Results

Progesterone

The mean concentrations of progesterone in maternal serum observed from Ist to 5th were 5.32±0.33, 8.26 ± 0.11, 7.70 ±, 0.23, 7.08 ± 0.35, 7.00 ± 0.32 respectively. The progesterone concentrations increased significantly (p=0.05) from Ist to 5th month pregnancy and decreased significantly (p=0.05) in 3rd month but decreased non-significantly from 4th to 5th. The progesterone concentration increased significantly (p+0.05) in between. The progesterone concentration increased significantly (p=0.05) in between groups (Khatoon, 2007).

Progesterone Receptor

The mean concentrations of progesterone receptor in uterine tissue from Ist to 5th month pregnancy serum samples were 3.81 ± 0.04, 4.33 ± 0.06, 4.38 ± 0.04, 4.09 ± 0.09, 3.86 ± 0.12 respectively. The mean concentration of progesterone receptor followed similar shift as in progesterone. The progesterone receptor concentration increased significantly (p=0.05) in between Ist and 2nd, Ist and 3rd, Ist and 4th and decreased significantly (p=0.05) in between 2nd and 4th, 2nd and 5th 3rd and 5th.

Estrogen Receptor

The mean concentrations of estrogen receptor in uterine tissue from GPI to Gp5 were 0.37 ± 0.01, 0.41 ± 0.01, 0.41 ± 0.01, 0.37 ± 0.02, 0.36 ± 0.03 respectively (Table: 4A.7). The estrogen receptor concentrations increased significantly (p=0.05) in between GpA and GpB, GpA and GpC and decreased significantly (p=0.05) in between GpB and GpD, GpB and GpE, GpC and GpD.

(9) Group nomenculture is as

GPA = Ist month of pregnancy
GPB = 2nd month of pregnancy
GPC = 3rd month of pregnancy
GPD = 4th month of pregnancy
GPE = 5th month of pregnancy

New Formula Invention

Prediction of age in days in relation to whole embryo/Crown-anus (CA) length (Wani-Khatoon Formula).

Sl. No.	Whole/CA Length in cm	Age in days
1	0.05	12
2.	0.07	14
3.	0.2	15.4
4.	0.4	18.0
5.	0.6	20.2
6.	0.8	23
7.	1.0	25.5
8.	1.3	29.5
9.	1.6	33.5
10.	1.8	36
11.	1.9	37.5

Morphological Changes

(i) Early Pregnancy

At 18 days of gestation, the right horn was larger than the left and right ovary contained corpus luteum. An elongated blastocyst was present with comma shaped embryo. The changes in the caruncles were same with previous sample .

At 20 days of gestation, the right horn was larger than the left and right ovary having the corpus luteum. An elongated blastocyst was present with embryo. Both caruncular and intercaruncular area gave live appearance. The centre of the caruncles was more reddish compared to other parts of the horns. Caruncles were swollen, slightly more larger than earlier and rounded.

At 23 days of gestation, the horns were asymmetrical and body of the uterus bulged dorsally. The left ovary had corpus luteum. An elongated blastocyst was present with embryo. The changes in the caruncle and intercaruncular areas were same with 20 days of pregnancy. The fitures depict the photographs of all events narrated above and hereinafter.

At 25 days of gestation, the right horn was slightly larger than the left. An elongated blastocyst was present with embryo. Both caruncular and intercaruncular area gave lively appearance. The centre of the caruncles had slight circular depression.

At 29 days of gestation right ovary had double corpus luteum and comma shaped embryos were present one in each of the horns. Changes in the caruncles were as like 25 days gestation. Stereomicroscopic view gave the distinct comma shaped feature of the embryo. Under microscope some other structures like somites, head, eye bud, tail bud and emerging blood vessels from hind gut could also be observed in the elongated blastocyst. Two vessels emerged from each of the embryos and entered into the either of the horns along with its ramification could also be observed grossly. From this stage onwards emerging of blood vessels from the hind gut of foetus became distinct grossly.

At 33 days of gestation right horn was more enlarged than the left. A curved embryo was present in the right horn, emerging of blood vessels from the hind gut of foetus became distinct grossly. Caruncles were flattened type.

At 36 days of gestation left horn was more enlarged than the right and the corpus luteum was present in the right ovary. Foetus was present in the left horn. At this stage the blastocyst was completely filled with fluid and whole blastocyst ramified with blood vessels and it was almost reddish in colour. From this stage head, eye bud and tail could be recognized grossly, this might be the beginning of foetal stage (Jainndeen and Hafez, 1984 and Hafez and Hafez, 2000). Circulation increased to the caruncle and intercaruncular areas. Caruncles were flattened; reddish colouration was present at the centre.

(ii) Mid Gestation

At 39 days of gestation left horn was more enlarged than the right and the corpus luteum was present in the left ovary. Foetus was present in the left horn. At this stage single layered reddish granules were present mostly in the membrane of the left side were nothing but the cotyledons, appeared for the first time. Cotyledones were arranged in single layered which was reddish in colour and granular. Ramification of blood vessels was present both the horns but was more prominent in the left side. Head, eye bud, limb and tail bud in the foetus could be observed grossly. Caruncles were flattened.

At 42 and 43 days of gestation right horn was larger than the left and the corpus luteum was present in the right ovary. Other changes were same as there was in 40 days of gestation except their size and shape of the foetus which were increasing.

At 44 days of gestation right horn was larger than the left and the corpus luteum was present on the right ovary and at around 45 days of gestation left horn was larger than the right and the corpus luteum was present in the left ovary. Other changes were almost same as was in 40 days gestation except, the size of the foetus which was larger and the umbilical chord was very distinct and showed total four numbers of blood vessels, two of them entered on either side of the horns. Cotyledons were multilayered and reddish in colour at this stage.

At 46 and 47 days of gestation right horn was larger than the left and the corpus luteum was present on the right ovary. From this stage onwards the cotyledons were raised type and the depression at the centre of the caruncles was increasing. The size of the cotyledons and caruncles increased with higher gestational age. The external body parts of the foetus became distinct from this stage onwards. Cotyledons could easily be separated from the caruncles by holding the uterine horn or by imposing slight pressure.

At 48 days of gestation, right horn was larger than the left and the corpus luteum was present on the right ovary. At this stage demarcation of placenomes was visible from outside of the uterine horns. Caruncular-cotyledonary attachment was strong that could not be separated easily and needed pressure for separation. The placentomes were cup shaped and cotyledons were raised type. From this stage onwards all the external body parts of the foetus developed properly and their size and shape increased as the age advanced.

(iii) Late Changes

From 80 days to 135 days the cotyledons were almost engulfed by the placentomes leaving no space at the centre of the caruncles. At this stage separation of cotyledones from the caruncles were very difficult but could be separated by imposing strong pressure.

The umbilical chords were consisting of four numbers of blood vessels out of which two vessels entered on either of the horns with elongated blastocyst. The sub branches developed from main two vessels and entered each of the cotyledons and ultimately connection developed between the cotyledons and the caruncles. Arrangement of blood vessels with the cotyledons were just like a full blooming tree, where blood vessels were just like branches of the tree and cotyledons were like blooming flowers at the terminal ends of the branches. As the time advanced the vessels were becoming thick and prominent.

Hormonal and Receptor Changes in Our Studies

The progesterone concentration increased significantly from GpA to GpB but decreased later on. Similarly, estrogens varied from GpA to GpE as in progesterone. The progesterone and estrogen concentration increased significantly with fluctuation between the groups. Rodrigues *et al.*, 1998, reported that serum levels of progesterone increased during gestation decrease abruptly only by two days before preterm. Hafez and Hafez, 2000 and Schular, 2002 suggested that both estrogen and progesterone is a placental hormone and ewe is a placenta depended animal (Hafez, 1987 and Hafez and Hafez, 2000). So, the ratio of estrogen and progesterone might increase as the gestation advanced. The estrogen and progesterone concentration in present study in non pregnant animal was 451.25 pg ± 9.53, and 1.2ng ± 0.05 respectively, where estrogen was much higher and progesterone was much lower in non pregnant animals. Estrogen and progesterone receptor increased upto Gp D and decreased in GpE but the differences were very less, they were almost like constant in between groups.

The estrogen and progesterone receptor concentration in present study in non pregnant animal was 0.55 ng ± 0.01, and 2.19 ng ± 0.08 respectively. Our study was supported by earlier reports given by Thomas and fuller (1995).

They have postulated that increased level of progesterone suppress both estrogen and progesterone receptor and increased level of estrogen stimulate both estrogen and progesterone receptor. Hewitt and Korach, 2000 postulated that the progesterone attenuates the proliferative effect of estrogen in the uterus, where as estrogen also induces the progesterone receptor.

The present results with respective to estrogen and progesterone in serum and those of estrogen and progesterone with reported by Hafez and Hafez, 2000, Cengiz *et al.*, 2000, Kurita *et al.*, 2001, Roth *et al.*, 2004. Differentiation and caruncular growth are known to be regulated by estrogen and progesterone (Schular, 2002). The endometrial receptivity changes both in the endometrium and placentation as seen morphologically with due evidences from histopathology could be properly explained in view of our studies on estrogen, progesterone and their receptor concentration estimated throughout the gestation in this study. Many questions regarding the role of ovine and caprine ovarian steroids in initiating endometrial receptivity as observed in this study could be easily explained in view of the review reported by Huct - Hudson *et al.*,1989, Wani, 1995, Wani, 1996, Wani, 2003, Wani, 2005, Wani, 2006 . Proliferation of uterine cells is influenced by E_2 and P_4 receptor . Our study corroborated and confirmed the studies of Lee and Dee Mayo, 2004. The ability of uterine caruncular epithelium to throw inter digital projection which fit into the evaginated projections of the cotyledones have been confirmed by ER and PR receptor studies in this investigation. This to a broader extend explains the concept of implantation as put forth by Lee and Dee Mayo, 2004.

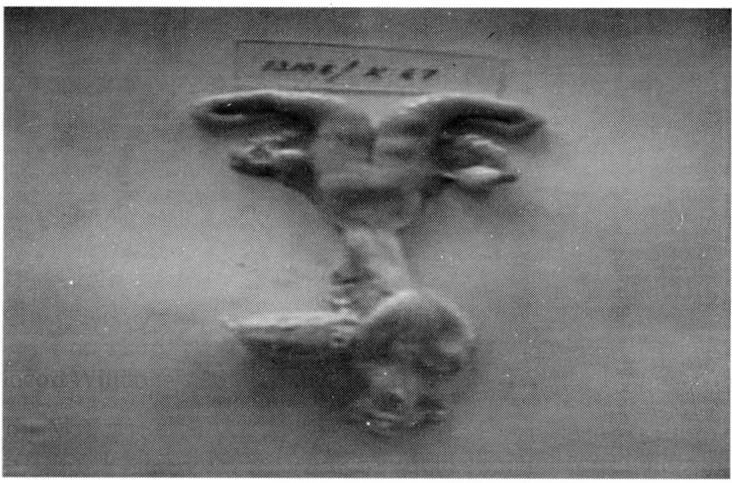

Plate 1. Showing non pregnant genital organ

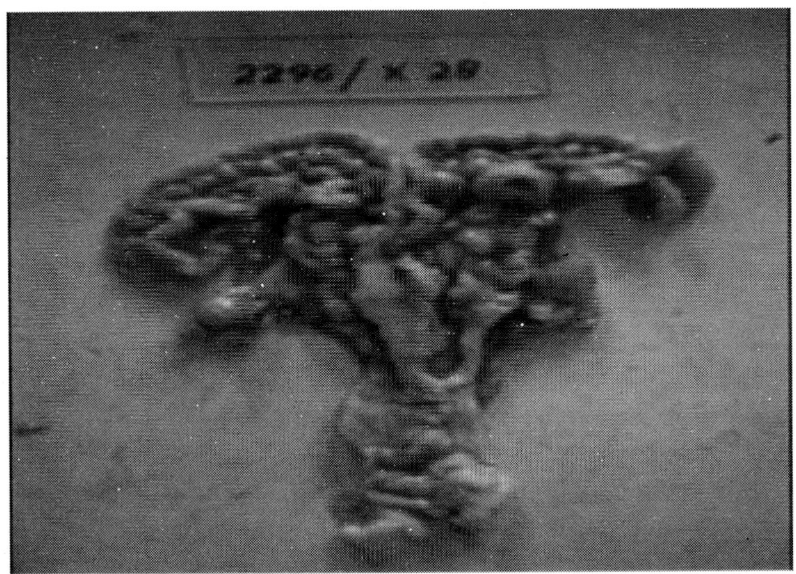

Plate 2. Showing caruncle in the non pregnant genital organ

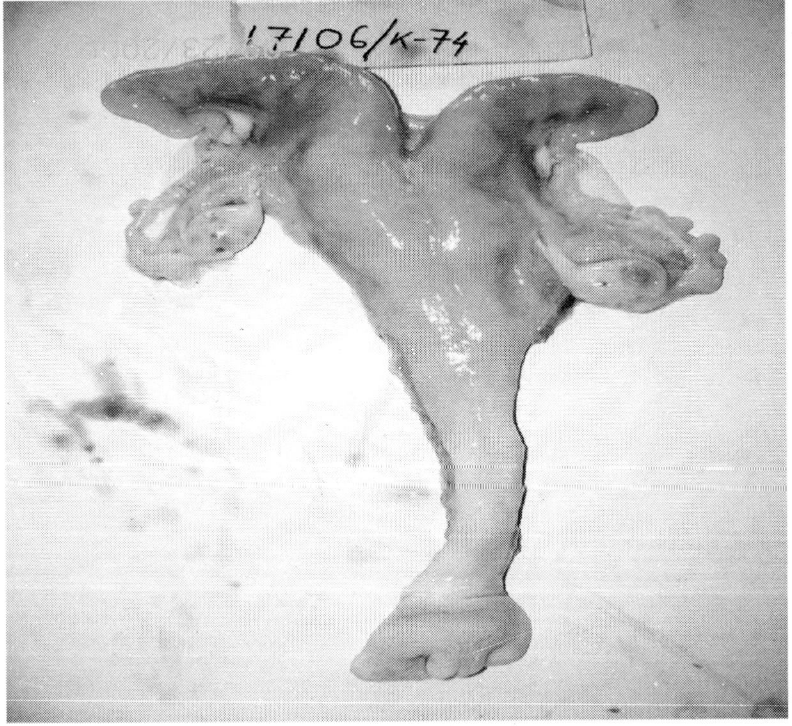

Plate 3. Showing Genital organ at 15 days of pregnancy

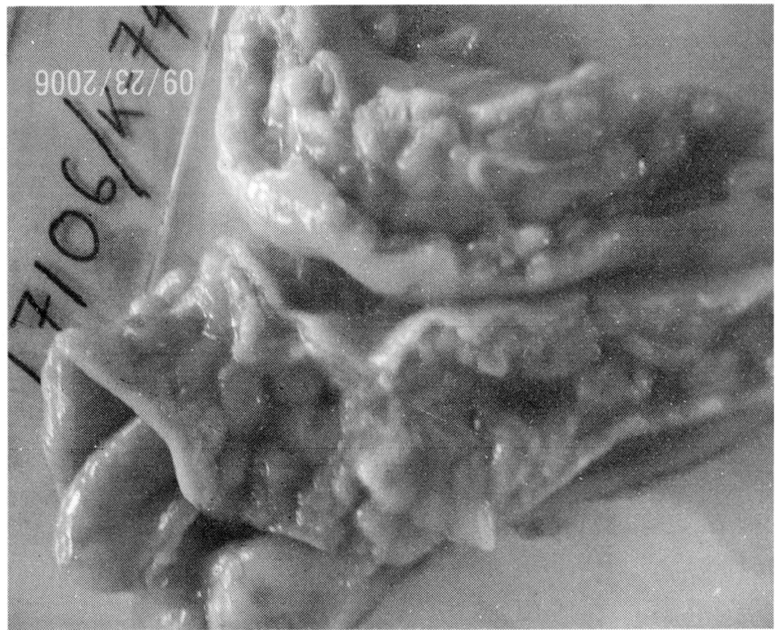

Plate 4. Showing elongated blastocyst in the uterine horns at 15 days of pregnancy

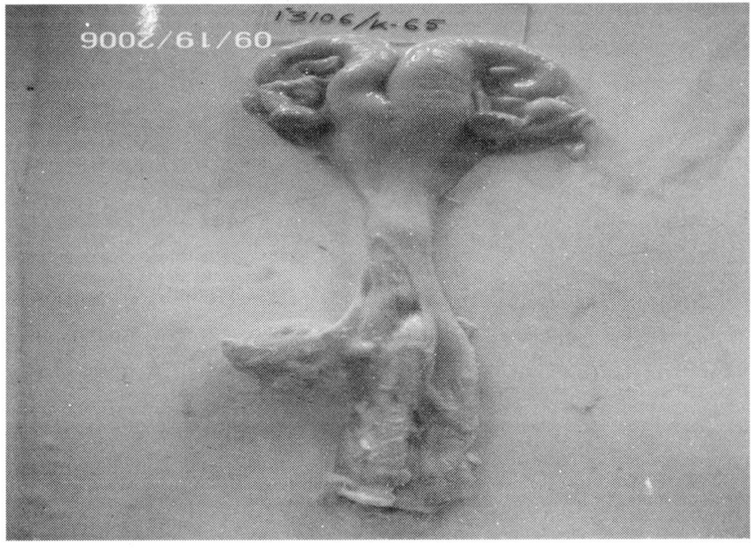

Plate 5. Showing genital organ of 18 days of pregnancy

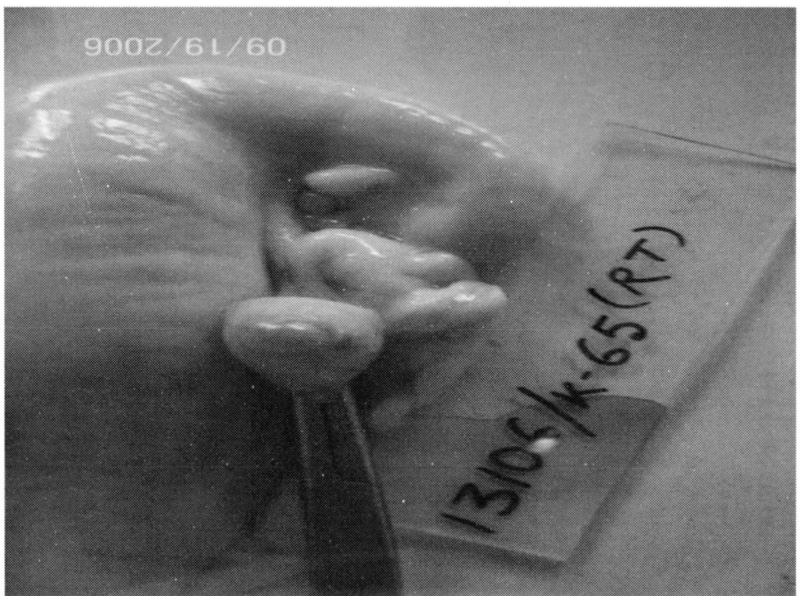

Plate 6. Showing corpus luteum at 18 days of pregnancy

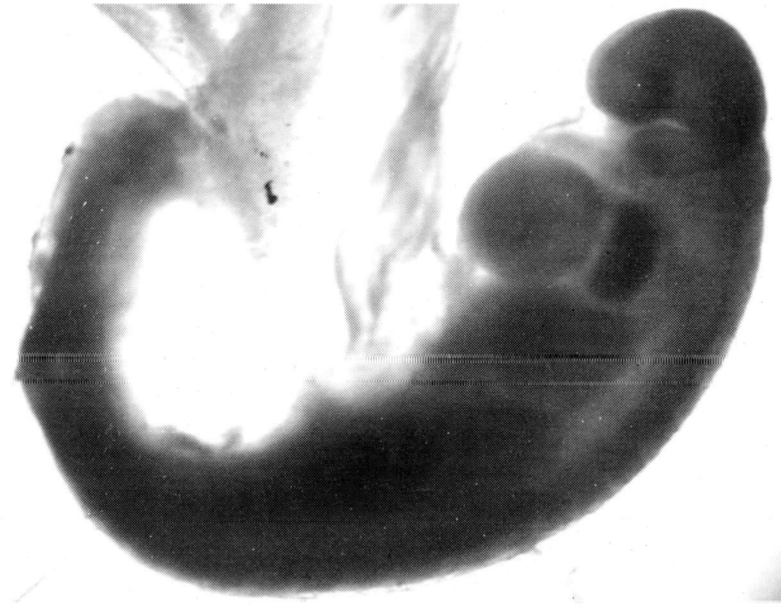

Plate 7. Showing microscopic view of 18 days embryo (2x magnification)

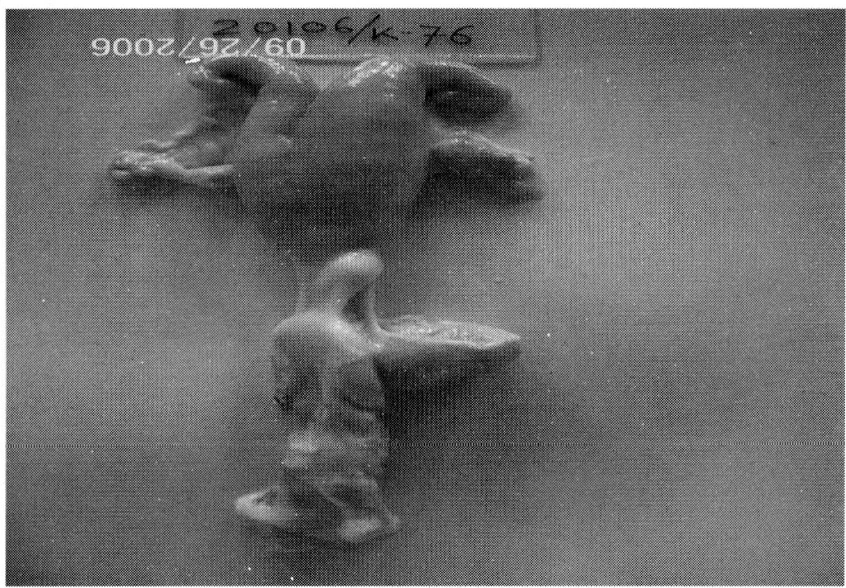

Plate 8. Showing genital organ of 20 days of gestation

References

A. Khatoon, G.M. Wani, A.R. Choudhury, M.Z. Khan, R.Islam and B.A.Pandit,(2007) Incidence of reproductive abnormalities in small ruminants. *The Indian Journal of Animal Reproduction*: 28(1):53-56; June 2007 www.buzzle.com .

Bhavsar, B. K., Pattabiraman, S. R. and Venkataswami, V. 1974. Endometrial changes during early pregnancy of Bubalus bubalis. *Indian Veterinary Journal,* 51(3): 175-176.

Dey, S. K. 2004. Focus on implantation reproduction, 128: 655-656.

Guillomot, M., Flechon, J. E. and Wintenberger Torres, S. 1982. Cytochemical studies of uterine and trophoblastic surface coats during blastocyst attachment in the ewe. *Journal of Reproduction and Fertility*. 65(1): 1-8.

Gordon Ian, 2004. In: Reproductive technologies in farm animals. CABI publishing.

Henricks, D. M and Harris, R. B. Jr. 1978. Cytoplasmic estrogen receptors and estrogen concentration in bovine uterine endometrium. *Endocrinology*, 103(1):176-185.

Hafez, E. S. E. and Hafez, B. 2000. Reproduction in farm Animals. Seventh Edition, Lippincott Williams & Wilkins, A Wolters Kluwer Company, pp. 126-139.

Huct-Hudson, M. Y., Andres, G. K. and Dey, S. K. 1989. Cell type specific localization of C-MYC-Protein in mouse uterus-modupition by steroid hormones and analysis of the pre implantation period. *Endocrionology*, 125: 1683-1690.

Hafez, E. S. E. 1987. In: Reproduction in farm animals, 5th edition, Lea & Febiger, Verghese Company

Kurita, T., Lee - Kijun, Saunders, P. T. K., Cooke, P. S., Taylor, J. A., Dahiya, R. and Cunha, G. R. 2001. Regulation of progesterone receptors and decidualisation in uterine stroma of the estrogen receptor- alpha knockout mouse. *Biology of Reproduction,* 64(1): 272-283.

King, G. J., Atkinson, B. A and Robertson, H. A. 1981. Development of the intercaruncular areas during early gestation and establishment of the placenta. *Journal of Reproduction and Fertility,* 61(2): 469-474.

Lee,K.Y and De-Mayo, F.J; 2004, Animal Models of Implantation Reproduction. 128, 679-695.

Roth, T. L., Bateman, H. L., Kroll, J. L., Steinetz, B. G. and Reinhart, P. R. 2004. Endocrine and ultrasonographic characterization of a successful pregnancy in a sumatran rhinoceros (*Dicerorhinus sumatrensis*) supplemented with a synthetic progestin. *Zoo- Biology,* 23(3): 219-238.

Schular, 2002. In: Reproductive Technology in Farm Animals. CABI publishing, pp. 237- 254.

Thomas E. Spencer and Fuller W. Bazer, 1995. Temporal and spatial alterations in uterine estrogen receptor and progesterone receptor gene expression during the estrous cycle and early pregnancy in ewe. *Biology of reproduction.* 53: 1527-1543.

Wani, G. M. 1995. Early diagnosis of pregnancy in surrogate mothers (Sheep and Goat). In National Symposium on Embryo Transfer and Genetic Engineering Technologies in Small Ruminants, May 7-8, CIRG, Makhdoom, Farah, India.

Wani, G. M. 1996. Embryo Biotechnology in Sheep and Goats. Publisher Valley Book House, University Road, Hazratbal, Srinagar, Kashmir.

Wani, G. M. 2000. Effect of ocyte harvesting techniques on invitro maturation and invitro fertilization in sheep. *Small Ruminant Research,* 36(1): 63-67.

Wani, G. M. 2002. Multipurpose Management/ Goat Husbandry. Encyclopedia, Dairy sciences, Editor Roginski, H, Academic Press, An imprint of Elsevier Science, London, Boston, Amsterdam, New York, Oxford paris, San Diego, San Francisco, Singapore, Sydney, Tokyo, II: 1259-1270.

Wani, G. M. 2003. Biotechnological advances in goats production. In *National Work-Shope on "Recent Advances in Goat Rearing"*. MPKKV, Agric.Univ. Rahuri, 29-30 December.

Wani, G. M. 2005. Advances in ovine and Caprine Reproduction, www.word-mart.com. Research Paper, 9: 1-25.

Wani, G. M. 2006. Biotechnological Advances in Goat Production. www.word-mart.com. Research paper, p. 10.

Wani G.M., (2007) Recent Trends in Policy Initiatives & Technological intervention Goat production as livelihood resource, presented at Sri Venkateswara Veterinary University, Tirupati w.e.from 20-22, June,2007

Wani G.M.,Khatoon A., 2006, Early embryonic development in sheep and goats www.buzzle.com http:/www.buzzle.com/edit orials/8-7-2007-104739 asp.

Wani, G.M and Sahni, K.L. 1980. An ultrasonic technique for detecting pregnancy in sheep and goats. *Veterinary Research Bulletin, 3:35-37.*

Wani, G.M. 1996. Embryo biotechnology in sheep and goats. *Published by Valley Book House, Srinagar- 190006.*

Wani G.M., A.Khutun, P. Goswami and M.Z.Khan (2006_b) Ovarian and foetal abnormalities in small ruminants . Pesented in National seminar on Innovations and Recent advances in reproduction for augmenting small ruminant production held at CSWRI, Avikanagar (India), December 28-30, 2006

Wani G.M., (2007) Recent Trends in Policy Initiatives & Technological intervention Goat production as livelihood resource, presented at Sri Venkateswara Veterinary University, Tirupati w.e.from 20-22, June,2007

Wani G.M.,Khatoon A., 2006 ,Early embryonic development in sheep and goats www.buzzle.com http:/www.buzzle.com/edit orials/8-7-2007-104739 asp.

□□□

Chapter 12

Ovarian and Foetal Abnormalities in Small Ruminants

G.M. Wani & G. Wani

Abstract

Small ruminant infertility patterns like cystic ovaries, luteal and follicular cysts are scantly reported in literature. Many reasons can be given, but the non-invisible rectal manipulations are restricted in small ruminants. We have tried to collect materials from slaughter house for both gravid and non-gravid uteri of sheep and goats . The breed, age or description of the animal was not known. The stage of pregnancy was calculated as per formula used by (Mufti et al, 2000). The basis of age determination was CAL (Crown-Anus-Length). Examination of these samples provided us excellent, specimens. The images of following reproductive structure along with histological evidences are presented and discussed in this paper. Structure of non-pregnant uteri, Asymmetry of early pregnant uteri, Cervical deviation of Sheep, Contour of uteri suspected of Brucellosis, Uterine contour carrying macerated foeti, Uterus carrying mummified foetus, Images of follicular cysts, Cystic ovary, and Images of luteal cysts etc.

Placental formation consists of Formation of caruncular part and Formation of cotyledonary part.Their union for development of umbilicus. The evidence of attachment of foetal-maternal placental components, which signifies "nidation" as misnomer in sheep or goats.

Gravidity, images from First month 0-30 days, Second month 31-60 days, 3rd month 61-90 days, 4th month 91-120, 5th month 121- term,

Histological evidences of Non-pregnant uteri, Corpus luteum Haemorrhagicum, Early Pregnant uteri, Double nuclei at development of cotyledons within foetal membranes Hybrid nuclei formation, as evidence of syncytia

Introduction

Foetal biology is a vast interesting subject. The recent studies of Erlebacher (2001), postulating why foetus is not rejected has revived the interest on foetal biology research. Uterine receptivity being still a biological mystery *(Makker and Singh 2006)* We have investigated it and few stimulated and innovative observations have been published. (*Wani et al 2006a, b&c Wani 2006*). The role of ovarian steroids in initiating pregnancy desired changes in endometeruim have been reviewed and are documented. (*Huet-Hudson, et al 1989*). The cognate nuclear receptors called ER,P4 receptors are playing a significant role (*Dey et al 2002*). Development factors like homes (Hox) gene and others regulating factors of implantation and conception have been reported. (*Tulac et al 2003 Wani et al 2006b).*

The futuristic technology knocking the doors of reproductive scientists is presumed as "*In Vitro Foetus Development*" (IVFD). Some scattered reports in this regard are coming up from Japan. What this new Buzz Word needs for successful demonstration is "development of conceptus" facts and figures. With Ethics restricting our biological live model use, we have no option but to study abattoir materials. In our quest for developing an (IVFD) modules we have been probing various gravid and non-gravid uterine samples from slaughter house. While the development of conceptus details are available in literature (*Wani and Buchoo, 1990, Mufti et al 2000, Wani et al 2006, Wani et al 2007*), very little information on various ovarian and foetal deformities are available in literature. As such we are presenting herewith some of our own observations on the topic. We hope it shall stimulate many more studies on the area, so that like ultrasonography, (*Wani, 1981*), laparoscopy (*Wani 1982*) we in India shall lead the researchers in IVFD too.

Materials and Methods

Deformities of ovarian and foetal development in sheep and goats were studied in abbatior materials. Reproductive tracts from slaughtered sheep and goats were obtained and carried to laboratory within 2-3 hrs in normal saline. The sample was examined, analyzed and their images were studied. Various ovarian and foetal deformities recorded are presented. The samples were processed for histological examinations. The structural and morphological changes observed and recorded were supported with histological and/or bacteriological evidences where necessary. All the samples were measured. The age of foetus was determined on the basis of crown-anus-length (*CAL*) as reported (*Mufti et al 2000*). The observations of the ovarian status, presence or absence of corpus luteum, (CL) status of CL, CAL, caruncle shape and contour at pregnancy, non-pregnancy, contour of normal, pathological ovaries, CL and uterine horn were recorded and are documented in this paper.

Results

The results obtained shall be presented as images of the various conditions observed as:

1) Non-Pregnant Uterus

Non pregnant uteri are without CL. The contour of uterus body, its horns and ovaries can be seen Fig 1, along with caruncle structure. They appear as solid, up right reddish projections.

2) Early Pregnant Uterus

Early pregnant asymmetrical uteri. Right ovary with corpus luteum Albicans (CLA) and left ovary with corpus luteum Haemorragicum *(CLH)*. Right ovary having 5 and left ovary with 8 follicles. The caruncles were hypertrophied, cup-shaped with mucin coated appearances Fig 2. Seen with OV.A & embryo too.

3) Cervix of Sheep

The cervical contour, rings its internal structure and deviation posing problems in AI can be seen in Fig 3.

4) Gravid Looking Uteri Containing Pus and Suspected of Brucellosis

The uterus was of irregular shaped with depressions. Right ovary had a hard CL which looked like a persistent CL . The CL was 1.1.x 0.8 cms thick at base and not as bright red as CLH. On opening uterus pyometra was observed. The material was suspected of Brucellosis and sent for identification to IVRI. The shape of uterus and PCL are shown Fig 4.

5) Mumified Foetus Case Fig.5

Left uterine horn bulged, Asymmetrical uterine horns, Right ovary with hard. PCL, left ovary with 5 follicles, on opening mummified foetus was observed.

6) Cystic Follicle Fig.6

Right and left ovaries fat overlapping and having cystic follicles Rt Sided cystic follicle 6.5 cms and Left sides 4.5 cms in diameter. Enlarged caruncles with distinct rows. Right side 36 nos and left side 37 Nos. counted.

7) Cystic Ovary Case Fig.7

Right ovary with 4 follicles less than 3 mm in diameter . Left ovary without follicle but completely cystic one both sides. Few hypertrophied caruncles only enlarged uterine horns 14-15 cms length and 7-8 cms breadth. No of caruncles Rt side 36 and left side 41 counted.

8) Persistant Corpus Luteum (PCL) Fig.8

Cervix closed like pregnancy, outside horns give a pregnant look.. Right ovary with enlarged hard corpus luteum. Uterine horn 24x13 cms on right side ipsilateral to PCL. Left horns 24x13 cms. Left ovary 1 large and 7 small follicles.

9) Pyometra/Macereated Foetus Fig.9

Right ovary with 2 cystic follicles. Left sided ovary with PCL. Uterine horns enlarged 21-24 cms length.

10) Early Pregnancy Gravid Uterine Horns

i)	Right sided foetus CAL 2.0 cms	Age 40 days	Fig.10
ii)	Foetal CAL 2.3 cms	Age 41 days	Fig.11
iii)	Foetal CA Length 3.2 cms	Age 43 days	Fig.12
iv)	CAL 10.4 cms Heart,Lungs	Age 58 days	Fig.13

Heart, lungs, liver, Kidney, Skull, Brain, Limbs all visible.

11.Advanced Pregnancy Details

(i) Right uterine foetus CAL 6.3 cms age 49 days Fig.14.

(ii) Placentomes visible from outside, Twin fetus CA 16.4 cms both age 70 days Fig.15.

(iii) Right uterine horns gravid CAL 27.5 age 94 days Fig.16.

12. Pregnancy

i. Right sided CAL 5.8 cms age 48 days Fig.17

ii. Caruncles with redish tinge Fig 18.

13. Trichomonus Foetus

Corn-flake like Pus on Left sided uterus Fig. 19.

14. Placental Formation

The formation of maternal caruncular portion of placenta consists of hypertrophy of the caruncles' The solid upward projected and pointed caruncles assume cup-shaped structural formation under the continued influence of progesterone secreted by CLH or even PCL. With developing conceptus and CLH, progesterone secretion stimulates mucin coated covering over wrapping the caruncles. This covering at attachment assumes angiogenesis. Thus, this covering of maternal caruncles contributes protective mechanism and avoids direct union of caruncles with cotyledons. The cotyledons grow within developing conceptus still within elongated blastocyst-fetal membranes. They get interwoven into a cable type networking system. Thus before these attachments, changes at cellular and molecular level occur in both caruncles and cotyledons to give what we know "Placentomes". These changes and their histological evidences could be seen in (Fig. 20).

15. Gravidity Images of the Foetus

The shape and structural topography of gravid uterine horns prenatal development and foetal fluids has been reported. (*Mufti* 1996, *Mufti et al,* 2000). We reproduce those observations in the figs 21-25 as per the months of gestation in sheep and goats.

16. Histological Evidences

The structural changes described above were confirmed by subjecting the morphological samples to histological investigations.. The histological differentiation of non-pregnant (NP) and pregnant (P) caruncles, NP and P uterine horns, CLH and PCL, cystic ovaries, and cystic follicles, binuclear and hybrid nuclei cells are presented in the figures Fig. 26-29. Some of the interesting features of this study is the confirmation of structural changes and differentiation made by us on union of the foetal and maternal tissues. The clear cut differentiation of cotyledons, caruncles and placentomes has been made, which support our earlier studies (*Wani et al 2006 abc*).

17. Histological Sections of Ovarian Deformities

The histological section of ovarian deformities in sheep and goats are poorly documented. They have been presented as described and shown in Figures cited above.

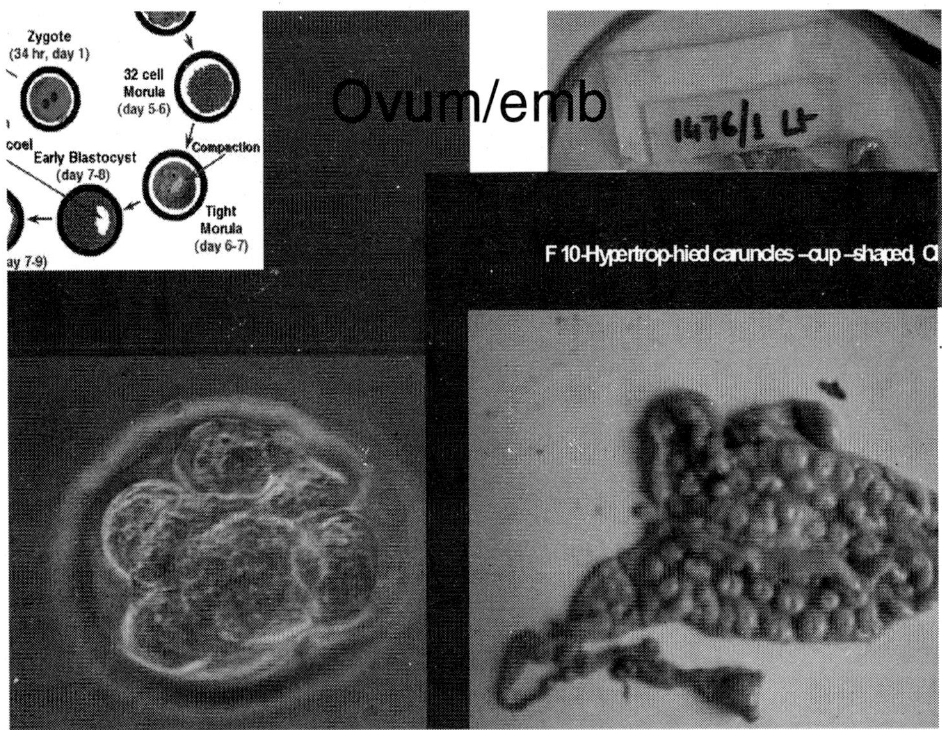

Fig. 1-2 : Shows non-pregnant caruncles, pregnant uterus caruncle ova and embryo images.

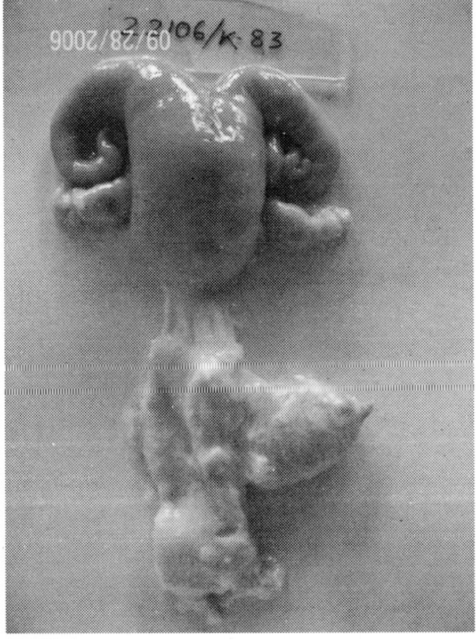

Fig. 3 : Cervical cotour and rings opened.

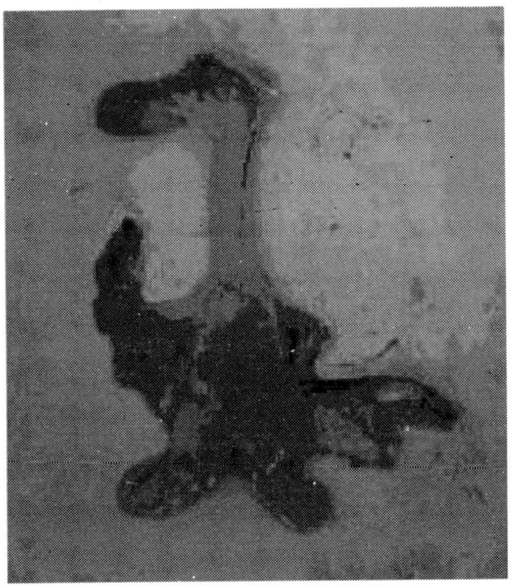

Fig. 4 : Pyomatra-Uterus containing pus suspected of brucellosis.

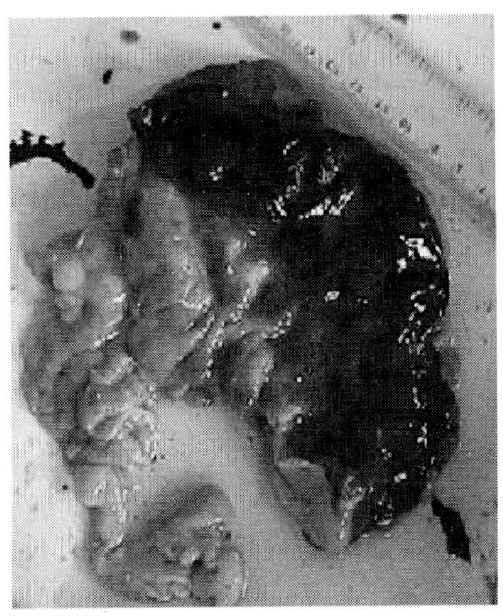

Fig. 5 : Mummified foetus below

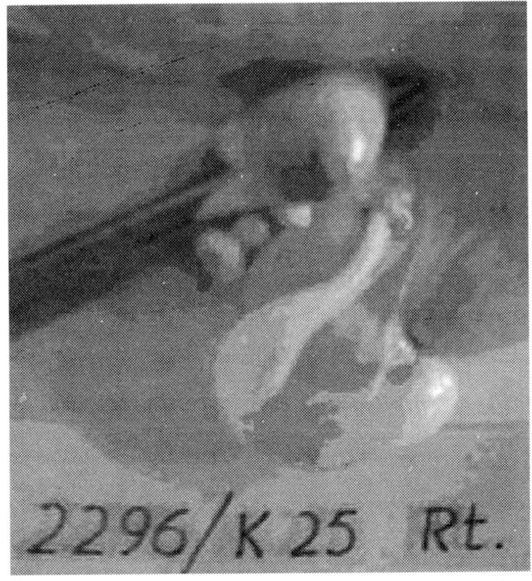

Fig. 6 : Cystic Follicles.

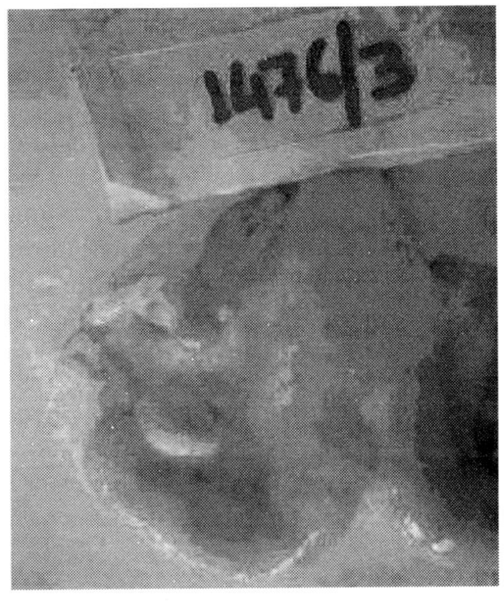

Fig. 7 : Cystic Ovary.

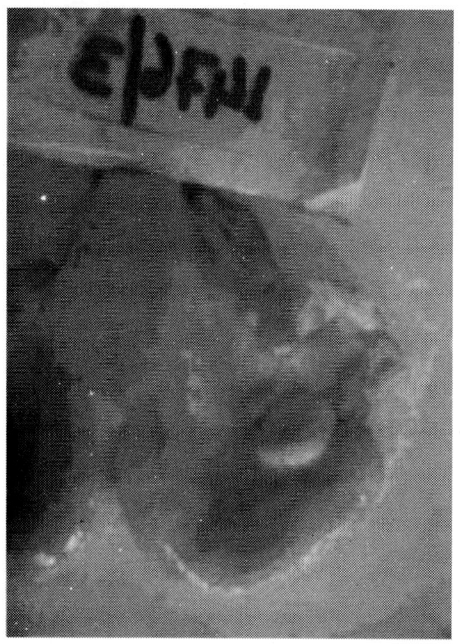

Fig. 8 : Persistent Corpus Luteum.

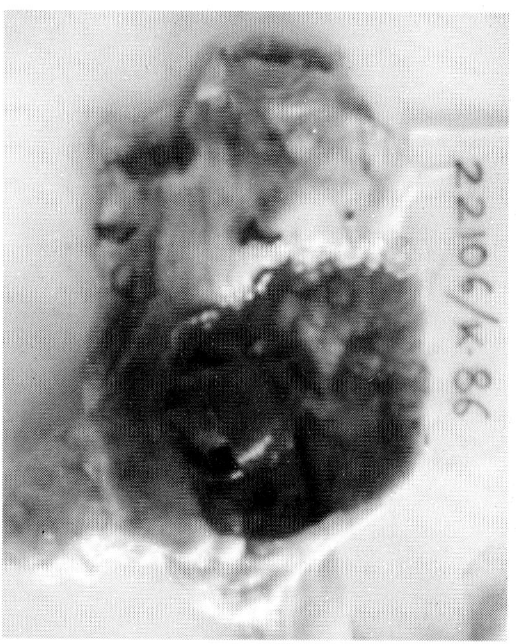

Fig. 9 Pyometra/Macereated Foetus.

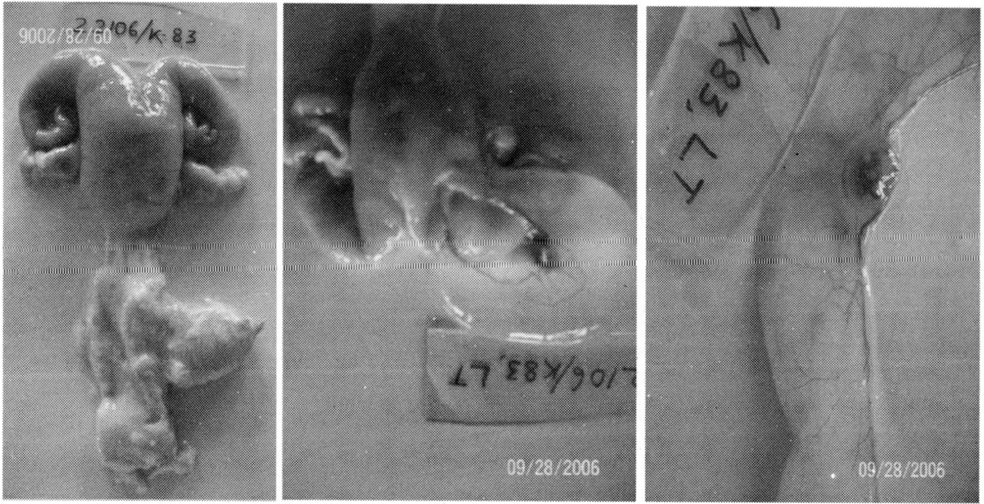

Fig. 10-13 : Pregnant Uteri

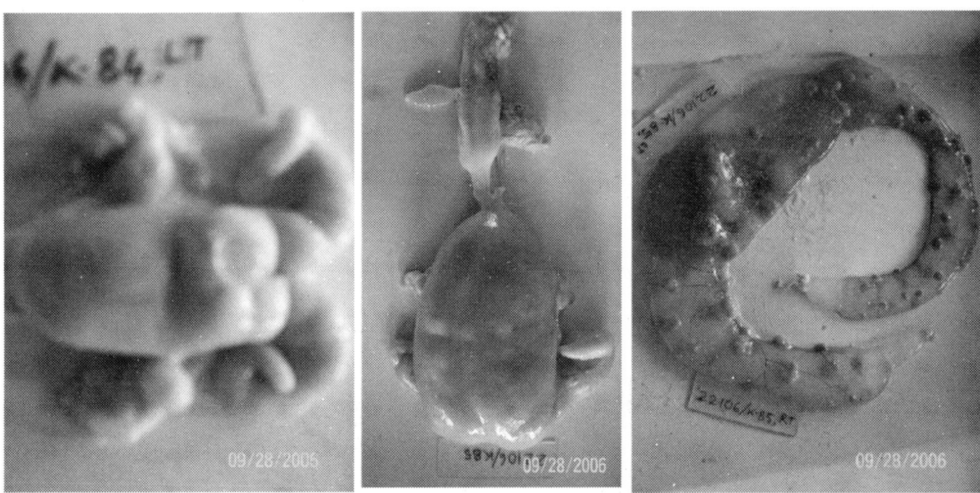

Fig. 14-16 : Advanced Pregnancy Figures.

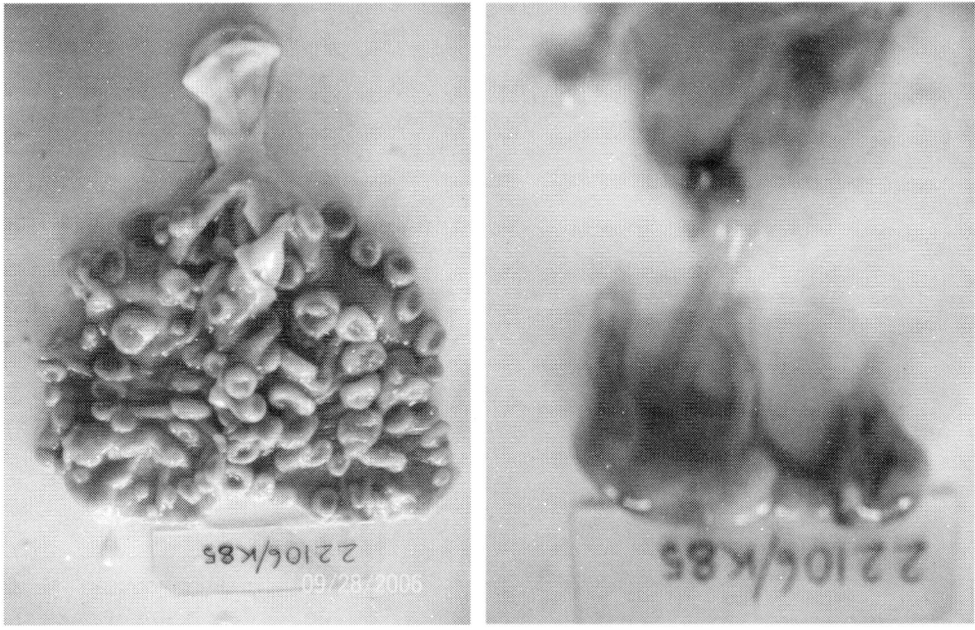

Fig. 17-18 : Pregnancy Plcentation.

Fig. 19 Tirchomonos Fetus Persisten Corpus Lutem Already Shown Refer Figure No 8.

Fig. 20 Placentation See Figure 18 and 17.

Fig. 21-25 Refer Various Figures of Pregnant Uteri in the Other Chapters and Among Other Figures Shown Above.

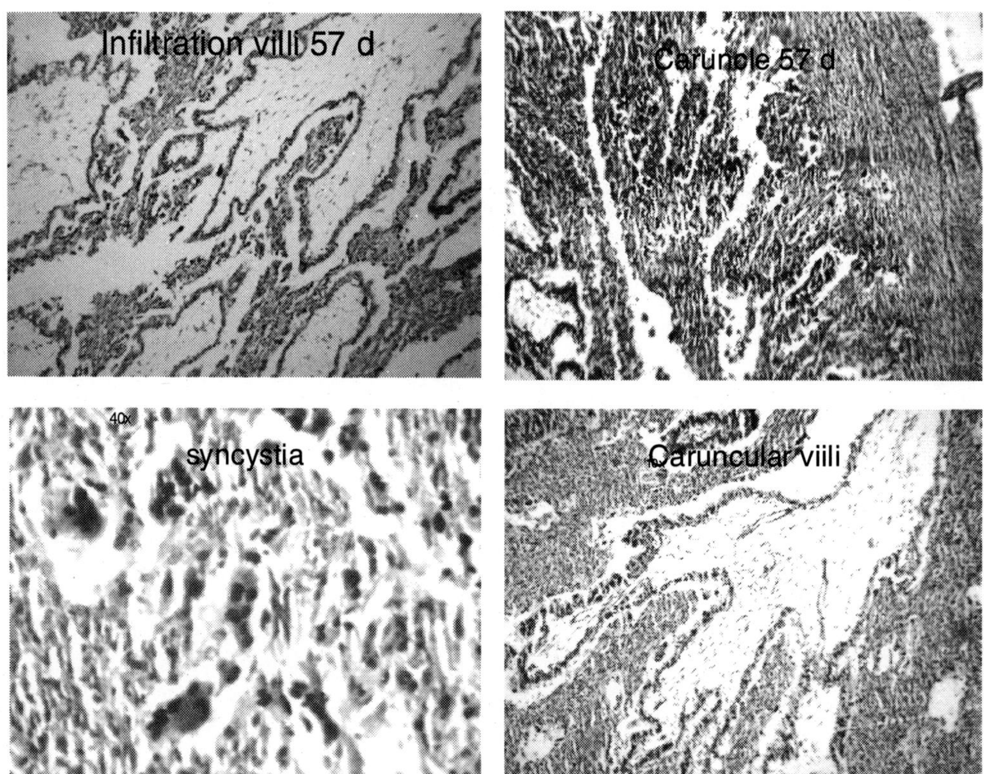

Fig. 26-31 Histological Evidences See Bellow and Refer Other Details in Other Chapters on Cervical Alterrations, Please.

References

Dey. S.K; Lim. H; Das. S.K; Reese. J; Paria.B.C.Daikoku.Ku and Wang.N.2002 Molecular Cues to implantation, *Endocrine Reviews*.25:341-373.

Huet-Hudson, M.Y; Andres, G.K and Dey, S.K.1989 Cell type specific localization of C-Myc-protein in mouse uterus-modulation by steriod hormones and analysis of the pre-implantation period. *Endocrinology*. 125:1683-1690.

Erlebacher A. 2001. Why is not foetus rejected? Cureent opinion in *Immunology*. 13: 590-593.

Mufti, A.M 1996. MVSc.thesis.Division of Animal reproduction, Gynaecology and Obstet, S.K. University Agric. Sci. & Technology, Shalimar- Srinagar. Advisor G.M. Wani

Mufti. A.M; Wani, G.M; Wani. N.A; Mir, M.M and Khan M.Z. 2000. Pre-natal development of ovine foetus. *Small Ruminant Research* 38(1):87-89.

Makker. A and Singh. M.M. 2006. Establishment of endometrial receptivity is still a biological mystery. *Med Res.Review*. May 18, 2006. E- Publication a head of print.

Tulac.S; Nayak.N.R; Kao. L.C; Von-Waes. M;Hang.J; Lobo.S; Germyer, A; Lessey.BA; Taylor.R.N;Suchanek.E. and Giudice.L.2003.Identificcation, Characterization and regulation of the canonical wnt segnalling pathway in human endometrium *J. Clinical. Endocroine and Metabolism* 88:3860-3866

Wani.G.M; A. Khatoon; P.Goswani; G.Wani and M.Z.Khan 2006a. Uterine changes under normal and bnormal corpus luteum in sheep and goats. Lead paper to be presented at XII Annual ISSAR Conference, at Mhow, MP Nov. 10-12, 06.

Wani, G.M.2006 Biotechnological advances in goat production www.word.mort .com. *Research Papers Page.10.*

Wani G.M. A. Khatoon, P.Goswami, Mir S; Darzi, M.M and M.Z.Khan 2006 b. Early embryonic development in sheep and goats. www.Buzzle.com.pet anaimals *.published on web 8.7.2006 visibility also on* www.Google.com.

Wani, G.M;P.Goswami; A.Khatoon; P.Wani and G.Wani (2006c). Foeto maternal attachment in small Ruminants. *Small Ruminant Research*. Submitted.

□□□

Chapter 13

Reproductive Health Management of Mammals

G.M. Wani & Talat Zahir

Fertility is the main economic factor in the livestock industry. Low fertility or loss of fertility results into tremendous loss to the entrepreneurs. The efficient reproductive management enhances the potential of livestock, mammal breeding and husbandry.

Abstract

Majority of world's goats live in Asia and Africa. Developing & poor economies have more goats sustaining their poorest of poor. This paper reviews the incidence of reproductive diseases in Africa (Nigeria) Europe (Cyprus) and India. Most common diseases inflicting reproductive losses are abortions, pregnancy toxaimia, milk fever in African sheep, having an incidence of 37, 25 & 25% respectively. The incidences of Dystocia, abortions, milk fever and Mastitis in goats under similar conditions range from 10-49%. Brucellosis is less common in sheep and goats than cattle. Chlamydia, leptospirosis, fever, rift valley fever, salmonella, toxoplasmosis, campylobacteriosis, bluetongue, viral border disease of new born lambs, white muscle disease are some of the common reproductive diseases inflicting economic losses to sheep and goat rearers. Some of the diseases have zoonotic importance too. Therefore, proper reproductive health management is "Key" to successful sheep and goat enterprise.

Better reproductive healthy management through proper sanitation, adequate feeding, preventive vaccination and seniotive diagnostic applications reduced some of the above disease incidences in Cyprus. The significant reduction in enzootic abortions. Blue tongue hydatidosis, brucellosis, and Q fever was recorded. Many reproductive deformities like foetal dystocia, foetal mummifications, foetal macerations, hydroallontois and hydroammions observed by author in J&K are reported with factual images. Some of the case histories reported from Rajasthan are also incorporated wherever necessary.

In order to reduce reproductive losses in sheep and goats, proper reproductive health management schedules need to be advocated. They can be divided into three categories.

1. Congenital defects like intersex and other genetic defects could be reduced by avoiding inbreeding and proper testing and selection of breeding rams and bucks.
2. Nutritional deformities could be reduced by applying supplementation schedules during breeding and pregnancy. The neonatal care and post-parturient management needs attention.
3. Hygienic and zoo-sanitation measures shall envisage proper examination of sires at breeding, disease prevention through vaccinations and hygiene. Easy pregnancy diagnosis using ultrosonography for pre-diagnosis of any foetal or pregnancy abnormalities like foetal death, amphisema, mummification, maceration etc.

Introduction

Reproduction in important for the survival of all be it humans or animals. No life without wife although a common solagan for human is true for all life on this Universe, be it insect or Elephant. The holy "Quran" says God created every living entity on earth in pairs. Thus reproductive management in general and reproductive health management is essential for conservation, propagation and sustenance of life on earth. Sheep and goats have an early domestication history. Domestic sheep (Ovis aries), order Artiodactyla and family Bovidae have been derived from ovis musimon, a species distributed across Mediterranean islands. Their domestication began some 10,000 years ago. Longevity differs among breeds but maximum life span is 19-20 years. Keisler (1999) in encyclopedia of Reproduction has given broadly reviewed global reproductive parameters of sheep and goats. The reader is also referred to review on sheep and goat production in general and goat management system in Encyclopedia of Dairy sciences, published by Academic Press. U.K in 2003 (Wani, 2003). Immunological aspects of sheep and goat health in general have been reviewed and depicted as html on http:/medicihne. ucsd.edu/cpa/sheep.html.2003.

Sheep and goats harbor Chlamydia psittaci and weak and dead foeti are born or aborted. Various diseases affecting reproductive health or sheep & goats have been reviewed (Smith et al, 1972). Various physiological data on sheep has been used to study human prenatal development. Many studies of value to control and cure reproductive losses in human were studied on sheep models. Like regulation of amniotic fluid in sheep suggested that (VEGF) vascular endothelial growth factor increases intramembranous abortion and micro vesicular transport. Similarly, technetium crossed ovine placenta more rapidly than smaller sodium ions. (Daneshmand et al 2003; Gilbert et al, 1996). These and other studies have revolutionized the health care of human and animal especially related to reproductive disease diagnosis and management.

The above short description and brief review envisages the importance of the topic under discussion. In order to reduce reproductive losses, disease, nutrition and breeding management schedules are to be advocated. It may need summing up information and methodology of clinical examination of reproductive organs, a checklist of physical signs of diseases associated with genital system. Besides clinical examination of goats their abnormalities needed to be discussed and imagined. For proper strategies of reproductive health management we need information on:

Incidence of reproductive disease and disorders. Diagnostic features of reproductive disorders. Management tools for prevention and cure of these diseases. This paper intends to review the data available on the topic and discussed it in the light of our own limited observations and profess a reproductive health management schedule from lamb/ kid born to lamb-kid born from her. Seed to seed plant or offspring to offspring cycle.

Broad Review of Facts

i. Common Reproductive Diseases

In Nov, 2006, a book detailing common reproductive diseases and disorders has been published by Saunders, edited by youngquist and Walter on ovine and caprine theriogenology. The common diseases listed are abortions, dystocia, listeriosis, mastitis, pregnancy toxemia, vaginal prolapse, ring womb, uterine prolapse, vibriosis (compylo bacteriosis), white muscle disease, toxoplasmosis, cache valley virus, Chlamydia, leptospirosis, Q fever, rift valley fever, salmonellosis, metritis, peritonitis, blue tongue, Fuzz lamb syndrome (Border lamb disease), foetal mummification, foetal macerations, cystic ovaries, persistant corpus luteum, trichomoniasis. We have also observed and reported Haenatic of foetal mummification in an ewe at CIRG, Makhdoom, first report of its kind in India (Wani et al, 1980) and a case of schitoschoma reflexus in an ewe at Srinagar. Sheep Breeding Farm, SKUAST (Wani et al., 2000). Besides a number of reports on reproductive diseases of sheep & goat from CIRG, CSWRI and other SAU have been reported and presented in various health related seminars or ISSGAPU in 1981, 1982, 1987, 1991, 2001, 2002, 2003, 2004, 2005, 2006 and published in their proceedings respectively.

ii. Incidence of Reproductive Diseases

A. (Europe) Brucellosis in Sheep & Goats has less occurrence than cattle in Europe (Cyprus). Its incidence is 0.0005% to 0.013% only. The engootic abortions, Q fever incidence in affected flocks were over 40%. The incidence of Hydatidosis in sheep was 0.79% but in goats it was negligible 0.09%, so one could profess goat meat safer than sheep, so far as hydatidosis as zoonotic disease is concerned. Blue tongue, toxoplasmosis incidence vary from 5-11% in sheep & Goats (Polydorou 1983). Chlamydia in sheep 11% and goat 9% is less than cattle (24%). Campylobacteriosis has an incidence of 1.2% in both sheep & Goats. Coxiella burnetic infections are 3 & 6% in sheep & Goats respectively. Toxoplasmosis is higher in sheep 23% than goats 7%, may be because of their clean and selective feeding habits. Listeria monocytogenes infection was very low in sheep (0.7%) than goats (2.2%) or cattle (4.6%). Brucellosis has host preference for cattle but sheep and goats too can harbour the infection (Luchsinger & Anderson, 1979).

B. (America) Blue Tongue was first reported from USA in 1948 in taxas sheep. Now virus has been detected in its 18 states. Initially its morbidity was 50-70% and mortality 20-50%. Subsequently with proper reproductive health measures its incidence has its incidence has been greatly reduced to minimal 1-2%. Cattle are the virus reservoirs. Culicoides is insect vector. Virus has been isolated from sheep kids (Melophagus ovinus) and cattle lice (Haematopinus eurysternus Denny). Dairy goats or Angora goats do not habour nutural Blue tongue. Vaccines have been produced to reduce its incidences. (Hourrigan and Klingsporn, 1975). Bluetongue can spread through winds as reported in cyprus (Pitzolis, 1987).

C. Global: Blue tongue occurance is traced to Europe, 125 years ago. It spread to all most all countries globally, through distribution and sale of European animals to Africa, Asia & Other countries. Blue tongue is spread by a biting gnats or midges, genus culicoides. (Walton, 2004). Caprine arthritis-encephalitis is seen often in new born kids but not before birth. The disease symptoms are similar to Visna in sheep, where Leuksencephalomyelitis occurs. (Adams & Crawford, 1980). DNA technology using 2 PCR tests for diagnosis of Paratuberculosis in goats was reported from Germany (Schroeder et al 2001). Flow cytometery using monoclonal antibody detector was used for diagnosis of paratuberculosis in Czech goats. (Faldyna et al, 1999). Toxoplasmosis in goats was reported from Malaysis. The incidence was 35% in affected herds incidence was higher 50% in small farms, Selanger and lowest (20%) in Melaka & Negri Sembilan farms. Incidence was more in Malay farmer flocks 40% than Chinese run farms 20%. No infection was detected in large commercial farms. (Dorry et al, 1993). A steering group on evolving national Livestock policy perspective in their report of 1996, describe, Blue tongue, Brucellosis, Chlamydia, Maedi-visna disease, (Sheep), mastitis affecting small ruminants, besides other diseases like John's disease goat pox, sheep pox, enterotoxaemia, foot and mouth disease, PPR, Rinder pest & pneumonia. On the basis of above facts a considerable economic loss is inflicted on national and international livestock economic through reproductive diseases. Therefore, it is very important to draw reproductive health management schedules to reduce the loss. In this regard we shall devote our attention in next few pages under following different heads:

1. Examination Methodology

Unless our vets, NGO, scientists and those working with sheep & goats are conversant with methods of detecting deformities it is difficult to identify and cure than. Thus examination of external and internal genital organs shall be discussed and demonstrated through slides and figures.

2. Investigative Technology

The technique used for clinical diagnosis of reproductive diseases and deformities shall be shown in slides under presentation and shall be discussed through figures and charts.

3. Projective Strategies

Reproductive health checklist shall be discussed to help farmers, NGO & others to operate efficient reproductive health management.

Examination Methodology

i. Male External Genitalia

Examination is performed under standing conditions. Visual and palpation of scrotum, spermatic cord may be made. Size, texture may hint towards abnormalities prolapse of intestine, presence of hernia or inguinal ring or ruptures may be looked for. Testes should be symmetrical, mobile, having an consistency like a ripe tomato. The total scrotal diameter in an adult Ram or back should be around 33 cms. Enlarged, painful or hard testicles indicate disease and thus need further laboratory investigation. Such Rams or bucks should be immediately isolated and not used for breeding till clearance from diagnostic examination.

Testicular atrophy & hernia may be indicated on examination physically or through ultrasonography. Infections epididimytis can be detected as enlarged epididymitis similarly asmetrical spermatic cords are on disease indication scars on scrotum detect casectomized ram and bucks. Examination of penis and prepuce in necessary. The method of its examination is shown in figs. Inflammation or ulceration may be seen. Many disease like Pizzle rot need to be scored oat by seminal and blood examinations. (Jackson and cockroft, 2002).

ii. Female Genital Tract

The examinaktion of udder, vulva, vagina and cervix becomes necessary. The method of examining udder is shown in Figs. Mastitis, gangrene, prolapse or vaginal secretion may reveal various disease conditions. Small necrotic plaques on vaginal wall be examined through head light and speculum. Locomotor problems of limbs and feet, lamness, nevous movements may reveal many diseases like border disease tremors in young lambs, paresis due to spinal cord abscess in 3 month old lambs, circling, head tilt conditions due to coenuriasis in 3month to 2 years age sheep and goat. Adult sheep nervous conditions include cervical subluxation (Paresis) as in listeriosis, brain abscess and scrapid (ataxia, pruritus, loss of weight). Metabolic disease include recumbancy, dullness and head pressing in pregnancy toxaemia, weakness, recombency and Haccid paralysis in hypocalcaemes, hyperasthesis, staggers in hypomognesaemia and kangaroo gait (ataxia).

Similarly intersexes in goats could be diagnosed by external examination of female genitalia, as shown in fig. Similarly pseudopregnancy and other genital tract defects or deformities could be detected via ultrasonography or hormonal, biochemical examination too. Radiograph or detection of estrone sulphate and progesterone in blood, milk may indicate pregnancy (wani, 1989). Similarly examination of vaginal discharge, cervix could help us to detect retained placenta, necrotic vaginitis. Acute or septic metritis, endometritis, chronic vaginitis, vaginal cysts or other deformities. Various shapes and sizes of these reproductive disorders, abnormalities will be presented as slides 1-29. Various lamb mortality in exotic sheep has been documented with managemental too for reduction (Suden et al, 1990).

Investigative Technology

a. Use of ultrasonography, laparoscopy and radiation tools have been used to explore reproductive diseases and disorders. Since their first use in India around 1978-82 (Wani, 1981, 1982) many groups have now emerged. Even our pioneering work at Makhdoom has been rated as pioneering, first of its kind by British & American Journal reviews notably theriogenology (1988) and 2002 website www.googli.com 21 years of ultrasound use. Now its use has been popularized and almost all veterinary college of India use it for research purpose. It is now time to conduct few diagnostic summer or winter school on it, so as to train atleast 10-20 veterinarian in each state for its use under farm conditions. Their use even in pre-slaughter animals may save economic resources and loss of lambs. We used many investigative technologies for small ruminants. Bachoo & Wani, 1991; Sudan et al, 1990; wani, 1981, 82; Wani et al, 1998 ; Wani & Bachoo, 1990. Wani & Bachoo, 1993, 1991; Wani, 1989.

b. Radio Immunoassays : This technology use have predicted pregnancy in goats, sheep by blood, milk examinations. Its use for diagnosis of foetal abnormalities needs to be explored. In collaboration with ultra-sonography, hormone, estrogen/progesterone values may help to diagnose foetal deformities, ovarian disorders or even placental defects. However, this technology has to be remodeled and a new area of clinical diagnostics in reproduction is on the walls in future. More landmarks as normal values during early, mid or late pregnancy images have to be corborated. Foetal dimensions, placements and growth factor study should be multiplied. The lone experiments at vety. College, Srinagar (Mufti et al, 2000) he fortified in all veterinary colleges as was done in ultrasonography. Various details of the internal structures of genitalia have been depicted by wani et al, 2006 a,b,c, 2007. For further use ahnd standardization of foeto-maternal interactions various articles of the author are available free of cost on www.wanigm.net or www.word-mart.com. Even clicking WANI.G.M on www.Googli.com search engine shall make available all these paper to the interested readers. (Wani, 1989; Sharief 1996).

Prospective Strategies

i. Even USA, National Academy has given green light for intensive study of small ruminant research on reproduction, which may serve as fertility control models in human too. Thus more funds and emphasis in research on basic, strategic and applied aspects is to be given in India too.

ii. An examination and investigative schedule for breeding rams needs to be formulated into a regulation. An organization as per Indian standard Institution needs to be established, which shall take all measures to safeguard our stock from emerging diseases through imports under WTO and others. Massive breeding ram investigations need to be advocated. Similarly, female before breeding or at culling need to be singled out.

iii. Common recommendations on the topic.

   a. Sheep and goat enterprise needs encouragement and adequate funds for establishing reproductive disease diagnostic centers at state / regional levels immediately. The emphasis on zoo-sanitation arising under WTO & import/ export restrictions needs these technical improvements. These labs could be located in veterinary colleges for survey and identification of these diseases. Use of advance techniques needs priority.

   b. Clinical aspects of Veterinary institutes including IVRI, NDRI, CSWRI or CIRG are poorly developed. ICAR should not forget that these institutions have public health importance. Their care and facilities should be of referral institutes like AIIMSW or other national or state Medical college. Even American NAS have now laid many emphasis on renewal of clinical diagnostic in small ruminant reproduction.

   c. Feeding mechanization, using cheap and easily available field staffs for sheep and goat wholesome and balanced feeds needs to be tested and used.

   d. A proper identification of the sheep and goat resources needs to be made using DNA finger printing and selection of our best breed semen, embryo needs propagation. It is prelude to seed our whole sheep & Goat population free of all diseases especially sexually transmitted diseases. A proper diagnostic checks are necessary for all imports or exports of semen or Embryos.

   e. The excellent facilities at CSWRI & CIRG need to be utilized as referral clinical diagnostic hospitals as in America, Europe and especially Germany. A sheep and goat clinic needs to be established in each SAU to serve as referral clinical at state level white as CIRG & CSWRI* should act as National referral institutes for Goat and sheep respectively at par with IVRI which looks after large animals.

   f. The global importance of sheep and goat meat being comparatively free of major zoonotic diseases threats needs moe emphasis on health, reproduction, physiology, nutrition, meat technology and extension of these technologies to poorer of the poor in this field. It is high time to re-organize these two species -specific institutes at par with IVRI & NDRI as deemed University as they are already serving as national institutes for sheep and goats.

References

Adams, D.S and Crawford, T.B, 1980. CAE(Caprine arthritis- encephalalitis)- a viral syndrome in goats. *Inht. Goat & Sheep Research* 1 (2): 168-172.

Bachoo, B.A and Wani, G.M, 1991. Induction of oestrus in corriedale ewes during anoestrus. *Indian J of Anim Sci* 61(4): 407-408.

Charke, G.L and Osburn, B.I, 1975. Border disease like syndrome found in California lambs. Proc. 18th Annual meeting of American Association of Vety. Lab. *Diagnostics* 1975: 303-325.

Danesmand, S.S; Cheung, C.Y and Bruce, A.A 2003. Regulation of amniotic fluid volume by intramembranous abortion in sheep. Role of passive permeability of vascular endothelial growth factor. *Amer. J. Obstet. Gynecol.* 1988: 786-793.

Dorny, P; Casman, C; Sani, R and Vercruysee, J. 1993. Toxoplasmosis in goats – a sero-epidemiological study in pennensular Malaysia. *Annals of Tropical Medicine and Parasitology.* 87(4): 407-410.

Faldyna, M; Toman, M and Pavlik, I, 1999. Leutocyte counts and Lymphocyte subpopulalzons in peripheral blood of pygmy goats from herd infected with Mycobacterium avium subspecies pseudotuberculosis veterinarrie-*Medicina.* 44(9): 259-262.

Gilbert, W.M; Newman, P.S; Eby-wilkens, E and Brace, R.A 2002. Technetium to 99 m rapidly crosses the ovine placenta and intramembranous pathway. *Amer. Obstet. Gynecol.* 175: 1557-1562.

Haurrigan, J.L and Klingsparn, A. L, 1975. Epizootology of Blue tongue- the situation in USA. *Australian Vet. Journal.* 1975; 51: 203-208.

Jackson, P.G. G and Cockroft, P.D, 2002. Clinical examination of farm animals. Backwell science publishing co, Australia. Website - www.blackwell- science.com. Keisler, D.H, 1999. Sheep and goats in Encyclopedia of Reproduction. E. Knobil and J.D Neill eds. Vol 4, pp 479-492. Academic press, san Diego, 1999.

Mattoo, F.A; Buchoo, B.A and Wani, G.M, 1989. Lambing process in ewes. *Int. J. Anim. Sci.* 4: 185-185.

Pitzolis, G. 1987. A review of Blue tongue in Cyprus with evidence of windborne spread vectors. Dept. of vet. Services, Ministry of Agriculture and Natural Resources, NICOSIA, Cyprus.

Polydorou, K. 1983. Cyprus Annual report of department of vet. Services. 1983-84. pp 224. Lushsinger, D.W and Anderson, R.K. 1979. Longitudinal studies of naturally acquired brucella abortion infection in sheep. *American. J. Vet. Res.* 40(9) : 1307-1312.

Schroeder, C; Seelinger, F; Gaede,W; Westermeier and Granter, M, 2001. Tierarztliche-Praxis- Ausgabe- G-Grosstier/ Nutztiere. 29(1): 19-26. Smith, HA; Jones, T.C and Hunt, R.D Veterinary Pathology. Lea & Febiger Phitadelphia, 1972.

Sudden, N.A; Risam,K.S; Mattoo, F.A and Wani, G.M, 1990. Lamb mortality in exotic sheep. *Indian Vet. J.* 67:51-54.

Walton. T.E. 2004. The history of blue tongue and current global review. *Veterinaria-Italiana* 2004; 40 (3): 31-38.

Wani, G.M and Bachoo, B.A, 1993. Prediction of ovarian response in sheep by laparoscopy. *Ind. J. Anim. Sci.* 63 (3): 300-300.

Wani, G.M and Buchoo, B.A, 1990. Laproscopic view of ovine conceptus. *Ind. J. Anim. Sci* 60 (5): 564-68.

Wani, G.M. 1981. Ultrasonic pregnancy diagnosis in sheep and goats. *World Review. Anim. Prod.* 17: 43-48.

Wani, G.M. 1982. Laproscopy in farm animals. World review of Animal Production. 18: 7-13. Wani, G.M. 1989. Plasma progesterone concentration during oestrous cycle in *Goat. IJAR.* 10(1): 44-46.

Wani, G.M. 1991. Oestrus activity in goats. World review of animal production (4): 73-7. Wani, G.M. 2003. Goat husbandry/ multipurpose management encyclopedia of dairy sciences. Eds. Elsevier Science ltd. PP 1259-1270.

Wani, G.M; Vihan, V.S and Sinha, N.K, 1980. Haematic type of foetal mummification in Muzaffarnagri ewes. *Indian Vet. Med. J.* 4 (March): 46-47.

Wani, N,A; Wani, G.M; Khan M.Z and Beig , M.A, 1998. Ultrasonic pregnancy diagnosis in Gaddi Goats. *Small Ruminant Research* 29: 239-240.

Wani, N.A; Wani, G.M and Bhat, A.S, 1994. Schistosoma reflexus in a corriedale ewe. *Small Ruminant Research.* 14: 95-97.

Wani, G.M and Bachoo, B.A, 1991. Detection of induced ovulation by laparoscopy in ewe lambs. *IJAR.* 12 (1) : 76-77.

Wani, G.M, 1989. Detection of foetal age in sheep and goats by ultrasonic technique. *Ind. J. Anim. Sci.* 59(12): 1525-1526.

□□□

Chapter 14

Pregnancy Diagnosis in Ruminants

G.M. Wani & P. Wani

1. Introduction

Rectal palptation in small ruminants is of little value due to the size of the pelvis. (Wani, 1981). The caudal artery monitoring, bloatment, non-return to oestrus, udder development and other tests tried have had little success, (Wani & Sahni,1980). The more recent interest in early pregnancy 3. diagnosis of small ruminants is of academic and economic importance (Mellado,2003). A highly valued zygote or embryo when transferred to a less valued surrogate mother (recipient) needs to be closely monitored and the early detection of conception helps in repeated use of baren females. Proper management of pregnant animals also prevents embryonic losses. The method applied should be safe to both offspring and dam and needs to be cheap and easily applied. A review of various methods and techniques used for early pregnancy diagnosis in small ruminants. (sheep and goats) is presented.

2.Early Signs of Pregnancy

2.1. Maintenance of a Functional Corpus Luteum

It was evident that conception prolongs the life of the CL and prolongation and maintenance of a functional CL is triggered by the developing conceptus.

These signals ensure the maintenance of the structural integrity of the CL. Corpus luteum produces progesterone, which maintains the uterine endometrium in a state permitting embryonic development, implantation and foetal-placental development (wani, 1984b). The formation and regression of the corpus luteum (CL) in Muzzaffarinagri ewes and Jamunapari goats was monitored at 3 days intervals for an entire oestruous cycle. Laparotomy and laparoscopic methods were used in these experiments.

The Endometrium undergoes tissue remodeling. This change in Extra cellular Matrix (ECM) components is needed for successful implantation. Cytokinens 8,18 and 19 have been detected in the caprine endometrium during early pregnancy using immunofluorescence. Thus the presence of these cytokinen at approximately day 15 post conception is indicative of pregnancy in goats.

The implantation process in goats starts around day 18 post mating. During this phase intense type I collagen staining was detected throughout the uterine caruncular and intracaruncular stroma. For embryonic trophoblastic adhesions with endometrium, local control of protease activity is suggested. (Guillomot, 1999).

The earliest signs of pregnancy is the non-regression of the cyclic CL, which can be observed by the following methods:

i. Laparoscopy and Laparotomy approximately day 18-25 post mating. (Wani, 1982, 1988, 1984b,Wani & Buchoo, 1990, Wani & Buchoo, 1993, Cuellar et al, 1990, Wani et al, 2003).
ii. Serum Progesterone values higher than 1 ng/ml e.g 2 to 3 ng/ml. (Wani, 1989; Shreif, 1997, Boscas et al, 2003, Al-Merestani et al, 1999, Zarkawiet et al, 1999). Diagnosis of Pregnancy accurately (100%) predicted on the basis of serum progesterone P4 values around 17-19 days post mating.
iii. Pregnancy associated ovine glycoproteins recorded approximately post mating indicate pregnancy in sheep. (Karen et al, 2003; Verberckmoes, et al, 2004) or secretion of 17 & 22-24 K Da proteins on day 17 post mating in the caprine conceptus. (Guillomot et al, 1998).
iv. Non-return to oestrus (Mellado, 2003)

Some of the other early pregnancy signs detected by various methods are set out in Table 1.

3 Non-Rejection of Early Conceptus

Progesterone maintains the uterine endometrium in a state which allows for embryonic development, implantation and foetal placental development.

Details of foetomaternal relationships have been described (Mufti, 1997, Mufti et al, 2000)and are shown illustrated in Fig 1to 5. The presence of an early conceptus prolongs the life of corpus- luteum. These pregnancy signals are secreted as proteins. (*Heap et al, 1990*). Some of these proteins have been identified as ovine Trophablast protein I (OTP-1) in sheep which prevents the release of PGF2 alpha and thus helps in the maintenance of the corpus luteum. In cyclic ewes (non-pregnant) PGF2 alpha pulses are released in response to oxytocin with receptors being in the endometrium. The earliest signal of pregnancy is detected by a marked reduction in the endometrial oxytocin receptor numbers. The OTP-1 may inhibit synthesis of endometrial receptors for oestrogen and oxytocin. This possibly prevents luteolysis and maintains the dominance of theuterus by progesterone which is pre-requisite for the establishment and maintenance of pregnancy. (Bretzlaft and Romano, 2001; Wani, 1996; Ala cam et al, 1988).

The expression of progesterone receptors (PR) in the caprine uterus markedly increases during the peri-implantation period and estrogen -(ER) receptors do not increase in relation to PR, thus signaling the non-rejection of the early conceptus. (Flores et al, 2001). Progesterone in milk too can be found during early fertilization and conception (Cough et al, 1989). Caprine H-type I antigen expression is unregulated during peri-implantation and progesterone P4 level stimulate it. It may be a useful marker to signal uterine preparations for receiving and retaining pregnancy in goats. (Powell et al, 2000). The caprine pregnancy related glycoprotein (Ca PAG) may help the conceptus to develop and is found around 18-19 day post mating . (Garbayo et al 2000). Endometrial tissue the undergoes remodeling to retain the conceptus in gravid small ruminant females. (Guillomot, 1999)

The dephosphorylated state of caprine uterine myocin in early pregnancy may help the conceptus to grow. Changes in the expression of native myocin, myosin heavy chains (MHCS) and myosin light chains (MLCS) were observed. (Kumar and Katoch, 1997).

For the development of the blastocyst, a proper uterine environment is essential. Besides the maintenance of the corpus luteum, production and availability of progesterone, the non-rejection of conceptus (blastocyst) is another critical feature of this period. The embryo produces interferons (embryo-IFN). This embryo IFN is homologous with - interfersons (? -IFN) and Ovine Trophoblast Interferons (OTI) of early pregnancy.

Purified OTP and recombinant OTP (r-oTP) produced in yeast exhibit antiviral activity and these r-OTP and OTP inhibit the release of endometrial PGF2 ? . This helps in the non-regression of the CL and indirectly maintains the early conceptus. Intra uterine r-OTP administered at a dose of 340 μg/ day for a week maintained the C.L in cyclic ewes for a month or so of. The inter- oestruos interval in 80% of the ewes was about a month or more. This dose r-OTP was as a effective as 14-16 day old conceptus. OTP was found to be immunosuppressive in several in-vitro and in-vivo assays. An assay on phytohaemagglutinin A revealed both OTP and r-OTP to be immunosuppressive. This was further verified by the inhibitory activity of r-OTP in Graft Versus Host Reaction. (GVH assays). Trophoblast interferons play a strategic role in the prevention of early pregnancy loss as it inhibits CD + blastogenesis. The role of CD + cells and as helper T lymphocytes and delayed+ Type hyper sensitivity mediators (DTHS) would explain this immunosuppressive rate of OTP. (ILeri et al, 1996; Karen et al, 2003; Wani, 1996).

4 Oestrogen: – Pregesterone Ratio (E:P Ratio)

The role of oxytocin in inducing uterine PGF2 alpha was discussed earlier. However, the release of PGF2 under the action of oxytocin depends on or is controlled by progesterone and oestradiol. It was further indicated that ewes with a high E:P ratio may generate stronger luteolytic signals. It was demonstrated that low progesterone and high oestradiol combination record the largest and sustained increase in PGF2 alpha following oxytocin injection. Trophoblast interferons act locally to suppress the uterine oxytocin receptors in sheep.(Karen et al, 2003).

5. Maternal Recognition of Pregnancy

The maternal recognition of pregnancy in sheep and cattle is centered around the production by the trophoblast of type I x interferon (tINF). This tIFN then suppresses uterine oxytocin receptor concentrations (OTr). The oxytocin receptor (OTr) occupancy is associated with oxytocin induced PGF2 alpha release. OTr inhibition may represent the principal antiluteolytic mechanism of tIFN and secretion of the conceptus secretory proteins or bovine recombinant IFN to the uterus reduces OTr. Concentrations in intact and ovarectionized steroid treated ewes . A relationship between the conceptus secretory proteins and the metabolic products and those in the peripheral blood of the dam exists. (Mufti; 1996; Mufti et al, 2000). There are conflicting reports making the action of oestradiol on oxytocin receptor concentration. (Powell et al, 2000). Trophoblastic cells contain interferon on day 14-17 after mating. During maternal recognition of pregnancy goat interferon was detected on day 18 post mating, its absence signifies pregnancy maintenance has been taken over by the corpus luteum. Thus a very thin line exists between maternal recognition of pregnancy and its maintenance or sustenance by the CL. (Gillomot et al, 1998).

6. The Reliability Pregnancy Tests

Various methods used for correctly predicting pregnancy in sheep and goats during gestation have been summarized in Table 2. The accuracy varies from 70 to100% with different ultrasonic equipment. Different models as well as principles involved have been extensively reviewed (Wani, 1991; Wani et al,1998) and other methods of pregnancy detection during this stage e.g serum progesterone determination, vaginal cytology, laparotomy, estrone sulphate are summarized (Table-2). Various techniques were also evaluated in assessing mid-gestation. The various pregnancy signs as quoted by various researchers using ultrasonography are summarized in Table 3. Of late certain anatomical features in the live, developing conceptus in vivo have been reported. This is reviewed & a summary is presented (Table 4). Various live foetal measurements like Biparietal diameter, Amniotic vesicle diameter, foetal radius and Tibia lengths are reviewed and shown (Table 5). Various pregnancy related images, histological sections and morphology of endometeruim have recently be published (Wani et al 2007, 2006 abc) where images are presented 6-15.

References

Alacam-E; Dinc-DA; Guler-M; Eroz-S;Sezer-AN, 1988. Veteriner-Fakultesi-Dergisi, Selcuk-Universitesi, 4(1),91-98.

Al-Merestani-MR; Zarkawi-M; Wardeh-M, 1999. Early breeding and pregnancy diagnosis in Syrian Awassi sheep yearlings. *Reproduction –in-Domestic-Animals*, 34(5), 413-416.

Blasco-I;Foklch-J; Echegoyen-E, 1989. *Tecnica-Economica-Agraria*, 20(82), 22-31.

Boscos, C.M; Samartzi, F.C; Lymberopoulos;A.G; Stefanakis, A; Belibasaki, S; 2003. Reproduction in domestic Animals. 38(3), 170-174. Assessment of progesterone concentration using enzyme immunoassay for early pregnancy diagnosis in sheep & Goats.

Bozkurt, T; Gundogan, M; Esen, F; Kul.S; 1999. Saglik- Bilimleri- Dergisi– Firat- Universitesi. 13(2),185-180. The use of B mode real time ultrasonography rectally for early pregnancy diagnosis in sheep.

Bretzlaff-KN; Romano-JE, 2001. Veterinary-clinics-of-North-America, Food-Animal-Practice, 17(2), 421-434. Advanced reproductive techniques in goats.

Calamari-CV;2001. Avaliacao-de-dois-metodos-de-diagnostico-precoce-de-gestacao-em-ovelhas:Ultrasonografia-transretal-e-detector-de-prenhez-para-pequenos-ruminantes - DPPR-80R.2001, 57pp. Evaluation of two methods for early pregnancy diagnosis inewe: transectal ultrasonography and pregnancy detector for small ruminant (DPPR-80®).

Chalhoub-M; Prestes-NC; Lopes-MD; Trinca-LA; Ribeiro-Filho-AL, 2001. Ars-Veterinaria, 17(1), 17-21. Pregnancy diagnosis and embryological/fetal quantification by transrectal ultrasonography in sheep.

Chen-Zao Ying; Lun-SiChun; Cao-Hong; Chen-Zy; Lun-SC; Cao-H, 1996. *Chinese Journal-of-Veterinary-Science- and-Technology*, 26(2), 12-15. Study on early pregnancy diagnosis of ewes by ultrasonic tomographic scanning

Cuellar-C; Fontanillas-JC; Palacio-M-del; Perez-Fuentes-J, 1990. *Avances-en-Alimentaction-y-Mejora-Animal*, 30(6), 251-257. A comparison of early pregnancy diagnosis in Manchega ewes using endoscopy or ultrasonics.

Chemineau, P; Gauthier, D; Poirier, J.C; Saumande, J. 1982. *Theriogenology*. 13 (3), 313-323. Plasma levels of LH, FSH, Prolactin, Oestradiol 17B and progesterone during natural and induced oestrus in the diary goat

Cuoghi, F; Castagnetti, G.M; 1989. Informatore- Agrario. 47, 27-32. Progesterone in milk- a method of early pregnancy diagnosis

Dickie-MB; Holzmann-A, 1992. *Journal of Veterinary-Medicine- Series*-A,39(7), 525-530. Investigations concerning the use of progesterone tests (Serozyme-Progesterone, Ovucheck) for pregnancy diagnosis of mountain sheep

Dionysius, D.A; 1991. *Australian. Veterinary Journal*.68(1), 14-16. Pregnancy diagnosis in diary goats using progesterone assay kits.

Flores, J.M; Sanchez, M. A; Nieto, A; Sanchez, B; Gonzalez, M; Garcia, P. 2001. *Theriogenology*. 56(2), 341-55. Delection of estrogen alpha and progesterone receptors and cell proliferation in the uterus during early pregnancy in goat.

Garbayo,J-M; Green,J-A; Manikkam,M; Beckers,J-F; Kiesling,D-O; Ealy,A-D, Roberts, R-M, 2000. *Mol-Reprod-Dev*. 57(4), their cloning, expression and evolutionary relationship to other PAG.

Garcia-A; Neary-MK; Kelly-GR; Pierson-RA, 1993.. Department of Veterinary Clinical Sciences, Purdue University, West Lafayette, in 47907, USA, 39((4), 847-861. Accuracy of Ultrasonography in early

Richardson, C 1972; Pregnancy diagnosis in the ewe. *Vet Record*. 90, 264- 275

Slosarz, P; Steppa,R;Gadek,A: 1999. *Medychna waterynaryjna* 55(10),686-688. The application of ultrasound for early pregnancy diagnosis in sheep

Shrief, M; 1997.. M.VSc Thesis, S.K. Univ.Agric.Sc & Technology, Shalimar, J&K India. Investigations on Caprine Ovarian activities during Oestrus cycle.

Tajik-P;Abbas-Veshkini;Sarang-Soroori, 2001. Veterinary-Conference-University of Stellenbosch, South-Africa. Measurement of different parts in Chall fetuses to determine the pregnancy age. *Proceedings-of-the-5th –International- Sheep.*

Taverne, M.A.M; 1991. Tievarztliche. Monatsschrift. 78(10),341-345. Application of two dimensional ultrasound in Animal Reproduction.

Verberckmoes, S; Vandaela, L; Cat,S.de; EL-Ameri, B;Sulon, J;Duchateau, L; Kruif, A.de; Beckers. J.F; Soom, A. Van; de-cat.S; de-Kruif, A;Van-Soom, A;2004. Vlaams-Diergeneeskundig- *Tijdschrift*. 73(2), 119-27. A new test for early pregnancy diagnosis in sheep: determination of ovine pregnancy associate glycoprotein (OVPAG) concentration means of a homologous radio immunoassay.

Wani, G.M, 1981. Ultrasonic pregnancy diagnosis in Sheep and Goats. *World Review of Anim. Prod.* 17, 43-48.

Wani,G.M, 1982;. *World Review of Anim. Prod.* 18,7-13. Laparoscopy in farm Animals.

Wani, G.M; 1991. Proc. Ist. National Seminar on small Ruminant Reproduction, CSWRI, Avikanagar, Rajasthan- India. Use of Laparoscope, ultrasound imaging in small ruminant Reproduction.

Wani-N.A; Shabeena-Mustafa; Misra-AK; Maurya-SN; Mustafa-S; 2003. Pregnancy diagnosis in farm animals - a review. *Indian- Journal-of - Dairy-Science* 56 (1),1- 8.

Wani, N.A; Wani, G.M; Mufti, A.M; Khan, M.Z. 1998. *Small Ruminant Research.* 29, 239-240. Ultrasonic pregnancy diagnosis in gaddi goats.

Wani, G.M; Sahni, K.L; 1980. *Vet. Research Journal* 2(1), 35-73. An ultrasonic technique for detection of pregnancy in Sheep and Goats

Wani, G.M; Sahni, K.L; 1981. *Indian J. Anim. Sci.* 51 (2), 194-197. Ultrasonic pregnancy diagnosis in ewes under tropical field conditions

Wani, G.M; Sahni, K.L; 1988. *Indian J. Anim. Sci.* 58(7), 802-804. Ovulation detection by Laparoscopy in Sheep

Wani, G.M; 1982.. *World Review of Anim. Prod.* 18(1), 7-13. Laparoscopy in Farm Animals

Wani, G.M; 1984a. Dr. Med. Vet.Diss.Tierartzlichen Hochscule, Hannover, Germany. Investigations on Embryo production and culture in goats.

Wani, G.M; 1984b. Ph.D thesis, Indian vet. Res. Inst, Izatnagar- U.P. India. Investigations on ovarian activity, ovulation and frequency diagnosis in sheep and goats.

Wani, G.M; 1989. *Indian Journal Animal Reprod.* 10(1), 44-46. Plasma progesterone concentrations during oestrus cycle in goats.

Wani, G.M, 1996. Embryo Biotechnology in sheep & goats. Published Valley Book House, University Road, Hazratbal, Srinagar.

Wani, G.M; Buchoo,B.A; 1990. *Indian J. Anim. Sci.* 60(5), 564-68. Laparoscopic view of ovine conceptus.

Wani, G.M; Buchoo, B.A; 1993. *Indian. J. Anim. Sci.* 63(3), 300. Prediction of Ovarian response in sheep by laparoscopy.

Zarkawi-M; Al-Merestani-MR; Wardeh-MF, 1999. *Small-Ruminant-Research,* 33(1), 99-102. Induction of synchronized oestrous and early pregnancy diagnosis in Syrian Awassi ewes, outside the breeding season.

□□□

Chapter 15

Future Research in Reproduction — Vision 2100

G.M. Wani, Paras Wani, Talat Zahir & Gazanfer Wani

Abstract

Endocrinological advances would need more research on hormone synergism, agonistic and antagonistic actions. Follicular atresia vis-in-vis intrinsic GnRH system would need more experimentation. The deterministic and permissive modes of GnRH action would be applied to reduce hyper stimulation of ovaries. The understanding of dominant follicular dynamics, shall help to harvest Follicular pool in thousands rather than in hundred at present. New biological sources of hormone from fish and other animals should find more clinical trials to popularize cheap and biological rather than chemical hormonal sources.

Sensitive disease diagnostics using RIA, REA, hormone detection kits, ultrasonography, laparoscopy, DNA finger printing, immunological and other psychometric techniques needs consideration. Gap junction and development of conceptus, foetal biological research will need emphasis by 2050. In vitro fertilization and embryo culture shall be replaced with in vitro implantation and in vitro foetal development. Foetomuscular development and foeto-maternal interactions will be more vivid and clear. Reproductive care and disease prevention shall need herbal drug stimulants. Such bio-plant stimulants and medicaments need innovative research attempts. Corpora luteal bio-sensitivity and its use in augmentation of reproduction needs trials. Fertility enhancement needs identification of desired fertility objectives. The problems and constraints in animal's reproduction have to be addressed by using modern reproductive technique to improve our traditional reproductive management.

Introduction

The use of artificial insemination (AI), embryo transfer (ET) have helped in elite gene propogation. "Dolley" the super sheep is known and many more genetic clones are in offering. Today our concern is bio-safety of germplasm and the animals products. With high risk of new disease emergence, all researchers are warned to use their materials with caution. The "mad cow disease" and heavy economic losses in U.K. and Europe are eye openers for all of us. The future of gene technology lies in bio-safety and fulfillment of international zoo-sanitation regulations. Disease free zones need more emphasis than producing high yielding livestock from disease infested area. Thus safe and quality animals material (germplasm) and products is the slogan for tomorrow. Zoo-sanitation certifications will demand more trained veterinarians and other para vet staff. This necessities review of human resource development too.

For production and utilization of elite gene pool of animals, we need to know available resources, both biological and material. For example, increasing meat/milk or fibre production from our small ruminants we need to know them better. We must analyse their gene pool both qualitatively and quantitatively as was done in goats (Gall el al., 1997) using DNA techniques. Once we know merits and demerits of our gene pool, could we preserve them or utilize them for further. The methods and postulations in this direction can be reviewed as :

Endocrinological Advances

Ovarian oogenic and endocrine functions are interlocked via Hypothalamo-Hypophysial Gonadal Axis (HHGA). The cyclicity and its behavioral overt and convert phases are intrinsic to these ovarian functions. They have been thus rightly named as " the Zeitgeber or biological clock (Baird, 1987). The existence of feedback loops of ovarian steroids with respective gonadotrphin is a biphasic phenomenon, where higher concentrations in blood have inhibitory or suppressive effects, low concentrations have stimulatory effects (Barrac Lough, 1973), Many reproductive mechanisms hang on changes in these ovarian steroids such as labour pains (Csapo, 1981), ovulatory failures (Snchez-criado et al., 1990). Interestingly these hormones interact to modify or sometimes antagonize each other. For example P4 appears to be potent antagonist of E2 at a combined ratio of 1:1 (Heap and Flint, 1987). The synergism, agonistic and antagonistic actions of these hormone combinations are the basis for all reproduction manipulations viz Control of oestrus, ovulation, conception, gestation, fertility, infertility, contraception and or abortions. Fish and sheep ovarian extracts were tested to find out species sensitivity. It seems it is possible to tap huge fish ovarian extracts for intra and interpolation in mammalian ovary (Najar, 1999). Coming years will see more attempts to use such resources to avoid side effects from chemical and synthetic analogues of hormones.

New Hormonal Roles - Follicular Atresia

GnRH – regulates gonadotrpin release from pituitary, but its continuous and prolonged use in natural form or as its potent agonist induces an inhibitory effect on ovarian steroidogenesis, follicular development and ovulation (Kledzik et al., 1978; Srivastawa et al., 1995). GnRH receptor gene encoding is expressed in the rat ovary (Kaiser et al., 1992). A new concept prevails today which postulates that an intrinisic GnRH system competes with receptor. Thus its biological response exist within the ovary. The follicular atresia is brought by this intrinsic GnRH system (IGS) (Erickson et al., 1994). A new vision of autocrine and paracrine secretion of ovary vis-in-vis GnRH becomes evident. Future modulations in correcting reproductive failure involving hormone therapy needs rational understanding of all endocrine, autorine and related control mechanisms. Ovarectomized ewes when implanted with E2 and P4 implants were maintained in an artifical follicular phase. This speaks of an endocrine model compatible to a natural sequencing hormones of desired levels (Krash and evans, 1996).

Intrinisic GnRH System (IGS)

Two theoretical methods called deterministic and permissive models have been put forth to explain role of GnRH (Karsch et al., 1977).

Deterministic Model

Deterministic model envisages that a surge-like increase in GnRH is needed to induce preovulatory LH surge. This model befits animals whose LH release and ovulation is either time bound (Copulation) or signal dependant (cervical signal, rabbit and neural signal Central circadian pacemaker in rata refers).

Permissive Models

Permissive models explain GnRH in relation to ovulatory LH sruge. Here feed back of steroids exert their influence on anterior pituitary gland. This is supported by evidences that LH surge continue in situations where GnRH secretion is apparently abolished. For example in monkeys with hypothalamic lesions, or in women with hypothalamic amenorrhea or congenital GnRH deficiency, normal LH surges occur (Martin et al., 1990).

Hormone Agonists/Antagonists

Hormone antagonist or antihormone are those compounds or elements which prevent an endogenous hormones from exerting its biological effect(s) (Vanlook, 1988), These antihormone need to be safe i.e. free from side effects. The hormone antagonists in principle acts in 3 ways.

i) Inhibition of hormone synthesis

ii) Neutralising hormone action by specific antibodies.

iii) Blocking the receptors of the hormone.

Use of antagonists, agonists have demonstrated their role in ovulation control. Their low histamine-releasing activity (Fraser, 1981) support their wider use which needs further extensive studies. More details are reviewed (Najar, 1999).

Hormones-Biological Sources

The synergism, agonistic or antignostic actions of steroid and other reproductive hormones are the basis of reproductive manipulations like ovulation and fertility control, conception, gestation, fertility, infertility and contraception (Wani, 2002b). Fish and sheep ovarian extracts were tested for interclass sensitivity. It was revealed that intra and interpolation in mammalian ovaries is possible using these materials (Najar, 1999). Future attempts to utilize these cheap and easily available biological materials for ovarian manipulation may reduce the use of high cost chemical and synthetic analogues (Najar, 1999).

The biggest problem in ovarian manipulation faced in the past was due to varied superovulatory response (wani, 1984). The follicular atresia was postulated to reduce this constraint if induced adequately. As our understanding of (IGS) intrinisic GnRH system was poor then, we could not use it to overcome our problems. Now we know follicular atresia is related to this IGS (Erickson et al., 1994) and therefore, our later manipulations incorporated GnRH with comparably better success in ovarian manipulations. A new vision of autocrine and paracrine secretions of ovarian GnRH became the subject of discussion. Estrogen alone can induce ovulation but for CNS (Central Nervious System) priming progesterone combination in desired ratio is essential (Evans et al., 1995). Thus future modeulations in correcting reproductive failures needs a proper understanding of hormonal inter and intra polation with physical, chemical, biological and neuroendocrine secretions. Pheromones and other psycoemotional factors too need an analysis.

In our experiments on infertility (anoestrus) in cows of Kashmir (District Budgam) we were puzzled to see better response with low doses (300 ug) of GnRH than the higher dose of 400 ug (Shah, 1996) . We were unable to explain our results until the recent studies on permissive and deterministic models of GnRH action came to light (Karsch and Evans, 1996,Karsch et al., 1997).

Herbal-Biostinulators

Pheromones, sire introduction effects, sensory reflexus and vast other stimulants increase receptivity as well as fertility in domestic animals. Sprays and herbal cure products will be in great demand to increase or decrease fertility in coming years. The use of stimulants or repellents as sprays and scents may be in need. Herbal cures for various reproductive wastage needs further research. The use of herbal products, fruits, nuts and medicaments in ancient Kashmiri culture after delivery, need to e looked at. The Cichorium spp. Saffron, Cofula anthemodies, Prunella vulgaris, Artemisa and other products in use during post delivery period needs scientific analysis.

The ovarian sensitivity decrease soon after partuition and its resurgence after 14 days in cows, 40 odd days in women and specific periods in other animals, reveal a lot which needs exploration to phase out various sequences of reproduction.

Corpora Luteal Biosensitivity

Corpus luteum growth and development is the first earliest pregnancy signal along with maternal recognition factors discussed (Wani, 1996). The corpus luteum development and receptivity to PGF2 alpha is now known. The ascending and descending corpora lutea are insensitive to PGF2 alpha. What makes them insensitive and what resurges their sensitivity need exploration. May be in future, CL insensitivity might have direct effects on survival of the conceptus. Does conceptus make CL more sensitive to PGF2 alpha? Or in other words, does the gravid CL and non-gravid CL behave similarly or differently to PGF2 alpha. Many such questions vis-à-vis the ovarian morphology, uterine environment and behavioral patterns need to be studied and co-related to develop a manipulative and curative safe system. In some species CL or pregnancy is necessary till delivery, in others it is needed only till placenta develops. Its complete genesis and biostimulation and bioregulatory roles are obscure and needs further research.

Dominant Follicle Dynamics

The dominant follicle dynamics may be in focus for ovulation control. Dominant follicle or would be bride at each ovulation appears around day 7 of cycle in women. It occurs much earlier on day- 1 to 0 in ruminants (cow, sheep,goats). This time point may be the focus to try future selective interferences in ovulation control. Antagonist use during this phase has shown decreasing levels of estrogens in serum thereby indirectly hinting at dominant follicle alteration, as it is one of chief source of this hormone during this phase (Fraser, 1987). Dominant follicle in its sojourn to freedom and release at ovulation seems to attain some specific qualities. Why? It alone releases itself of bondage's in ovary (ovulation) and rest of its competitors are hindered through the process as atretic follicles. What make this dominant follicle to attain these qualities among a vast pool of follicles needs further study. In coming 50 years one shall possibly understand and interlock a pre-determined messenger system triggering dominance in one or few follicles destined to become the new offspring of that species. This system when decoded shall offer new opportunities in controlling ovulation both for contraception (women) or augmentation (animals). New genepool of dominant follicles from dead high yielding animal ovaries would help to propagate and multiply the elite. Over stimulatory effect during superovulation could be avoided by use of antagonists as dominant follicles seem to evade suppression, thus follicular selection will prevent excessive stimulation and cystic follicles. Follicular fluid constituents containing occyte awaiting for biochemical signals to resume (meiosis) growth and development to ensure timely and healthy ovulation need more exploration (Lenton et al., 1988). The specific signals triggering occyte release might help in in vitro occyte maturation success rate.

Development of Conceptus-In Vitro Foetus

The lean meat production needs more studies on development of conceptus. We know that prenatal development undergoes in three stages. The early pregnancy signals skeletal & nervous tissue growth, on priority. During mid-gestation the hind and fore-limbs develop. During late gestation muscular development along with peripheral extremities including hair develop. In order to increase lean meat production, research in needed to investigate relations between hard and soft tissues. Maternal and foetal interactions needs a probe. Some basic studies (Mufti, 1996) indicated prospects of manipulating lean meat production (Wani, 2001, 2002 a) by bio-stimulators during late gestation, which may help, the production of muscular lambs/ kids at birth, besides, increasing birth weight. More details have been described (Wani, 2001, 2002a) Fig.1-5). More studies in future are needed to reveal the exact mechanism involved in critical transition of power control from maternal to embryonic genome. Application of recent techniques of reverse transcription, PCR and others may reveal the details (Schultz et al., 1990).

Slaughter House Follicle Pool

A good number of our endangered species die in wild climatic conditions are hard and harsh. Under temperate climatic conditions dead bodies are entrapped in snow for long. It gives a hope to revive their follicular pool. Our studies on resumption of life using follicles from slaughtered sheep and fertilizing them with epididymal semen was an pioneering attempt in this direction (Wani, 1996) (Fig 6-8). The protein content of unfertilized ovum is around 100 mg in rabbit, 27 mg in mouse (Brinister, 1967, 1971). The small genetic material needs extensive analysis to unveil vast informations in its codans and anticodans. Further ovum maturation or resumption of meiosis is initiated during birth or shortly before birth and thus the pool of follicles remains fixed at birth. Therefore, foetal ovary has to be the subject of further manipulations especially in view of its immunological insensitivity (Guraya, 1985).

Gamete Biosafety

Bacterial counts in semen in unprecedented numbers effect semen fertility (Kher, and Dholokia, 1985), The contaminational hazards have increased due to environmental pollution, water and air contamination. Above all the development of resistance of known bateria and other micro-organism to antibiotics (Gupta and Maurya, 1993). We have examined the semen of about 40 bulls maintained at the Frozen semen bank, Rambirbag, Kashmir for bacterial loads in semen and tested antibiotic sensitivity (Koul, 2002).

Presence of bacteria in large numbers in semen, effects its quality and fertility (Kher and Dholokia, 1985). The source of contamination of semen vary from inflammatory foci and lesions on penile apparatus, contaminations on equipments, appliances, diluters, extenders air and laboratory environment. These contamination hazards have increased with environmental pollution. Many bacteria having developed resistance of routine antibiotics used (Gupta and Maurya 1993) in semen extenders has made inseminations more critical and vulnerable. More studies in this direction are needed to develop a bio-safe and infection free germplasm pool of frozen and fresh semen in future.

Reproductive Physiology

Reproduction is a complex phenomenon, starting with oogenesis as early as 135th day post mating. By 50-100th day post conception (mating) sheep and goat conceptus in-utero shows follicular growth. This follicular growth and development upto primary oocycte stage in some mammals ceases at birth (Wani, 1996). The further growth and maturation of primary occytes is triggered at puberty i.e. at the onset of first sexual activity. The release of ova i.e. ovulation follows. During this proves of folliculogeneis and ovulation, followed by fertilization and development of conceptus, the male and female gametes undergo various morphological and physical changes. They pass through various barriers, reservoirs, under the influence of bio-physical, hormonal,morpho-chemical mechanisms (Wani, 1996, Wani and Buchoo, 1990, Wani, 1984a and b, Wani et al.,1986). More in-depth studies on barrier mechanism vis-à-vis morphological changes during different hormonal milieu is needed.

Foetal Muscular Development

Genetic,hormonal and environmental factors influence foetal growth and development. The proper in-utero maturation and growth of foetus signify proper care and management of gravid nannies. The pre-natal development of foetus is exhibited at birth by birth weight. The head and skeletal tissues grow and mature earlier, followed by fore and hind limbs. The muscular development of foetus occurs during late gestation. Thus, during early pregnancy nervous and skeletal tissues are having priority for nutritive requirements in-utero(Musti, 1996). Maternal nutrition is important for growth and development of meat in off spring in-utero. Muscular growth stimulants may thus help production of quality meat lambs or kinds in future.

Foeto-Maternal Interaction

The role of endometrial stromal cells during pre-implantation stage needs further research. Some studies on ultrastructure of endometrium (Brinsfield et al., 1974) are still not sufficient to answer the questions. What is the role of immune cells and endometrial cells in building of the immunological barrier to protect the foetus. Foetal biology is abarren field of research in the world. After few initial steps by Dr. Moris Yound in UK and D. Srivastava in India, only a scattered information is available on the subject (Wani and Buchoo, 1990; Mufti, 1996). Much is still to be unveiled by 2050. The placental hormone influence mammary development and milk production of the dam. This needs further research by 2050. Foetal membranes serve as Chief source of nutritive exchanges. The arrangement of cotyledons, caruncles, positioning of single and twin conceptus, indicate complex mechanisms with common roots. (Mufti et al., 2000) These mechanisms of foeto-maternal interactions and relationships emphasize proper foetal number diagnosis and corresponding materno-nutritional management strategies.

Small Ruminant Reproduction

There is a need of ovulation, pregnancy and foetal number detection by methods described (Wani, 1981, 1982; Wani and Sahni, 1980, 1981; Wani and Buchoo, 1991, 1993). These reports demonstrated successfully results ofr ovulation detection, multiple ovulation predictions, early pregnancy and foetal numbers. The induction of multiple ovulation, in-vitro fertilizations, embryo collections, embryo culture, conservation, micromanipulation, splitting, cloning and sexing are some of the reproductive high-tech applications used to increase reproduction rates (Wani 1996; Wani et al., 1990; Wani et al., 2000 a,b; Buchoo et al., 2000). These techniques have but a little application under field conditions in small ruminants especially under extensive management system. However, their use has already revolutionized propagation of elite germplasm and can be used under buck mother farms to produce high quality sires. The futuristic attempts in this direction should include in vitro post implantation and foetal development experiments. These future manipulations at organogenesis level may help to induce desired quality changes in offspring so that dressing percentages could be increased from present 40-50 per cent to 55-60 per cent or more.

Reproductive Management

The reproductive management strategies to enhance meat production could safely be divided into two kinds:

i. Traditional management: Animal are still managed under zero-input/ migratory /tribal or extensive management system. The application of above described methods will be limited. They may indirectly help in this system by better sire availability.

ii. Modern management: Zero-grazing, stall-fed or semi-intensive, commercial goat farming system may use the above methods more and more to produce high quality germplasm. They may use variety of improved reproductive technique or reported (Wani,1996; Wani et al., 200a; Wani et al., 1998; Mufti et al., 1997; Wani and Sahani, 1981; Wani et al., 1986, 1987, 1988; Wani and Buchoo, 1991; Buchoo and Wani, 1991; Wani, 1989, Wani, 1995; Wani, 1984a and b; Wani and Sahani, 1980a).

Enhancement of Fertility

Desired fertility objectives have been discussed (Wani, 2001). Increased prolificacy, early rebreeding, safe/clean parturition, set-time animal crop, elite gene introduction are some of the desired fertility enhancement goals. The constraints in achieving desired objectives in small ruminant livestock sector are the problems of inbreeding, reproductive losses (Wani, et al., 1980; 1981; 1994), pre-pubertal mating, indiscriminate or illicit mating, venereal diseases, nutritional and other stresses. These above referred factors reduce fertility status of our ruminants especially under transhumane migratory systems.

These constraints mentioned could be overcomed by adopting enhancement strategies (Wani, 2001), which include exchange of sires by tribal, Gujjar, Bakarwals and Chopans to reduce inbreeding. This could be ensured through legislature or forced castrations and introduction of sires from outside the flock. The pregnancy detection methods described (Wani, 1981) and later on used extensive in J&K under farm condition (Wani et al., 1998; Wani, 1989) needs adoption as a routine in each flock for better care of offspring and dam.

Nutritional Bio-Conversion

The nutritional status of dams before mating could be improved by grazing them in pastures having 4-5 cm sward height herbage, yielding 1500kgs and above of dry matter per hectare . Such pasturing shall increase ovulation. Reduced sward height pastures needs supplementation and introduction of nutritional block at least 4-5 weeks before mating. Thus proper feeding 100 days prior to mating and parturition may help in healthy and heavy weight lamb/kid crop.

The pasture nutrient analysis and its supplementation as per requirement may help to avoid losses due to such deficiencies as cobalt, iron, phosphorus etc. which effect herbage and through herbage the animal health. The sire fertility test and use of disease free semen is the foremost requirement towards increased productivity from small ruminants.

The nutrition plays important role in maintaining fertility status of animals. With decrease in land for fodder cultivation, alternate sources and technologies to utilize agricultural waste for animal food is needed. The biotechnological intervention in converting crop residues and other agri-horticulture waste into animal feed needs further research.

Conclusion

Critical gaps identified are poor-zoo-sanitary conditions and sufficient human resources. Both these area need emphasis. Development and application of accurate, sensitive and timely disease diagnosis and surveillance techniques especially in the field of animal reproduction should invite more research proposals in future. This will ensure biosafe livestocks and germplasm. Gene propogation of elite germplasm should ensure increase in per animal production. This would means more production from less livestock, thereby relieving soil, water and environmental pressures. Endocrinological advances should aims at replacing chemical hormones with biological hormones, which could be cost-effective and without side effects. The use of fish, sheep and other ovarian extracts need further research. Follicular atresia, intrinsic GnRH system, hormone agonists and antagonists should attract more funds and research attempts. The use of herbal stimulators and medicaments in caring and prevention of reproductive diseases and incapacities needs immediate attention. More innovative approaches using these indigenously available materials need to be prioritized. Copora lutea biosensitivity, development of conceptus, follicular dynamics, gamete safety procedures, foeto-maternal interaction, stromal cell studies would be future areas of frontline researches in animal reproduction. A policy in view of majority 85% of our livestock production systems being small holder enterprises. Research and extension establishments and NGO in animal reproduction needs reorganization to serve as a protective armour around small producers and farm women. Blending of advanced technology available as embryo transfer, AI, blastomer splitting, DNA micro-injections, DNA finger prints, gene mapping, DNA hybridization, ovulation control, in vitro fertilization, foetal monitoring, ultrasonography with traditional modes of livestock production system needs further innovative attempts. This should aim to further fortify our cost-efficient livestock productivity.

Hitherto untapped areas of wild and zoo animal conservation, protection and propagation, pet animal care, wildlife retrieval and referral clinics are the need of the hour. Nutritional requirement during foetal life, prenatal and neonatal life needs more emphasis along with care and management of dams during mating, pregnancy and parturition. Increased prolificacy, early rebreeding, set-time parturitions are some of the desired fertility enhancements goals. The problems of inbreeding, reproductive losses, pre-pubertal and illicit mating, nutritional stresses could be overcome by adopting fertility enhancement procedures like sire exchanges, pregnancy detection, better nutrition before mating, during pregnancy and after parturition. Extension and information technology missions for animal reproduction are to be intensified especially in rural India. Scientist Industry interaction, technology interaction under rural sectors and linkages between research, extension, NGO and farmers needs priority by 2050.

References

Barraclough, Charles A. 1973. Sex steroid regulation of reproductive neuroendocrine processes.In :Handbook of Physiology, Section 7-*Endocrinology* Vol.II-Female

Reproductive system, Part ! American Physiological Society, Washington, printed in USA by Waverly Press, Inc., Baltimore, Maryland-21202.

Boshier, D.P. Fariclough, R.J. and Holloway, H.1987. *J. Reproduction Fertility,* 79:569-573

Brinsfield, T.H., Fisher, M and Hawak, H.W.1974. *J.Anim.Sci.*39:565-574.

Buchoo, B.A. and Wani, G.M.1991.Induction of oestrus in Corriedale ewes during anoestrus. *Indian Journal of Animal Science,* 60(4) : 407-408.

Buchoo, B.A.,Wani,N.A.,Khan,M.Z and Wani, G.M.2000.Laparoscopic detection of Ovulation and pregnancy in different sheep breeds. *Indian Journal of Animal Sciences* 70(2):61-62.

Csapo, A.I. (1981). Force of labor in :Principles and practice of Obstetrics and Perinatology. Vol.2 (L.Iffy and MA Kaminetzky, eds.), pp.761-799,John Wiley and Sons, New York.

Erickson, G.E.,Lid, Sadrkhanloo R.; Liu X-J, Shimasaki,S. and Ling, N. (1994).

Extrapituitary actions of gonadotropin releasing hormone in stitulation of insulin-like growth factor-binding protein and atresia. *Endocrinology* 135:1365-1372.

Evans, N.P., Dahl, G.E., Maugar, D. and Karsch, F.J.1995. Estradiol induces both qualitative and quantitative changes in the patterns of GnRH secretion during the pre-surge period in the ewes. *Endocrinology* 136: 16-3-1609.

Fraser, H.M. (1981). Effect of oestrogen on gonadotrophin release in stumptailed monkeys (Macaca arctoides) treated chronically with an agonist analogue of LHRH. *Journal of Endocrinology* 91:525-530.

Fraser, H.M. (1987). LHRH antagonists and female reproductive function. In: LHRH and its analogs: Contraceptive and Therapeutic Applications. Part 2 (BM Vickery and JJ Nestor Jr.eds.) pp.227-244,M.T.P. Press, Lancester.

Gall, C., Gelderman, H. and Wani, G.M. 1997. Analysis of genetic resources of goat breeds in Kashmir. Final Report of Research Co-operation between SKUAST and University of Hohenheim, Germany.

Gupta, R.S. and Maurya, S.N. 1993.Microbial flora of bovine semen and their antibiotic sensitivity pattern. *Indian Journal of Animal Reproduction,* 14(3):107-108.

Guraya, S.S. (1985). Biology of ovarian follicles in mammals, Springer-Verlag (Publishers), Berlin.

Heape, R.B., Wang, M.W.,Whyte, A. and Taussing, M.J. 1990. In:Preoceedings of the international Conference on Frontiers in reproductive Physiology, AIIMS, New Delhi, 8-10 November, 1990 L17.

Kaiser, H.B.; Zhao, D.; Cardona, G.R, and Chin, W.W.(1992). Isolation and characterization of cDNAs encoding the rat pituitary gonadotropin-releasing hormone receptor. Biochemical and Biophysical Research Communications 189:1645-1652. Karsch, Fred J.; Bowen, Jennifer M.; Caraty, Alain; Evans, Neil P, and Moenter, Suzanne M.(1997) Gonadotropin-Releasing Hormone Requirement for Oculation. Ovulation. *Biology of Reproduction* 56:303-309.

Karsch, Fred, J. and Evans, Neil P.(1996) Feedback actions of estradiol on GnRH secretion during the follicular phase of the estrous cycle. *Acta Neurobiol. Exp.* 56:715-725.

Kher, H.N and Dholakia, P.M. 1985. Bacteriological studies of bovine semen. *Indian Journal of Animal Reproduction,* 5(1):78-84.

Kledzik, G.S.; Cusan, L.; Auclair, C.; Kelly, P.A. and Labrie F. (1978). Inhibitory effects of luteinizing hormone releasing hormone agonist on rat ovarian LH and follicle stimulating hormone receptor levels during pregnancy. *Fertility and Sterility,* 29:560-564.

Koul, N.A.2002 Bacterial flora of bovine semen and their antibiotic sensitivity pattern. M.V.Sc thesis, SK. University of Agricultural Sciences & Technologyof Kashmir, Srinagar-191 121.

Lenton, E.A.; King, H.; Thomas, E.J.; Smith, S.K. MeLachlan, R.I.; MacNeil, S. and Cooke, I.D. (1988). The endocrine environment of the human oocyte. *J.Reprod.Fert.*82:827-841.

Martin, K.; Santoro, N.; Hall J.; Filicori, M.; Wurman, M. and Crowelery, W.F. Jr. (1990). Management of oculatory pulsatile gonadotropin-releasing hormone. *J. Clin Endorcinol and Metab.* 17: 1081A-1081G.

Mufti, A.M.1996.Studies on morphological characters and some metabolic constituents at different stages of development of ovine foetus. M.V.Sc. thesis, Sher-e-Kashmir University of Agricultural Sciences and Technology, Shalimar, Srinagar.

Mufti, A.M., Wani, N.A.; Wani, G.M., Mir, M.M. and Khan, M.Z,1997 Superovulatory response in sheep during different months of breeding season.*Small Ruminent Research,* 25:181-184.

Najar, A.M.(1999). Effects of fish ovarian extracts on mammalian ovary, Ph.D. Thesis of University of Kashmir, Srinagar-190006.

Sanchez-Criado, J.E.; Bellido, C.; Galiot, F.; Lopez, F.J. and Gaytan, F. (1990). A Possible dual mechanism of the anovulatory action of antiprogesterone RU 486 in the rate. *Biology of Reproduction* 42:877-886.

Schultz, G.A., hahnel, Ann., Watson, A.J and Hojan, A. 1990. In:Proceedings of the international Conference on Frontiers in Reproductive Physiology AIIMS, New Delhi, 8-10 November, 1990. L20.

Shah, N.A. 1996. Effect of gonadotrophin releasing hormone (GnRH) on anoestrous crossbred cows under temperate climatic conditions. MVSc Thesis submitted to Sher-e-Kashmir University of Agricultural Sciences & Technology, Srinagar.

Srivastawa, R.K.; Krishna, A. and Sridaran R. (1995). Ovarian development, steroidogeness and gonadotrphin secretion in response of long-term treatment with a gonadotrophin-releasing hormone agonist in the rat. *Journal of endocrinology* 146:349-357.

Van Look, P.F.A. (1988). Hormone fertility reguation at the crossroads. A short review of the potential of hormone antagonists . In Proceedings of a symposium on the Use of Hormone Antagonists for Fertility Regulation organized by the Indian Society for the Study of Reproduction and Fertility and the World Health Organisation's Special Programme of Research, Development and Research Training in Human Reproduction on November 4, 1988 at Mumbai, India. Edited by Chander P.Puri and Paul, F.A.Van Look, Printed by Good Print, Dr.Ambekar Road, Parel Mumbai-400012 (India).

Wani, G.M. 1998.Ultrasonic pregnancy diagnosis in Sheep and goats. *World Review of Animal production,* 17:43-48.

Wani, G.M. 1982. Laparoscopy in farm animals. *World review of Animal Production,* 18: 7-13.

Wani, G.M. 1984a. Vet suche Zur Embryo Production and Kultur bei Ziege. Dr. Med.Vet Dissertation, Tieraertzlichen Hochschule, Hannover, Germany.

Wani, G.M. 1984b. Investigations on ovarian activity, ovulation and pregnancy diagnosis in sheep and goats. Ph.D. Thesis, IVRI, Rohikhand University, Indian Veterinary Research Institute, Izatnagar, U.P.

Wani, G.M. 1986. Embryo transfer in goats. 6th National Congress on Animal Reproduction, Guwhati 28-30th November, 1986.

Wani, G.M. 1989.Detection of foetal age in sheep and goats by ultrasonic technique. *Indian Journal of Animal Science,* 59(12):525-30.

Wani, G.M. 1989. Detection of foetal age in sheep and goats by ultransonic technique. *Indian Journal of Animal Science,* 59(12):1525-26.

Wani, G.M. 1989. Ovulation detection by laparoscopy in goats. *Indian Journal of Animal Science.* 59(12):1527-28.

Wani, G.M. 1995. Early diagnosis of pregnancy in surrogate mothers. National symposium on embryo transfer and genetic engineering technologies in small ruminants. May 7-8, 1995, Makhdoo, Farrah. India.

Wani, G.M. 1996. Embryo biotechnology in sheep and goats. Published by Valley Book House, Srinagar- 190006.

Wani, G.M. 2001. Enhancement of chevon production through reproductive management. In: Proceedings of IV National Seminar of Indian Society for Sheep and Goat Production & Utilization (ISSGPU), held at Bharathiar University, Coimbatore from November 9-11, 2001.pp.60-68.

Wani, G.M. 2002. Goat Husbandry/ multipurpose management. Encyclopedia of Dairy Sciences. Edited by Roginishi, Hubert, Enquay, Jhon W. and Fox patrict, Academic Press, London, NW 17B4MS192.

Wani, G.M. and Buchoo, B.A.1990. laparoscopic view of ovine conceptus. *Indian Journal of Animal Science*, 60(5):564-568.

Wani G.M. and Buchoo, B.A. 1991. detection of induced ovulation by laparoscopy in ewe lambs. *Indian Journal of Animal Reproduction*, 12(1):76-77.

Wani, G.M. and Sahni, K.L. 1980. Ultrasonic pregnancy diagnosis in sheep and goats. 2nd All Indian Symposium on Animal Reproduction, Banglore, pp. 1-5.

Wani, G.m. gelderman, H. and Hahn, J. 1986. Brunstsychronization bei Ziege. *Zuchthygiene*, 21: 1-2.

Wani, G.M., Gelderman, H. and Hahn, J. 1990. Superovulations during early luteal phase in goats. *World Review of Animal Production*, 29(2):41-43.

Wani, G.M., Risam, K.S. and Nowshahri, M.A. 1988. Effect of oestrus synchronization on lambing in corriedale ewes. *Indian Journal of Animal Science*, 58(7):800-801.

Wani, G.M. 1989.Plasma progesterone concentrations during oestrus cycle in goats. *Indian Journal of Animal Reproduction*, 19(1):44-46.

Wani, G.M. Sadat Rehman and Shah Masood 2000. Higher Education and Human resource Development: Knowledge Perspectives in Current Millennium. Paper presented at the Invernational Conference on Higher Education for Human development, organized by Association of Indian Universities at New Delhi on February 6 22-24, 2000.pp.239-247.

Wani, G.M., Sinha, N.K. and Khan, B.U. 1987. Oestrus synchronization with progestagens in Muzzaffarnagri ewes. *Indian Journal of Animal Science*. 57(12)

Wani, G.M., Sinha, N.K. and Sahni, K.L, 1980. Haematic type of foetal mummification Muzaffernagri ewes. *Indian Veterinary Medical Journal*, 4:46-47.

Wani, G.M, Sinha, N.K. and Sahni, K.L. 1980. Note on gestational oestrus in the Muzzafernagri ewes. *Indian Journal of Animal Science*, 50(11):1009.

Wani, G.M and Sahni, K.L. 1980. An ultrasonic technique for detecting pregnancy in sheep and goats. *Veterinary Research Bulletin*, 3:35-37.

Wani, G.M., Sinha, N.k. and Sahni, K.l. 1980. Note on seasonality of breeding behaviour in Jamunapari goats, *Indian Journal of Animal Science*, 50:1153-55.

Wani, G.M., Sinha, N.k. and Sahni, K.L. 1981. Incidence of abortionsand stillbirths in Muzzaffarnagri and crossbred ewes. *Indian Veterinary Journal*, 58:558-561.

Wani, G.M., Tripathi, S.S. and Saxana, V.B. 1980. Sialoproteins in the cervicaql mucus of repeat breeding cows during oestrus. *Indian Veterinary Medical Journal*, 4:49-53.

Wani, N.A., Wani, G.M., Mir, M.M. and Beg, M.A. 1998. Ultrasonic pregnancy diagnosisin Gaddi goats. *Small Ruminant Research*, 29: 239-240.

Wani, G.M. (2002b). Advances in animal Reproduction in: Proceedings of NationalSeminar on "Recent Research Trends in Life Sciences" , (August 26-28, 2002) organized by PG Department of Envirornmental Science and Centre of Research for Development, The University of Kashmir, Srinagar- 190006.

□□□

Chapter 16

Wild Animals Damages to Agrarian Economy

G.M. Wani & Talat Zahir

The paper describes the damages caused by wild animals on agrarian economy. This study looks into various aspects of damages by elephants, birds and other species.

Introduction

Wild animal-human conflicts have started since beginning of human era from Adam and Eve. This conflict of wilderness made man to hide in caves and he was called as "Cave man". Slowly, with his advancement it is he who invented Axe and other weapon in stone and iron ages to frighten the wild animals, initially. Later on he hunted them to save himself. This feeling of uncertainty and fear of wild animals and wilderness reduced with the invention of fire. He made sharp weapons of bones and iron. These initial weapons were the beginning of this conflict, Animal human conflict.

Thirty thousand years ago, the human population rose to 6 million. They were still hunters. With the invention of fire, he set fire a vast majority of sanctuaries, which scared wild life and they migrated from his neighborhood. Many forests, hills and difficult terrains were still beyond the reach of man three thousand years before, although human population has increased to 60 million. Man has already started primitive agriculture. He had made his terrains and wild life scared by him left his close habitats and searched for fresh abodes. Man by now had lust for fur, horns, ornaments and other forest resources. He invented many means to frighten whole wild life. He became a "Danger "not only for wild animals but for his own species, environment and eco-biodiversity.

Three hundred years ago he industrialized crop production and produced enough food, for nearly 600 million people. This continued and from 30 year now he is feeding 6000 million people. Today we have a global food security for 7.5 billion. Human food security gains resulted in reduction of all other wild species, thus, the origin of wild- life- human conflict is the lust of man for more food, more luxury or sometimes fun for hunting or fur. This reduced wild life reserves and now a open conflict came into existence.

Many man-wild life conflicts have been reported from Gir forests of Gujarat, Rajaji National Park in UP and many other states where forest lands became cultivable lands. Elephants, wild boars, monkeys, squirrels, deer , birds like crows, parakeets, wild dogs, jackals, gaur, sambur, langure, fowls, pea cocks, neilgai, Hippos, biats, blackbirds, rodents, wild pigs, feral species, primates, beetles, foxes, pigeons, feral hogs and a variety of other species damage crops. The carnivores even attack human too. These attacks are for search of food or their loss of habitat. Many such accidents came to be known in Jammu and Kashmir. The bear leopod and other wild animals are reducing in number. Their habitats used by human. The human causalities are due to carnivorous species, but herbivores inflict economic and human losses too. The crop damages by wild life has been the new threat to agricultural productivity throughout the world. This also concerns us in Asia and India. This review is aimed to find out:

1. Extent of damage to crops.
2. Nature of crop damage
3. Ways and means to prevent these losses.
4. A strategic planning to drive a line between wildlife conservation and farm economics especially in India.

Executive Abstracts and Strategic elementaries: [EASE]

1. Importance

Species causing crop damages ranges from elephant's wild birds, monkeys, squirrels, deer, parakeet, wild dogs, foxes, deer and many others like Neilgai. On an average this damage to crops by wild animals amounts to U$ 961 per hectare. It is much more than an Indian Farmer earns from a hacter annually. Therefore, by these estimates, the damages are spectacular and economically important.

2. Human Elephant Conflicts (HEC)

i) Economic Losses

Among elephants crop raiding is common. The crops near their home ranges are damaged more. Elephants damaging crops had twice big ranges than those who did not damage the crops. Thus, more proximity of the crops near their ranges are prone to crops damages. Train accidents instigate elephants more to crop or human damages. Indonesia saw more frequent raids of elephants on crops. Human elephant conflict (HEC) is frequent and poses serious challenges in Africa.

Both male groups and family group attacks have been observed. HEC losses in West Bengal were worth 3.2 croses of Rupees. This damage occurred in 3368sq kms.radius. The numerical number of elephants was 62 only. Assam observes damages to the woodcutters by elephants Asian wild Elephants raid and damage crops in herds of 10-13 individuals or big herds comprising 50-74 elephants. In Darjeeling district alone over an area of 200 kms. East to west, in last two decades 277 houses were demolished by elephants, killing 66 people in 5 districts. As a result of this conflict 23 elephants lost their lives. In 2001, economic loss of the order of US fifty thousand dollars was estimated to be inflicted by elephants. This scenario necessitates comprehensive measures to be taken to lessen these damage. The review of the measures, around the world suggest following few studies to be undertaken and resultant measures to be applied to lessen these damages.

ii) Mitigation of Losses

Methods developed and used to mitigate crop damages by elephants consisted of frightening of animals by drum beating, firegracks or even air gun fires. Guarding the crops by fences or even using electric fencing or raising poles and wiring may be effective. Chemical based deterrents, and electric fencing have been found to reduce crop damages. High voltage electric fencing using energizers in west have prevented wild animals damaging crop but this preposition may not be applicable in areas where even habitations do not have access to electricity. However, this method may act as a temporary boundary separating wild and domestic habitats.

This could prevent intrusions from sly vatic to domestic foci. Guarding fields, digging trenches, modifying cropping patterns have a possibility of reducing man elephant conflicts or Human elephant conflicts. A 30% open space between two habitats may help to avoid HEC. This means a distance to be maintained between cropping fields and elephant habitates. A proper investigation on these lines is needed. A grid based geographical information system (GIS) with a 25Km2 resolution may help to have cost effective data source to analyze these situations. There is urgent need of identifying spatial predictors of HEC. On the basis of this study one can suggest or plan mitigation strategies, early warnings of attacks, use of barriers and deterrents. The utility of the methods could be assessed for land use and livelihood strategies to limit HEC.

3. Dear-Nilgai Damages

A survey of 2500 farms in UK revealed that 69% (n=192) cereal crops were damaged by deer. This damage costed £500 per annum per farm. Damages varied with deer density. In USA too deer damages comprised of crop loss, landscape damages, car accidents and property damages. Nilgai damages alongwith deer damages are common in India too. Tree cover of Acacia in the area is generally used as shelter by Nilgai. We have encountered huge crop damage in Mathura (CIRG) and nearby area by Nilgai visits. Grazing and browsing of Nilgai inflict losses on farms. This is regarded as a mammalian crop threat by the farmers. This behaviour of their inflict ozone injury to the young sapling, so precious for the growth of trees in Indian semi-arid farms.

Corn damages by deer in USA amounts to 6.6.% per hectare. In an area where a farmer owned 125 hectare on an average 55 hectares were used to sow corn. There is a 6.6% loss of the product which is a huge economic loss. Similar damages by deer in Ontario, Canada, Portugal, Japan, South America and other parts of the world have been reviewed. In Virginia too a study involving 1506 agricultural producer farmers revealed 58% of them experiencing deer damages. Ways and means to prevent such attacks on crops have been reviewed. Most of the possible measures to be adopted are similar to these described for elephants (HEC).

4. Other Mammal and Bird Damages

Monkey damages maize, sweet potato and other crops. It is suggested to reduce or change cropping pattern or alternating with non-agricultural activities near location of monkey habitats. Various methods are reviewed. Bat damage, Hygo crop damages in Japan, Grape wine damages by bats in Andra Pradesh, India have been studied. Crop economic loss assessment has been reviewed in these pages. Clover rather than grass can reduce wild mammal damages. Alley cropping of black walnuts helped to save soyabean and maize crop damages. Pesticide damages wildlife and minimize their attacks, enclosures and other electric fencing help to avoid such damage. Can one use harmful means to save crops, need a suitable strategy to preserve ecology and biodiversity. Both crops and wild animals need security and conservation. A management strategy is needed to safeguard human and wildlife equally. A policy framework is envisaged.

Blackbird crop damage in USA amounts to 5-8 million dollars. A considerable sum of crores of rupees have been estimated to be damages in the form of crop, human and property losses by wild animals, birds and other rodents in India. Many measures to reduce these losses are needed and have been reviewed.

Review

1. Wild Animals Damaging Crops

To elephants wild boar, porcupine, rheus macaque (Macaca mulatta), hoary-bellied squirrel, barking deer (Muntiacus muntjak), red-breasted parakeet (Psittacula alexandri), and wild dog are wild animals damaging crops.

Methods developed and being used to mitigate man-wildlife conflict include, frightening the animals; guarding the crops; and using some sort of scarecrow and traps (Miah et al 2001).

Accurate estimation of crop damage by wildlife (raccoons, white-tailed deer, and coyotes) often requires labour-intensive sampling procedure. Variable area transect (VAT) sampling has been identified as a potential labour-saving alternative to quadrat sampling (Engerman et al 2002).

2. Wild Life Human Conflicts

Agricultural landowners suffer in the form of damage to crops, livestock, and other property. Some wildlife agencies maintained abatement and compensation programmes. A model of deer-inflicted crop damage used to facilitate agency decisions regarding deer densities and distribution, abatement use, and to forecast compensation. The model is applied to field-level compensation claims in Wisconsin, USA. The results are consistent with theory, (Yoder J, 2002).

3. Crop Economic Loss Due to Wild Life

People's perceptions were discerned through participatory discussions covering 419 households distributed in 10 villages in the buffer zone.

Traditional uncodified rights of local people were substantially reduced through policy interventions set in since 1860s. Local people as well as tourists have been excluded from the core zone covering an area of 625 km2 since 1982. Deterioration of rural economy due to damage to crop and livestock by wildlife and, termination of opportunities of income from wild medicinal plant resources and tourism in the core zone were the key negative impacts of conservation policy felt by more than 90% of respondents.

Mean annual economic loss per household was estimated as Rs.1285, Rs. 1195 and Rs.156 due to damage caused by wildlife to food crops, fruit trees and beehives, respectively, Rs. 1587 due to ban on collection of wild medicinal plants for marketing and Rs.7904 due to ban on tourism in the core zone. The Reserve authority granted compensation for livestock killed by wildlife but it was hardly 5% of the market value of killed livestock as assessed by the people. People did not appreciate much the present benefits from the reserve management in the form of wages for carrying out afforestation work, partial compensation for livestock depredation and availability of solar power devices, wool, and spinning devices.

Approximately 95% respondents identified empowerment of local people in respect of realizing income from timber from dead/diseased trees in community forests, income from medicinal plants in buffer zone and opening of the core zone for tourism as potential development options. Improvement in rural economy, the prime concern of local people, has not received as much attention as legal enforcement of protection by the reserve management. There is a need for developing policies and management actions that serve the economic interests of local people together with enhancement of environment conservation goal (Maikhuri et al 2001).

4. Crop Selection

Damage was less (34%) in experimental carrot as multiple crop than carrots as only crop planted (62% damage). Staggered plantings of canola, which continuously produced flowers, was the most effective lure crop of the green manure crops we tested. Carrot producers should use electric fences or 2.4-m woven wire fences, perhaps combined with staggered canola plantings, to reduce carrot depredations (Schwab et al 2001). The colver, rather than fertilized grass, is more effective cover crop on AFAs, against damage by migratory geese. The sward should be managed to encourage clover growth, which would probably involve frequent cutting but no fertilizer. Further research is needed on clover replenishment rate over the winter season and possible benefits of clover leys to other wild life (McKay et al 2001). Alley cropping of black walnuts and percent with maize and soyabean rotations may avoid will animal damages. Tree -crop ratio of 1:10 may help(Gcdsey, 2000).

5. Elephant-Human Conflicts

Conflicts between elephants (Elephas maximus) and human occur in Rajaji National Park (RNP), Uttar Pradesh, one of eleven reserves designated in India, to conserve Asian elephants. Elephant-human. The conflicts in RNP from 1996 to 1999 were studied, and all human and elephant deaths or injuries caused by conflict were recorded. The impact of human colonies on elephant movement was studied in 18 villages along 17 km of the sourthern boundary of the study site and 4 village in the Chilla-Motichur corridor. Three male and four female elephants were radio tracked for 1-2 years. Primary conflicts included crop raiding, competition between humans and elephants for vegetation, and elephant mortality due to train accidents. Adult males that raided crops had home ranges twice as large as adult males that did not raid crops. Elephants only damaged crop of fields that occurred within their home ranges. Field trails of chemical based deterrents and electric fencing should be tried to reduce crop damage. Train speeds need to be reduced to prevent accidental elephant mortality (Williams et al 2001).

Wild Elephant Damages

A rapid village and field assessments, data survey showed. Elephants raided crops at a rate of 0.53 elephants per day in Indonesia. The frequency of crop raiding was related to vegetation type along the park border, the size and presence of rivers, and the distance to the park's Elephant Training Centre (ETC), which houses about 150 captive elephants. Wild elephants damaged at least 450000 m2 of maize, rice, cassava, beans and other annual crops, and close to 900 coconut, banana and other perennial trees in the area surveyed. Elephants killed or injured 24 .Villagers try to reduce elephant damage by guarding fields, digging trenches between the park and their fields, and modifying their cropping patterns. Elephants-human conflict decreases the probability of support from local people for conservation efforts. The approaches are suggested consist of elephant trenches, electric fences, external support to affected villages, and compensation to villagers for any damage caused (Nyhus et al 2000).

This study explores land use conflict in south east Kajiado District, Kenya. The results of household surveys conducted with farmers and herders in 1977 and 1996 to examine changes in land management strategies are compared. The conflict reflects ongoing competition over access to scarce land and water resources between herding, farming and wildlife are the reason of damage of crops. This man-animal conflict needs understanding the conditions that have created the present conflicts (Compbell et al 2000).

It is, therefore, suggested that 30% open space be used as a basic division for stratifying thickets into low -use and high-use categories for deer density estimation. The proportions of each type could be derived from grid-square measurements of aerial photographs (Latham J 2000).

Human Elephant Conflict

Human-elephant conflict (HEC) in Africa occurs wherever these two species coincide, and poses serious challenges to wild life managers, local communities and elephants alike. Mitigation requires a details understanding of underlying patterns and processes. Although temporal patterns of HEC are relatively predictable, spatial variation has shown few universal trends, making it difficult to predict where conflict will take place. Crop raiding was further subdivided into incidents involving only male elephants or family groups. A relatively fine-resolution, systematic, grid -based method was used to assign the locations of conflict incidents, and spatial relations with underlying variables were explored using correlation analysis and logistic regression. Crop raiding was clustered into distinct conflict zones. Both occurrence and intensity could be predicted on the basis of the area under cultivation and, for male elephant groups, proximity to major settlement. Conversely, incidents human injury and death were less predictable but were correlated with proximity to roads.

A grid-based geographical information system (GIS) with a 25km2 resolution utilizing cost-effective data source, combined with simple statistical tools, was capable of identifying spatial predictors of HEC, At finer resolutions spatial autocorrelation compromised the analyses. Synthesis and application. These results suggest that spatial correlates of HEC can be identified, regardless of the sex of the elephants involved. Moreover, the method described here is fully transferable to other sites for comparative analysis of HEC. Using these results to map vulnerability will enable the development and deployment of appropriate conflict mitigation strategies, such as guarding, early warning systems, barriers and deterrents. The utility of such methods and their strategic deployment should be assessed alongside alternative land-use and livelihood strategies that limit cultivation within the elephant range (Sitati et al 2003). Human elephant conflicts (HEC) in west Bengal was an economic loss worth 3.2 crores. This much damage occurred in 3368 sq km radius inhibiting 62 elephants (Singh et al 2002).

6. Kerala Survey

Crop damage by wild animals in Kerala, India, was studied from 1993 to 1996. Data were collected from the offices of the Kerala Forest Department, field survey and from the intensive study area at Marayur, Idukki District. Forty-five species of crops were destroyed by wild animals in Kerala, the species commonly destroyed by wild animals were paddy, coconut plam , plantains , cassava , arecanut, coffee, oil plam , pepper , jack tree, mulberry and manago. The main animals involved in crop damage were elephant (Elephas maximus), gaur (Bos gaurus),sambar (Cervus unicolor), wild boar (Sus scrofa), bonnet macaque (Macaca radiate), common langur (Presbytis entellus), blacknaped hare (Lepus nigricollis) and pea fowl (Pavo cristatus). Among these, elephants and wild board gave maximum damage . Of the total compensation claimed by the farmers, only 8.2% was sanctioned by the Kerala Forest Department.

The highest crop damage (30%) was recorded from the forest ranges coming under the Northern Circle: pinapple (47%), sweet potato (47%), tapioca (42%), alocasia (39%), beans (25%) and plantains (23%) recorded highest percentage of damage. In the intensive study area at Marayur, 28 species of crops were damaged and highest damage was during the summer months. At maximum damage was due to elephant (72%) followed by gaur (62%) , sambar (17%) and wild boar (16%). Tiger (Panthera tigris), panther (leopard) (Panthera pardus) and wild dog (Conine alpinus) were the main cattle lifters in the state. A total of 31 deaths and 64 injuries caused by wild animals were recorded from the state during the period 1983 to 1993. Thirteen indigenous methods used for controlling the crop damage had been identified. High voltage electric fencing using energizer was effective for stopping elephants and other herbivores from entering the agriculture fields. Crop damage is found to be linked to the cropping pattern and location of the agriculture fields. Short term and long-term measures needed to prevent the crop damage are discussed *(Jayson EA,1999).*

7. Bird Crop Damages

Use of non-lethal method to avoid crop damages by bird have been reported. Blueberry damages by cedar waxwings (Bombycilla cedrotun were minimized. (Avery et al 2002).

8. Wild Bird Damage

In the northern Great Plains of USA, conflicts between red-winged black birds (Agelaius phoeniceus) and sunflower (Helianthus annus) growers have intensified since the late 1960s due to the expanded commercial production of sunflowers. We studied the potential population effects of the removal of up to 2 million red-winged blackbirds annually under a 5 year programme of baiting during spring with DRC-1339 (3-choloro-4 methalalanine) treated rice. They also examined whether lethal control, in combination with current levels of breeding habitat management, would be cost effective in decreasing depredation of sunflower crops during late summer. They evaluated the cost benefit ration for 4 culling scenarios involving (1) variable annual cullus, not exceeding 2 million birds, with and without density compensation (i.e., a positive density-dependent response) on adult survival and (2) culls of 2 million birds annually with and without density compensation.

We constructed a red-winged blackbird population model represented as an age-based matrix and calibrated to stable growth. We assumed a total population of 27 million birds on 1 April (week 1), representing the red-winged blackbird breeding population staging in eastern Southern Dakota and migrating into North Dakota.Under each culling scenario, we reduced the stable red-winged blackbird population (Equally for females and males) and project the population through week 23 of the annual cycle (2 Sep). We then evaluated the associated costs of the management relative to potential sunflower crop losses, assuming $0.07 in damage per bird and 4% loss to other factors. Variable annual culls, likely the more biologically realistic model scenarios, yielded mean annual removals of 1 240 560 (SE=12 328) birds with density compensation and 1 231 620 (SE=28 811) birds without density compensation,, with cost benefit ratios of 1:2.3 and 1:3.6, respectively.

Annual intrinsic rates for the model population over the 5 year period ranged from 1-4 to 4.8%. Considering potential variability in the effectiveness of the cull and the combination of direct and indirect costs, we contend that the realized benefits to sunflower growers by lethal control of red-winged blackbiards via spring baiting, in combination with current nonlethal management efforts, would likely be negligible (Blackwell et al 2003).

The efficacy of hydrolyzed casein (HC) and retail products that contain HC in reducing deer damage to trees and shrubs was determined in a field experiments conducted in USA during 2004-05. The results of the experiments indicate the suitability of HC as a deer repellent. Technical grade HC completely eliminated browsing damage to evergreen shrubs (Gaultheria shallon) and conifers (Thuja plicata). Retail sources of HC were not as effective as the pure hydrolyzed protein (Kimball et al 2005).

9. Blackbird Damages

The economic impact of blackbirds can be severe to rice producers in the United States. One approach to managing this damage is the application of bird-deterrent chemical to the crop. Previous pilot trials suggested that caffeine offered potential as a safe, economical bird repellent. In this study, cage feeding trials with female red -winged blackbirds and male brown headed cowbirds confirmed that a treatment rate of 25000 ppm caffeine on rice seed reduced consumption as much as 76%. Trials with mixed species blackbirds flocks in a 0.2-ha flight pen resulted in just 4% loss of caffeine-treated rice compared to 43% loss of untreated rice. Field trials of a 10 000 ppm caffeine treatment in Louisiana revealed > 90% of caffeine-treated rice seed remained unconsumed on days 2 and 3 of the study whereas blackbirds consumed > 80% of the untreated seed. As a rice seed treatment to deter blackbirds, caffeine appears to be effective, economical and environmentally safe, although additional aquatic toxicity testing is desirable. Improvements in formulation will be needed to make the compound practical for general agricultural spray applications and to extend the adherence of caffeine to rice seeds in field conditions ((Avery et al 2005).

10. Deer Damage

A questionnaire was distributed to over 2500 farms to know damage caused by lowland deer to crops, trees and vegetation. Results from the questionnaire showed that 69% (n=1192) of responding farmers had deer on their holdings and that Roe and Fallow were the most frequently seen species. On those farms with deer present, cereals were the most commonly damaged crop (44%), but only 15% of these farmers claimed that the annual cost of damage to cereals exceeded £500 each year for the whole farm. Validation assessments were based on two visits to assess deer damage to the crop, with a deer species/density assessment during the March assessment and an assessment of grain yield and quality during the August assessment. Respondents were generally accurate in the density and species of deer reported. The percentage of the farm suffering damage attributable to deer was very variable, generally being higher at the first assessment than the second. The figures calculated for yield loss were generally low, Farmers were poor at estimating the economic impact of deer damage when compared to validation data, but a number of parameters may have changed in the two years between the questionnaire distribution and validation, including changes in deer density, crop rotations, and the marked drop in grain prices, which may account for some of the inaccuracies. There were no statistically significant relationships between deer damage assessments and yield loss, either for individual species or both species combined. The relationship between Roe deer damage at the harvest assessment and Roe deer density was significant (Post et al, 2001).

Wildlife managers must consider the public's preferences for wildlife population levels when determining management policies. 849 farmers, hunters and the general public of Maryland. USA, were surveyed in 1996 to determine their preferences for increasing, maintaining, or decreasing deer population numbers. Using a random utility theoretic framework, the factors that explain preferences such as residential location, socioeconomic characteristics, landscape damage, agricultural yield loss and vehicle accidents were analyzed. Results suggested that the majority of people benefit from deer and want to keep deer population at current levels. Other characteristics such as age, income, education, and residential location have minor or no impact on preferences. Property damage, crop loss, landscape damage, and car accidents appear to be the biggest concerns (Curtis et al 2001).

11. Sika Deer Population in Japan

Sika deer Cervus nippon population in eastern Hokkaido, Japan, increased rapidly during 1990-1998. This increase appeared to have halted in 1999-2000, probably due to increased hunting and nuisance control. The period of rapid increase was associated with a disproportionately rapid increase in compensation paid for deer damage to crops. We studied changes in diet during 1990-2000, as reflected by stable isotope ratios of C and N in tooth collagen. We hypothesized that isotope ratios would demonstrate dietary shifts related to population levels and/or time, and that shifts in isotope ratios would be consistent with increasing individual consumption of pasture grasses at higher population levels, delta 13C isotope ratios of tooth collagen in 3 year-old sika indicated a diet dominated by C3 plants throughout the period, and that forage species other than pasture grasses and dwarf bamboo Sasa nipponica (the main crop and woodland understory plants, respectively) were important elements. There was a significant decline in the delta 13C isotope ratio during 1990-2000 in both males and females, delta 15N values showed no trend with time for males, but increased over time in females. Indices of population (Sightings per Unit Effort, SPUE) were negatively correlated with female delta 13C, and positively correlated with female delta 15N, values indicating a shift in diet over the period, especially among females. This shift may be related to population and/or offtake levels, in particular the rapid increase in female offtake for nuisance control and hunting during the period. The data are consistent with a relative increase in pasture grass consumption per individual at higher population levels, however, other explanations of the data are equally plausible. Possible dietary changes, and other factors, influencing the observed shift in isotope ratios are discussed. Although statistically significant, the magnitude of dietary shifts nevertheless appeared small, and did not provide evidence which would justify modifying the current policy, of limiting crop damage through managed population reduction to about 25% of peak levels (Halley et al 2006).

12. White-Tailed Deer Damages

White-tailed deer (Odocoileus virginianus) may cause more damage than any other species of wildlife. These damages include crop loss, automobile and aviation collisions, disease transmission, environmental degradation, and destruction of ornamental plantings. One practical method of controlling deer damage is the use of exclusionary fences. The relatively high cost of labor and materials required to build effective fences has limited most applications to the protection of orchards, vegetable farms, other high -value resources, and mitigation of human health and safety risks. Improvements in fence technology resulting in less expensive, yet effective fence have expanded the use of fence to manage damage caused by deer. Fence typically installed to manage white-tailed deer damage include wire or plastic mesh, electrified high-tensile steel wire, and electrified polytape or polyrope fence. They reviewed the scientific literature on fencing to determine which fence designs would be the most effective for excluding deer in a variety of situations (VerCauteren et al 2006).

The installation of fences to protect agricultural products, natural resources, or other areas from deer (Odocoileus spp.) can be expensive and potential benefits of fencing are difficult to quantify. A rational method is needed to help evaluate whether fencing can be cost effective and which fence designs will be optimal for particular applications. They describe an interactive, dynamic simulation model that conducts economic analyses and predicts economic benefit associated with fences for crops relative to area and perimeter of protected plot, value of crop, percentage of crop damaged by deer annually prior to fencing, efficacy of fence, and costs of fence materials and labor. Users of the model can easily adjust these variables to fit their individual situations and needs. By running a series of simulations, model users can answer questions related directly to fence efficacy and cost-effectiveness (VerCauteren K et al 2006).

13. Corn Damage by Wild Life

Corn damages in USA were estimated at 6.6 per hac due to wild life. The white tailed deer was the wild animal responsible for loss. The average hacters owned by farmers were 125 out of which 55 hectares were sown corn. (Tzilkowsi et al 2002).

14. Deer Damage

Deer (Odocoileus spp.) can cause substantial damage to agricultural crops, resulting in economic losses for producers. They developed a deer activated bio-acoustic frightening device to reduce white-tailed deer (O, virginianus) damage in agricultural fields. The device considered of an infrared detection system that activated an audio component which broadcast recorded distress and alarm calls of deer. They tested the device against unprotected controls in cornfields during the silking-tasseling stage of growth in July 2001. The device was not effective in reducing damage: track-count indices (F1,4=0.02), corn yield (F1,9=1.27, P=0.289), and estimated damage levels (F1,10=0.87, P=0.374) did not differ between experimental and control fields. The size (F2,26=1,00,P=0.380), location (F2,25=0.39, P=0.684), and percent overlap (F2,25=0.20,P=0.818) of use-areas of radiomarked female deer did not differ between during and after treatment periods. They concluded that the deer-activated bio-acoustic device was not effective in protecting cornfields in this study; however, the device may be more effective in small areas such as gardens or for high value crops that do not grow tall enough to offer protective cover (Gilsdorf et al 2004). White-tailed deer (Odocoileus virginianus) cause millions of dollars of damage to agricultural crops annually. They tested the effectiveness of propane exploders and Electronic Guards (Pocatello Supply Depot, Pocatello, Id). For reducing deer damage in corn fields during the silking-tasseling stage of growth. Track-count indices (F2,7=0.70,P=0.532), corn yield (F2,6=0.14, P=0.873), and estimated damage levels (F2, 12=1,45 P=0.272) did not differ between experimental and control fields. The size (F2,11=0.08,P =0.924), Location (F2,9=0.30, P=0.750), and percent overlap (F2,9=0.46, P=0.644) of use -areas of radiomarked female deer in the vicinity of experimental fields did not differ among before, during and after 18 day treatment periods. In a related study, we placed propane exploders in cornfields within use-areas of 12 radiomarked female deer. The deer did not react appreciably to the devices; the size (F2, 17=0.08, P=0.921), location (F2,22=1.37, P=0.275), and percent overlap (f2,10=0.47, P=0.636) of deer use areas did not differ among before, during, and after 14 day treatment periods. They conclude that propane exploders and Electronic Guards have limited potential for reducing deer damage to corn at the silking-tasseling stage (Glisdorf et al 2004.

A welfare measure for wildlife damage to Ontario (Canada) field crop producers during the 1998 was. The welfare measure presented in this study provides a more accurate picture of losses from wildlife damage to agricultural. Other damage estimates based on yield loss overstate the damage since benefits from wildlife are netted out. Results for the Ontario field crop producers indicate that the magnitude of the difference between the value of the yield loss and the welfare measure of damage is approximately 50%. This difference indicates that most farmers were willing to tolerate the wildlife damage they experienced (Heigh et al 2001).

15. Grapevine Damages

In field trials in 1999-2000 in Gundla Pochampally, Andhra Pradesh, India , the incidence of damage to green grapevines was studied. Visits to the vines by bats begain around 45 minutes after sunset and foraging continued until 1 hour before sunrise. Damage occurred to ripe fruit only, and increased around harvesting time, percentage damage ranged from 0 to 100% (Bhargavi et al, 2001).

16. Hippo Damage

The introduction to the paper described the various ecological effects known to result from grazing, movement along paths, and wallowing by the common hippo. Hippopotamus amphibious. The study reported was carried out at Kainji Lake National Park., Nigeria, during the dry season periods {of 1991 and 1992}. The method adopted by Agnew, A.D.Q. (East African Wildlife Journal (1966) 4, 38-46) was used to assess hippo foraging footprints at three hippo pool sites. A total of 32 footpaths were enumerated out of which 18 were located at Kaii hippo pool site, while the frequency of utilization of paths was also higher for this site relative to those in other area. The upstream-downstream trend in hippo occupancy of dry season water pools could expose the hippo to crop damage conflicts at the peripheral areas.

17. Hygo-Japan-Mammal Crop Damages

Mammals inhabiting Hyogo consist of seven orders, 17 families and about 40 species. Except for Lagomorpha and Artiodactyla, the remaining five orders among them include species which need some protection and they total about 55% of all species excluding extinct, introduced and feral species. Ecological information in Hyogo prefecture has been accumulated in few protection-required species there is no recent information of spatial distribution on Oriental water-shrew, Japanese noctule-bat and Japanese dormouse; and little information on Japanese shrew, Japanese horse-shoe bat, Japanese large - footed bat. Schreiber's bent-winged bat, Japanese tube-nosed bat, common parti-colored bat, Japanese squirrel, Japanese small flying-squirrel, Japanese giant flying-squirrel, smith's vole harvest mouse and Japanese badger. Damage and population managements is also necessary in sika deer and Japanese wild boar, to reduce their crop—damaging, and comprehensive management in Japanese black bear, an endangered local population, to prevent human-bear fatal accident. Habitat alteration due to human activities, however, has affected the population sizes and spatial distribution of all these, mammals in Hyogo. Habitat management has priority over damage. Fundamental and applied scientific studies and understanding of ecology and wildlife management science needs promotion by citizens (Mitani M 2000).

18. Monkey Damage

Forty-seven property owners in Entebbe, Uganda were questioned about vervet monkey activities on their property. The objective was to investigate the interactions between humans and vervet monkeys in an agricultural area adjacent to a forest zone. Other studies have reported that farms located within 300 m of a forested boundary probably incur the greatest risk of crop-raiding. Two other factors that may influence susceptibility to vervet crop-raiding were also examined: the types of crops grown and the types of direct preventive measures used. The effect of these two factors on vervet crop-raiding is not straight forward. However, the distance a property is located from the forest edge is an important factor influencing vervet crop raiding. Surveyed gardens 200 m from the forest edge received significantly less crop -raiding than farms located 100 or 50m (P=0.040,< alpha=0.05). In this study, 8 out of 13 farmers (61.5%) said that maize, sweet potato, or both, were most likely to be damaged by the vervets. Other crops singled out by the farmers are receiving damage by the vervets were bananas, mangoes, groundnuts and yam. They suggest that the development of non-agricultural activities on land directly adjacent to forested areas may reduce vervet crop -raiding by deterring vervets from traveling greater distances from the forest edge due to increased obstacles or risks (Saj et al 2001).

19. Bait Damage

The longevity of zinc phosphide (ZP) on whole wheat bait was determined at the end of the "dry" and "wet" seasons, is Western Australia.. While the total rainfall during the two trials was 74 mm and 155mm, substantial loss of ZP was recorded only after significant rainfall events. Irrespective of season, the loss of ZP from bait applied in bait stations was minimal. The maximum recorded loss was 17% and this occurred after 21 days' exposure during the wet season where the bait stations were placed in-crop. Nevertheless, regardless of the application method, sufficient ZP always remained on the wheat bait. Theoretically it was lethal to rats for at least 8-14 days (Twigg et al., 2001).

20. Venezuela Experience

In Venezuela, lethal control of wintering Dickcissels (Spiza Americana) is considered a threat to the species survival. To help farmers protect their rice and sorghum crops from by Dickcissels and to minimize the killing of large numbers of these birds, alternative non-lethal crop protection measures are needed. To that end, the responses of captive Dickcissels to three bird-repellent chemicals (anthraquinone,methyl anthranilate and methiocarb) applied to rice seed were evaluated. In one-cup feeding trials, treatments of methiocarb (0.05% g/g, applied as Mesurol 75% wettable powder) and anthraquinone (0.5%, applied as Flight Control) reduced consumption of rice by 70% relative to pretreatment consumption. Other anthraquinone treatments (0.05,0.1%) and methyl authranilate (0.05%) were inrffective. In two -cup trials, with untreated millet as the alternative food, consumption of rice treated with 0.05 and 0.1% anthraquinone was reduced by 90% relative to pretreatment levels. Overall, Dickcissels responded to the repellents similarly to the red-winged blackbird (Agelaius phoeniceus). Because Flight Control has been used successfully to reduce blackbird use of rice fields in the USA, the prospect is good for successful reduction of damage to repening rice by Dickcissels in Venezuela, particularly if repellent use is coupled with the establishment of alternative feeding sites (Avery et al 2001). Deer selected carrots over all green manure crops.

21. Nilgai Damages in India

Crop-damage by nilgai has been widely reported from India. Are give Nilgai is for increasing in this region. Lack of natural predators, deforestation overgrazing and the protection of these animals from Hindu communities are reasons for their overpopulation. Tree cover of Acacia are generally used by nilgai as a day time shelter but not food, therefore it goes for crop-raiding in the late evening and at night, jumping across 6-7 feet high stone wall, barbed fencing and fences of dead or live thorny plant material and any other fencing/barrier made to protect the crop-. Due to habit of both grazing as well as browsing they devore every kind of farm species (both rabi and kharif crops). It has been observed that eating less but destroying more by trampling and causing damage are therefore regarded as serious mammalian crop pest and farmers wants to get ride of this unconventional pest. The farmers chase them away by just following them by making loud sound by crackers or air gun fires, following through tractors, empty tin or dried pumpkin filled with small stones and connected with strings. Technically, carrots(enclosures), trenching or power fencing are suggested to mitigate the crop damage. Secondly, animals could be translocated to wildlife sanctuaries from the sites they seen overcrowded or severe crop raiding problems (Goyal et al 2000).

22. Pesticides and Wildlife

A range of monitoring activities has shown impacts of anticholinesterase pesticides on UK wildlife, and continued risks are evident from laboratory and field experiments, together with the scale of use in the field. Along with other broad spectrum insecticides, many organophosphates have adverse direct effects on non-target arthropods in farmland, and so are likely to contribute towards indirect effects of pesticides on farmland biodiversity. The anticholinesterase insecticides have both lethal and sublethal effects on aquatic wildlife, however the history of recent incidents of damage to river ecology following the wider use of synthetic pyrethroid sheep dips, illustrates the need to consider the implications of changes in the use of alternative products when reviewing these insecticides (Burn 2000).

The use of anthraguinone-based flight control and methyl anthranilate-based non toxic avian foraging repellent we used to avoid crop damage by sandhill cranes. Thought both repellents were effective at deterring cranes from treated corn, neither has been tested on corn under field conditions. (Blackwell et al 2001). Oak seedlings were scientifically raised. Seedling mortality and wild life browse damages were minimal when certain herbicide mixture was used. Biologically and aesthetically, the procedure was extremely successful (Ezell et al 1999). Pre-commercially thinned (forests are less prone to moose damage (McLaren et al, 2000).

23. Pesticide Use in Conflicts

Pesticides can cause damage to man and beneficial organism. Some sub-lethal effects of pesticides were studied in birds with a view to identifying characteristic biochemical responses that may be useful for the monitoring of exposure to sub-lethal levels in the field. Pesticides were used; demeton-S-methyl, (DSM), chlorpyriphos, chlorfenviphos, triazophos, pirimicarb, methiocarb and permethrin. Blood was collected before dosing, and 2,6,24,48 and 72 hours after the treatment from the brachial vein of birds. Enzyme, activities were assayed in the plasma or serum samples obtained. The assays used were GOT,MHD, GDH, SDH,GAMMA GT and ChE. The results showed an increase in plasma and serum GOT and gamma-GT levels were found in all animals treated with the previous pesticides. The level of ChE increased in birds after treatment with permethrin. It was concluded that the pesticides cause structural and functional changes in the liver and also, the measurement of the previous parameter activities may be useful for assessing exposure and sub-lethal effects of pesticides on the wildlife (Dahamna et al 2004).

24. IPM and Crop Losses

The queensland sugar, industry has recently implemented a comprehensive integrated pest management (IPM) system to minimize crop losses from two antive rodent species, Rattus sordidus (canefield rat) and Melomy burtoni (climbing rat). These species inflicted approximately $25M of damage in a major outbreak in the 1999-2000 seasons. Both of these rodents are listed as common wildlife under the schedules of the queensland nature Conservation (Wildlife) Regulation 1994. The IPM programme is based on understanding the ecology and biology of each species. It incorporates a large-scale monitoring programme aimed at providing early warning of imminent rodent build up to avert major outbreaks. The industry has also developed a memorandum of understanding with Queensland State Government, which delivers on the industry's pest management needs, while providing an improved system of accountability for the taking of two of Queenslands native wildlife species. The consensus reached between the cooperating parties (The Bureau of sugar Experiment Stations, CANEGROWERS, regionally-based Cane Productivity Services, and the outcomes can be negotiated between rural industry and environmental interests (Hunt et al 2004).

In Areas around Lake Mburo National Park ,large wild animals wander in close proximity to human settlement. This poses serious conflict in terms of crop damage. The integration of conservation with other land uses is difficult where densely settled agricultural land surrounds a protected area potentially containing problem animals, as is the case for several parks in Africa and Asia. The intensity of crop raiding was quantified through the use of random crop quadrants/plots and area estimation techniques in a portion of raided fields. The animal species concerned were documented from observations, footprints and any other marks left behind. Three variables were tested as predictors of damage; human population density, distance from the park boundary and season. In this study data is presented regarding crop loss in the different seasons of the year, analysis of crop damage variation and animal species involved in crop loss. A diverse assemblage of animals foraged on subsistence crops and analysis of crop damage revealed significant crop depreciation by wildlife (Kagoro et al 2004).

25. Hawai Pest

The apple snail, P.canaliculata, is an aquatic freshwater snail native to South America, Originally imported to Hawai'i as pets for the aquarium trade, they were soon introduced into wetland plots known as "lo's" where taro (Colocasia esculenta), an economically and culturally significant crop, is grown. Some individuals reasoned that the snails, being edible, could be harvested as food, and that raising the snails along with the taro in the "Io's" would provide income supplemental to the taro harvest with minimum additional input. This introduction of snails into the taro "lo'i" however, proved to be a disaster. Farmers failed to take into account the voracity , reproductive potential, and rapid growth of the snails.

Because of the ideal conditions in the taro "lo'i", the snails multiplied rapidly and fed heavily on the taro shoots and corns in many cases, destroying a complete crop before harvest time. Hindisight has shown that the snails are dissipated via the irrigation system throughout the "lo'i" and then spread to the surrounding wetland areas. Large breeding populations are now established in wetland areas on the islands of Hawai'e, O'ahu, Kaua'I, and Maui. Some of these wetlands are wildlife preserves with state and federal mandates that restrict the potential methods of eradication. Background information is provided on both P.canaliculata and taro to fully explain the challenges and opportunities that this situation presents (Tamaru et al 2006).

26. Chemical Repellents

Chemical feeding repellents applied to ripening sunflower might help reduce blackbird (lcteridae) damage, which is a chronic agricultural problems from seed information harvest. However, cost are high to develop and register new repellents for agricultural use. In 2003 and 2004, we evaluated feeding repellency of 8 pesticides registered by the Environmental Protection Agency for use in sunflower. Caged red-winged blackbirds (Agelaius phoeniceus) were fed unshelled sunflower seeds treated with the following pesticides: 5 pyrethroid insecticides, an organochlorine, an organophosphorus, and a gungicide. Compared to untreated reference groups, feeding rates were reduced for 4 of the 5 pyrethroid insecticides. Only the organophosphorus (chlorpyrifos), however, significantly decreased feeding rates. More research on repellency effects of this product in field efficacy trials is probably warranted based on the results of our cage experiments. Depending on timing of application, registered insecticides with blackbird feeding repellency could provide supplemental economic benefits to sunflower producers through dual purpose use (Linz et al 2006).

27. Persistent Organic Pollutants (POPS)

Persistent organic pollutants (POPs) have spread throughout the global environment to threaten human health and damage ecosystems, with evidence of POPs contamination in wildlife, human blood, and breast milk documented worldwide. Based on data from the US Food and Drug Administration, this article provides a brief overview of POPs residues in common foods in the

USA food supply. The analysis focuses on 12 chemical compounds now targeted for an international phase out under the Stockholm convention on POPs. The available information indicates that POPs residues are present in virtually all categories of foods, including baked goods, fruit, vegetables, meat, poultry, and dairy products. Residues of five or more persistent toxic chemicals in a single food item are not unusual, with the most commonly found POPs being the pesticides DDT (and its metabolites, such as DDE) and dieldrin. Estimated daily doses of dieldrin alone exceed the US Environmental Protection Agency and US Agency for Toxic Substances Disease Control reference dose for children. Given the widespread occurrence of POPs in the food supply and the serious health risks associated with even extremely small levels of exposure, prevention of further food contamination must be a national health policy priority in every country. Implementation of the Stockholm Convention will prevent further accumulation of persistent toxic chemicals in food. Early ratification and rapid implementation of this treaty should be an urgent priority for all governments (Schafer et al 2002).

28. Netherland Experiences

Traditionally, pink-footed geese Anser Brachyrhynchus wintering in Denmark, the Netherlands and Belgium have used the Danish sites only during mild winter, rapidly moving southwards in case of cold spells. Since the 1980s, an increasing number of geese have remained on the Danish wintering grounds despite cold spells, foraging on pastures Because winter wheat represented a reliable and profitable food source even in sever winter, the recent change in Agricultural practice has enhanced the development of a new wintering strategy of pink-footed geese, allowing a northward expansion of their winter range. Potentially, this will increase the crop damage conflict and may lead to further population growth (Therkidsen et al 2000).

Enclosure trials near Huron, CA in the San Joaquin Valley from 12 to 23 January 1999, determine the efficacy of Flight Control TM (50% anthraquinone) and Mesurol R (75% methiocarb) in preventing horned lark damage to lettuce seedlings. Flight control TM (FC) and Mesurol R were evaluated as foliar sprays at application rates of 2.79 and 2.27 kg ha-1, respectively. Horned lark damage to lettuce seedlings treated with anthraquinone was greater (p=0.015) than for methiocarb R, 60 versus 20% , respectively, and seedlings in control plots were 100% destroyed. While this level of damage is probably unacceptable to lettuce growers, it should be remembered that the enclosure situation caused an artificially high bird pressure on the crop. Further studies in open fields under a more normal bird pressure are warranted (York et al 2000).

29. Ozone Injury

Incidence and severity of visible foliar ozone injury on cutleaf coneflower (Rudbeckia laciniata L.) and crown-bread (Verbesina Occidentalis Walt). Were determined. It is thus a matter of consideration that zone injury may harm vegetation harmed by browsing or even cutting. Ozone injury was greatest on the lower leaves for both species sampled with over 95% of the injured leaves occurring on the lower 50% of the plant. This is the first report of foliar ozone injury on these plant species in situ, in the Park, illustrating the great variability in symptom expression with time, and within and between populations (Chappelka et al 2003).

30. Protected Arrears and Humans

Knowledge of conflicts between people and protected areas is required for the design of sustainable conservation strategies for the management of most protected areas. The study identifies the causes of conflicts between local people and the Benous Wildlife Conservation Area (BWCA), which includes the Benous National Park, In northern Cameroon. Informal interviews and questionnaires were administered to 114 households in three communities, and to 17 park staff and 7 professional hunting guides from July -October 1997. Crop damage affected 86% of the surveyed household, with 31% of crop income lost on average, and with the damage varying significantly between communities. Elephants, baboons, patas monkeys. Warthogs and green parrots accounted for 97% of crop damage, with the staple food maize and millet being most affected. Of the respondents 27% experienced livestock depredation, with 18% of livestock income lost on average.

The civet cat was the main predator. The involvement of local people in illegal activities, their lack of access to natural resources, and damage by wildlife were identified as principal causes of conflicts. Local people, park staff and professional hunting guides had diverse and differing perceptions about the causes of the conflicts and made various suggestions for reduction of wildlife damage including animal scaring and controlled shooting. We conclude that, under current wildlife policy, conflict between people and BWCA (Bonous Wildlife Conservation area) is difficult to resolve. To reduce conflicts and promote sustainable conservation, we suggest co-management of wildlife involving all stakeholders, establishment of crop damage control teams, and promotion of tangible benefits to local people. There may be a requirement for site-specificity in management strategies (Weladji et al 2003).

31. Low Technology Use to Avoid Damages

It is suggested that an integrated, community-based, low technological approach will be the most sustainable solution to this conflict (Osborn et al 2003). Blackbirds (lcteridae) annually damage US$5-8 million of ripening sunflower in the northern Great Plains. Baiting blackbirds with avicide-treated rice during spring migration might reduce the regional breeding populations. birds can be successfully baited with avicide-treated rice placed in corn stubble (Linz et al 2003). Plant debris accumulation is viewed as a key factor determining small mammal abundance and potential damage in low-till agricultural ((Stermer et al 2003) areas.

The projected total value of crop yield losses due to wildlife damage for buffer zone villages located in Garhwal Himalaya in about Rs.5 38 620 (US$15 389). Besides food grain, horticultural crops i.e apple, also suffered. Major wildlife agents responsible for crop damage were wild board, bear, porcupine, monkey, musk deer and partridge (Chokor). Monkey and wild boar alone accounted for about 50% to 60% of total crop damage in the study village. Goat and sheep are the major livestock killed by leopard. The total value of livestock losses at prevailing market rates is about Rs. 10 24 520 (US$ 29 272) in the study village. Due to existing conservation policies and laxity in implementation of preventive measures, the problems for local inhabitants are increasing. Potential solutions discussed emphasize the need to undertake suitable and appropriate protective measures to minimize the crop losses. Change in cropping and crop composition, particularly cultivation of medicinal plants (high value low volume crops), were also suggested. Besides, fair and quick disbursement of compensation for crop loss and livestock killing need to be adopted. Local people of the buffer zone area already have a negative attitude towards park/reserve establishment due to socio-political changes inducing major economic losses and this attitude may lead to clashes and confrontations if proper ameliorative measures are not taken immediately (Rao et al 2002).

32. Rodent Control by Capsaicin

Capsaicin, present in most hot sauces and salsas, are responsible for the "hot" sensation in many spicy foods. At high doses, capsaicin cause significant discomfort upon contact with the sensitive tissues of the mouth and throat of mammals. By applying a capsaicin containing paste to the outside of maple tree sap collection, tubing, operators hope to minimize rodent (primarily red squirrel, Tamiasciurus hudsonicus) gnawing damage to the tubing. However, some operators and sap processors have expressed concern regarding the potential migration of capsaicin through the tubing and into the tree sap, leading to contaminated maple syrup. To address these concerns, we filled a variety of new and used sap collection tubing with maple sap. Plugged the ends, and coated the tubing with a commercially available capsaicin-based rodent repellent paste. Following storage, the contents of the tubes were carefully removed and subjected to a solid-phase extraction clean-up process. Capsaicins in the sap were then quantified by high performance liquid chromatography/fluorescence detection. Results indicate that polyethylene tubing was more resistant to capsaicins migration than was polyvinyl tubing. While capsaicins were detected in the sap, the perfected levels in syrup would be below the human taste threshold (Johnston et al 2002). Wildlife management in Australia with crop losses by rabbits and foxes is discussed (Kidd 2002).

33. Rice-Damage in Portugal

Rice fields are preferred feeding areas of mallard, Anas platyrhynchos, in central Portugal. Rice consumption by mallards prior to crop harvest causes conflicts between farmers and waterfowl managers. They analysed the food habits of mallrads collected from rice fields and quantified the impact that mallards had on rice production. Oesophagi, including proventriculi (n=52)< were obtained from hunter-shot mallards, from 15 August to late September of 1993, 1994 and 1995, on a 600-ha farm included in the main feeding area of the lower Mondego population. Point counts were performed to estimate the numbers of mallards using the area. The diet included rice seeds (51.2% dry matter), weed seeds of rice fields (23.6%, dry matter), and animal matter (25,1% dry matter). By means of the diet composition, we calculated that each mallard needed to ingest 93.4 g (dry matter) of food daily including 47.8 g of rice, to meet its energy requirements. The estimated mallard population using the area was 1600 ducks. They consumed approximately 0.2% of the 4000 kg/ha minimum rice production. Therefore, the decrease in rice production caused by mallard was insignificant, which suggests that the existence of mallards on rice fields is compatible with rice production (Rodrigues et al 2002).

An effective, economic , and environmentally safe bird repellent is needed to reduce blackbird (Icterinae; such as Agelaius phoeniceus, Quiscalus quiscula, Q. Major, Q. mexicanus and Molothrus ater) depredations to newly planted rice. We evaluated Flight Control TM, a 50% anthraquinone produce, as a seed treatment for newly planted rice in Louisiana, USA, during March 1998. We treated rice seed with Flight Control at a 2% (g/g) concentration (1% active anthraquinone) the day of planting. This concentration reduced the number of blackbirds (P=0.0003) using treated fields and blackbird damage to rice seed (P=0.0124). The chemical concentration of anthraquinone on rice seed averaged 0.79% (SE=0.06%) at planting, 0.39% (SE=0.04%) at day 1,0.34% (SE=0.05%) at day 3, and 0.41% (SE=0.06%) at day 5 post-planting. Rice seedling counts were similar between treated and untreated enclosures, suggesting that Flight Control had no phytotoxic effect to rice seed. Our results showed Flight Control to be an effective blackbird repellent that warrants further developments as a management tool to reduce blackbird damage to newly planted rice and other agricultural commodities (Cummings et al 2002).

34. South American Conflicts

The coypus (also called nutria) has been introduced from its origins in South America to every continent except Australia and Antarctica. While perceived in some regions as a valuable resource, in most regions the animals are considered a pest species. Copypus have caused damage to water control structures, crops, and marsh systems and are considered a disease host. Eradication efforts have met with varying degrees of success. For those efforts to be successful, the coypus populations must be isolated to prevent reintroduction, harsh winters are necessary to reduce their populations, and intensive trapping efforts mist be sustained until the last coypu has been removed (Carter et al 2002).

35. Virginia Reports

White-tailed deer (Odocoileus virginianus) populations in Virginia and other eastern states have grown noticeably over the last decade, elevating concern about negative human-deer interactions. To evaluate these concerns among a subset of Virginia (USA) stakeholders, we sent a questionnaire to 1506 Agricultural producers and homeowners to determine their perceptions about deer damage and management during 1995 (adjusted response rate=52%). Most respondents (58%) experienced deer damage to their crops or ornamental plants in 1995, but producers experienced more damage than homeowners. An individuals experience with damage during 1995 strongly influenced his opinion about deer and their management. Respondents experiencing severe damage were more likely to consider deer a nuisance and support dramatic reductions in Virginia deer herd. Deer density influenced agricultural producers' perceptions regarding deer damage and their desire for future population management. Our observations may serve as a foundation for future efforts to quantify cultural carrying capacity (CCC) and incorporate it into management (West et al 2002).

36. Forest Management Practices

Initially, elephant caused considerable damage to man and crops. With the introduction of Joint Forest Management practices, development programmes including Project Elephant scheme are implemented mostly through people's participation in Koundinya Wildlife Sanctuary. Man-elephant conflict mitigation, viz., ex-gratia and compensation for losses, preventive measures initiated by the management of protected Areas, raising fodder plots, improvement of water facilities, construction of pulsative power fence, communication network, awareness campaigns, and change in crop pattern, and their impact are discussed (Prasad et al 2002).

An evaluation of 49 existing power fences in Northern Begal that have been established from 1982 to 1997 to keep elephants (Elphas maximus) out from the plantation, human habitations in various ranges and tea gardens and crop fields, was carried out. Fences varied from single strand to 6 strands but majority were in the category of 2-3 strands. There was a general flaw of species-species-specific design for elephants in terms of energizing the surroundings of the fence posts, which exposed them for their vulnerability to uprooting or pushing over. Corroding GI wires and improper insulation materials were the reasons for power loss varying from 0.5-7.3 KV between source to the end . In 18% of the cases (n=9) poor battery maintenance was the reason for low power output to shock the elephants. The energizer output at negative and positive terminals varied from 1.6-9.6 KV and 1.2-8.5 kV, respectively. The leaking at earthling points varied from 0-1.9kV, which is quite substantial. . The technical and maintenance problems were invariably acute in nearly all fences yet 26% were effective in output terms. The authors recommend initiation of stakeholder participation of installation and maintenance of fences, as large number of poorly maintained fences could be sources of learning for elephants to breach the fence line repeatedly. Important recommendations for design, maintenance and periodic training need to frontline staff and stakeholders are suggested (Chuhan et al 2002).

37. Rodent Damages

The mountain beaver (Aplodontia rufa) is a rodent species endemic to the Pacific Northwest and portions of California. The herbivore inflicts millions of dollars of damage annually to forest seedling plantations. Currently, extensive trapping prior to planting is the most reliable method for reducing damage. With increasing restrictions placed on trapping, forest resource managers need alternative tools to minimize forest damage. This study assessed the potential of four toxicants registered for underground use to control mountain beaver; zinc phosphide, diphacinone, chlorophacinone, and strychinine. Zinc phosphide and strychnine are acutetoxicants, whereas diphacinone and chlorophacinone are anticoagulants. Anticoagulants prevent the recycling of vitamin K in the body, which inhibits the production of clotting factors. Efficacy varied among treatments. Zinc phosphide and strychnine were avoided by mountain beaver, Pre-baiting marginally increased acceptance of strychnine, but did not alter mountain beaver acceptance of zinc phosphide. Diphacinone and chlorophacinone were both readily consumed, but only chlorophacinone was 100% effective after a 14 -day baiting regime.

Subsequently, we tested the effects of diet on the efficacy of diphacinone by varying the availability of food containing citamin K, the anticoagulant antidote. Restricting access to potential sources of Vitamin K appeared to increase efficacy. They conclude that anticoagulants hold some promise as additional tools for managers to reduce mountain beaver populations with chlorophacinone showing the most promise. However, limitations to anticoagulant baits include the necessity of long -term bating (greater than 10 days), a possible decrease in toxicity if baits contact moisture, and potential primary hazards (Arjo et al, 2004}.

38. Proximity to Human Habitation Hazards

In areas around Lake Mburo National Park, large wild animals wander in close proximity to human settlements. This poses serious conflict in terms of crop damage. The integration of conservation with other land uses is difficult where densely settled agricultural land surrounds a protected area potentially containing problem animals, as is the case for several parks in Africa and Asia. The intensity of crop raiding was quantified through the use of random crop quadrants/Plots and area estimation techniques in a portion of raided fields. The animal species concerned were documented from observations, footprints and any other marks left behind. Three variables were tested as predictors of damage; human population density, distance from the park boundary and season. In this study, data is presented regarding crop loss in the different seasons of the year, analysis of crop damage variation and animal species involved in crop loss. A diverse assemblage of animals foraged on subsistence crops and analysis of crop damage revealed significant crop depreciation by wild life (Kagoro et al 2004).

The Queensland sugar industry has recently implemented a comprehensive integrated pest management (IPM) system to minimize crop losses from two native rodent species. Rattus sordidus (Canefield rat) and Melomys burtoni (Climbing rat). These species inflicted approximately $25 M of damage in a mjor outbreak in the 1999-2000 seasons. Both of these rodents are listed as common wildlife under the schedules of the Queensland Nature Conservation (Wildlife) Regulation 1994. The IPM programme is based on understanding the ecology and biology of each species. It incorporates a large-scale monitoring programme aimed at providing early warning of imminent rodent builup to avert major outbreaks ((*Hunt et al 2004).*

39. Control Measures

Recent advances in weed control technology have changed how vegetational diversity can be managed in agroecosystems. This 2 year study (2001 and 2002), in Delaware, USA) assessed the impact of altering the timing of herbicide application in herbicide (glyphosate)-resistant field maize (cv.Dekalb 626RR) on the European maize borer, O.nubilalis, and on the beneficial insects that prey on O. nubilalis egg masses, Treatments in this study were intended to allow weed presence in the field for varying periods of time, but the control of the weeds before crop yield was negatively affected. Effects on O.nubilalis oviposition were assessed directly by counting egg masses and indirectly, through assessment of the damage done to corn stalks resulting from O. nubilalis or the beneficial insects that prey on its egg masses. End -of-season stalk comparisons showed no differences in O. nubilalis infestation levels among the treatments. Predation on sentinel egg masses showed few significant differences among treatments, and predator densities were only rarely significantly different by treatment and showed no evident trends. This data suggest that under the conditions of this experiment, manipulating herbicide applications to minimize O.nubilalis damage is not a viable management technique in field maize (Wilson et al., 2004).

40. Impact of Control Measures

Data from an attitudinal questionnaire survey of 202 households in four villages are used to examine local perceptions of wildlife crop-damage in terms of relative impact and which wildlife species were responsible . We explore the influence of wildlife crop-damage on attitudes to the adjacent game reserve. Data on the frequency of crop-damage events and estimated severity impact, recorded during a 6 month programme of crop -damage monitoring in one of the survey villages, are used to described on-farm patterns of crop-damage Comparison of the two data sets indicates a disjunction between the nature of the wildlife crop -damage, conflict as perceived by local villagers, and as it actually occurs in the study area *(Gillingham et al 2003*).

In the northern Great Plains of USA, Conflicts between red-winged blackbirds (Agelaius phoeniceus) and sunflower (Helianthus annuus) growers have intensified since the late 1960s due to the expanded commercial production of sunflowers. The potential population effects of the removal of up to 2 million red-winged blackbirds annually under a 5-year programme of baiting during spring with DRC-1339(3-choloro-4-methalalanine)-treated rice. It was examined whether lethal control, in combination with current levels of breeding-habitat management, would be cost-effective in decreasing depredation of sunflower crops during late summer. They evaluated the cost -benefit ratio for 4 culling scenarios involving (1) variable annual culls, not exceeding 2 million birds, with and without density compensation (i.e., a positive density-dependent response) on adult survival ; and (2) culls of 2 million birds annually with and without density compensation. They constructed a red winged blackbird population model represented as an age-based matrix and calibrated to stable growth. A total population of 27 million birds on 1 April (week 1), representing the red-winged blackbird breeding population staging in eastern South Dakota and migrating into North Dakota. Under each culling scenario, they reduced the stable red-winged blackbird population (equally for female and males) and projected the population through week 23 of the annual cycle. They evaluated the associated costs of the management relative to potential sunflower crop losses, assuming $0.07 in damage per bird and 4% loss to other factors. Variable annual culls, likely the more biologically realistic model scenarios, yielded mean annual removals of 1240560 (SE=12328) birds with density compensation and 1231620(SE=28811) birds without density compensation, with cost benefit ratios of 1:2.3 and 1:3.6, respectively. Annual intrinsic rates for the model population over the 5-year period ranged from 1.4 to 4.8%. Considering potential variability in the effectiveness of the cull and the combination of direct and indirect costs, we contend that the realized benefits to sunflower growers by lethal control of red-winged blackbirds via spring baiting. In combination with current nonlethal management efforts, would likely be negligible (Blackwel et al 2003).

41. Community Dependence on Wild Life

A study was conducted in December 1999 in four villages in Assam, India to: identify the nature of dependence of local communities on the forest resource; evaluate the pressures on the forest, especially due to wood collection; evaluate the socio-economic status of the villages; and recommend solutions based on community participation. The villagers were 100% dependent on the forest for fuelwood and grazing, but there was only a 2% dependence for non-timber forest products. All four forest villages faced problems of crop damage by wild animals, particularly by elephants {Elephas maximus). The field visits and interviews revealed that 73% of the woodcutters or fuelwood collectors were men while only 10 and 16% were women and children, respectively. The majority of the woodcutters belonged to the 16 to 30 years old age group and most woodcutters collect wood on a weekly-basis. It was also found that the woodcutters collect 48.5 of wet wood and 35.5% of regenerating trees, which is the major factor leading to the destruction of the forest. Some of the problems related to the community-based conservation approach include lack of communication with villagers, lack of coordination between departments, difficulties due to delays in salary payment, lack of field equipment, non-involvement of grassroots people in decision making, lack of communication between villagers and the district administration, short-term strategies, and lack of alternative livelihoods (Dilip et al 2003).

The Indian wild pig (Sus scrofa cristatus) was identified as the major crop pest in this area, accounting for 99% of the crops damaged The actual loss was estimated at 257.19 kg ha-1 accounting for approximately 7% of the actual produce. The electric fence was not effective in preventing crop raiding by the wild pig. Extent of damage might be a factor of time spent in the paddy field, suggesting that wild pigs might raid paddy fields for habitat requirements rather than for nutritional requirements. It is vital to understand crop raiding patterns prior to the implementation of control strategies (Jeyasingh et al 2003).

Bahundangi Village Development Committee (BVDC) was selected for crop damage assessment by Asian Wild Elephants (Elephas maximus). Out of 9 wards of BVDC, 8 wards were selected for household survey. Regression analyses were carried out to define crop raiding based on the distance of wards and wildlife habitat. Paired T-tests were carried out to compare the economic loss in 1999 and 2001. The number of the small and big herds were composed of 10-13 and 50-74 elephants, respectively. The big herd raided BVDC and Darjeeling district in India, whereas the small herds raided areas from Jhapa to Udayapur districts (about 200 km from East to West). Within one and half decades, elephants demolished 277 houses and killed 66 people in the 5 districts; 23 elephants were killed. The damage due to elephants vary within wards. Economic loss of US$ 54567.00 in 1999 and US$ 33669.00 in 2001 were estimated. Farmers of VDC spend US$ 175.90 to US$ 228.60 to deter elephants. The paper suggests compensatory regulations as a solution to the problems (Yadav BR, 2003). Net costs of wildlife damages are reviewed (Rollin et al 2004, Potter, 1998).

42. England Scenario

Characteristics of damage to agricultural land caused by a small feral population of wild board Sus scrofa in southern England were studied. Fifty-eight reports of rooting damage to grassland and one of rooting in a cereal crop were recorded between December 1997 and November 2002. Most reports (74%) concerned damage in January, February or March. Damage mainly occurred in fields adjacent to woodland, but there was no evidence of selection between different grassland types. The severity of damage was assessed and potential options for reducing damage in the future are suggested (Wilson C.J, 2004).

43. Sterilization Control

A graphical and quantitative frame work is described linking production yield, damage by vertebrate pests, their abundance and density-dependent responses to imposed sterilization. Compensatory responses of yield to pest damage and pest populations to fertility control are described as part of the modeling. The focal relationship is that between yield and the proportion of a pest population permanently sterilized, which is shown to be generally positive though the form of the relationship varies with differing assumptions. Compensatory responses to pest damage of production systems, such as crops livestock, trees or fish, generate non-linear responses between yield and pest damage, and yield and pest abundance. Compensatory responses by a pest population generate linear or curved relationships between abundance and the proportion of females sterilized. The model is illustrated using data from empirical studies, especially of European rabbits in Australia and New Zealand. It is recommended that the framework be evaluated on wild population of vertebrate pests (Hone J 2004).

44. Animal VS Wild Life Damages

The cost of living with wildlife are assessed using Namibian subsistence farmer's willingness to pay (WTP) for deterrents to attack on crops and livestock as a measure of damage costs. The household survey conducted for this assessment covered 165 households from different agro-ecological zones and villages. A utility -theoretic approach jointly estimates household WTP for deterrent programmes in two "currencies", maize and cash. This has a double payoff. Use of a non-cash staple increase respondent comprehension and provide more information about preferences, improving the accuracy of results. The household shadow value of maize is also identified. Significant costs from living with elephants and other types of wildlife are demonstrated. Compensation for farmers may be warranted on equity and efficiency grounds.

Uncontrolled domestic cattle generate even high costs to farmers than wildlife, highlighting the need to clarify property rights among these farmers (Sutton et al 2004).

45. Chemical Control of Rats

Replicated fields trials were conducted to compare the efficacy of zinc phosphide, strychnine and chlorpyrifos for the control of house mice (Mus domesticus) infesting recently sown wheat in South Australia. Bait was prepared using whole-wheat grain or grain-based pellets and broadcast into the crops at 1 kg/ha. Treatment with zinc phosphide reduced mouse number by 98%. Two treatments with strychnine baits, applied 11 days, apart, also reduced mouse number by 98% with no evidence of bait aversion in mice that survived the initial treatment. On the basis of these and other published results, zinc phosphide is considered an effective alternative to strychnine for control of house mice in cereal crops. Chlorpyrifos baits reduced mouse number by less than 10%. The trial began too late in the growing season to prevent substantial mouse damage to seed (Mutze et al 2004).

46. Wildlife Migration

This study examines the role of cultivated food plants and their seasonality in the diet of wild boars, determine the extent and damage of crop raiding by wild boars, lists the species of crops, vegetables, fruits and flowers affected by the wild boards, estimates the economic losses and threats to the livelihood of farmers living in a around Kumbhalgarh Wildlife Sanctuary , Rajasthan, India and looks at the various strategies employed by the people in the conservation and management of wild animals (Chhangani et al 2004).

This study develops a model for wildlife migrating seasonally between a conservation area and a neighbouring area. When the wildlife is outside the conservation area, harvesting take place by a group of small-scale farmers. The local people have two motives for harvesting to get rid of "problem" animals as roaming wildlife destroys crops and agricultural products; and to hunt for meat and trophies. Depending on the specification of the property rights, the harvesting is legal or illegal. It is demonstrated that it is far from clear which of the two property rights regimes give the highest wildlife abundance. Hence, contrary to what is argued for in the literature, handing the property rights over to the local people does not automatically mean more wildlife and more "sustainable" resource utilization. The reason lies in the nuisance motive for harvesting. The exploitation under the two different property rights regimes are illustrated by numerical calculations that represent the exploitation of the wildebeest population in the Serengeti-Mara ecosystem (on the border of Tanzania and Kenya) (Johannesen et al 2004).

47. Conservation and Conflicts

Conflict with human over livestock and crops seriously undermines the conservation prospects of India's large and potentially dangerous mammals such as the tiger (Panthera tigris) and elephant (Elephas maximus).This study, carried out in Bhadra Tiger Reserve in South India, estimates the extent of material and monetary loss incurred by resident villagers between 1996 and 1999 in conflicts with large felines and elephants, describes the spatiotemporal patterns of animal damage, and evaluates the success of compensation schemes that have formed the mainstay of loss-alleviation measures. Annually each household lost an estimated 12% (0.9 head) of their total holding to large felines, and approximately 11% of their annual grain production (0.82 tonnes per family) to elephants. Compensations awarded offset only 5% of the livestock loss and 14% of crop losses and were accompanied by protracted delays in the processing of claims. Although the compensation scheme has largely failed to achieve its objective of alleviating loss, its implementation requires urgent improvement if reprisal against large wild mammals is to be minimized. Furthermore, innovative scheme of livestock and crop insurance need to be tested as alternatives to compensations *(Madhusudan 2003)*.

48. Fire Wood and Conflicts

The study examined how multiple land use and activities such as firewood collection, livestock grazing, collection of non-timber forest products, fishing and hunting vary among villages and the losses households have incurred from wildlife raiding crops and preying on livestock within the Bhadra Wildlife Sanctuary. A household level social survey was conducted during 2002 in all 16 villages in Bhadra, and results are described from 6 villages. Regardless of socioeconomic status and their ability to afford alternatives, all households collected firewood. Livestock were grazed primarily in the forest. Three commercial non-timber forest products were collected and sold to traders or directly; seegakai (Acacia concinna), soap nut (Sapindus emarginatus) and wate huli (Artocarpus locoocha). The collection of non-commercial non-timber forest products appeared to be driven by their distribution and ease of extraction. Fishing was an important source of protein in most households. Although illegal in Bhadra, hunting did occur, but boar was extremely high in all villages. Fuelwood, fodder, fish and non-timber forest products are major natural subsidies for household in villages in Bhadra. Their current availability, unrestricted access, ease of extraction and market price make them appealing options to villages and therefore vulnerable to over–exploitation. While benefiting socio economically from access to forest resources, these villages suffer from serious human-wildlife conflicts unlike those in villages further from the protected area (Karanth K.D, 2003).

49. Weed Control Technology

Recent advances in weed control technology have changed how vegetational diversity can be managed in agroecosystems. Ths 2 year study (2001 and 2002, in Dealware, USA) assessed the impact of altering the timing of herbicide application in herbicide (glyphosate)- resistant field maize (cv. Dekalb 626RR) on the Eureopean maize borer, O. nubilalis, and on the benefical insects that prey on O.nubilalis egg masses,Treatments in this study were intended to allow weed presence in the field for varying periods of time, but the control of the weeds before crop yield was negatively affected. Effects on O.nubilalis oviposition were assessed directly by counting egg masses and indirectly, through assessment of the damage done to corn stalks resulting from O. Nubilalis infestations. The impact of weeds on the main generalist predators in this system was quantified through direct counts and predation trials on sentinel egg masses. End of season stalk comparisons showed no differences in O. nubilalis infestation levels among the treatments. Predation on sentinel egg masses showed fe significant differences among treatments, and predator densities were only rarely significantly different by treatment and showed no evident trends. This data suggest that under the conditions of this experiment, manipulating herbicide applications to minimize O. nubilalis damage is not a viable management technique in field maize (Wilson et al 2004).

50. California Reports

The mountain beaver (Aplodontia rufa) is a fossorial rodent species endemic to the Pacific Northwest and portions of California. This herbivoure inflicts millions of dollars of damage annually to forest seedling plantations. Currently, extensive trapping prior to planting is the most reliable method for reducing damage. With increasing restrictions placed on trapping, forest resource managers need alternative tools to minimize forest damage. This study assessed the potential of four toxicants registered for underground use to control mountain beaver, Zinc phosphide, diphacinone, chlorophacinone, and strychnine. Zinc phosphide and strychnine are acute toxicants, whereas diphacinone and chlorophacinone are anticoagulants. Anticoagulants prevent the recycling of vitamin K in the body, which inhibits the production of clotting factors. Efficacy varied among treatments. Zinc phosphide and strychinine were avoided by mountain beaver. Pre-baiting marginally increased acceptance of strychnine, but did not alter mountain beaver acceptance of zinc phosphide. Diphacinone and chlorophacinone were both readily consumed, but only chlorophacinone was 100% effective after a 14-day baiting regime. Subsequently, they tested the effects of diet on the efficacy of diphacinone by varying the availability of food containing vitamin K, the anticoagulant antidote. Restricting access to potential source of vitamin K appeared to increase efficacy. They conclude that anticoagulants hold some promise as additional tools for managers to reduce mountain beaver populations with chlorophacinone showing the most promise. However, limitations to anticoagulant baits include the necessity of long-term baiting (greater than 10 days), a possible decrease in toxicity if baits contact moisture, and potential primary hazards (Arjo et al 2004).

51. Tanzanian Study

This paper presented an analysis of perceived patterns of wildlife crop-damage in relation to an on-farm assessment of damage in an area bordering the Selous Game Reserve (SGR) in south-eastern Tanzaina. Data from an attitudinal questionnaire survey of 202 households in four villages are used to examine local perceptions of wildlife crop-damage in terms of relative impact and which wildlife species were responsible. They explore the influence of wildlife crop-damage on attitudes to the adjacent game reserve. Data on the frequency of crop-damage events and estimated severity of impacts, recorded during a 6 month programme of crop -damage monitoring in one of the survey villages, are used to describe on-farm patterns of crop-damage. Comparison of the two data sets indicates a disjunction between the natures of the wildlife crop-damage conflict as perceived by local villagers, and as it actually occurs in the study area. This disjunction is discussed in relation to the effect of extreme damage events on local people's views, the opportunity costs involved in guarding farm plots against crop-damage, and the tenure arrangements for wildlife that define the relationship with the state wildlife management authority (Gillingham et al 2003).

52.Asian Wild Elephants

Bahundangi village Development Committee (BVDC) was selected for crop damage assessment by Asian Wild Elephants (Elephas maximus). Out of 9 wards of BVDC, 8 wards were selected for household survey. Regression analysis were carried out to define crop raiding based on the distance of wards and wildlife habitat. Paired T-tests were carried out to compare the economic loss in 1999 and 2001. The number of the small and big herds were composed of 10-13 and 50-74 elephants, respectively. The big herd raided BVDC and Darjeeling district in India, whereas the small herds raided areas from Jhapa to Udayapur districts (about 200 km from East to West). Within one and half decades, elephants demolished 277 houses and killed 66 people in the districts; 23 elephants were killed. The damage due to elephants vary within wards.

Economic loss of US$ 54 567.00 in 1999 and US$ 33 669.00 in 2001 were estimated. Farmers of VDC spend US$ 175.90 to US$ 228.60 to deter elephants. The paper suggests compensatory regulations as a solution to the problems ((Yadav 2003).

53. Reserve Reports

Crop raiding patterns by wildlife at 10 villages along the eastern boundary of the Kalakad-Mundanthurai Tiger Reserve (KMTR) in Kerala, Southern India were studied from December 1998 through March 1999. A total of 121 farmers were interviewed, and 39 instances of fresh crop raids were observed. Of these , 35 were on paddy fields, 3 on banana plantations and one on sugarcane. The Indian wild pig (Sus scrofa cristatus) was identified as the crop loss caused by this species was quantified. The effectiveness of an electric fence to prevent crop raids by wild pig was also assessed, comparing (1) mean group size of pigs, (2) extent of damage in m2, (3) actual and potential loss, and (4) frequency of wild pig raids. The actual loss was estimated at 257.19 kg ha-1 accounting for approximately 7% of the actual produce. The electric fence was not effective in preventing crop raiding by the wild pig. The number of wild pigs was not correlated with the extent of damage. Extent of damage might be a factor of time spent in the paddy field. Suggesting that wild pigs might raid paddy field for habitat requirements rather than for nutritional requirements. It is vital to understand crop raiding patterns prior to the implementation of control strategies (Jeyasingh et al 2003).

A study was conducted in December 1999 in four villages in Assam, India to: identify that nature of dependence of local communities on the forest resource, evaluate the pressures on the forest, especially due to wood collection; evaluate the socio-economic status of the villages and recommend solutions based on community participation. The villagers were 100% dependent on the forest for fuelwood and grazing, but there was only a 2% dependence for non-timber forest products. All four forest villages faced problems of crop damage by wild animals, particularly by elephants (Elephas maximus). The field visits and interviews revealed that 73% of the woodcutters or fuelwood collectors were men while only 10 and 16% were women and children, respectively. The majority of the woodcutters belonged to the 16-to 30 year old age group and most woodcutters collect wood a weekly basis. It was also found that the woodcutters collect 48.5 of wet wood and 35.5% of regenerating trees, which is the major factor leading to the destruction of the forest. Some of the problems related to the community based conservation approach include lack of communication with villagers, lack of coordination between departments, difficulties due to delays in salary payment, lack of field equipment, non-involvement of grassroots people in decision making, lack of communication between villagers and the district administration, short -term strategies, and lack of alternative livelihoods (Dilip et al, 2003).

54. German Reports

Increasing quantities of specialized crops such as vegetables and fruit are being grown on a field scale in Germnay. This provides an attractive new food supply for game and other wildlife, and associated damage can be substantial. In a recent case heard by the provincial court in Verden, south-east of Bremen, the financial loss following fallow deer grazing on asparagus was assessed at some euro 72,000 over 10 years. This exceeded the annual stalking rent more than tenfold. Under existing laws, shooting tenants and associations are required to compensate farmers for game damage to field crops and to ensure protection of high-risk areas that form apart of shooting grounds. This, however, applied only to common crops such as cereals. In the case of specialized crops the grower is responsible for protecting them from wildlife and incurs the whole damage risk. If the holding does not include registered shooting grounds the grower has not shooting rights of his own. At present the legal definition of crops as "field" or "garden" is insufficiently precise. Asparagus, soft fruit, and vegetables have historically been classed as garden crops but are not longer exclusively grown as such, and clarification of their status is now needed (Haarstrich 2005).

55. India Reports

This paper analyses the economics of biodiversity conservation in the context of a tropical forest ecosystem in the Western Ghats region of India, where coffee is the main competitor for land use. Using primary data covering a cross-section of coffee growers, the study notes that the opportunity costs of biodiversity conservation in terms of coffee benefits foregone are quite high. Even after including external costs due to wildlife damages and defensive expenditure to protect against wildlife, the NPVs and IRRs from coffee for all lad holding groups were high. Including external costs, these NPVs across different land holding groups ranged between Rs. 17 thousand to over Rs.106 thousand per acre at 12% discount rate, and the IRRs between 16.6% and 23%. Even if the expected benefits were to decrease by 20% and costs rise by a similar proportion, still the IRRs from coffee were quite high (19.5-20.1%). The study notes that the external costs accounted for between 7% and 15% of the total discounted costs of coffee cultivation, and smaller holdings proportionately incurred high external costs as compared to large holdings. The study also notes high transaction costs incurred by the growers to claim compensation for wildlife damages. Notwithstanding these disincentives, the study notes that the local community had a positive attitude towards biodiversity conservation and were willing to pay in terms of spending time for participatory biodiversity conservation. Taking elephants, a keystone and threatened species in Asia and the study region, for the contingent valuation survey, the study notes that the respondents are willing to spend 25.8 human days per household annually which works to over Rs.6003 per household per annum in terms of the income foregone. They also preferred a decentralized government institution for participatory biodiversity conservation (Ninan et al, 2005).

56. Uganda Reports

Crop raiding by primates in particular and wild animals in general is a significant source of people forest conflict around the Budongo Forest Reserve, Uganda. Crop loss to wild animals undermines local support for conservation efforts in this area. Patterns of primate crop raiding were studied over a period of 14 months in six villages (five adjacent to the Budongo Forest Reserve and one that is approximately 3,500 m from the forest edge). Data were collected via a questionnaire survey. Additional information was obtained from the relevant local government offices. Chimpanzees Pan troglodytes, baboons Paio anubis, other monkey, bush pigs Potamochoeus procus and porcupines Hystrix cristats were reported by farmers to be the major causes of crop losses by wild life. Of farmers, 73% reported suffering crop damage caused by primates, and 79% considered baboons to be the most destructive of all crop raiding species. Drought, insect pests, poor sowing, plant disease and accidental fires were other sources of crop losses to farmers, though the risk of crop damage particularly by primates is perceived as the most serious potential cause of losses. Using chimpanzees as a case study, patterns of crop damage across the year are compared with seasonal fluctuations in availability of wild foods (Tweheyo et al 2005).

57. Non-Lethal Methods of Prevention

There are increasing demands, for non-lethal methods of resolving foraging conflict between people and a wild range of wildlife species. Badgers make good models for studying human-wildlife conflict resolution, and they epitomose the circumstances driving research in this field; they cause millions of pounds worth of crop damage each year in England and Wales, and yet they are protected legally from culling. In anticipation of future field conditioning trials, we set out here to identify the minimum concentration of the repellent ziram for inhabiting badger damage to maize. They believe this to be the first study to investigate the dose dependent effects of a repellent on a wild free -ranging population of mammals. They adapted a multi-choice design to test the acceptability to free ranging badgers of maize cobs treated topically with a range of ziram concentrations (0.5.3% w/w). We used video surveillance to obtain detailed behavioural observations of individually identifiable animals. The seven badgers that ate maize during the trial did not discriminate between treatments on the first night on which they fed. However, on subsequent nights, most of these badger either failed to return to experimental sites or returned without feeding. These individuals visited and fed throughout the study, and demonstrated a clear negative dose-dependent response that reached plateau at a ziram concentration of 1.3% (w/w). This effect was evident in both individual behavioural data and the proportion of cobs suffering badger damage . Our aim was to identify an optimal (threshold) concentration of ziram for deterring badger feeding on maize; we conclude that maize cobs should be treated at approximately 1.3-1.5% 9w/w), by topical application of a 10% (w/v) ziram paste. A field experiment of similar design could provide a model starting point for research with other species, repellents and food substrates (Baker et al 2005).

58. Ecosystem Services

The ecosystem services provided by mangroves are often ignored in the ongoing process of mangrove conversion. Services provided by the Bhitarkanika mangrove ecosystem in India and estimated cyclone damage avoided in three selected villages, taking the cyclone of 1999 as a reference point , were valued by assessing the socio-economic status of the villages, the cyclone damage to houses, livestock, fisheries, trees and other assets owned by the people, and the level and duration of flooding. Eleven variables were used to compare damage in the villages, one protected by mangroves, one unprotected by mangroves, and the third possessing an embankment on its seaward side. Attitude surveys were carried out in 10% of the households in 35 villages located in the Bhitarkanika Conservation Area to assess local peoples perceptions regarding the storm protection function of mangroves and their attitude towards mangrove forests generally. In the mangrove protected village, variables had either the lowest values for adverse factors (such as damage to houses), or the highest values for positive factors (such as crop yield). The loss incurred per household was greatest (US$ 153.74) in the village that was not sheltered by mangroves but had an embankment, followed by the village that was neither in the shadow of mangroves or the embankment (US$ 44.02) and the village that was protected by mangrove forests (US$ 33.31). The local people were aware of and appreciated the functions performed by the mangrove forests in protecting their lives and property from cyclones, and were willing to cooperate with the forest department in mangrove restoration (Badola et al 2005).

59. Willow Decline

Populations of braver and willow have not thrived in riparian environments that are heavily browsed by livestock or ungulates, such as elk. The interaction of beaver and elk herbivory may be an important mechanism underlying beaver and willow declines in this competitive environment. They conducted a field experiment that compared the standing crop of willow three years after simulated beaver cutting on paired plants with and without intense elk browsing (~85% utilization rate). Simulated beaver cutting with intense elk browsing produced willow that was small (biomass and diameter) and short, with far fewer, but longer, shoots and a high percentage of dead biomass. In contrast, simulated beaver, cutting without elk browsing produced willow that was large, tall, and leafy, with many more, but shorter, shoots (highly branched) and a lower percentage of dead biomass. Total stem biomass after three years was 10 times greater on unbowsed plants than on browsed plants. Unbowsed plants recovered 84% of their per-cut biomass after only two growing seasons, whereas browsed plants recovered only 6%. Thus, the interaction of beaver cutting and elk browsing strongly suppressed the standing crop of willow. They predict that a lack of willow suitable as winter food for beaver can cause beaver populations to decline, creating a feedback mechanism that reduces beaver and willow populations. Thus, intense herbivory by ungulates or livestock can distrupt beaver willow mutualisms that naturally occur in less competitive environments (Baker et al 2005).

60. Conservation Preservation

In an efforts to attenuate human-wildlife conflict and promote conservation of charismatic megafauna, compensation programs for wildlife damages have been implemented in many countries. Compensating pastoralists and farmers for damage caused by wildlife reduces bunting pressure on wild animals populations . However, it can also lead to a decrease in efforts to prevent damage and exacerbate conflicts with wildlife. Furthermore, compensation programs increase the return to agriculture and can therefore be viewed as a subsidy toward crop and livestock production. Such subsidies can trigger agricultural expansion (and habitat conversion), an inflow of agriculture producers, and intensification of agricultural production. Each of these impacts is shown to have potentially adverse effects on the wildlife population that compensation intendeds to favor. In some circumstances, the net effect on the wildlife stock could be negative. This calls for a careful assessment of local ecological and economic conditions before compensation is implemented. Incentive mechanisms that are directly tied to conservation outcomes (e.g. payments to locals based on the size of the wildlife population) should be considered instead of compensation programs(Bulte et al 2005).

This paper examines the environmental and economic consequences of introducing a programme to compensate peasants of a small economy for the (crop) damage caused by wildlife . It shows that the widely held belief that compensation induces wildlife conservation may be erroneous. In a partially open economy, compensation can lower the wildlife stock and result in a net welfare loss for local people. In an open economy, compensation can trigger wildlife extinction and also reduce welfare. The conditions leading to a reduction of the wildlife stock are identified and the implications for current and planned compensation programmes are discussed (*Rondeau et al 2003*).

61. Beaver Mountain, Utawah

In 1994, ground fire ignited in forests of Douglas-fir, Pseudotsuga menziesii var, glauca, on Beaver Mountain, Utawah, USA. The Douglas-fir beetle, Dendrocotonus, pseudotsugae, attack a range of moderately fire-injured host conifers in 1995 . Logistic regression models run for 1995 data illustrated that one year after the fire event, the Douglas-fir beetle selected and attacked large-diameter Douglas-fir with 60-80% bole char,60-80% crown volume scorch, and 50-70% probability of mortality due to fire. In 1996, beetle preference shifted to smaller-diameter trees with lighter fire injury, because most large, fire damaged conifers were colonized by beetles in 1995. Although beetle populations did not reach outbreak proportions outside the fire boundary, host selection shifted to green trees in 1997 along the burn perimeter. Log linear analysis indicated that increased brood production was conditioned by increased diameter and moderate fire damage to the trees (Cunningham, 2005).

62. European Range

Since 1980, populations of wild board (Sus scrofa) have increased over the species' entire European range. This increase had led to conflicts because wild boars cause crop damage amounting to several million U.S dollar every year. Wildlife management agencies promote and financially support 3 major methods to reduce the loss (1) intensive harvest, (2) supplemental feeding in forests to bait animals for easier shooting and to distract them from agricultural fields, and (3) building electrical fences around crops at risk. Our objective was to investigate how effective these methods were in reducing field damage by wild boars. Based on data from 44 hunting territories in the Canton Thurgan, Switzerland, we related damage frequency to harvest success, supplemental feeding, and fencing effort by means of 2 multiple regression analyses. The analysis of mean damage frequency among territories (averaged over 3 year) and changes in damage frequency within territories from 1994 to 1996 showed that only hunting reduced damage by wild boars. Because our results question the effectiveness of wild board management practices and wild board populations and damage are increasing throughout Europe, we suggest that control efforts and funds be reconsidered. Because only hunting seems to clearly reduce wild board damage, we suggest more emphasis be put on the development and introduction of new harvest models among local hunting teams (Geisser et al 2004).

63. Sustainable Wildlife Management

For the sustainable management of small mammals, fertility control may be used in the future. Little is known about what proportions of females need to be sterilized to achieve an impact on population size and what compensatory processes may act on the population level. We tested that impact of surgical sterilization of zero, 25, 50 and 75% of females on the population dynamics and demography of enclosed populations of rice field rats (Rattus argentiventer) and damage to rice crop. Sterilizing 50% of female founder (6 of 12) decreased population size at the end of the breeding season by about 50%. We used a simulation model, based on the breeding biology of the rice field rat in the field and in the control enclosures, to generate the expected dynamics of the enclosure populations. The results suggested that compensation occurred in the enclosures where 75% of female founder (9 of 12) had been sterilized. We detected a slight tendency for 50% higher numbers of recent uterine scars in fertile founder females in the 50% and 75% treatments versus 25% treatments and controls ($p=0.198$). The primary demographic mechanism for compensation was higher survival of young rats in enclosures where 75% of females were sterilized. However, compensation only partially offset the decrease in population size. We found no convulsive evidence that the reproductive output of F1 generation females was higher when large proportions of female founders were sterilized. Early in the breeding season, the per capita damage to rice plants in populations without sterilized rice field rats was increased. Our results suggest that the sterilization of.50% of females in rice field rat populations can reduce rat population growth and rat damage to rice crops ((Jacob et al 2004).

64. Farm Management Practices

The impacts of a range of farm-magement practices on house mouse (Mus domesticus) populations were tested in a large replicated field study in a complex irrigated farming system in southern New South Wales, Australia. An advisory panel, made up of farmers, extension officers, industry representatives and scientists developed a series of best practice farm management actions to minimize the impact of mice. Twelve experimental sites were split into six treated sites, where farmers were encouraged to conduct the recommended practices, and six untreated sites, where farmers were encouraged to conduct the recommended practices, and six untreated sites, where farmers conducted their normal farming practices. Mouse abundance was generally low to moderate for the 4 year project (5-60% adjusted trap success). They found significant reductions in population abundance of mice on treated sites when densities were moderate, but no differences when densities were low. Biomass of weeds and grasses around the perimeter of crops were significantly lower on treated sites because of application of herbicide sprays and grazing by sheep. We could not detect any significant difference in mouse damage to crops between treated and untreated sites; however , levels of damage were low (<5%). Yields of winter cereals and rice crops were significantly higher on treated sites by up to 40%. An analysis of benefits and costs of conducting, farming practices on treated sites compared with untreated sites howed a 2:1 benefit to cost ration for winter cereals, 9:1 for rice and 4:1 for soybeans (Brown et al 2004).

65. Physical and Biotic Effects

Physical and biotic effects of straw mulch applied at two different times to potatoes and watermelons were assessed in field trials over 2 yr. Plots with straw mulch generally and lower soil temperatures and higher soil moisture than control (weedy, no straw) plots. When straw was applied at planting weeds were suppressed, whereas straw applied after cultivation 4 wk after planting had less effect on weeds. In 2000, potatoes with straw at planting had fewer potato leafhoppers, Empoasca fabae (Harris) (and less associated plant damage) and more colonizing Colorado potato beetle, Leptinotarsa decemlineata (Say), adult than the other treatments. Subsequent Colorado potato beetle egg mass and larval numbers, however, were not higher in this treatment. Subsequent Colorado potato beetle egg mass and larval members, however, were not higher in this treatment, possibly because of the higher numbers of predators in these plots as assessed by pitfall beetle adults, but by midsummer there were more Colorado potato beetle larvae in the control plots than in the straw plots, again possibly due to differential predation. The few pests observed in the watermelon plots were not affected by the straw at planting (in which weeds were suppressed) in 2000. In 2001, when melons were planted earlier in the spring, melon plant biomass at 4 wk was lower in plots with straw at planting than in the other treatments, probably due to reduced soil temperatures. Thus, for potatoes, straw may be useful to control weeds and enhance predator numbers, but insect pests may still require control. In melons, straw should be used only if the crop is planted after the soil is sufficiently warm (Jhonson et al 2004).

66. Damages in England

Characteristics of damage of agricultural land caused by a small feral population of wild board Sus scrofa in Sourthern England were studied. Fifty-eight reports of rooting damage to grassland and one of rooting in a cereal crop were recorded between December 1997 and November 2002. Most reports (74%) concerned damage in January , February or March. Damage mainly occurred in fields adjacent to woodland, but there was no evidence of selection between different grassland types. The severity of damage was assessed and potential options for reducing damage in the future are suggested ((Wilson 2004).

This study models net welfare impact on producers who receive utility from on-farm wildlife populations that are not costlessly diposable. Wildlife damage levels where net benefits are zero indicate producers' maximum willingness to pay for on-farm wildlife. An empirical model is developed and applied to 1998 data for 241 field crop producers in Ontario, Canda., Results suggest that the net welfare loss from damage is approximately half of the value of the yield loss for those with damage.In aggregate, however, on farm wildlife generates net benefits to producers that outweigh costs by about 10:1. The distribution of net benefits is highly skewed across producers (Rollins et al 2004).

67. Field experiences

Field experiments were used to test whether intercropping sunflower (Helianthus annus) in organic vegetables would (i) attract insect -eating birds and encourage them to (2) forage in greater numbers and (3) for more time in cropped fields. Cropped areas with sunflower treatments of one or two rows per 0.4 ha exhibited significantly greater mean abundance of insectivorous birds than did control plots, across a variety of crop types. Additionally, both mean numbers of individual birds foraging on insect prey and mean insect foraging time per hour in crops were significantly greater in plots with sunflower rows than without. Birds actively pursuing prey in study plots consumed economically important pest species and did not damage crops during the study. The addition of sunflower intercrops proved to be an effective habitat modification for augmenting avian insectivore numbers and insect-foraging time in organic vegetables (Jones et al 2006).

68. Bhuton Reports

The attitudes and feelings of concerning conservation policies and wildlife conflict affect their behaviour, and understanding this is important in involving local people in conservation planning and decision making processes. This paper examines these important issues in Jigme Singye Wangehuck National Park, located in central Bhutan. A survey of 274 household was carried out to asses farmers' perceptions of the influence of Park management policies and protection regimes on traditional resource uses, and their attitudes towards the Park and conservation policies set forth in the Forest and Nature Conservation Act of 1995, integrated conservation development programmes (ICDPs) and wildlife conservation, and determine how demographic and socioeconomic variables influence these attitudes. Among local farmers, 52.2% disliked the Park and the Conservation Act, and 67.5% supported exterminating problem wildlife. Negative attitudes were linked to loss of resource use right, livestock depredation and crop damage, lack of compensation strategies and exclusion of farmers from the Park;s planning processes. However, 76.3% of the respondents appreciated the Park's development programmes, the positive attitudes associated with an expectation that significant economic benefits would be available from ICDPs sponsored by the Park. Empowerment of local communities associated with monetary benefits from non-timber forest products and compensation for loss of crops and livestock were emphasized by more than 70% of the respondents. These attitudes were related to age and literacy of the respondents, number of livestock owned and size of land holdings. Though important to the future of Bhutan's parks, study results also have wider applicability to conservation professional throughout the developing world for resolving human land use conflicts and involving local farmers in the protection of nature (Wang et al 2006).

In the protected areas of Bhutan, crop damage by wildlife impacts the livelihoods of local agropastoralists. We interviewed a stratified -random sample of 274 farmers living in the Jigme Singye Wangchuck National Park in an attempt to better understand and manage growing concerns. Most of the farmers interviewed suffered major financial losses annually due to crop damage (rice, wheat and maize) by wild pigs (Sus scrofa), barking deer (Muntiacus muntjak), macaques (Macaca mulatta), and sambars (Cervus univolor). All respondents reported crop losses to wild animals, and wild pig was the most common cause (97%). Farmers responded by implementing nonlethal methods such as guarding, fencing and performing religious rituals to protect their crops. Growers blamed the park's conservation policies for the high level of reported losses. Crop damage, especially by macaques, increased after establishment of the park in 1993 and implementation of the Forest and Nature Conservation Act of 1995. Management actions should focus on increasing tolerance among farmers and reducing severe losses caused by wild pigs and macaques (Wang et al 2006).

69. Mayamur Experiences

Declaration of forests as Protected Areas, implementation of Wildlife Protection Act, denial to/restriction on natural resources to the depending people coupled with degradation/fragmentation of habitat lead to conflict between interests. A questionnaire survey was conducted among the villagers of the selected settlement in Marayur Forest Range for information on matters related to socio-economic status and wildlife conservation. About 10% of the household in each settlement were selected randomly. Most of the people (77.78%) have their own lands and the major source of income is cultivation. According to the respondent's paddy, sugar cane, plantain and vegetables were the highly raided crops by wild animals, Gaur, Elephant and Wild boar were held responsible by the respondents for most of the damagers to crops. Guarding combined with ordinary fencing was the protection method employed by majority of the cultivators in Marayur. The reasons for crop damage and compensation details were discussed (Veeramani et al 2004).

70. Japanese Experiences

Direct sowing of rice is increasing in Japan as a means of reducing farm labour. Seed damage by birds, including the Oriental Turtle Dove Streptopelia orientalis, has, however, prevented the spread of direct sowing. An experiment with five replications was conducted from 1 May to 12 September 2000 in an outdoor aviary to determine whether damage by Oriental Turtle Doves could be reduced by providing barley as an alternative field. The aviary was divided with nets into four plots and assigned to four treatments of 2x2 about feed provision and experimental period. In the experiment, the alternative feed was either "provided" or "not provided",, while during the experimental period, the treatments were assigned as either "early" (doves were introduced into a plot right after the sowing and kept there until germination) or "late" (doves were introduced on the day of germination and kept there for the same number of days as in the "early' treatment). Damage was evaluated by comparing the number of seedlings with that in enclosures set inside the plots. In the "early" treatment, the number of seedlings in the alternative feed treatment did not differ from that in the enclosures, but in the non-alternative feed treatment, the number of seedlings was significantly fewer than that in the enclosures. In the "late treatment, the number of seedlings did not occur after germination. These results showed that the damage by doves ceases at germination, and can be reduced by providing an alternative feed (Yamaguchiet al 2006).

71. Bolina Reports

Damage caused by wildlife to human economic activities represents a serious problems to conservation efforts. Conflict resolution efforts are particularly complex in Bolivian protected areas, which generally harbour human populations that settled there before the establishment of the protected area. They carried out a study that estimated wildlife damage to crops using paired enclosures and open (control) fields, in a protected area in Bolivia, that was primarily montane forest. We compared the performance (proportion of attacked plants and total crop yield) of three crop species within and outside of the enclosures, Simultaneously, we recorded the wildlife species that visited crop fields by means of track plots, and identified their share in the damage. The damage was significantly higher in control fields for two out of three crop species. Total yield was higher, though not significantly, in exclosures for all three crop species. Brown agouties (Dasyprocta variegate) was the most frequent wildlife species in crop fields and was responsible for some crop damage; however, collared peccaries (Tayassu tajacu) caused complete crop loss in two cases. We discuss our results from a broad management perspective and propose that conservation efforts, especially in Bolivia and other third world countries, should consider both applying methods to reduce wildlife damage and direct or indirect economic compensation. The latter could be achieved through the commercial use of non-threatened local wildlife species (Perez et al 2006).

72. Fox Damages

Flying foxes come into conflict with fruit growers when they raid fruit crops in response to limited native food resources. A standard technique used by fruit growers to deter flying-foxes has been to shoot the animals as they enter the orchard. This activity has been regulated by the National Parks and wildlife service (NPWS) in New South Wales, Australia, since the protection of flying-foxes in 1986. A NPWS policy was developed on this issue in 1998. It provides for licences to be issued to harm a limited number of protected species by gunshot only. The policy does not cover threatened species. The recent change in the conservation status of the Grey-headed flying –fox (Pteropus poliocephalus) from protected to vulnerable means that the NPWS policy must be adapted to address damage mitigation of a threatened species. The NPWS has considered a number of alternatives in its management strategy, including ending licenses harm to flying foxes , issuing licences under the Threatened Species legislation, and encouraging farmers to prepare property management plans. The preferred NPWS position is to continue to issue licences under the National Park and Wildlife Act 1974 for a further 3 years, during which time farmers are expected to take all reasonable means to adopt non lethal deterrent strategies. Important aspects of this new policy was ensuring that licensed damage does not constitute a threat to the long term survival of the Grey headed flying fox, and improving compliance with the policy so that all harm may be quantified and the impact on the state-wide population properly assessed and managed (Walples K, 2002).

73. Madeira Island, Portugal Reports

This study was conducted over a two year period in the north west of Madeira Island (Macaronesian Atlantic Islands, Portugal). It investigates the relationship between fruiting patterns in the indigenous forest and the use by the Madeira laurel pigeon of forest habitats and cultivated areas. Habitat use observations were carried out using line-transects in forest sites where the availability of fruit (the most important element of the diet) was assessed. The use of agricultural fields was measured indirectly by the assessment of the damage caused to crops. The findings provided strong evidence that bird movements and shifts in abundance are related to fruit stage and availability. The use of fruits is mostly opportunistic but when many species are available, selection was observed. From an ecological perspective, our findings indirectly confirm a high concordance between pigeon diet, studied by microhistological techniques, and habitat use. The use of crops also seems to be mostly opportunistic and is governed by the birds' movements within the adjacent forest. It is hypothesized that fruit phenology will influence the use of agricultural areas only to the extent that it governs such movements; there is no strong evidence that crops are attacked only when the availability of natural foods is low. Our results suggest that the current management strategy for the forest is adequate for the conservation of this pigeon but that the conservation value of forest edges should be redefined. It is also proposed that crop damage may be reduced by identifying the proximate factors, which lead to the use of specific fields (Oliveira et al 2006).

74. Feral Hog (Sus Scrofa) Invasions

A field study was conducted in Texas,USA, in 2003 to determine the consequence of feral hog (Sus scrofa) invasions in several ecoregions of Texas. The observations, experiences and action of landowners and managers concerning feral hogs on their property were examined by using purposive sampling of landowners and managers who fit one or more 3 selection criteria. Landowners and managers were either sent a self administered, mail -out questionnaire or given a copy of the questionnaire. during pesticides applicator workshops. There were 775 survey participants. The effective response rate from those landowners and managers who received a mailed questionnaire was 62% (n=294). Nearly all (95%, n=491) of the pesticide applicator workshop participants turned in a completed questionnaire. Sampling error based on the farms (includes ranches) in Taxas and in each region was + or-3%, alpha=0.05. The majority (74%)of respondents were ranchers and 18% were farmers. Most respondents felt that feral hogs came from neighbor's property and were an agricultural pest. Rooting, wallowing and crop damage were the major forms of damage caused by feral hogs. The average economic loss due to hog damage, over the lifetime ownership of the land by the respondent, was $7515.Hog control was an incidental process. The average cost for hog control over the lifetime ownership of the land by the respondent was $2631. There was strong support for programmes related to feral hog management and control, but only half of the survey participants responded to the question. The average quiz score of 11.5 indicated that respondents could correctly respond to, 50% of the 26 questions. Region was found to have an effect on all questions tested except one. Management implications included the need for educational programmes about feral hogs, how landowners can make better use of feral hogs on their property, ongoing education efforts about feral hogs and the impact of this study on the public policy and decision making process (Adams et al 2005). Compared to a survey conducted in 1982, 17 of the 32 (53%) populations are extinct or have dropped to levels so low that local hunters have failed to encounter the species in recent years. This indicates a rapid population decline. They hypothesize that this loss is primarily caused by a decline in suitable habitat, especially stands of teak Tectona grands forest or similar forest plantations, and by high hunting pressure. Competition from and hybridization with the Eurasian wild boar Sus scrofa may be further threats to S.verrucosus. Rapidaction is needed to prevent extinction in the wild. We recommended effective protection of selected S.verrucosus population, lobbying of the Government to give protected status to S.verrucosus, conducting ecological research and investigating crop damage issues, and establishment of conservation breeding programmes (Semiadi et al 2006).

75. Nursery Management Practices

Integrated Pest Management is a system that combines cultural, biological and chemical technologies to reduce insect, fungal and weed populations to levels below those that result in economic damage. Nursery managers in the sourthern. United states currently use many practices to control pests of southern pine seedlings. Over the last three decades, improvements in chemical, cultural, and biological pest control practices have increased seed efficiency (defined as the number of plantable seedlings produced divided by the number of pure live seed sown) and reduced the percentage of production costs associated with pest control. As crop values increase, the economic thresholds for applying control measures decrease. However, since the statistical power of most trials in bareroot nurseries is low, the likelihood of experiments that detect "real" treatment difference (e.g those that consistently increase seed efficiency to the point, where economic returns are affected) will be low. This paper describes some current practices in southern pine nurseries and provides some economic injury levels for various pest control treatments(South et al 2006).

76. Low Forest Damages

During the past several decades, an overaboundance of sika deer has caused dramatic increase in agricultural and forestry damage, and has severely affected natural vegetation and ecosystems in many areas of Japan as a result of heavy foraging pressure. In National Parks and nature reserves, where conservation of natural environments is the highest priority, managing overabundant deer is an imminent and major concern.. Sika deer, which were once locally extirpated, re-colonized the Shiretoko Peninsula in the 1970's. The population has been increasing since the mid 1980s, resulting in chronic effects on the natural vegetation of forests and grasslands. Although severe climatic conditions and scarcity of food resources during the winter may regulate the deer population in Shiretoko National Park, the deer are able to sustain a high density due to the high female survival rate as well as a high intrinsic rate of increase. The population can be reduced only by heavy snowfalls that are severe enough to kill females.

Therefore, natural processes alone are unable to alleviate the impacts of deer on vegetation. Unfortunately, from the currently available data, we cannot determine whether the irruption of sika deer in the NP is caused by socioeconomic activities or by ecological process. It is imperative to establish a comprehensive management plant for sika deer that considers differences in land use status and conservation purposes both in Shiretoko NP and in peripheral areas. Also, vegetation monitoring is essential for evaluating the extinction risk of plant species and communities. If habitat restoration and population control of sika deer are to be implemented in and around the NP, these measures must be closely linked with deer and vegetation monitoring results to adaptively manage sika deer and ecosystems in Shiretoko NP (Tokida et al 2004).

An account is given of the emerging vegetation-raiding status of golden langur (Trachypithecus geei), one of the endangered and restricted range species in fragmented forest pockets in western Assam, India. It was found that the golden langur causes maximum damage to Gmelina arborea trees planted by cillagers, as this crop is one of its natural food item. Although the damage caused at this stage is not considerable and has yet to cross the economic threshold level, this emerging ecological problem needs special attention from the point of view of future conservation and management efforts *(Rekha et al 2005).*

77. Strychnine Baiting

Underround strychnine baiting is a common technique used to suppress pocket gopher (Thomomys sp). Populations until seedlings are established. Non-targets risks from underground baiting can still occur if strychnine-baited animals die aboveground. Accordingly, the fate of strychnine-baited carcasses of four small mammal species, Thomomys, Mus, Microtus, and Peromyscus, were documented during four trials on the Rogue River National Forest. We found no difference among species and type of damage, either insect or predator, during the four trials. However, survival differed between species in two trials where fewer Thomomys carcasses survived until completion of the study. We also examined the risk to tertiary non-targets from insects collected at the sites. Strychnine concentration among insect groups from treated carcasses differed with Diptera larvae and ants consistently containing higher concentrations. Even with high strychnine concentration (0.2756 mi cro g/g), risk assessments showed negligible tertiary risks from consuming strychnine-laced insects (Arjo et al 2006).

78. Canadian Experiences

In many agricultural regions of Canada, the most common remnant natural areas are field boundaries, these habitats being linear features or narrow areas located beside cropland. Boundaries are often perceived to harbour noxious weeds, insects and birds that could potentially damage crops or interfere with crop production. Therefore, boundary habitat may be degraded by pesticides, fertilizers, tillage, wind and water exposure, excessive burning, haying and grazing. One conservation objective is to work with land owners to retain and protect existing boundaries, a goal that could be achieved more readily with evidence of benefits and practical ways of managing field margins. Direct services provided by boundary habitats include control of soil and water erosion , protection (e.g, from agro-chemical) of surface water used by livestock and people, and provision of forage for livestock through grazing or haying. Boundaries serve as refugia for plants, insects or other animals that are either neutral or beneficial to agriculture. Native plants often are more common farther from field edges and in habitats abating pastures and hayfields, whereas weeds are more abundant in boundaries adjacent to intensively managed agricultural fields, possibly as a result of competitive advantages or outright loss of native species created by disturbance and agrochemical use. Wildlife has been studied in several countries and under different agricultural settings, but survival and reproductive rates of animals occurring in boundaries are not well known. Relationships between boundary width, height and composition and wildlife value, carbon storage, and protection of surface water are poorly quantified. Answers to these questions will help land owners, conservation agencies and policy-makers make better decisions about sustainable farm practices (Clark et al 2005).

79. Wild turkey Losses

The wild turkey (Meleagris galopavo) repatriation in the USA and Canada has been a major achievement in wildlife management. This success has led to increasing wild turkey populations and calls to wildlife professionals complaining about wild turkeys causing agriculture damage. A mail survey was distributed to state, federal and provincial agencies that respond to wildlife crop damage complaints. Survey recipients were asked to report the number of crop damage complaints and their severity received from farmers, how many of these complaints were investigated and by what means, how often damage was confirmed when investigated, the severity of actual damage observed and how frequently the damage was caused by other species. They wanted to compare farmer's perceptions of crop depredation levels to actual damage and the economic impact relative to specific agricultural crops. A total of 23 different crops were reported as having confirmed damage caused by wild turkeys. In the majority of these cases (93%) damage was reported as light. In 45% of inspected cases, investigators reported that most of the damage (76-100%) was caused by another species. Our results provide wildlife managers with an understanding of the nature of crop depredation relative to the wild turkey (Teftt et al 2005).

80. Mycoaster Coypus Damages

Introduced coypus (Mycoastor coypus) (nutria) damage crops and natural vegetation and undermine riverbanks. Voypu was controlled using case traps to protect vegetation in 3 wetlands in the park of the rivers Po and Orba in northwestern Italy. A field experiment was conducted in a natural reserve in Valenza (NRV), Italy, from 1998 to 2000, to control coypu. The park wardens in NRV were involved in approximately 18 man days of work annually and the cost was 1518 Euros ($ 1867) per year. From 2001, a controm programme was conducted in 3 areas. Removal dednsities (0.06- 1.3 animal/ha) sugeested that 2 trapping periods conducted annually were adequate to slow down coypu population increase. The recovery of yellow waterlily (Nuphar lutea) in 3 plots and the colonization of new ponds by vegetation indicated that coypu were maintained at a level sustainable for vegetation development (Bertolino et al 2005).

81. Cold Temperatures and Wildlife

Cold temperatures and wildlife (e.g deer) predations are the two reasons for seed loss at Vallonia Nursery in Indiana, USA. The use of wheat straw mulch in reducing damage to seed from cold temperatures, loss of seed to wildlife predation and delayed emergency in early spring when frost is likely to occur, was evaluated. Experiment showed that a combination of cover crops (rye and wheat) and straw mulch effectively protect autumn-sown hardwood seeds (e.g Quercus alba, Carya, Juglans and Corylus) from cold temperature damage and predation. Detailed steps on the proper use of mulches are discussed. A tabulated list of hardwood seed species susceptible to cold injury when autumn planted at Vallonia Nurse is given. These species were grouped into three: very susceptible, moderately susceptible and generally not susceptible to cold injury. The advantages of removing straw mulch with fire are briefly discussed. Finally, the costs and benefits of using straw mulches are analysed and discussed (Wichman et al 2005).

82. Visual Monitoring

In 2001 and 2002, the elephant, Loxodonta Africana, herds of Alfakoara in the hunting Zone of Djona (Zone Cynegetique de la Djona, ZCD) in Benin were visually monitored to study their migration, and 500 pellets were collected to analyse their diet. A socio-economic survey covering the village of Alfakoara and its surroundings was achieved near three different socio-professional categories (farmers, ranchers/transhumant producers, and hunters). During the May-September rain season and the October-December crop-ripening season, the elephant herds invaded the community lands, increasing their home ranges outside the reserves. The diet included 71 vegetation species during the rain season and 30 during the dry season. During both seasons the most eaten species were: shrubs (Piliostigma thonningii, Strychnos spinosa), small trees (Cassia sieberiana) and trees (Acacia sieberiana, Balanites aegyptiaca, Diospyros mespiloformis, Adansonia digitata). The results of the survey show that the human-elephant relations in the ZCD were relatively satisfactory, despite damages to crops (mainly corn, zea mays, followed by cotton, Gossypium sp., and to a lesser extent, groundnut, Arachis hypogaea, and millet, Panicum miliaceum), because the riparian populations of Alfakoara are conscious that the elephants are part of their "patrimony", and that they must be conserved at all costs. The keen interest of the populations to adhere to the forest management through the AVIGREF (Association Villageoise de Gestion des Reserves de Faune) and during rescuing operations of accidentally -stuck elephant calves, testifies their effective will to conserve the fauna in general, and particularly the elephants of the site, However, the incessant damage cause by the elephants in the fields of the riparian population and the physical risks they face are also factors that cause irritation among the inhabitants. These factors thus may modify the otherwise serene human-elephant rapports in this zone (Imorou et al 2004).

83. Myanmar's Biodiversity

An effective protected area system is essential for the long-term conservation of Myanmar's biodiversity. This study examined the attitudes of 2915 residents in 97 communities around three protected areas (Pas) in upper Myanmar:Alaungdaw Kathapa National Park in the western mountains, Htamanthi Wildlife Sanctuary in the hills bordering the Chindwin and Uru rivers, and Chattin Wildlife Sanctuary in the central dry zone. Logistic regression indicated a positive attitude toward the Pas was most highly correlated with a perception of conservation benefits and benefits resulting from management of the areas. Attitude was also significantly correlated with a perception of extraction benefits, conflicts with PA staff and crop damage by wildlife. Socioeconomic variables were less powerful than perceptions in predicting attitude and, unlike perception, their effects varied among the areas. The much greater effect of perceptions, especially positive ones, on people's attitudes indicates that understanding perceptions is important to improving the relationship between local residents and these PAs. This finding underscores the fact that a focus on conflicts to understand people's attitudes towards Pas may undervalue or miss critical positive perceptions that people hold. Understanding local residents' perceptions of Pas makes possible the creation of strategic, place-based management strategies that build on people's positive perceptions and mitigate their negative perceptions (Allendorf et al 2006).

84. Japanese Macaques

The current crop damage by Japanese macaques (Macaca fuscata) was examined and the control methods that are employed were evaluated. The frequency of crop damage was highest in winter, followed by summer, autumn, and apring. Wild Japanese macaques primarily consumed leafy and stem vegetables (e.g .Welsh onion and spinach) in winter and spring, fruit vegetables (e.g pumkin and cumcumber) in summer, and fruits (e.g. Japanese persimmon and Chinese gooseberry) in autumn. In winter, the distance between the forest edge and farmland areas receiving crop damage increase, and the maximum distance recorded was 180 m. Japanese macaques also fed on crop residue and waste in winter. These observations suggest low food availability in the forest habitat; thus, to reduce crop damage in winter, food availability in the forest habitat should be augmented. Moreover, to reduce crop damage in farmland, it is necessary to properly dispose of raw crop waste and crop residue, and it is critical to educate the local communities. There was no management strategy to limit crop damage in most areas in which damage occurred. Damage prevention, by constructing net walls to exclude macaques, was exercised at only eight sites (2.9%). However, of the 47 days when wild Japanese macaques were observed in residential areas or surrounding farmland, residents chased them off on only 18 days (38.3%), and the average number of people who participated in this artificial exclusion management procedure was only 1.61 per event. Therefore, it is necessary to establish a cooperative management system that includes the participation of women from local communities to reduce crop damage (Yoshida et al 2006).

85. Transhumance in the Area

This study aimed to evaluate the status of transhumance in the area of the"W"park. It showed that the movement routes lead to the territories located in or bordering the Sahelo-Soudanian zone or the bordering villages, and either lead into the W park, or crossed it in the direction of the South-Borgou (Benin) and Togo. The occupation of pastoral areas in the peripheral zones by crop fields and game reserves, hunting zones and the concern of avoiding browsing damage to standing crop on one hand the fields, the abundance of the feed resources and the "freedom of access" on the other hand made W park the principal transhumance zone destination of cattle. This illegal cattle grazing in the park is considered to be incompatible with the objectives of the wildlife conservation. It causes the degradation of the habitat, a competition with wild animals for feed competition and a health risk between cattle and wildlife. In order to reduce the recurring conflicts which oppose the transhumant herders, the foresters and the farmers, a process of dialogue between all involved was implemented that led to the development of a regional strategy for transhumance management around the W park (Kangone et al 2006).

86. Willow (Salix) Damages

Willow (Salix) communities are important components of riparian ecosystems. However, browsing by livestock and wildlife species can negatively impact willow size and abundance, and make restoration efforts difficult. A common solution has been fencing of affected willows to exclude ungulates, but fencing is expensive and may not complement desirable land management strategies. An alternative to fencing is the use of structures that limit access to streamside willows, without excluding ungulate access to the entire riparian zone. We examined the use of felled western juniper trees (Juniperus occidentalis Hook) placed over streamside willow shrubs. Four replicates of felled western juniper treatments (covered) and nonconvered treatments were applied to a 1.2 km length of stream in southeastern Oregon. Willows (,2m) within treatments areas were censused, tagged, examined for evidence of browse-use, and measured for maximum height during August 2002, before treatment. Posttreatment measurements were made in August and October 2003 . Results indicate that by August 2003 (posttreatment) the average growth of willows in covered treatment was 25 cm (48%) greater than in noncovered treatments. By October 2003 (posttreatment), more shrubs were browsed in noncovered (84%) than covered (39%) treatments. Our data suggest that covering small willow shrubs (< 2m tall) with felled western juniper is an effective deterrent to browsing (Matney et al 2005).

87. Man-Wildlife Conflicts

A case study is presented on the man-wildlife conflicts brought about by the interface of wildlife habitats and a human-use dominated landscape around the Gir Protected Area in Gujarat, India. Information collected from approximately 100 villages around the protected area include; economic status; fuelwood and resource use' grazing' mining activities and people's perceptions. Also included are recommendations on sustainable utilization and management of peripheral forests ;human injured and livestock kills by large fields; minimizing crop damage problems; and conflict zone (Sinha et al 2004).

Summary and Conclusion

The species causing crop damage include wild boar, porcine, rheus macaque (Macaca mulatta), Hoary-bellied squirrel, barking deer (Muntiacus muntjak), red-breasted parakeet (Psittacula alexandri), and wild dog. On average, the worth of damage per ha of crops US$961.82. Methods developed and being used to mitigate man-wildlife conflict include, frightening the animals; guarding the crops; and using some sort of scarecrow and traps (Miah et al 2001).

Primary conflicts included crop raiding, competition between humans and elephants for vegetation, and elephant mortality due to train accidents. Adult males that raided crops had home ranges twice as large as adult males that did not raid crops. Elephants only damaged crop fields that occurred within their home ranges. Field trails of chemical based deterrents and electric fencing should be tried to reduce crop damage. Train speeds need to be reduced to prevent accidental elephant mortality (Williams et al 2001). Population reduction as a possible component of integrated bird management strategies is useful future research is proposed (Avery et al 2002). Accurate estimation of crop damage by wildlife often requires labour-intensive sampling procedure. Variable area transect (VAT) sampling has been identified as a potential labour-saving alternative to quadrat sampling (Engerman et al 2002). The development of non-agricultural activities on land directly adjacent to forested areas may reduce crop –raiding . (Saj et al 2001). There is a need for developing policies and management actions that serve the economic interests of local people together with enhancement of environment conservation goal (Maikhuri et al 2001). Carrot producers should use electric fences or 2.4-m woven wire fences, perhaps combined with staggered canola plantings, to reduce carrot depredations (Schwab et al 2001).The colver, rather than fertilized grass, is more effective cover crop on AFAs, against damage by migratory geese. The sward should be managed to encourage clover growth, which would probably involve frequent cutting but no fertilizer. Further research is needed on clover replenishment rate over the winter season and possible benefits of clover leys to other wild life (McKay et al 2001). Elephants-human conflict decreases the probability of support from local people for conservation efforts. The approaches suggested consist of elephant trenches, electric fences, external support to affected villages, and compensation to villagers for any damage caused (Nyhus et al 2000). It is, therefore, suggested that 30% open space be used as a basic division for stratifying thickets into low –use and high-use categories for deer density estimation. The proportions of each type could be derived from grid-square measurements of aerial photographs (Latham J 2000). Because winter wheat represented a reliable and profitable food source even in sever winter, the recent change in Agricultural practice has enhanced the development of a new wintering strategy of pink-footed geese, allowing a northward expansion of their winter range. Potentially, this will increase the crop damage conflict and may lead to further population growth (Therkidsen et al 2000). Horned lark damage to lettuce seedlings treated with anthraquinone was greater (p=0.015) than for methiocarb R, 60 versus 20%, respectively, and seedlings in control plots were 100% destroyed. While this level of damage is probably unacceptable to lettuce growers, it should be remembered that the enclosure situation caused an artificially high bird pressure on the crop. Further studies in open fields under a more normal bird pressure are warranted (York et al 2000).Incidence and severity of visible foliar ozone injury on cutleaf coneflower (Rudbeckia laciniata L.) and crown-bread (Verbesina Occidentalis Walt), were determined Ozone injury was greatest on the lower leaves for both species sampled with over 95% of the injured leaves occurring on the lower 50% of the plant. This is the first report of foliar ozone injury on these plant species in situ, in the Park, illustrating the great variability in symptom expression with time, and within and between populations (Chappelka et al 2003).

A grid-based geographical information system (GIS) with a 25km2 resolution utilizing cost-effective data source, combined with simple statistical tools, was capable of identifying spatial predictors of HEC. It is suggested that an integrated, community-based, low technological approach will be the most sustainable solution to this conflict (Osborn et al 2003). Blackbirds (lcteridae) annually damage US$5-8 million of ripening sunflower in the northern Great Plains. Baiting blackbirds with avicide-treated rice during spring migration might reduce the regional breeding populations. birds can be successfully baited with avicide-treated rice placed in corn stubble (Linz et al 2003).

Plant debris accumulation is viewed as a key factor determining small mammal abundance and potential damage in low-till agricultural ((Stermer r RT et al 2003). The projected total value of crop yield losses due to wildlife damage for buffer zone villages located in Garhwal Himalaya in about Rs.5 38 620 (US$15 389). Besides food grain, horticultural crops i.e apple, also suffered. Maximum damage. Major wildlife agents responsible for crop damage were wild board, bear, porcupine, monkey, musk deer and partridge (Chukor). Monkey and wild boar alone accounted for about 50% to 60% of total crop damage in the study village. Goat and sheep are the major livestock killed by leopard. The total value of livestock losses at prevailing market rates is about Rs. 10 24 520 (US$ 29 272) in the study village. Due to existing conservation policies and laxity in implementation of preventive measures, the problems for local inhabitants are increasing. Potential solutions dicussed emphasize the need to undertake suitable and appropriate protective measures to minimize the crop losses. Change in cropping and crop composition, particularly cultivation of medicinal plants (high value low volume crops), were also suggested. Besides, fair and quick disbursement of compensation for crop loss and livestock killing need to be adopted. Local people of the buffer zone area already have a negative attitude towards park/reserve establishment due to socio-political changes inducing major economic losses and this attitude may lead to clashes and confrontations if proper ameliorative measures are not taken immediately (Rao et al 2002). Results for the Ontario field crop producers indicate that the magnitude of the difference between the value of the yield loss and the welfare measure of damage is approximately 50%. This difference indicates that most farmers were willing to tolerate the wildlife damage they experienced (Heigh et al 2001).

Capsaicin, present in most hot sauces and salsas, are responsible for the "hot" sensation in many spicy foods. At high doses, capsaicin cause significant discomfort upon contact with the sensitive tissues of the mouth and throat of mammals. By applying a capsaicin containing paste to the outside of maple tree sap collection, tubing, operators hope to minimize rodent (primarily red squirrel, Tamiasciurus hudsonicus) gnawing damage to the tubing.Capsaicins in the sap were then quantified by high performance liquid chromatography/ fluorescence detection. Results indicate that polyethylene tubing was more resistant to capsaicins migration than was polyvinyl tubing. While capsaicins were detected in the sap, the perfected levels in syrup would be below the human taste threshold (Johnston et al 2002). Wildlife management in Australia with crop losses by rabbits and foxes is discussed (Kidd 2002). Rice fields are preferred feeding areas of mallard, Anas platyrhynchos, in central Portugal. Rice consumption by mallards prior to crop harvest causes conflicts between farmers and waterfowl managers. Point counts were performed to estimate the numbers of mallards using the area. The diet included rice seeds (51.2% dry matter), weed seeds of rice fields (23.6%, dry matter), and animal matter (25,1% dry matter). By means of the diet composition, we calculated that each mallard needed to ingest 93.4 g (dry matter) of food daily including 47.8 g of rice, to meet its energy requirements. The estimated mallard population using the area was 1600 ducks. They consumed approximately 0.2% of the 4000 kg/ha minimum rice production. Therefore, the decrease in rice production caused by mallard was insignificant, which suggests that the existence of mallards on rice fields is compatible with rice production (Rodrigues et al 2002).

Reference

Adams, C.E et al, 2005. Regional perspectives and opportunities for feral hog management in Texas. *Wildlife Society Bulletin* 33(4): 1312-1320.

Allendorf T et al., 2006. Community attitudes toward three protected areas in Upper Myanmar (Burma). *Environmental Conservation* 33 (4): 344-352.

Arjo W.M and Nolte D.L, 2004. Assessing the efficacy of registered underground baiting products for mountain beaver(Aplodontia rufa) control. *Crop Protection* 23 (5) 425-430.

Arjo W.M et al, 2006. Potneital non-target risks from strychnine containing rodent carcasses. *Crop Protection*: 25 (2):182-187.

Arjo WM; Nolte DL, 2004. Assessing the efficacy of registered underground baiting products for mountain beaver (Aplodontia rufa) control. *Crop Protection*, 23:5,425-430.

Avery, M.L et a l2005.Caffeine for reducing bird damage to newly seeded rice. *Crop Protection* 24 (7) 651-657.

Avery, ML, Levery DJ(ed); Silva WR (ed) and Galetti-M. Behavioural and ecological consideration for managing bird damage to cultivated fruit, 2002, Third International Symposium Workshop on Frugivores and Seed Dispersal, Sao Pedro, Barzil 6-11 August 467-477.

Avery, ML, Tillman, EA and Laukert CC., 2001. Evlauation of chemical repellents for reducing crop damage by Dickcissels in Venezuela. *International Journal of Pest Management*, 47:4,311-314.

Azell, AW, Lowery, J, Leopoid, B, Minogue, PJ and Haywood, JD , 1999. Use of imazapyr injection to promote oak regeneration and wildlife and improvement in bottomland hardwood stands. Proceedings of the tenth Biennial Southern silvcultural Research Conference, Shreveport, Louisiana, USDA No.SRS-30,151-153.

Badola R and Hussain S.A., 2005. Valuing ecosystem functions: an emprirical study on the storm protection function of Bhitarkanika mangrove ecosystem, India. *Environmental Conservation* 32 (1): 85-92.

Baker B.W et al, 2005. Interaction of beaver and elk herbivory reduce standing crop of willow.*Ecolbgical Application* 15 (1) 110-118.

Baker et al, 2005 . A dose response trial with ziram treated maize and free ranging European Badgers Meles meles. *Applied Animal Behaviour Sci.* 93(3/4): 309-321.

Bertolino S et al, 2005. Effectiveness of coypu control in small Italian wetland areas. *Wildlife Society Bulletin* 33 (2): 714-720.

Bhargavi, Srinivasulu, Srinivasulu-C and Srinivasulu, B, 2001. Magnitutde of depredation on grapes by short-nosed fruit bats Cynopterus sphinx Vahl, 1797 in Secunderabad, India, *Current Science* 2001, 80:1,14-15.

Blackwell B.F et al 2003. Lethal control of red winged blackbirds to manage damage to sunflower: an economic evaluation. *J. of wildlife Management* 67 (4) 818-828.

Blackwell BF, Huszar E, Linz GM, and Dolbeer RA., 2003, Lethal control of red winged blackbirds to manage damage to sunflower, an economic evaluation. *Journal of Wildlife Management* 67:4 818-828.

Blackwell, BF, Helon, DA and Dolbeer, RA, 2001, Repelling sandhill cranes from corn, whole kernel experiments with captive birds, *Crop Protection* , 20:1,65-68.

Brown P.R et al, 2004. Can farm management practices reduce the impact of house mouse populations on crops in an irrigated farming system. *Wildlife Res.* 31 (6): 597-604.

Bulte E.H and Rondeau D., 2005. Why compensating wildlife damages may be bad for conservation. *J. of Wildlife Management* 69(1) 14-19.

Burn, AJ, 2000, Anticholinesterase pesticides and wildlife in the UK., The BCPC Conference Pests and disease volume 3 Hotel Brighton UK, 13-16, 1117-1124.

Campbell, DJ, Gichohi, H, Mwangi, A and Chege, L., 2000. Land use conflict in Kajiado District Kenya., Land Use Policy, 17:4,337-348.

Carter Jand Leonard BP, 2002.A review of the literature on the worldwide distribution, spread of, and efforts to eradicate the coypu (Myocastor coypus). *Wildlife Society Bulletin* 30: 1, 162-175.

Chappelka, AH, Neufeld, HS, Davison, AW, Somers, GL, Renfro, JR and Skelly, JM., 2003. Ozone injury on cutleaf coneflower (Rudbeckia laciniata) and crown-beard (Verbesina occidentials) in great Smoky Mountains National Park. Symposium held in conjection with the 34th Air Pollution Workshop, USA 125,1,53-59.

Chauhan NPS, Sushant Chowdhury and Chowdhury S., 2002. Evaluation of electric fences for their efficacy in controlling elephant damage in Northern West Bengal and Suggesting Improvements., *Indian Forester* 2002, 128:2, 179-188.

Chhangani, A.K and Mohnot S.M. 2004 Crop raiding by wild boar (Sus scrofa) in and around Aravalli, and its management in Rajasthan, India. *Tigerpaper* 31(2):1-5

Clark R.G et al, 2005. Living on the edge field boundary habitats , biodiversity and agriculture, Field boundary habitats and disease management, 113-133.

Cummings JL,Avery ML,Mathre, O, Wilson EA, York DL, Engeman RM Pochop PA, and Davis, JE Jr. 2002, Field evaluation of Flight Control TM to reduce blackbird damage to newly planted rice.*Wildlife Society* 2002, 30:3, 816-820.

Cunningham C.A et al, 2005. Attrack and brood production by the Douglas-fir beetle (Coleopetera : Scolytidae) in Douglas-fir. *Western North American Naturalist* 65 (1): 70-79.

Cutis, J and Lynch, L ,2001. Explaining deer population preferences : an analysis of farmers, hunters and the general public. *Agricultural and Resource Economic Review*, 30:1,44-55.

Dahamna S et al, 2004. Biochemical indicators of hepatotoxic effects of pesticides. *Communications in Agri and Applied Bio. Sci* 69 (4):821-828.

Dilip Chetry et al 2003. Community based conservation approach around Nameri National Park, Assam India. *Tigerpaper* 30 (3):16-19.

Dilip Chetry, Rekha Medhi, Bhattacharjee PC, Chetry D and Medhi R 2003, Community based conservation approach around Nameri National Park, Assam India., *Tigerpaper* 30:3, 16-29.

Doney,J, and Packer,JJ, 2001. The impact of deer on agriculture, results of a questionnaire survey and subsequent validation, Deer Journal of the British 12(2) 98-104 Engeman, RM and Sterner, RT, 2002 , A comparison of potential labour saving sample methods for assessing large mammals damage in corn. *Crop Protection* 21:2,101-105.

Geisser H and Reyer H.U, 2004, Efficacy of hunting, feeding and fencing to reduce crop damage by wild boars. *J. of Wildlife Management* 68 (4): 939-946.

Gillingham S, and Lee PC. People and protected areas: a study of local perceptions of wildlife crop damage conflict in an area bordering the Selous Game Reserve, Tanzania. *Oryx*, 37:3, 316-325.

Gillingham S and Lee PC, 2003. People and protected areas: a study of local perceptions of wildlife crop damage conflict in an area bordering the Selous Game Reserve, Tanzania. *Oryx* 37 (3) 316-325.

Gilsdorf J.M et al, 2004. Propane exploders and Electronic Guards were ineffective at reducing deer damage in corn fields. *Wildlife Society* 32(2) :524-531.

Gilsdorf J.M et al. 2004 Evaluation of a deer activated bio-acoustic frightening device for reducing deer damage in corn fields. *Wildlife Society* 32 (2): 515-523.

Godsey, LD,2000. An economic analysis of black walnut alley cropping in NE, Temperature. *Agroforester*, 8,4 ,8-9.

Goyal, SK and Rajpurohit, LS, 2000, Nilgai, Boselaphus tragocamelus- A mammalian crop pest around Jodhpur. Uttar Pradesh. *Journal of Zoology*, 20:1,53-59.

Haarstrich R.J, 2005. Compensation for damage by game to specialized crops. *Agar and Umweltrecht;* 35(4): 109-112.

Halley, D. et al, 2006. Variation in stable isotope ratios of carbon and nitrogen in Hokkaido sika deer Cervus Nippon during 1990-2000. *Wildlife Biology* 12(2): 211-217.

Heigh L, Rollins K, and Kanetkar V., 2001. An appropriate welfare measure of wildlife damage. Deptt. of Agri,Econ. and Business Univ. of Guelph No.02-04, 35pp. Hone-J, 2004. Yield compensation and fertility control : a model for vertebrate pests. *Wildlife Research* , 31 4:357-368.

Hunt W et al , 2004. System policy and infrastructure for rodent management in the Queensland sugar industry. Conference of the Australian Society 4-7 May, 2004, 1-8.

Hunt W, Dyer B, Kerkwyk, RE, Marohasy J, Thompson, J and Hogarth DM., System policy and infrastructure for rodent management in the Queensland sugar industry.

Imorous S.A.G et al 2004. The human elephant (Loxodonta africana). *Game and Wildlife Sci.* 21(4) 553-569.

Jacob. J et al, 2004. The impact of sterilized females on enclosed populations of rice fields rats. *J. of Wildlife Management* 68 (4): 1130-1137.

Jakson, EA,1999. Studies on crop damage by wild animals in kereala and evaluation of control measures. *KFRI, Research Report* No.169,iii + 48 pp.

Jeyasingh P.D and Priya Davidar, 2003. Crop depredation by wildlife along the eastern boundary of the Kalakad Mundanthural Tiger Reserve, Southern India. *J. of the Bombay Natural History Society* 100 (1): 38-45.

Jeyasingh PD, Priya Davidar, Davidar P, 2003. Crop depredation by wildlife along the eastern boundary of the Kalakad Mundanthurai Tiger Reserve Southern India. *Journal of the Bombay Natural History Society* 100:1,38-45.

Johannesen A.B and Skonhoft A,. 2004 Property rights and natural resource conservation. A bio-economic model with numerical illustrations from the Serengeti-Mara ecosystem. *Environment and Resource Economic* 28:4: 469-488.

Johnson J.M et al, 2004. Effects of straw mulch on pest insects, predators and weeds in watermelons and potatoes. *Environmental Entomology* 33(6): 1632-1643.

Jones G.A and Eieving K.E, 2006. Intercropping sunflower in organic vegetables to augment bird predators of arthropods. Agri. Ecosystems and Environment 117 (2/3): 171-177. *Journal of Environmental Manager* 66:3,317-327.

Kagone H et al, 2006. Pastoralism and protected area in West Africa. *Bulletin of Animal Health and Production in Africa* 54(1): 43-52.

Kagoro Rugunda G., 2004. Crop raiding around Lake Mburo National Park Uganda. *African Journal of Ecology* 42:1,32-41.

Kahoro Rugunda, 2004. Crop raiding around Lake Mburo National Park, Uganda. *African J. of Ecology* 42 (1): 32-41.

Karanth K.D, 2003. Forest use and human wildlife conflicts in Bhadra wildlife Sanctuary, Karnataka, India, Tropica Resources Bulletin of the Yale Tropical Resource Instt. 22:48-58.

Kidd H, 2002. Wildlife management in Australia, Pesticide outlook, 13:6, 249. Kimball B.A et al 2005. Hydrolyzed casein reduces browsing of trees and shrubs by white tailed deer. *Hort Science* 40 (6):1810-1814.

Latham J., 2000. Use of thicket stages of Scottish Conifer Plantations by red and roe deer in relation to openness., *Forestry Oxford*, 73:4,403-406.

Linz G.M et al., 2006. Evaluation of registered pesticides as repellents for reducing balckbird (leteridae) damage to sunflower. *Crop Protection* 25 (8) 842-847.

Linz GM, Knutsen GA, Homan HJ and Bleier WJ, 2003. Baiting blackbirds (lcteridae) in stubble grain fields during spring migration in South Dakota. *Crop Protection* 2003 22:2 261-264.

Madhusudan M.D. 2003 .Living amidst large wildlife Livestock and crop depredation by large mammals in the interior villages of Bhadra Tiger Reserve, South India, *Enviornment Management* 31(4):466-475.

Matney C.A et al., 2005. Use of felled junipers to protect streamside willows from browsing. *Rangeland Ecology and Management* 58(6):652-655.

McKay, HV, Milsom, TP, Feare, CJ, Ennis, DC. O, Connell, DP and Kaskell DJ, 2001, Selection of forage species and the creation of alternative feeding areas for dark bellied Brent geese Branta bernicla Bernicla in Sourthern UK coastal areas, *Agriculture -Ecosystems and Environment*, 84:2,99-113.

MCLaren, BE; Mahoney, SP, Porter, TS, Oosenbrug, SM.,2000. Sapatial and temporal patterns of use by moose of pre-commercially thinned, naturally-regenerating stands of balsam fir in central Newfoundland., *Forest Ecology and Manager* 133.3, 179-196.

Miah, MD, Rehman, ML and Ahsan, M.F 2001. Assessment of crop damage by wildlife in Chunati Wilklife Sanctuary, Balgladesh. *Tigerpaper*, 28(4), 22-28.

Middelmann M, 2006. Warning Get friendly. Acta Horticulturae (716): 69-70. Mitani M, 2000, Present situation of wild mammals and problems of wildlife management in Hyogo prefecture; a review. *Human and Nature* No,11,43-59.

Muir JP, Ocumpaugh WR, and Read JC, 2003., Spring forage yield and nutritive value of Texas black medic accessions. *Agronomy Journal* 95:4, 908-912.

Mutze G, and Sinclair R, 2004.Efficacy of zinc phosphide, strychnine and chlorpyrifos as rodenticides for the control of house mice in South Australian cereal crops. *Wildlife Research* 31 (30:249-257.

Nigam BC, 2002. Elephants of Jharkhand-increasing conflicts with man. *Indian Forester*, 128:2,189-196.

Ninan K.N. and Jyothis, 2005. The exonomics of biodiversity conservation: a study of a coffee growing region in the Western Ghats of India., *Ecological Econ.* 55 (1) 61-72.

Nyhus, PJ, Tilson, R and Sumianto, 2000. Crop raiding elephants and conservation implications at Way Kambas National Park, Sumatra, Indonesia. *Pryx* 34:4,262-273.

Oliveria P et al,2006. The influence of fruit abundance on the use of forest and cultivated field habitats by the endemic Madeira laurel pigeon Columba trocaz implication of conservation. *Biological Conservation* 130 (4): 538-548.

Onyeanusi AE, Ambudode, FO, Oduwaiye, EA(ed), Obiaga, PC, Abu, JE, 1997. Common hippopotamus (H.amphibius Linn 1758) foraging movements and their crop damage potential at the peripheral Ecological environment of Kainji Lake National Park, Nigeria. Annual Conference of the Forestry Association of Nigeria held in Ibadan, Oyo State, Nigeria 191-197.

Osborn FV, and Parker GE., 2003. Towards an integrated approach for reducing the conflict between elephants and people : a review of current research. *Oryx*, 37:1,80-84.

Perez E and Pacheco L.F, 2006. Damage by large mammals to subsistence crops within a protected area in a monane forest of Bolivia. Crop Protection 25 (9):933-393.

Post, LS, Schlarbaum, SE, Barber, LR, Tolman, DF; and Cecich, RA, 2001. Bifenthrin (Capture (R) reduces Curculio spp. (Coleoptera, Curculioriae) damage in northern red oak acorns. *Journal of Entomological Science*,36:2, 222-225.

Prasad NS, and Reddy KS. 2002. Man elephant conflict and mitigation-Koundinya wildlife Sancturay, Andhra Pradesh. *Indian Forester*, 2002, 128:2, 137-144.

Rao, KS, Maikhuri RK, Nautiyal S and Saxena KG., 2002. Crop damage and livestock depredation by wildlife a case study from Nanda Devi Bioshphere Reserve, India. Rekha Medhi et al, 2005. A note on crop raiding by golden langur in fragmented forests in Western Assam, India. *Tigerpaper*: 32(3): 23-27.

Rodrigues D, Figueiredo M and Fabiao A., 2002. Mallard (Anas platyrhynchos) summer diet in central Portugal rice fields. Game and Wildlife Science 2002 19:1,55-62. Rollins et al 2004. Net costs of wildlife damage on private lands, *Journal of Agri.Econ.*Univ.of Nevanda Reno 1664 USA.

Rollins K et al, 2004. Net costs of wildlife damage on private Lands. *J. of Agri. and Resource Economics* 29 (3): 517-536.

Rondeau D and Bulte E, 2003. Compensation for wildlife damage habitat conversion, species preservation and local welfare. Working paper Resource and Environ. University of Victoria (2003-01): 36 pp.

Saj.TL, Sicotte.P, and Paterson, KD. The conflict between vervet monkeys and farmers at the forest edge in Entebbe, Uganda. *African Journal of Ecology*, 39(2) 195-199.

Schafer KS and Kegley, SE., 2002, Persistent toxic chemicals in the US food supply. *Journal of Epidemiology and Community Health* 56:11,813-817.

Schwab ,FE, Pittoello,FG and Simon,NPP, 2001. Relative palatability of green manure crops and carrots to white-tailed deer. *Wildlife Society* 29,1,317-321.

Semiadi G and Meijaard E, 2006. Declining populations of the Javan warty pig Sus verrucosus. *Oryx* 2006 40(1): 50-56.

Seward N W et al., 2004. Feral swine impacts on agriculture and the environment. *Sheep and Goat Res.J.*19:34-40.

Singh, AK, Singh RR, Sushant Choedhury and Chowdhury S., 2002. Human elephant conflicts in changed landscapes of South West Bengal, India. *Indian Forester*, 128:10,1119-1132.

Sinha S.P et al, 2004. Man-animal conflicts in and around protected areas. *Tigerpaper* 31 (3): 27-32.

Sitati, NW, Walpole, MJ, Smith RJ, and Leader Williams N., 2003. Predicting spatial aspects of human elephant conflict. *Journal of Applied Ecology*, 40: 4, 667-677.

South D.B and Enebak S.A, 2006.Integrated Pest Management practices on Southern pine nurseries. *New Forests* 31(2): 253-271.

Sterner RT, Petersen BE, Gaddis SE, Tope KL and Poss DJ,2003,. Impacts of small mammals and birds on low tillage, dryland crops. *Crop Protection*, 22:4,595-602.

Sutton W et al, 2004. A new approach for assessing the costs of living with wildlife in developing countries. Working Paper Deptt. of Agri. And Res.Economic Univ. of California (04-001).

Tamaru C.S et al., 2006. Control of the apple snail, Pomacea canaliculata, in Hawai'i challenge or apportunity. Global advances in ecology and management of golden apple snails 459-473.

Tefft B.C et al.,2005. Assessment of crop depredation by wil turkeys in the United States and Ontario, Candada. *Wildlife Society Bulletin* 33 (2): 590-595.

Therkidsen OR, and Madsen, J. , 2000. Energetics of feeding on winter wheat versus pasture grasses; a window of opportunity for winter range expansion in the pink footed goose Anser brachyrhynchus. *Wildlife Biology* 6:2,65-74.

Tokida , K et al 2004. A deer management approach to promote ecosystem management in National Parks. *Japanese J. of Conservation Ecology* 9(2): 193-202.

Tweheyo et al, 2005. Patterns of crop raiding by primates around the Bundongo Forest Reserve, Uganda. *Wildlife Biology* 11(3):237-247.

Twigg, LE, Martin, GR, Wilson, N, Goddard, D; Watkins, R, and Armstrong, PJ. 2001.Longevity of zinc phosphide on wheat bait in tropical Australia. *Wildlife Research*, 28:3,261-267.

Tzukjiwsju WM, Brittingham MC, Lovallo MJ.,2002, Wildlife damage to corn in Pennsylavania : farmer and on the ground estimates. *Journal of Wildlife Management* 66:3,678-682.

Veeramani A et al, 2004. Socio economic status of cultivators and their interface with wild animals: a case study of Marayur Forest Range, Kerala. *Indian Forester*: 130 (5): 513-520.

VerCauteren K.C et al 2006. A simulation model for determining cost effectiveness of fences for reducing deer damage. *Wildlife Society Bulletin*: 34(1): 15-22.

VerCauteren K.C et al 2006., Fences and deer damage management : a review of designs and efficacay. *Wildlife Society Bulletin* 34(1):191-200.

Walples K, 2002. Review of the NPWS policy on the mitigation of commercial crop damage by flying foxes. Managing the grey headed flying fox-as-a threatened species in NSW 39-46.

Wang S. W et al 2006. Farmer attitudes towards conservation in Jigme Singye Wangehuck National Park, Bhutan. *Enviornmental Conservation* 33 (2) 148-156.

Wang S.W et al 2006. Farmer perceptions of crop damage by wildlife in Jigme Singye Wabgehuck National Park, Bhutan. *Wildlife Society Bulletin*: 34 (2): 359-365.

Weladji RB, and Tchamba MN., 2003. Conflict between people and protected areas within the Benoue wildlife Conservation Area, North Cameroon. Oryx, 37:1, 72-79.

West BC and Parkhurst JA, 2002. Interaction between deer damage, deer density and stakeholder attitudes in Virginia., *Wildlife Society Bulletin* 30:1,139-147.

Wichman J et al, 2005. Straw mulch prevents loss of fall sown seeds to cold temperatures and wildlife predation.*Native Plants Journal* 6(3): 282-285.

Williams, AC, Johnsingh, AJT and Krausman, PR 2001, Elephant human conflicts in Rajaji National Park,northwestern India. *Wildlife Society Bulletin* 29:4,1097;23.

Wilsin C.J., 2004. Rooting damage to farmland in Dorset, southern England, caused by Feral Wild Board Sus Scrofa. *Mammal Review* :34 (4): 331-335.

Wilson A.P et al, 2004. Effects of varying weed communities in corn on European corn borer Ostrinia nubilalis (Hubner) . *Environment Entomology* 33 (2):320-327.

Wilson AP, Hough Goldstein JA, Vangessel, MJ and Pesek JD., 2004. Effects of varying weed communities in corn on European corn borer, Ostrinia nubilalis (Hubner) (Lepidoptera: Crambidae), oviposition and egg mass predation. *Environmental-Entomology* 2004, 33:2,320-327.

Wilson CJ, 2004. Rooting damage to farmland in Dorset, southern England caused by Feral Wild Board Sus Scrofa.*Mammal Review*, 34(4) 331-335.

Yadav B.R, 2003. Assessment of crop damage by Asian wild elephants in Jhapa District, Nepal. *Banko Janakari* 13 (1): 50-56.

Yadav BR,2003. Assessment of crop damage by Asian Wild elephants in Jhapa District Nepal. *Banko Janakari* 13:1, 50-56.

Yagi H et al, 2004. Farmland classification method considering of controlling crop damage by wildlife. *J. of Rural Economic* 76 (Suppt): 342-347.

Yamaguchi Y and Yoshida H., 2006. Alternative feed as a means of reducing damage to direct sown rice by the Oriental Turtle Dove, Streptopelia orientalis. *Japanese J. of Ornithology:* 55(1): 1-6.

Yoder, J., 2002, Estimation of wildlife inflect property damage and abatement based on compensation program claims data. *Land, Economics* 78:1,45-59.

York, DL, Cummings, JL, Engeman, RM and Davis JE,Jr.2000., Evaluation of Flight Control TM and Mesurol R as repellents to reduce horned lark (Eremophila alpestris) damage to lettuce seedlings. *Crop Protection* 13,201-203.

□□□

Chapter 17

Nuclear and Embryo Manipulation Techniques

G.M. Wani, P. Wani & G. Wani

Abstract & Introduction

Past century saw many techniques like micro-surgery of embryos, embryo co-culture, in vitro fertilization and production of monozygotic twins, chimeras besides cryo preservation and vetrification techniques for preserving goat embryos. The recent micro injection techniques made possible cloning and DNA transfer and production of animal clones like modern day "Dolley". Various reproductive and genetic engineering techniques like DNA finger printing gene mapping and recumbent DNA technologies are in active use.

Formation, growth and maturation of female gametes begins during embryonic stage and continues after birth evincing climax at ovulation (Wani, 2002). The single cell released from ovarian follicle at ovulation in an "Egg". In most of the mammals it is a "secondary oocyte", when first polar body is already released at primary oocyte stage. In most of the mammals, the second meiotic division is arrested at metaphase and proceeds further only after sperm penetration. The number of primodial follicles in a mammalian dam far exceeds its production of offspring even after use of modern day techniques mentioned above.

Biotechnology has been used for various purposes in biomedical sciences which stands reviewed Wani (1998), Wani (2003).

1. Augmentation of Primodial Germ Cell

The number of primordial germ cells in an ewe and goat is around 86,000 and cow nearly 1 lakh. Out of them hardly 100-200 grow to mature in life and only a few are fertilized and develop into a conceptus. The gap between potential pool of oocyte and those available for fertilization is large. Trapping this pool conventionally means using 0.2% of available genetic material for manipulation by known applicable techniques of superovulation, synchronization and assisted reproductive techniques reviewed and reported (Wani, 1998, 2003).

The varied ovarian reaction, inadequate knowledge on atresia and mechanism of favourable status of maturing follicle and final selection of the particular oocyte for release and its fertilization make genetic selection much variable and distant relatively in terms of number of follicle which do degenerate at each atresia. Thus, the intricate mechanism of hormonal play in induction of ovulation or maturation of oocytes needs more awareness and thus signifies more research (Wani, et al 1997).

In this context the neuroendocrine signal which triggers massive and sustained GnRH surge may be the switch inducing ovulation (Karsch et al., 1992). Few questions can be asked & needs answer in our todays review:

Does recruitment of follicle have any relationship with preovulatory GnRH secreting neurons? Similarly, Is the post-dated GnRH surge, involving morphological changes in neuron (Karsch and Evans, 1996) related to morphological adjustment, recruitment for follicle maturation and its subsequent release at ovulation. The 40 fold increase in GnRH over and above baseline values during surge period thought to be dynamic switch control of ovulation in sheep needs further elucidation, which demands more studies on morphological aspects of follicular maturation, selection and ovulation, especially in the light of deterministic and permissive models of GnRH and LH surges vis-à-vis folliculogenesis and ovulation in goats (Karsch, 1997).

Therefore, our innovative vision should be to remodel our techniques incorporating GnRH signals and oestradiol modulation for exploiting primodial germ cell pool to mature more oocytes in vivo or in vitro. This is what stem cell technology offers. If by 2150 A.D. we are able to use even 10%

of available genetic primodial germ cell pool of an improved mammalian species, it would be a landmark in the biological discoveries. This would produce more from less and less numbers. This is what we all strive for. Keep the number of goats as minimal as possible. Increase their productivity many fold, so that we have enough milk, meat and fur available for growing population at minimum costs. This is what we call sustainable goat production system.

2. Embryo Production Techniques

Embryo production using techniques of synchronization, super-ovulation, embryo collection, embryo evaluation, grading culture, manipulation, preservation and transfer along with modern useful techniques of in vitro maturation of oocytes and in vitro fertilization, along with recent diagnostic and explorative use of ultra sounds and laparoscopy have been discussed. (Wani, 1984, 1992, 1995, 1998, 2002, 2003).

3. Embryo Evaluation and Grading Techniques

Various scales for grading mammalian embryos are reported (Wani, 1992) and are also detailed (Wani, 1996) However, what we need is a numerical grading mechanism for embryos as in sperms. The examination procedures of embryos shall be simplified for use in the field conditions in future.More intensified research on grading of embryos and oocytes are to be searched out. This is particularly important now, because of recent aging effect reported on "Dolley".

4. Determination of Embryo Viability

Fluorescent technique for determining living embryo cells got a boost by revealation that staining with Fluorescin - diacetate (FDA) dye does not effect further development. Biochemical methods based on glucose uptake; oxygen uptake or enzymic releases have poor practical applicability. More refined techniques predicting embryo viability have to be searched, especially for use under field conditions.

5. Embryo Transfer Techniques in Goats

Surgical techniques (Wani, 1984) have been replaced with laparoscopic and trans-cervical non surgical methods (Hays, 1998). More simplification and search is needed to make whole set of techniques operative under farm conditions to produce future breeding sires from superior sire mothers (donors). More research on trans-cervical embryo transfers in recipients needs encouragement. A few innovative beginnings for ultrasonically guided laparoscopic collection of oocyte in being attempted and planned.

6. Embryo Manipulation Techniques

These techniques reviewed and improved by our colleagues in the developed countries need to be operative under field conditions in developing country where most of the goats are reared. (Wani, 1996). Manpower or expertise is available; funds are needed to intensify perfection of these techniques.

7. *In Vitro* Fertilization and Prior Sexing of Embryos

Some of the techniques already operative (Jufen et al., 1991; Wani, 1996). We have used in vitro fertilisation techniques in sheep with a unique distinction. We used semen from slaughtered rams for fertilization of oocytes recovered from slaughtered ewes (wani et al 2000, 1999, Wani 1996). This promises its potential use for regeneration of rare wild animals. The slaughter house pool shall provide valuable elite germ plasm even after their death. This is essential to conserve wild goat ancestors and best producers like Cherion, Ibex etc.

8. Cryo Preservation and Vetrification Techniques

Various culturing techniques like embryo co-culture, cryo preservation and vetrification needs more emphasis. This has been added with Cryopreservation of whole ovary and its use for oocyte maturation, fertilization and transfer. (Wani, 2003).

9. Ultrasonography and Laparoscopy

Prior indications of early pregnancy through ultrasonography, laparoscopy(Wani, 1981, 1982) have further improved and were reviewed (Wani, 1995). Early indication of pregnancy limits reproductive losses and promises efficient production systems (Wani,1995, Wani et al, 1998; Wani, 2005). New vistas in this arena have been opened up. In India our humble start in late seventies (Wani, 1981) has now advanced to detection of implantation of embryo too). We could moniter embryonic development and implantation through ultrasounds. The prospects of in vitro implantation and development of conceptus needs basic investigation. We tried to set up basic land marks in this regard. (Mufti et al 2000). The whole implantation concept has been rediscovered. A new invention defining the role of foetal and maternal compartments has been published. (Wani, et al, 2006, 2006a, 2006b, 2006c and 2007). A few new founded real photographs could be seen I the photogalary of this book.

10. Production of Desired Chimeras and Clones

Production of chimeras and clones having high yielding genes through aggregation of cells from two or more pre-implanation embryos may enhance Goat production in 21st century. The addition or modification of DNA related to production traits may help in production of high yielding goats replacing Dolly with "Trolleys of DNA chips" on sale like computer chips in future. New pre-implantation signals hinted (Heap et al., 1990) and reviewed (Wani, 1995) may open new progressive research phase involving more and more reproductive scientists to unveil the hidden truth (Al-Quran, Suratul Mominoon, 12-134)

Inclusion of specific aminoacids in the culture media has helped to grow embryos beyond 8 cell stages. Now more advances in culturing ovine embryos by co-culture or group culture have been known. More search on rate of involution, development of concepts in vivo and in vitro implantation and transfer are some of the immediate research themes to be searched in near future (Wani, 1998, Wani 2002, Wani 2001).

11. Foetal Maternal Relationship

Various biochemical and morphological investigation on foetal growth and development has been jointly explored by SKIMS and SKUAST (Mufti, 1996). This is an innovative search to unveil various mechanisms involving in vitro implanation and transformation of blastocyst into foetus in future. More funding is needed for investigative and innovative projects on foeto-maternal associations such as reported by Mufti et al (2000), wani et al 2007, 2006, abc.

12. DNA Finger Printing

Genotype environment interactions are closely related to mammalian reproduction. Knowledge of genetic constitution and embryonic differentiation is a pre-requisite for breed conservation and improved selective breeding. Since, DNA of an individual is not directly influenced by environmental conditions except loci coding for T-cell receptors, antibodies and others; short term environmental influences like season, feeding and management may not influence. DNA sequences based on this central theme, an investigation was undertaken in collaboration with University of Hohenheim Germany. The study was financed by German Research Foundation. The project has ended, but its savings has enabled us to institute an fellowship to be called Gall, Geldermann, Wani fellowship for goat reproduction. Under this project, blood samples from Pashmina and Bakerwal Goats were analysed.

DNA was extracted from leucocytes with phenol-chloroform. Different regions of the expressed DBR gene and the DBR pseudogene were amplified by PCR. Fragment lengths of the entire exon 2 and the microsatellite were analysed, using automatic DNA sequencer, Automatic Laser Fluorescence (ALF) and by Single Strained Confirmation Polymorphism.

Mannalian Foeto-Maternal Attachment - A New Invention

The blood of mother and foetus never mixes freely in ovine or caprine foetus. The fertilized zygote develops into a morula which decends and reaches uterus in 3-4 days. The presence of corpus Luteum haemorrhagicum (CLH) and its subsequent growth ensures progesterone secretion.

Endometrial recieptivity was morphologically documented in this study. The solid tooth shaped projections of non-pregnant uterus change into mucin coated cup-shaped caruncles. Conco mitant to these changes blastocysts elongates. Within elongated blastocysts cotyledons develop which fix into caruncles to form the placenta. The unique feature presented in this paper is recognition of non-self (foetus) by self (endometrium). We present here the following observations".

1. The caruncles in non-pregnant uterine horns consist of solid projection arranged in rows. They are red and upright ville-like. No Corpus Luteum Haemorrhagicum (CLH) was present.
2. In the presence of CLH these caruncles change in form, colour and structure. They are cup-shaped , mucin coated, whitish in colour and give an impression as if a whitish cover or membrane overwrap them.
3. In the meantime blastocyst elongates and covers the whole uterine horn. Within the blastocysts or now we can call it an elongated ballon shaped foetus floating over hypertrophied caruncles. Within this developing foetus sac small round, red shaped ball develop and it is called cotyledons. These cotyledons further fix themselves into cup-shaped caruncles. The formation of placentomes begins both within foetal membrances and endometruim cover over cotyledons and caruncles respectively, before union. Thus an indirect attachment of cotyledons and caruncles occurs which is separated by different membranous coverings described. The mechanism of such attachment and a system of manifold protective covers prevent direct contact of foetus and maternal tissues. The minute details of attachments are supported by histological nuclear unions too. To the best of our knowledge, it is the first finding of the kind being published in literature.

References

Al-Quran, Surat Mominoon, 12:134. Young, L. E, 2002, Somatic cell nuclear transfer,419, 6907, 583-586

Anderson.G.B; Murray J.D and Medrano.J.F, 1996, Expression of a bovine k-CN cDNA in the mammary gland of transgenic mice utilizing a genomic milk

Casey.D.C and White.K.L, 1990, Incorporation and expression of a neo gene after Cloning and sequence analysis of the cDNA for a protein from coccidioides Colman.A and Garner.I, 1995, The transgenic mammary gland as a bioreactor:

Dugger.K.O; Villareal.K.M; Ngyuen.A; Zimmermann.C.R; Law.J.H and Galgiani.J.N, expectations and realizations, 6,107.

Gall, C; Gelderman, H; Wani, G.M. 1997. Analysis of Goat Genetic Resources of goat breeds in Kashmir. Final report of research co-operation between SKUAST and University of Hohenheim, Germany.

Hays, B.L. 1998. Non-surgical embryo collection in the goats. Proceedings of 11th International Congress on Animal Reproduction and AI, Dublin 2: 165.

Heap, R.B; Wang. M.W; Whyte, A and Taussing, M.J. 1991. Significant of peri-implantation signals. L. 17. Proceeding Int. Conf. on Frontiers in reproductive Physiology, Nov. 8-10, 1991 at AIIMS, New Delhi.

Jufen, Q; Zhiming, H; Yong, Z and JianChen , W. 1991. Invitro capacitation of ejaculated spermatozoa and invitro fertilization in dairy goats. *Theriogenology*. 35(1):219.

Karsch, F.J and Evans, N.P. 1996. Feedback actions of oestradiol on GnRH Secretion during follicular phase of the oestrus cycle. *Acta. Neurobiol. Exp.* 56:715-725.

Karsch, F.J; Bowen, J.M; Caraty, A; Evans, N.P and Moenter, S.M. 197. Gonadotrophin-releasing harmone requirements for ovulation. *Biology of Reproduction*, 56h:303-309.

Karsch, F.J; Moenter S.M. and Caraty, A; Moenter, S.M. 1997.The neuroendoerine signal for ovulation.*Animal Reproduction Science*, 28:329-341.

Maughan .P.J; Philip.R; Cho.M.J; Widholm.J.M and Vodkin.L.O, 1999, Biolistic.

Mufti, A.M. 1996. Studies on morphological characters and some metaolic consititutes at different stages of development of ovine foestus. MVSC Thesis, SKUAST, Srinagar.

Mufti, A.M; Wani, G.M; Wani, N.A, and Khan, M.Z. 2000. Prenatal development of ovine foetus. *Small Ruminant Research*, 38: 87-89.

Mufti, A.M; Wani, N.A; Mir, M.M; Khan, M.Z and Wani, G.M. 1997. Superovulatory response in corriedale ewes during different months of breeding season. *Small Ruminant Research* 25:181-84.Protein gene as an expression cassette, 5 (4), 271-79 transfer of Caprine hematopoietic cells using a modified retro virus, 68,2,449-53.

Pawshe, C.H; Totey, S.M and Jain, S.K. 1994. Effect of the methods of recovery of goat Oocytes for invitro maturation and fertilization. *Theriogenology*, 42:117-125

Wani , G.M. 1996. Embryo biotechnology in sheep and goats. Valley Book House University Road, Srinagar.

Wani, G.M. 1981. Ultrasonic pregnancy diagnosis in sheep and goats. *World Review of Animal Production*, 17:43-48

Wani, G.M. 1982. Laparoscopy in Farm animals. *World Review of Anim. Production*, 18: 7-13.

Wani, G.M. 1984. Investigations Caprine Embryo production and culture. Dr. Med. Vet.Thesis Vet Univ, Hannover, Germany.

Wani, G.M. 1992. Advanced Techniques in Animal Reproduction. SKUAST Press, Shalimar No.75/250/12/92.

Wani, G.M. 1995. Early diagnosis of pregnancy in surrogate mothers (Sheep and Goats).National Symposium on Embryo Transfer and Genetic Engineering Technologies in Small Ruminants; May 7-8, 1995, CIRG, Makhdoom, Farah, India.

Wani, G.M. 1998. Role of assisted reproductive technologies in augumentation of sheep production. Lead paper presented on National eminar on "Strategies for maximizing production and efficiency in sheep", held at CSWRI, Avikanagar, Rajasthan, April, 1998 PP.18-22.

Wani, G.M. 2001. Enhancement of chevon Production through reproductive management proceedings IV National Seminar of Indian Society for Sheep and Goat Production and utilization (ISSGPU) Nov 9-11, 2001, Bharathiar University, Coimbatore, Tamilnadu, India.

Wani, G.M. 2002. Research in Animal Reproduction-Vision 2050. Proceedings XVII Annual Convention of ISSAR and National Symposium, Nov.14-16, 2002, IVRI, Izatnagar U.P

Wani, G.M. 2003 a. Biotechnological advances in Goat Production. National Workshop on "Recent advances in Goat rearing", MPKVK, Agric. Univ. Rahuri, 29-30 Dec, 2003.

Wani, G.M. 2003. Goat Husbandry- multipurpose management. Encyclopedia of Dairy Sciences, edited by Roginishi Hubert, Euquary, J.W., Fax, P. Academic Press London Vol.2 PP 1259-1270.

Wani, N.A, and Wani, G.M. 2003. Use of epididymal spermatozoa from slaughtered sheep for invitro fertilization of invitro matured Oocytes. *Indian.J.Anim.Sci*, 73:143-146.

Wani, N.A; 1996. Invitro maturation and invitro Gertilization of ovine Oocytes MVSc Thesis, SKUAST.

Wani, N.A; Wani, G.M; Khan, M.Z and Salahuddin, S.2000. Effect of Oocyte harvesting techniques on invitro maturation and invitro fertilization in sheep. *Small Ruminant Research.* 36: 63-67.

Wani, N.A; Wani, G.M; Khan, M.Z and Siddiqui, M.A. 1999. Effect of different factors on the recovery rate of Oocytes for invitro maturation and invitro-fertilization. *Small Ruminant Research* 34:71-76.

Wani, N.A; Wani, G.M; Mir, M.M and Beg, M.A. 1998. Ultrasonic pregnancy diagnosis in Gaddi Goats. *Small Ruminant Research* 29:239-240.

□□□

Chapter 18

Early Embryonic Development in Sheep and Goats

G.M. Wani, A. Khatoon, P. Gowsomi, S. Mir & M. Darzi

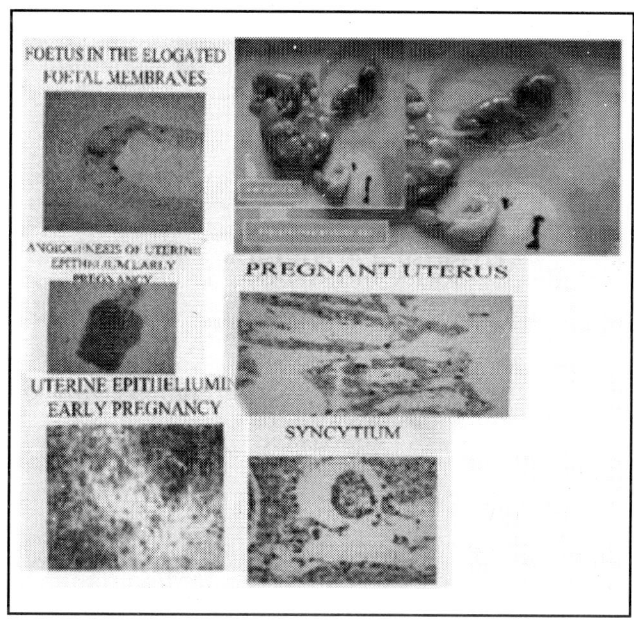

1. Role of corpus luteum in initiating caruncle and cotyledon development. 2. Syncytia formation and blastocyst elongation. To investigate Binucleate and hybrid nucleate cell formatioin in placeta. To gather information on signaling pathways and co-relate it with visual post-mortem samples. All studies on abattoir materials would be collaborated through in vivo-ultrasonic and electronic images to be posted latter, as work is in progress.

Introduction

Sheep uteri consist of a large number of caruncles. These caruncular areas are the sites of implantation. (Gray et al, 2002). Ovine preimplantation conceptus undergoes elongation. The elongated blastocyst extends itself both in gravid and non-gravid uteri. Uterine epithelium under goes through different changes prior to implantation. What is the nature of these changes and how they can be morphologically recognized is still obscure. The placentomes serve as immunological barrier membrane and mediate gas and micro-nutrient exchange. The changes under the influence of corpora haemorrhagicum is the begining of immunological protective cover, which prevents rejection of the cenceptus as a foreign body or influences maternal recognition of pregnancy (Wani, 1996).

This early development mechanism is our primary concern to be investigated in this study. Perhaps this invisible mechanism was populated by Corner back in 1947. He said, "Anterior chamber of eye is superior to womb for nourishing early embryo except that the ovarian hormones make it a well cherished superior chamber for survival and growth of the embryo."

Attempts to describe this superiority of "Womb" have remained limited to endocrinological and hormonal explanations and explorations. This investigation attempts to unravel the morphological and cytological aspects to explain the mechanism of protection, placenta development, immunological barrier and implantation signals in ovine conceptus. Paper presented on 2nd J&K Science Congress, held at Kashmir University in July, 2006.

Objectives

The embryonic diapauses model of delayed implantation in lactating mothers was thought to give important clue on the subject. (Lopes et al 2004).. Morphological and cytological studies with endocrinological and other physiological aspects explanations is not available to find basic answer to many questions like:

1. Role of corpus luteum in initiating caruncle and cotyledon development.
2. Syncytia formation and blastocyst elongation.
3. Binucleate and hybrid nucleate cell formation. As such this study has investigated.
    a) The local uterine vascular permeability and receptivity of uterine epithelium and its morphological and cellular visualization. It was the primary focus for this investigation.

b) This study shall gather information on signaling pathways and co-relate it with visual post-mortem and slaughter house samples. All these studies on abattoir materials would be collaborated through in vivo-ultrasonic and electronic images to be posted in literature later, as work is in progress.

Brief Review

Endometrial receptivity is stimulated by ovarian steroids. The role of ovine and caprine ovarian steroids in initiating endometrial receptivity have been reviewed by various workers.(Huct-Hudson et al, 1989, Wani, 1995, 1996 Wani, 2003 , Wani 2005, Wani, 2006). Proliferation of uterine cells is influenced by E2 and P4. These hormones have their cognate nuclear receptors in ovine/ Caprine uteri called E2 receptor (ER) and P4 receptor (PR). {Lee, and De Myo, 2004). This intrinsic ability of uterine epithelium has been reviewed (Dey et al 2004).

The intercaruncular regions filled with uterine glands in sheep develop during Ist month post Lambing. A simple tubular luman is replaced with aglandular, carcuncle inhabited or projected uteri as shown in our prdiminary investigations Fig.1 Wani et al 2006 and others too {Wiley et al, 1987). These uterine glands fail to develop if pre-natal withdrawal of P4 at birth was removed. This mean at birth the P4 supplies to new born ewe lamb hinders uterus glandular formation due to lack of progesterone stimuli. In an overetomized new born ewe lamb, implants of 19 norprogestin, failed to develop uterine glandular system. Thus "Progesterone knock out" at birth stimulates glandular formation during first month of newborn ewe lamb. These experiments on prolonged use of P4 stimulation in neonate ewe lambs has lead to a model commonly called as "Ovine uterine gland Knockout (UGKO)." Animal model. (Spencer et al, 1999). This model has been used to identify gene expression in endometrial epithelium (Gray et al 2002). Gene expressions under the stimuli from P4 are essential for implantation and maintenance of pregnancy. Various other factors having significance for inducing implantation are calcitonin, Indian hedge Log (IHA) Ambhiregulin etc (Lee and Dehayo, 2004).

Various growth factors like epidermal growth factors and their family members like TGF2 (Transforming Growth factor 2) released during pre-implantation, heparine binding EGF (HB-EGF) related to blastocyst attachment in mice and other factors like vascular endothelial growth factors (VEGF) reviewed by Wani, 2006 and described by Tamada et al 2002, Paria et al 2001, Tischer et al, 1991 need exploration in dertermining ovine implantation mechanism. Various studies in humans (Mueller et al, 2000, Wol in Wang et al 2003 Row et al 2003) could help to develop ovine-implantation-mechanism model envisaged in this study. The role of cytokines (Hilton, 1998) like Leukemia inhibitory factors (LIF) and others like interleukins-II; colony stimulating factor-1 (CSF1) and Modulators cell adhesions Mucin I (MucI) are important biochemical parameters to be probed for explaining ovine implantation model. Developmental factors like Homesbox (Hox) genes, Wnt genes could help to understand regulatory and signaling processes involved in ovine implantation. Many such studies have been undertaken in mice or primates. (Tulac et al 2003) but remain obscure in ovine and human uteri. This implantation mechanism knock out is the primary objective of this study. Once understood embryonic mortality and fertility improvement in livestock shall be easier. It may help to devise side affect free biological control of human fertility. Its economic importance can be counted in millions of dollars and human welfare, thus this study shall have both economic and human welfare prospective. The following innovative observations accredited to our group are detailed as follows. This innovative lead paper shall describe and show our pioneering and original results. To the best of our knowledge no such observations on foetal development are available in literature.

As we all know sheep and goat fertilized eggs develop into a Morula (8 cells) within 2-3 days post coitus. The figures shows the oocytes and morula of sheep cultivated in our laboratory (Wani, 1996, Wani et al, 2000}. The corpus luteaum Haemorrhagium (CLH) can be seen figure. The early uterine epitheliaum under the influence of P4 secretions from functional corpus luteaum can be seen figure. This is a clear angiogenic activity in early pregnancy and indicates preparation of the uterine epithetheluim to accommodate the incoming conceptus. On the other side of the figure marked A is another figures showing functional CL at mark "B" figure shows the persistent corpusluteum (PCL). The hypothetical attachment site in sheep can be seen figure. These are the elongated foetal blastocyst. The foetus lies within amniotic sac. The reddish areas are the implantation site as observed in our investigation. An external view of 25 & 57 day old foetus within maternal uterus can be seen figure. The raised nodes on the large sac(A) are developing. The caruncle in a non-pregnant uterus can be seen figure.

There is a wide confusion among biologists about the three terms, caruncles, cotyledons and placentomes. Our observations are the first to show a vivid distinction between the three. Caruncle can be seen as solid projections on the surface of the uterine epithelium as shown figure. They may be hypertrophied in some disease conditions or during normal pregnancy of early 2 week as shown figure. The early foetal development in an elongated blastocyst before attachment can be seen figure. The histological evidence of the union of foetal placental membranes with the uterine caruncle can be established only when the foetal cotyledons as shown figure get attached with maternal cornucle Fig. 8. This results into an intimate association with the mother in thè form of placentomes as show figure.

The development of the foetus during first two weeks takes place within the foetal membranes as shown figure. The corresponding uterine epithelial changes during first week of gestation can be increased blood supply to the uterus as shown figure. The corresponding histological evidences of syncytium development in early pregnancy can be seen figure. These findings shall be discussed in the presentation.

References

Corner, G.W. 1947. The hormones in human reproduction. Ed; Princeton N.J.Princeton University Press.

Dey.S.K.2004. Focus on implantation reproduction 128: 655.

Gray, C.A; Barghardt, R.C; Johnson, G.A; Bazer, F.W. and Spencer, T.E 2002. Evidence that absence of endometreal gland secretions in uterine gland Knock-out ewes compromises conceputs survival and elongation Reproduction. 124:289-300

Hilton, D.G. 1998. Leukemia inhibitory factor in sytokines, PP 277-296 ed. AR-Mirc-Sluis. New York, Academic press.

Huct-Hduson, M.Y.Andres, G.K. and Dey, S.K. 1989. Cell type specific localization of C-Myc-Protein in mouse uterus-modulation by steroid hormones and analysis of the pre-implantation period. *Endocrinology.* 125 :1683-1690

Lee, K.Y. and De Mayo, F.J.2004. Animal models of implantation. *Reproduction.* 128: 679-695.

Lopes, K.Y. and De-Mayo, F.I. 2004. *Animal Models of Implantation Reproduction* 128: 679-695

Mueller, M.D; Vigne,J.L; Minchenko, A;Lebovic, D.I, Leitman, Dc and Taylor, R.N. 2000. Regulation of vascular endothelial growth factor (VEGF) gene transcription by E2 receptors alpha and Beta. *PNAS.* 97: 10972-10977.

Rowe,A.J; Wulff. C. and Fraser, H.M. 2003. Localization of MRNA for vascular endothelial growth factor (VEGF), angiopoitins and their receptors during peri-implantation period and early pregnancy in marmosets (Callitrix Jacchus) *Reproduction.* 126:227-238.

Spencer, T.E. Stagg, A.G; Joyce, M.M; Jenster, G;Wood, C.G; azer, W.F. Wiley, A.A. and Bartol, F.F. 1999. Discovery and Characterization of endometrial epithelial messenger ribonucleic acids using the ovine gland knok out model. *Endocrinology,* 140:4070-4080.

Tamada, H; Tsubutani, D; Kawate, N; Inaba, T; Matsuya ma, S; Imakawa, K; Sakai, S; Christenson, R.K. and Sawada, T. 2002. Detection of transforming growth factor alpha and epidermal growth factor receptor RNA immunohistochemical localization of their proteins in the ovine uterus during early implantation period Histochemical localization of their proteins in the ovine uterus during early implantation period. *Histochemical Journal.* 34:383-390.

Tulac, S;Nayak, N.R; Kao, L.C; Van Waes, M; Huang, J; lobo. S; Germyer; A, Lessey, B.A, Taylor, R.N, Suchanek, E and Giudice. L.C.2003. Identification, Characterization and regulation of the canonical Wnt segnalling pathway in human endometerium, *J. Clinical, Endocrine and Metabolism* 88:3860-3866.

Wang, W;Li, Q; Lin, H; Yu, X Qian, D; Dai, J; Daun, E and Zhu, C.2003. Expression of vascular endothelial growth factor and its receptors in the rhesus monkey (Muscaca mulatto) endomenterium and placenta during early pregnancy. *Molecular Reproduction and Development.* 65:123-131.

Wani, G.M. 1995 Wani, G.M. 1995. Early diagnosis of pregnancy in surrogate mothers (Sheep & Goats). National Symposium on Embryo Transfer and Genetic Engineering Technologies in Small Ruminants, May 7-8, CIRG, Makhdoom, Farah, India

Wani, G.M. 1996. Embryo Biotechnology in Sheep and Goats. Publisher Valley Book House, University Road, Hazratbal Srinagar - Kashmir

Wani, G.M. 2005. Advances in ovine & Caprine Reproduction www.Word-mavt. Com.Summaries free. *Research paper* 9 : 1-25.

Wani, G.M.2003. Biotechnological advances in goats production National Workshop on "Recent Advance in goat Rearing. MPKKV, Agric. Uni, Rahuri, 29-30 De, 2003

Wiley, A.A; Bartol, F.F. and Barron, D.H.1987. Histogenesis of the ovine uterus. *Journal of Animal Science.* 64:1262-269.

□□□

Chapter 19

Mammalian Foetal Attachment

G.M. Wani, P. Gowsomi, A. Khatoon, P. Wani & G. Wani

CLH/SECTION

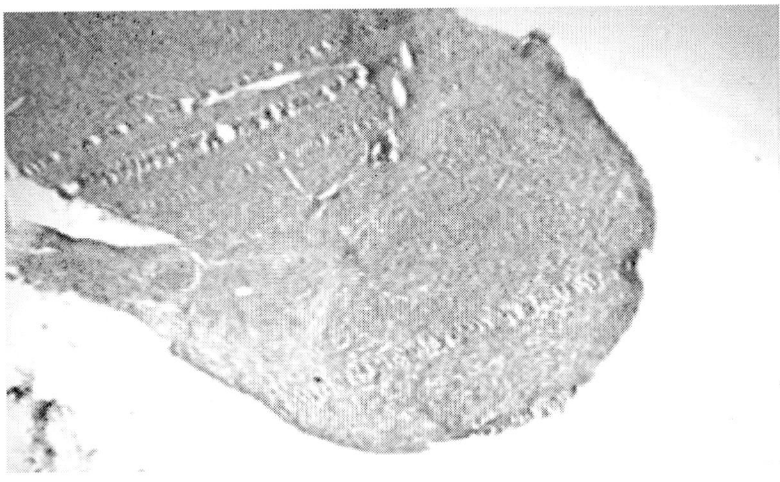

Key words : Placenta, Cotyledons, caruncles, sheep, mammals, goat, pregnancy, foetus.

* Director Extension Education, SKUAST-K Shalimar Srinagar

Corresponding author :E.mail: wanimohyuddin@yahoo.com. Website.www. wanigm.net

** Bolan Medical College, *** Hamdard University

Abstract

The blood of mother and foetus never mixes freely in ovine or caprine foetus. The fertilized zygote develops into a morula which decends and reaches uterus in 3-4 days. The presence of corpus Luteum haemorrhagicum (CLH) and its subsequent growth ensures progesterone secretion.

Endometrial recieptivity was morphologically documented in this study. The solid tooth shaped projections of non-pregnant uterus change into mucin coated cup-shaped caruncles. Conco mitant to these changes blastocysts elongates. Within elongated blastocysts cotyledons develop which fix into caruncles to form the placenta

The unique feature presented in this paper is recognition of non-self (foetus) by self (endometrium). We present here the following observations"

1. The caruncles in non-pregnant uterine horns consist of solid projection arranged in rows. They are red and upright ville-like. No Corpus Luteum Haemorrhagicum (CLH) was present.
2. In the presence of CLH these caruncles change in form, colour and structure. They are cup-shaped , mucin coated, whitish in colour and give an impression as if a whitish cover or membrane overwrap them.
3. In the meantime blastocyst elongates and covers the whole uterine horn. Within the blastocysts or now we can call it an elongated ballon shaped foetus floating over hypertrophied caruncles. Within this developing foetus sac small round, red shaped ball develop and it is called cotyledons. These cotyledons further fix themselves into cup-shaped caruncles. The formation of placentomes begins both within foetal membrances and endometruim cover over cotyledons and caruncles respectively, before union. Thus an indirect attachment of cotyledons and caruncles occurs which is separated by different membranous coverings described.

The mechanism of such attachment and a system of manifold protective covers prevent direct contact of foetus and maternal tissues. The minute details of attachments are supported by histological nuclear unions too. To the best of our knowledge, it is the first finding of the kind being published in literature.

Introduction

Mammalian uterine structure and its attachment with foetal tissues differs in various species and is well described. Ruminants like sheep, goats and cows have uterine epithelial projections called caruncles which act as anchors for foetal membrane attachment. Ovine pre-implantation conceptus under goes elongation. The role of corpus luteum, caruncle, cotyledon development has been recently elaborated. (Wani et al, 2006).

Pregnancy starts with intimate association between mother and foetus. This was thought to be a maternal controlled mechanism, however now we know the formation of this association is a joint foeto-maternal interaction. Gray et al 2005, Spencer et al 2004). Prevental development landmarks during different stages of mammalian pregnancy have been reported. (Mufti et al, 2000). However, the mechanism that co-ordinates, orients and initiates union of foetal cotyledons and caruncles is still obscure. The uterine epithelial angionesis, increased vascularity and various molecular changes have been documented in mammalian models (Lee and Demyo, 2004), various genes, transcription processes and vascular stimulators have been postulated. The structural changes in Endometrium and foetal membranes before actual union is unknown.

Foetus separates from mother through placenta. Gaseous nutrient and waste materials are exchanged but the earlier coating and the umbilicus development is white, shining devoid of any blood or colouration (Ain 2003). The midgestation placental attachment via umbilicus is still white shinning Wani et al 2006). The advanced connections are blood stained as shown Mufti, 1996 Mufti et al 2000). The blood of the mother never mix especially during first trimester of ovine or caprine gestation (Raven 2002).

The placental barrier comprises of trophoblast cells which serve as protective barrier I, for non-rejection of the foetus is know (Billington, 2003). Does endometrial barrier II too exists ? or does the barrier I is supported by double coat of elongated blastocyst existing prior to attachment. The exact protective barrier is still obscure. This paper investigated the possible morphological union of the foetal and endometrial parts. The histological evidences were explored to support the morphological findings.

Materials and Methods

Pregnant ovine or caprine Genital tract samples were collected from slaughter house. The breed, age and breeding status of these animals was not known. The genital tract samples after slaughter were carried to laboratory in thermos flask.

The genital tract were cleared of fat and examined for presence of functional Corpora lutea Haemorrhagicum (CLH). The uterine horns were opened and status of caruncles was photographed in both the uterine horns. It was campared with caruncles of uterine horns having no corpus luteum. The status of gravidity was determined by measuring Crown Rump. Length(CRL) and calculating age of foetus as described. (*Mufti 1996).* Morphological appearance of CLH, caruncles and uterine endometrium was photographed. A portion of the ovary, CLH endometrium, caruncles was sectioned and processed for histological examinations. The morphological and histological observations of animals having no CLH, corpus luteum Albicon (CLA) were compared. On the basis of CLA or CLH presence the differentiation of non-pregnant and pregnant uterine morphology was described with circumstantial evidence of histological sections. The development of cotyledon, their fusion with caruncles and development of placentomes were studied and documented.

Results

The images of CLH, caruncles in non-pregnant animals and early pregnancy are shown in Fig 1-3. The histological view of the above three conditions is shown in Fig 4-6. The hypotrophy of the caruncles uner progesterone influence can be seen Fig 7-8. These figures of the caruncles before attachment indicates receiptivity changes in caruncles.

Similar changes in elongated blastocyst occur, when foetus has an extra-membranic cover. The formation of foetus, foetal membrances, cotylodons and umbilical connections can be seen in figures 9-10. These images clearly indicate formation of interconnected red ball as cotyledons within the elongated foetal membranes. It is the extended balloon shaped foetal Sac, which as such fuse or attach within the cupshaped caruncles. (Fig. 11). The whole union is indirect through protective coverings which consist of:

i. Caruncular mucin coated covering - Fig 6
ii. Angiogenic covering over this white membrancous covering. - Fig 8
iii. Foetal membrance covering foetus & foetal cotyledous interwoven through blood vessels. - Fig 10.
iv. Fusion of the cotyledons and caruncles indirectly with their respective membrances inbetween them as shown –Fig 11. Thus the foetal and maternal tissues in sheep and goat(mammalian species) studied never directly meet, but have indirectly connection which forms the basis of non-self as self recognition by the mother. This is the basis of immunoprotection too as confirmed by double nuclei and hybrid nuclei formation as shown in Fig 12 & 13.

Discussions

Establishment of endometrial receptivity is still a biological mystry (Makker and Singh 2006). We present hereby images (1-3) of CLH, caruncle in NP and Early pregnant uteri along with their histological evidences images (4-6). These findings clearly show that caruncles acquire structural changes before embayo reaches uterus. These changes occur due to presence of CLH which secretes progesterone. The life of CLH beyond the cycle end depends on the presence of conceptus.The fusion of binucleate cell trophoblast with caruncular epithelial cells forms syncytia. These syncytia secrete placental lactogen (Bowen, 2000).

The changes as seen in the caruncles and endometrium image, are receptivity changes before attachment. The attachment process begins when foetus is already differentiated into organs. We noticed the development of umbilicus, which never attaches to the endometrium at any specific point. The ballon like foetal sac with foetus and foetal membranes is fully recognizable pre-implantation or attachment. The confusion within scientific faternity on usage of term implantation in ovine or caprine is hereby cleared. It is exactly an attachment. The attachment site isnot pre-determined. All the cotyledons join together and net work of connection within all cotyledons is formed before attachment. At attachment all these indivisual unions of caruncles and cotyledous make placentomes. These placentomes are inter-connected and all connections web together as umbilicus. Thus unblical card is a wire-net-connections of placentomes which fuse together like a telephone cable, connecting individual connections to the main point called Umblicus. The earlier reports of attachment around day 18-20 could not be confirmed in this study as 34 day foetus was still within elongated foetal memberanes. We noticed loose attarchment around 48-50 days in sheep foetus. The recieptivity of the endometrium consists of hypertrophied caruncles, membranous structures wraped over caruncles. It is the beginning of Ist Maternal component of placental barrier hither to unknown, but confirmed by histological examination, as no binuclear or hybrid nuclear cells were observed. The thickening of myometrium and caruncular epithelium as shown do propose structural changes akin to pregnancy. The binuclear cells in the foetal placental membrane and later hybrid multinuclear syncytia in placenta (caruncles + cotyledons) were confirmed. Our postulation and hypothesis as confirmed is by present observations of triple membranous barrier cannot be compared, due to absence of information in the literature. To that extent these findings are new addition to knowledge, which need further confirmation through biochemical and hormonal estimations under investigation.

These findings confirm the earlier studies of Billington, 2003 regarding non-rejection of foetus. He describes outer layer of trophoblast as barrier I. To our understanding three such barriers function as inner and outer layer of trophoblast and mucin coated membrance wrap covering of hypertrophied cup-shaped caruncles before attachment.

Thus the beginning of non-self as self is this barrier I over caruncles including their desensitization under progesterone phase. The two other barriers are trophoblastic layers of the elongated blastocyst, which as such wraps within, uterus and *the invagination of cotyledons within cup shaped caruncles as shown in the images. This could possibly explain the immunogenic tolerance for foetus as postulated (Thellin 2000) and (Jer Zak,2002).*

Early pregnancy CLH and slight asymetrica luterine horns

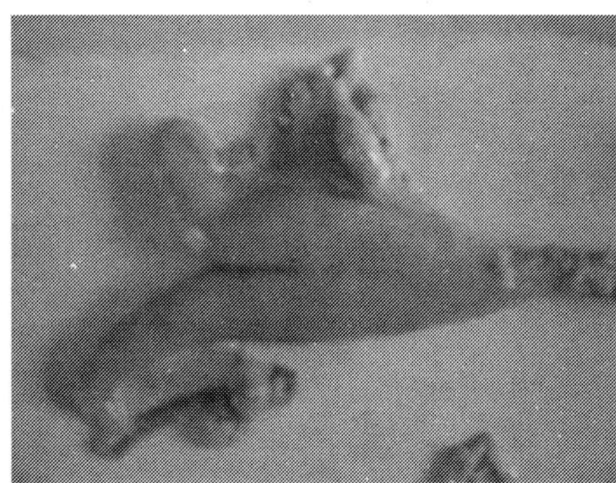

Fig. 1

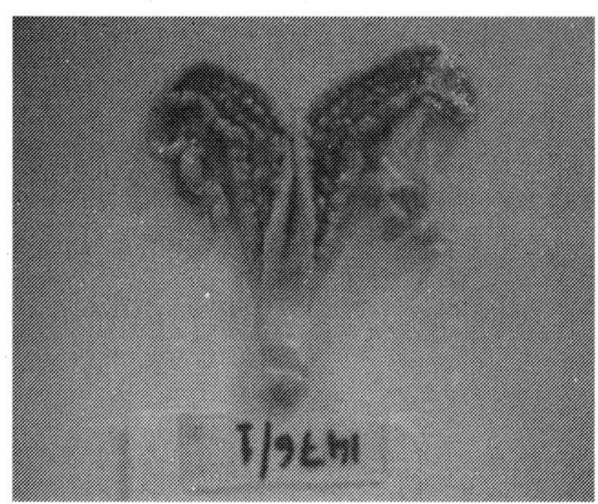

Fig. 2

Pre-implantation caruncular changes

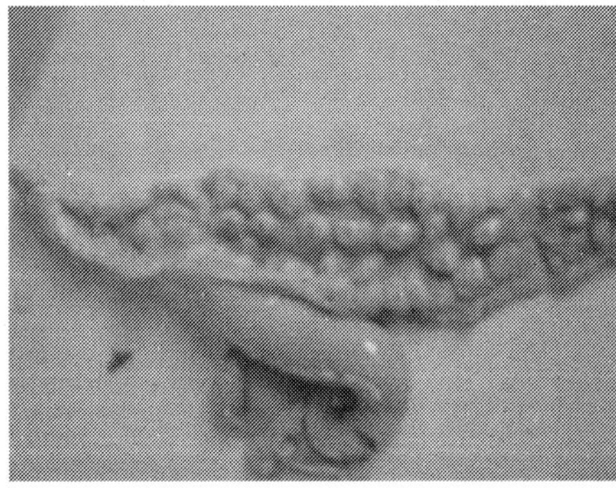

Fig. 3

CLH/SECTION

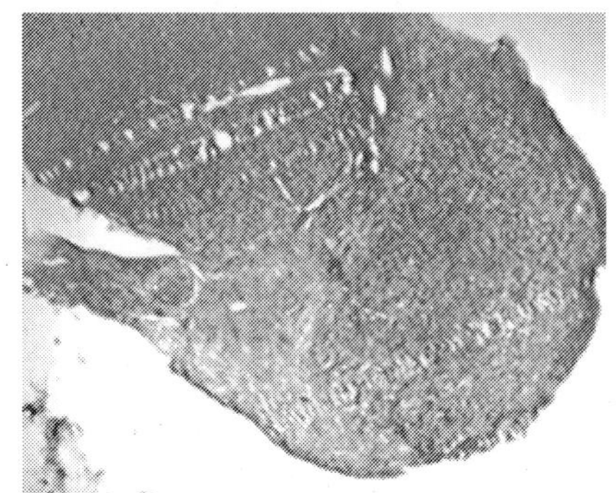

Fig. 4

contd....

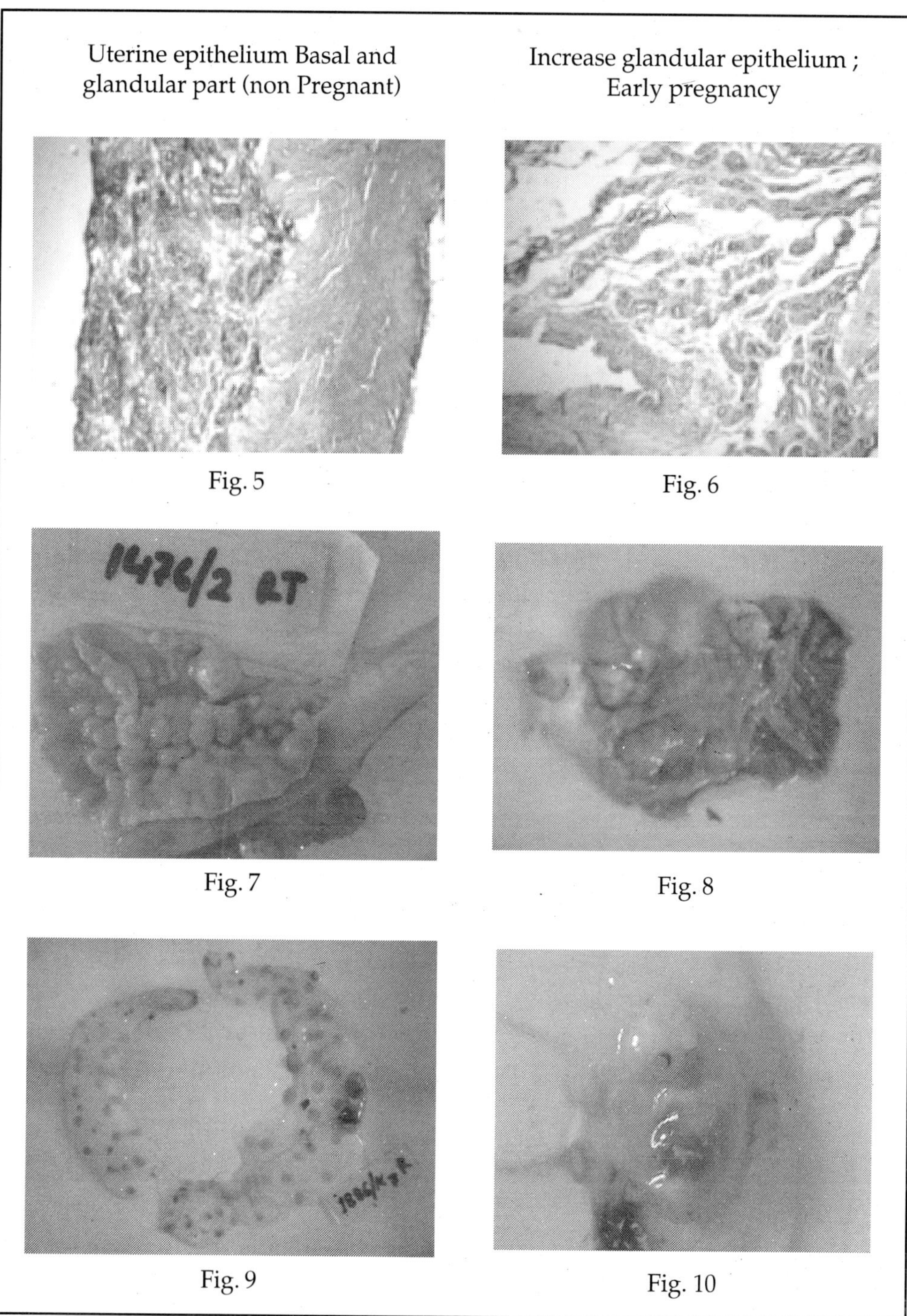

Uterine epithelium Basal and glandular part (non Pregnant)

Increase glandular epithelium ; Early pregnancy

Fig. 5

Fig. 6

Fig. 7

Fig. 8

Fig. 9

Fig. 10

contd....

Higher magnification; elongation of nucleus

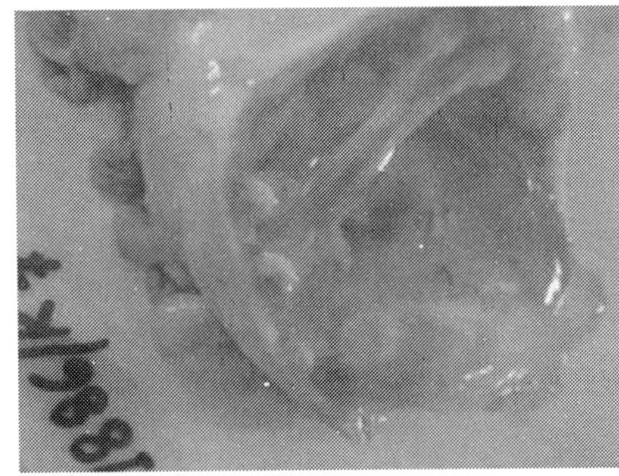

Fig. 11

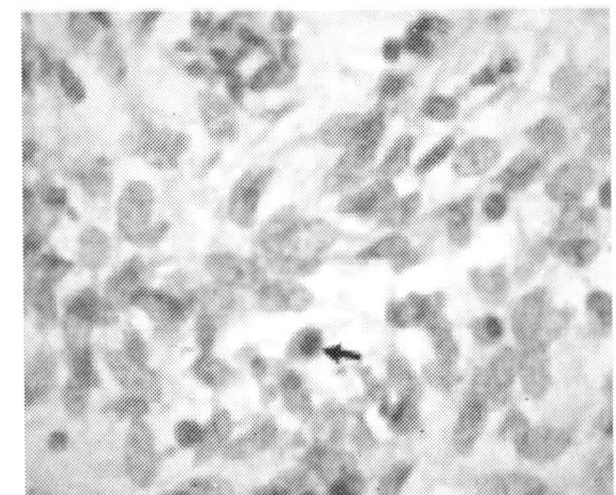

Fig. 12

Syncytia

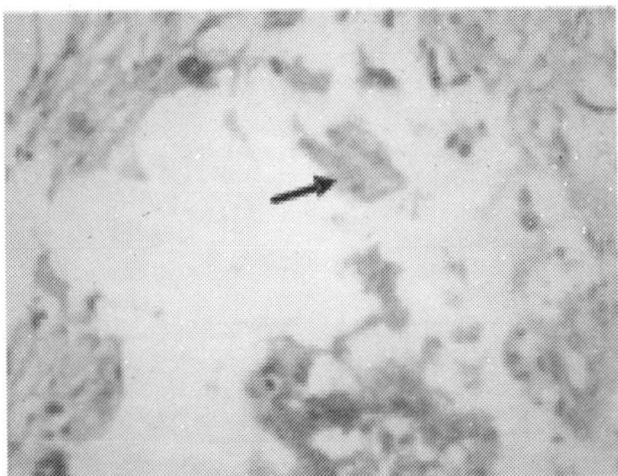

Fig. 13

Chapter 20

Uterine Physiology and Foetal Development

G.M. Wani, P. Gowsomi, A. Khatoon, P. Wani & G. Wani

This paper presents results on clinical aspects of embryonic and foetal development. It describes and answers some questions on implantation, foetal development.

ABSTRACT

This paper presents our own results on clinical aspects of embryonic and foetal development in Small Ruminants. We shall be presenting a review and our results with discussions on:

i) Estrogen-progesterone milieu in non-pregnant and pregnant animals during different (pregnancy).

ii) The new concepts of Estrogen: Progesterone receptors shall be compared in non-pregnant and pregnant animals.

iii) These hormonal values shall be co-related with morphological and physiological development of caruncles, blastocyst, foetus, with histopathological and actual visual appearances. This shall help clinico-reproductive diagnostics in small ruminants.

Introduction

Embryonic losses in sheep and goats are very high. About (22-44%) embryos are lost during early pregnancy, maximum losses occur during pre-implantation period (Wani, 2006). Various biochemical, nutritional and genetic causes have been discussed (Lee and De Mayo, 2004 and Wani, 2005). Ovine pre-implantation blastocyst undergoes elongation for up to 30 days and feeds on uterine milk through histotrophs (Wani et al., 2006). Uterine epithelium undergoes various changes. The morphological nature of these changes is not recorded. The changes in caruncles and endometrium are the beginning of immunological protective cover which prevents rejection of conceptus (Wani, 1996). Therefore, this study was planned to find out morphological, physiologic and histological changes. They were viewed with biochemical changes too.

Review

Uterine Physiology

The uterine endometrium of pregnant females is dominated by progesterone which stimulates development of glandular epithelium that becomes highly secretory and produces histotrophe which is essential for conceptus development. Placental production of progesterone is known to increase markedly between 70th and 100th days of gestation in the ewe after the placenta has reached its maximum size (Gordon, 2004). The corpus luteum of pregnancy is essential throughout pregnancy in all farm animals except, ewe and mare as placenta produces sufficient progesterone to maintain pregnancy in these species. Endometrial oestrogen and plasma estrogen levels were also inversely correlated with plasma P4 levels. Thus, P4 levels influences ER levels, the hormone may make the tissue less responsive to ovarian oestrogen at such times as the luteal phase of the cycle and during pregnancy (Henricks and Harris, 1978). Therefore, the concentrations of Estrogen Progesterone receptors have an importance for pregnancy.

Estrogen/Progesterone Receptors: The concentration of oestradiol cytosolic rcceptors (ERc) and oestradiol nuclear receptors (ERn) in non placental and placental myometrium was low in early and mid pregnancy, increasing in late pregnancy and was highest at parturition. The levels of progesterone cytosolic receptor (PRc) in non placental regions were low in early pregnancy, increased in late pregnancy but declined during parturition whereas progesterone cytosolic receptor (PRc) in placental regions and progesterone nuclear receptor (PRn) in placental and non placental regions showed reverse character.

Thomas and Fuller (1995) suggested that uterine estrogen receptor (ER) and progesterone receptor (PR) gene expression is regulated in a tissue and cell. In ovine uterus, estrogen increases both estrogen receptor (ER) and progesterone receptor (PR) expression, where as progesterone (P) inhibits estrogen receptor (ER) and the expression of its own receptor. Further they have recorded that alterations in tissue and cell specific expression of uterine estrogen receptor (ER) and progesterone receptor (PR) not only depend on circulating steroid hormone concentrations but also depend on locally produced, paracrine acting factors. Serum progesterone rapidly rises following formation of corpus luteum. Timing of implantation may be regulated by length of time the uterine endometrium is exposed to progesterone stimulation. The conceptus can not attach to the uterine epithelial surface until the large glycoprotein called MUC-1 has been removed. Over 8-10 days of exposure to progesterone, nuclear progesterone receptors in the uterine epithelium down-regulate leading to a loss of the direct effect of progesterone on this cell type. Since epithelial synthesis of MUC-1 is stimulated by progesterone, loss of progesterone receptor from the uterine epithelium would reduce MUC-1 production and open a receptive state for conceptus attachment (Hafez and Hafez, 2000). Endometrial Receptivity: Endometrial receptivity is stimulated by ovarian steroids. The role of ovine and caprine ovarian steroids in initiating endometrial receptivity has been reviewed by various workers (Huct-Hudson et al., 1989, Wani, 1995, Wani 1996, Wani, 2003, Wani 2005 and Wani, 2006). Proliferation of uterine cells is influenced by estrogen (E2) and progesterone (P4). These hormones have their cognate nuclear receptors in ovine / caprine uteri called E2 receptor (ER) and P4 receptor (PR). (Lee and De Mayo, 2004). This intrinsic ability of uterine epithelium has been reviewed (Dey, 2004). IFN-tau is believed to act by down-regulating oestrogen receptor thus preventing the oxytocin receptor and regulate the progesterone secretion (Gordon, 2004). Serum progesterone concentration remained at luteal levels 1.5 ± 0.5 ng/ml for the first two months of pregnancy and then gradually increased in pregnant Sumatran rhinoceros. However, progesterone decreased almost to luteal levels during the 5th months after that it increased again and eventually reached peak concentration (13.3 ± 1.9 ng/ml) shortly before parturition (Roth et al., 2004). Histological changes in the female genital organ: Extensive erosion of pseudostratified columnar epithelium of the uterus occurred at the time of invasion of binucleate cells of the trophoblast (Bhavsar et al., 1974).

On day 18th the presence of blastocyst reduced the height of the uterine epithelium, on day 20th the trophoblast was starting to contact with maternal epithelium; while on day 21st and 30th attachment became more complex. King et al. (1981)

This review suggests that biochemical changes occur in the composition or distribution of the trophoblastic cell coat during the process of blastocyst attachment in the ewe (Guillomot et al., 1982). Upregulation of progesterone receptors (PR) in endometrial stroma was mediated by estrogen signal which induces stromal cells to differentiate into decidual cells (Kurita et al., 2001).

Our Own Results

Progesterone : The mean concentrations of progesterone in maternal serum observed from Ist to 5th were 5.32±0.33, 8.26 ± 0.11, 7.70 ±, 0.23, 7.08 ± 0.35, 7.00 ± 0.32 respectively. The progesterone concentrations increased significantly (p=0.05) from Ist to 5th month pregnancy and decreased significantly (p=0.05) in 3rd month but decreased non-significantly from 4th to 5th. The progesterone concentration increased significantly (p+0.05) in between. The progesterone concentration increased significantly (p=0.05) in between groups (Khatoon, 2007).

Progesterone receptor : The mean concentrations of progesterone receptor in uterine tissue from Ist to 5th month pregnancy serum samples were 3.81 ± 0.04, 4.33 ± 0.06, 4.38 ± 0.04, 4.09 ± 0.09, 3.86 ± 0.12 respectively . The mean concentration of progesterone receptor followed similar shift as in progesterone . The progesterone receptor concentration increased significantly (p=0.05) in between Ist and 2nd, Ist and 3rd, Ist and 4th and decreased significantly (p=0.05) in between 2nd and 4th, 2nd and 5th 3rd and 5th.

Estrogen receptor : The mean concentrations of estrogen receptor in uterine tissue from GPI to Gp5 were 0.37 ± 0.01, 0.41 ± 0.01, 0.41 ± 0.01, 0.37 ± 0.02, 0.36 ± 0.03 respectively (Table: 4A.7). The estrogen receptor concentrations increased significantly (p=0.05) in between GpA and GpB, GpA and GpC and decreased significantly (p=0.05) in between GpB and GpD, GpB and GpE, GpC and GpD.

(9)		Group	nomenculture	is	as
GPA	=	Ist	month	of	pregnancy
GPB	=	2nd	month	of	pregnancy
GPC	=	3rd	month	of	pregnancy
GPD	=	4th	month	of	pregnancy
GPE	=	5th	month	of	pregnancy

New Formula Invention

Prediction of age in days in relation to whole embryo/Crown-anus (CA) length (Wani-Khatoon Formula).

Sl. No. Whole/CA Length in cm Age in days		
1	0.05	12
2.	0.07	14
3.	0.2	15.4
4.	0.4	18.0
5.	0.6	20.2
6.	0.8	23
7.	1.0	25.5
8.	1.3	29.5
9.	1.6	33.5
10.	1.8	36
11.	1.9	37.5

Morphological Changes

(i) *Early Pregnancy* : At 18 days of gestation, the right horn was larger than the left and right ovary contained corpus luteum. An elongated blastocyst was present with comma shaped embryo. The changes in the caruncles were same with previous sample.

At 20 days of gestation, the right horn was larger than the left and right ovary having the corpus luteum. An elongated blastocyst was present with embryo. Both caruncular and intercaruncular area gave live appearance. The centre of the caruncles was more reddish compared to other parts of the horns. Caruncles were swollen, slightly more larger than earlier and rounded.

At 23 days of gestation, the horns were asymmetrical and body of the uterus bulged dorsally. The left ovary had corpus luteum. An elongated blastocyst was present with embryo. The changes in the caruncle and intercaruncular areas were same with 20 days of pregnancy. The fitures depict the photographs of all events narrated above and hereinafter.

At 25 days of gestation, the right horn was slightly larger than the left. An elongated blastocyst was present with embryo. Both caruncular and intercaruncular area gave lively appearance. The centre of the caruncles had slight circular depression.

At 29 days of gestation right ovary had double corpus luteum and comma shaped embryos were present one in each of the horns. Changes in the caruncles were as like 25 days gestation. Stereomicroscopic view gave the distinct comma shaped feature of the embryo. Under microscope some other structures like somites, head, eye bud, tail bud and emerging blood vessels from hind gut could also be observed in the elongated blastocyst. Two vessels emerged from each of the embryos and entered into the either of the horns along with its ramification could also be observed grossly. From this stage onwards emerging of blood vessels from the hind gut of foetus became distinct grossly.

At 33 days of gestation right horn was more enlarged than the left. A curved embryo was present in the right horn, emerging of blood vessels from the hind gut of foetus became distinct grossly. Caruncles were flattened type.

At 36 days of gestation left horn was more enlarged than the right and the corpus luteum was present in the right ovary. Foetus was present in the left horn. At this stage the blastocyst was completely filled with fluid and whole blastocyst ramified with blood vessels and it was almost reddish in color. From this stage head, eye bud and tail could be recognized grossly, this might be the beginning of foetal stage (Jainndeen and Hafez, 1984 and Hafez and Hafez, 2000). Circulation increased to the caruncle and intercaruncular areas. Caruncles were flattened; reddish coloration was present at the centre.

(ii) *Mid Gestation :* At 39 days of gestation left horn was more enlarged than the right and the corpus luteum was present in the left ovary. Foetus was present in the left horn. At this stage single layered reddish granules were present mostly in the membrane of the left side were nothing but the cotyledons, appeared for the first time. Cotyledones were arranged in single layered which was reddish in color and granular. Ramification of blood vessels was present both the horns but was more prominent in the left side. Head, eye bud, limb and tail bud in the foetus could be observed grossly. Caruncles were flattened.

At 42 and 43 days of gestation right horn was larger than the left and the corpus luteum was present in the right ovary. Other changes were same as there was in 40 days of gestation except their size and shape of the foetus which were increasing.

At 44 days of gestation right horn was larger than the left and the corpus luteum was present on the right ovary and at around 45 days of gestation left horn was larger than the right and the corpus luteum was present in the left ovary. Other changes were almost same as was in 40 days gestation except, the size of the foetus which was larger and the umbilical chord was very distinct and showed total four numbers of blood vessels, two of them entered on either side of the horns. Cotyledons were multilayered and reddish in color at this stage.

At 46 and 47 days of gestation right horn was larger than the left and the corpus luteum was present on the right ovary. From this stage onwards the cotyledons were raised type and the depression at the centre of the caruncles was increasing. The size of the cotyledons and caruncles increased with higher gestational age. The external body parts of the foetus became distinct from this stage onwards. Cotyledons could easily be separated from the caruncles by holding the uterine horn or by imposing slight pressure.

At 48 days of gestation, right horn was larger than the left and the corpus luteum was present on the right ovary. At this stage demarcation of placenomes was visible from outside of the uterine horns. Caruncular-cotyledonary attachment was strong that could not be separated easily and needed pressure for separation. The placentomes were cup shaped and cotyledons were raised type. From this stage onwards all the external body parts of the foetus developed properly and their size and shape increased as the age advanced.

(iii) *Late Changes :* From 80 days to 135 days the cotyledons were almost engulfed by the placentomes leaving no space at the centre of the caruncles. At this stage separation of cotyledones from the caruncles were very difficult but could be separated by imposing strong pressure.

The umbilical chords were consisting of four numbers of blood vessels out of which two vessels entered on either of the horns with elongated blastocyst. The sub branches developed from main two vessels and entered each of the cotyledons and ultimately connection developed between the cotyledons and the caruncles. Arrangement of blood vessels with the cotyledons were just like a full blooming tree, where blood vessels were just like branches of the tree and cotyledons were like blooming flowers at the terminal ends of the branches. As the time advanced the vessels were becoming thick and prominent.

Hormonal and receptor changes in our studies: The progesterone concentration increased significantly from GpA to GpB but decreased later on. Similarly, estrogens varied from GpA to GpE as in progesterone. The progesterone and estrogen concentration increased significantly with fluctuation between the groups. Rodrigues et al., 1998, reported that serum levels of progesterone increased during gestation decrease abruptly only by two days before preterm. Hafez and Hafez, 2000 and Schular, 2002 suggested that both estrogen and progesterone is a placental hormone and ewe is a placenta depended animal (Hafez, 1987 and Hafez and Hafez, 2000). So, the ratio of estrogen and progesterone might increase as the gestation advanced. The estrogen and progesterone concentration in present study in non pregnant animal was 451.25 pg ± 9.53, and 1.2ng ± 0.05 respectively, where estrogen was much higher and progesterone was much lower in non pregnant animals. Estrogen and progesterone receptor increased upto Gp D and decreased in GpE but the differences were very less, they were almost like constant in between groups.

The estrogen and progesterone receptor concentration in present study in non pregnant animal was 0.55 ng ± 0.01, and 2.19 ng ± 0.08 respectively. Our study was supported by earlier reports given by Thomas and fuller (1995). They have postulated that increased level of progesterone suppress both estrogen and progesterone receptor and increased level of estrogen stimulate both estrogen and progesterone receptor. Hewitt and Korach, 2000 postulated that the progesterone attenuates the proliferative effect of estrogen in the uterus, where as estrogen also induces the progesterone receptor.

The present results with respective to estrogen and progesterone in serum and those of estrogen and progesterone with reported by Hafez and Hafez, 2000, Cengiz et al., 2000, Kurita et al., 2001, Roth et al., 2004. Differentiation and caruncular growth are known to be regulated by estrogen and progesterone (Schular, 2002). The endometrial receptivity changes both in the endometrium and placentation as seen morphologically with due evidences from histopathology could be properly explained in view of our studies on estrogen, progesterone and their receptor concentration estimated throughout the gestation in this study. Many questions regarding the role of ovine and caprine ovarian steroids in initiating endometrial receptivity as observed in this study could be easily explained in view of the review reported by Huct-Hudson et al.,1989, Wani, 1995, Wani, 1996, Wani, 2003, Wani, 2005, Wani, 2006 . Proliferation of uterine cells is influenced by E2 and P4 receptor . Our study corroborated and confirmed the studies of Lee and Dee Mayo, 2004. The ability of uterine caruncular epithelium to throw inter digital projection which fit into the evaginated projections of the cotyledones have been confirmed by ER and PR receptor studies in this investigation. This to a broader extend explains the concept of implantation as put forth by Lee and Dee Mayo, 2004.

□□□

Chapter 21

New Embryonic Concepts

G.M. Wani, P. Wani & G. Wani

Abstract as Introduction

Ovulation, its process and mechanism has been discussed (Wani, 1996). Here we shall discussed recent advances in ovulation process. They include ovarian angiogenesis and biochemical basis of corpur luteum formation, embryonic development blocks which hampered embryonic development beyond 8 cell stage in the past has been over comed. How it was achieved shall be disussed. New advance and terms used in literature like Genome actiation Gap-junction and blastocal formation shall be xaplained pre-implantation embryos and biochemical characteristics shall be discussed, however morphological events shall be depicted in some othe chapter , Embryo biosafety and early pregnancy factors has been reviewed in this chapter. Use of anthelmintics like Benacil and tetramisole orally results in toxicity during pregnancy. The embryotoxicity was 100% during 9-18 days post coitus and abnormal lambs delivered by ewes dosed during pregnancy ranged from 34-57% (Grunzhinskas, 1984).

1. Ovarian Angiogenesis

Angiogenesis is a process of vascular growth. The ovarian angiogenesis increases new blood vessels in ovary. They supply nutrients and hormones to the follicles. Thus angiogenesis promotes follicular growth and corpusluteum formation (Tamanini & Ambrogi, 2004). The developing follicles have endothelial cells. They form a network in the theca layer. Growth factors mainly produced by granulose cells as VEGE (Vascular endothelial growth factor) and bFgF (basic fibroblast growth factor) act as stimuli for promoting follicular growth and antrum formation (Ziecik, 2004; xiao et al, 2001).

After ovulation, newly formed blood vessels cross basement membrane between theca and granulose cell layers and undergo a rapid growth to form corpus luteum. The luteal vascular growth varies and angiogenesis and angioregression determine corpusluteum growth and regression. Thus in future the improvement in angiogenic development of ovary may help in reproductive performance (Tamanini and Ambrogi, 2004).

These Hemodynamic changes in ovarian and corpora luteal walls was studied using colour Doppler ultrasonography. The flow of blood is assessed (Acosta and Miyamoto, 2004). This confirms our earlier studies using Laparoscopy for ovulation (Wani & Sahni, 1988, Wani, 1989 Sharief, 2000).

2. Embryo Block

Embryo blocks action of antibodies and becomes a homograph and not a heterograph, eight cell embryo block was very common block in mammalian embryonic development. Sheep and goat embryos did suffer from this block too. Development beyond 8 cell was diffcult in vitro. Use of special medias have overcomed this problem The 8-cell embryo block in development is overcomes by various growth factors incorporated in culture media. The HB-EGF (heparin-binding epidermal growth factor like growth factor) improves the development of 8-cell embryo to the blastocyst stage. This factor is helpful in preimplantation embryo development. (Tamada et al, 1999). embryo, makes direct contact with uterine lining only for a short period and thereby thrives on it indirectly through placenta via umbilical cord. It remains protected in an immunologically protected chamber the amnionic sac. Feotal development and biochemical constituents of foetal fluids and morphological development features observed in our studies are published . (Mufti, 1996, Mufti et al, 2000, Wani, 2004, 2005). Embryo development can be viewed in web images too. Early embryo mesoderm cells called somites differntiatic into myoblasts-resulting into muscle and skeletal tissues. Various (MRF) Muscle Regulatory Factors in foetus determine the muscular growth and were professed to be used in future research. Their exploitation would bring muscular lambs at birth. (Wani 2001). Many advanced aspects of myoblast .Determination flow have been documented. Huskey, 2001. (Robert.J.Huskey, myoblast determination flow interent explorer- Huskey, R.J. 2001). Myoblast Determination flow Microsoft. Internet Explorer.

Tight junctions prevent ion and water movement between cells and Na/K pumps move into the centre of the blastocyst resulting in to osmotic pressure difflections and formation of blastocoel. Plasmogen and its activator results into enzymatic digestion of zona-pellucida during this process. (www.wisc.Edie/ansici-repro/Lec-18 or search Embryo block sheep Metacrawler- Microsoft Internet explorer).

3. Genome Activation

Activation of embryo is detected by pronuclear formation, cleavage and blastulation. Addition of vitamins during oocyte maturation helps subsequent blastulation and embryo viability (Bormann et al, 2003). Pregnancy associated glycoproteins (PAGS) are antigens synthesized in ruminant trophoblast. They enter maternal blood stream and thus signal early pregnancy (Wani, 2005). Invitro growth of granulosa cell complexes isolated from cryopreserved ovine ovary has been reported (Newton et al, 1999). These cells have a capacity to produce immunereactive inhibin with FSH, (Compell et all, 1996). The ovine Booroola fecundity gene (Fec B) so widely known has been linked to markers from a region of human chromosome 4q. Thus linking research in ovine or caprine reproduction for improvements or control of reproduction in humans. The VEGF factors (Vascular endothelial growth factors) characterized and expressed in ovine corpus luteum could be of immense "healing touch" value if properly innovated to use in repairs of various body injuries. Ovine FSH has been of value in developing long-term contraceptive vaccine in bonnet (Redmer etal, 1996).

4. Gap-Junction Network

Mammalian oocytes are surrounded by follicular cells. A communication system exists between these oocytes and follicular cells. This system of communication is called Gap -Junction network. This system provides transfer of nutrients and signals between follicular cells and oocytes. The proteins of connexin family form these building blocks. Oogenesis, primodial follicle activation and subsequent ovulation is the result of these and some other proteins factors. They are e-kit receptor, transforming growth factor beta (TGF beta) and bone morphogenic proteins. Progressive morphological changes in the oocytes result into structural rearrangements. These generate energy. The increased energy facilitates oocyte development and ultimately determine its capacity and competence to dominate and ovulate (Fair, 2003).

Pregnancy recognition during day 12-22 in sheep and 15-21 in goats is made on the basis of interferon tau a member of type I interferon. It is the first protein secreted by trophoectoderm of the ruminant conceptus. It blocks the uterine luteolytic mechanism and is responsible for the sustenance of corpus Luteum during early pregnancy days. It acts on the uterine epithelial cells by suppressing their ability for transcription of gene for propogation or activation of oestrogen or oxytocin receptors. It also induces several proteins responsible for embryo implantation.

5. Per-implantation Embryo Nutrition

The most important nutrients required by the developing mammalian preimplantation embryo are carbohydrates, anino acid which provide energy. It also prevent cellular stress. Quantification of embryo energy metabolism is thus necessary. An attempt in this direction has been shown in figures. (Jamada, 1998).

The method of assessing preimplantation embryo viability by the self-refrencing electophysiological technique is possible. This is an non-invasive measurement. The technique measures the gradients of calcium, potassium, oxygen, and hydrogen peroxide (Gardner, 1998).

Protein, prostaglandins and steroid synthesis in caruncular and intercaruncular endometrium in sheep has been estimated. It was determined that attachment occurs around day 15-16 which however was not confirmed by our recent studied (Wani et al 2996, 2007) (Findly et al,Comparing invitro and invivo embryonic development in ruminants is possible by use of immunocytochemistry, confocal laser scanning miscroscopy and fluorescence techniques. By combination of these techniques it has been possible to demonstrate that:

1. RNA gene activation is similar
2. Embryonic nuclear development is similar.

However, embryo produced through nuclear transfers are associated with increased apoptosis, in the inner cell mass of the blastocysts. Besides embryo produced in-vitro by nuclear transfers or otherwise are associated with mixoploidy. Mixoploidy means embryos with abnormal haploid or polyploidy cells besides normal diploid cells. (Cinobell et al, 2003).

Laparoscopy in sheep and goats has been used for ovulation detection, embryo aspiration and diagnosis of pregnancy. (Wani, 1982 Wani and Buchoo, 1990, 1997, 2000) Embryo recovered through laparoscopy in Boer goats have a of 62% and were economically viable. (Besenfelder et al, 1994)'

6. Embryo Biosafety and Disease Risk

Scrapie, an ovine and caprine transmissible spongiform encephalopathy is widely spread among sheep populations in many European Countries. As it is known that susceptibility to scrapie is determined genetically, breeding programmes aiming at providing scrapie- resistant flocks have been established - (sipos et al, 2002). The pregnant endometrium secreted a number of neutral -to-acidic proteins which constituted, in part, the histotroph. A vitamin A-transport protein, RBP was identified in cultures, conditioned by the endometrium at days 13-21 of pregnancy. The endometrium is the only source of retinol for embryonic tissues. The uterine RBP appears to transport retinol locally toward embryonic tissues. Secretion of RBP by caprine endometrium at days 13, 15 and 21 of pregnancy suggested that retinol played an important role in conceptus development during the periattachment period of early pregnancy - (Lue et al, 2002) (Young et al 2000, 2004).

Caprine oviduct epithelial cell (COEC) are commonly used in vitro goat embryo production protocols to stimulate early embryonic development. These feeder cells are usually collected from slaughterhouses from unknown serological status animals. In many regions of the world, issues derived from the resource may be contaminated with CEV (Caprine Arthritis Virus) CAEV (Caprine Arthritis Encephalitis Virus) and the use of such material in invitro fertilization systems may contribute to transmission of this pathogen to the cultured embryos and dissemination via embryo transfer (ET). The use in invitro fertilization systems of coculture feeder cells that support efficient replication of CAEV(Caprine Arthritis Encephalitis Virus) to high titers could represent a serious risk for permanent transmission of virus to the cultured embryos and to the surrogate dam involved - (Lamara et al, 2002).

7. Early Pregnancy Factors

Pregnancy - associated glycoproteins (PAGs) are antigens synthesized in the superficial layers of the ruminant trophoblast. Initially, they were identified either as proteins differed into the maternal bloodstream (where they have applications in pregnancy diagnosis+PAG or as molecules binding to the LH receptor (PAG2). In this study, double radial immunodiffusion was used to test the ability of antisera raised against different PAG molecules (bovine,

ovine and caprine) to react with placental extracts from non ruminants (rabbit, cat, mouse, pig and wild pig) and ruminants (cow, ewe and goat) Placental extracts from all nonrumiants tested except rabbit reacted with anti bovine. PAG2 (anti- boPAG2). Extracts of ruminant placentas react with different antisera, confirming the expression of various PAG molecules. According to the time at which the placentas were collected (early or middle pregnancy), the reaction differed as regards the thickness, position and number of precipitation lines, suggesting that PAG expression varies as pregnancy progresses. Bos indicus and Bos Taurus placental extracts exhibited different reactions with anti-boPAG2, a single precipitation line in the former case and two lines in the latter. This suggests differential expression of boPAG2related glycoproteins in these two subspecies - (EL-Amivi, et al, 2003).

References

Acosta,T.J. and Miyamoto,A. 2004 Vascular control of ovarian function: ovulation, corpus luteum formation and regression. *Animal Reproduction Science*. 82/83: 127-140.

Argov, N. and Sklan,D. 2004 Expression of MRNA of lipoprotein receptor related protein 8, low density lipoprotein receptor and very low density lipoprotein receptor in bovine ovarian cells during follicular development and corpus luteum formation and regression. *Molecular Reproduction and Development*. 68(2): 169-175.

Bormann,C.L; Ongeri,E.M. and Krisher,R.L.2003 The effect of vitamins during maturation of caprine oocytes on subsequent developmental potential in vitro. *Theriogenology*. 59(5/6):1373-1380.

Battye, K.M; A.J; O-Neill-Co. and Evans, G. 1991 Production of platelet-activating factor by the pre-implantation sheep embryo. *Journal of Reproduction and Fertility*. 93(2): 507-514.

Buchoo, B.A and Wani, G.M.(1991). Induction of Oestrus in Corriedale ewes during anoestrus. *Indian J.Anim Sci* 60(4):407-408.

Campbell, B.K. and Baird, D.T. 2001 Inhibin A is a FsH-responsive marker of granulose cell differentiation which has both autocrine & paracrine actions in sheep. *Journal of Endocrinology*. 169: 333 345.

Campbell,B.K; Webb, R; Dobson, H. and Dobers Ka, C. 2003 Reproduction in Domestic Ruminant. Symposium on Reproduction in Domestic Ruminants held in Crieff, Scotland - U.K. August 2003 PP 103-116. *Reproduction Supplement Series*. 61:103-116.

Evans,A.C.2003 Characteristics of ovarian follicle development in domestic animals. *Reprod. Domest. Anim*. 38(4):240-6.

El-Amri,B; Sousa,N.M.De; Mecif,K; Desbuleux,H; Banga-Mboko,H. and Beckers,J.F. 2003 Double radial immunodiffusion as a tool to identify pregnancy- associated glycoproteins in ruminant and nonruminant placentae. *Theriogenology*. 59(5/6):1291-1301.

Fair,T. 2003 Follicular oocyte growth and acquisition of developmental competence. *Animal Reproduction Science*. 78(3/4): 203-216.

Fakowska, Podstawka,M; Wrona,Z. and Krakowski,L. 2002 Biological role of interferon tau. *Medycyna Weterynaryjna*. 58(8):568-571.

Findlay, J.K; Ackland, N; Burton, R.D; Davis, A.J; Walker, F.M.M; Walters, D.E. and Heap, R.B.1981 Protein, prostaglandin and steroid synthesis in caruncular and intercaruncular endometrium of sheep before implantation. *Journal of Reproduction and Fertility*. 62(2):361-377.

Gardner,D.K.1998 Changes in requirements and utilization of nutrients during mammalian preimplantation embryo development and their significance in embryo culture. *Theriogenology*. 49(1):83-102.

Gruzhinskas,V.A. 1984 Effect of the anthelmintics Benacil and tetramisole on prenatal and postnatal development (sheep, swine, cattle, horse). *Veterinariya-Moscow*, USSR. (11):61-62.

Juengel,J.L; Nett,T.M; Anthony,R.V. and Niswender,G.D. 1997....on expression of mRNA encoding insulin-like growth factor 1 and growth hormone receptor in the ovine. *J.Reprod.Fertil.*

Kaloo, F.A (2002). MVSc Thesis, SKUAST-K.

Kaul, N.A (2002) MVSc Thesis, SKUAST-K.

Kawate,N; Morita,N; Tsuji,M; Tamada,H; Inaba,T. and Sawada,T.2000 Roles of pulsatile release of LH in the development and maintenance of corpus luteum function in the goat. *Theriogenology*. 54(7):1133-1143.

□□□

Chapter 22

Early Pregnancy Signs

G.M. Wani, P. Wani & G. Wani

1. Introduction

Rectal palpation in small ruminants is of little value due to the size of the pelvis. (Wani, 1981). The caudal artery monitoring, bloatment, non-return to oestrus, udder development and other tests tried have had little success, (Wani & Sahni,1980). The more recent interest in early pregnancy 3. diagnosis of small ruminants is of academic and economic importance (Mellado,2003). A highly valued zygote or embryo when transferred to a less valued surrogate mother (recipient) needs to be closely monitored and the early detection of conception helps in repeated use of baren females. Proper management of pregnant animals also prevents embryonic losses. The method applied should be safe to both offspring and dam and needs to be cheap and easily applied. A review of various methods and techniques used for early pregnancy diagnosis in small ruminants. (sheep and goats) is presented.

2. Early Signs of Pregnancy

2.1. Maintenance of a Functional Corpus Luteum

It was evident that conception prolongs the life of the CL and prolongation and maintenance of a functional CL is triggered by the developing conceptus.

These signals ensure the maintenance of the structural integrity of the CL. Corpus luteum produces progesterone, which maintains the uterine endometrium in a state permitting embryonic development, implantation and foetal-placental development (wani, 1984b). The formation and regression of the corpus luteum (CL) in Muzzaffarinagri ewes and Jamunapari goats was monitored at 3 days intervals for an entire oestruous cycle. Laparotomy and laparoscopic methods were used in these experiments. The Endometrium undergoes tissue remodeling. This change in Extra cellular Matrix (ECM) components is needed for successful implantation. Cytokinens 8,18 and 19 have been detected in the caprine endometrium during early pregnancy using immunofluorescence. Thus the presence of these cytokinen at approximately day 15 post conception is indicative of pregnancy in goats.

The implantation process in goats starts around day 18 post mating. During this phase intense type I collagen staining was detected throughout the uterine caruncular and intracaruncular stroma. For embryonic trophoblastic adhesions with endometrium, local control of protease activity is suggested. (Guillomot, 1999).

The earliest signs of pregnancy is the non-regression of the cyclic CL, which can be observed by the following methods:

i. Laparoscopy and Laparotomy approximately day 18-25 post mating. (Wani, 1982, 1988, 1984b,Wani & Buchoo, 1990, Wani & Buchoo, 1993, Cuellar et al, 1990, Wani et al, 2003).

ii. Serum Progesterone values higher than 1 ng/ml e.g 2 to 3 ng/ml. (Wani, 1989; Shreif, 1997, Boscas et al, 2003, Al-Merestani et al, 1999, Zarkawiet et al, 1999). Diagnosis of Pregnancy accurately (100%) predicted on the basis of serum progesterone P4 values around 17-19 days post mating

iii. Pregnancy associated ovine glycoproteins recorded approximately post mating indicate pregnancy in sheep. (Karen et al, 2003; Verberckmoes, et al, 2004) or secretion of 17 & 22-24 K Da proteins on day 17 post mating in the caprine conceptus. (Guillomot et al, 1998).

iv. Non-return to oestrus (Mellado, 2003).

3. Non-Rejection of Early Conceptus

Progesterone maintains the uterine endometrium in a state which allows for embryonic development, implantation and foetal placental development. Details of foetomaternal relationships have been described (Mufti, 1997, Mufti et al, 2000) and are shown illustrated in Fig 1to 5. The presence of an early conceptus prolongs the life of corpus- luteum. These pregnancy signals are secreted as proteins. (Heap et al, 1990). Some of these proteins have been identified as ovine Trophablast protein I (OTP-1) in sheep which prevents the release of PGF2 alpha and thus helps in the maintenance of the corpus luteum. In cyclic ewes (non-pregnant) PGF2 alpha pulses are released in response to oxytocin with receptors being in the endometrium. The earliest signal of pregnancy is detected by a marked reduction in the endometrial oxytocin receptor numbers. The OTP-1 may inhibit synthesis of endometrial receptors for oestrogen and oxytocin. This possibly prevents luteolysis and maintains the dominance of theuterus by progesterone which is pre-requisite for the establishment and maintenance of pregnancy. (Bretzlaft and Romano, 2001; Wani, 1996; Ala cam et al, 1988).The expression of progesterone receptors (PR) in the caprine uterus markedly increases during the peri-implantation period and estrogen -(ER) receptors do not increase in relation to PR, thus signaling the non-rejection of the early conceptus. (Flores et al, 2001). Progesterone in milk too can be found during early fertilization and conception (Cough et al, 1989). Caprine H-type I antigen expression is unregulated during peri-implantation and progesterone P4 level stimulate it. It may be a useful marker to signal uterine preparations for receiving and retaining pregnancy in goats. (Powell et al, 2000). The caprine pregnancy related glycoprotein (Ca PAG) may help the conceptus to develop and is found around 18-19 day post mating . (Garbayo et al 2000). Endometrial tissue the undergoes remodeling to retain the conceptus in gravid small ruminant females. (Guillomot, 1999)The dephosphorylated state of caprine uterine myocin in early pregnancy may help the conceptus to grow. Changes in the expression of native myocin, myosin heavy chains (MHCS) and myosin light chains (MLCS) were observed. (Kumar and Katoch, 1997).For the development of the blastocyst, a proper uterine environment is essential. Besides the maintenance of the corpus luteum, production and availability of progesterone, the non-rejection of conceptus (blastocyst) is another critical feature of this period. The embryo produces interferons (embryo-IFN). This embryo IFN is homologous with - interfersons (? -IFN) and Ovine Trophoblast Interferons (OTI) of early pregnancy. Purified OTP and recombinant OTP (r-oTP) produced in yeast exhibit antiviral activity and these r-OTP and OTP inhibit the release of endometrial PGF2 ?. This helps in the non-regression of the CL and indirectly maintains the early conceptus. Intra uterine r-OTP administered at a dose of 340 μg/ day for a week maintained the C L in cyclic ewes for a month or so of. The inter- oestruos interval in 80% of the ewes was about a month or more. This dose r-OTP was as a effective as 14-16 day old conceptus. OTP was found to be immunosuppressive in several *in-vitro* and in-vivo assays. An assay on phytohaemagglutinin A revealed both OTP and r-OTP to be immunosuppressive. This was further verified by the inhibitory activity of r-OTP in Graft Versus Host Reaction. (GVH assays). Trophoblast interferons play a strategic role in the prevention of early pregnancy loss as it inhibits CD + blastogenesis. Therole of CD + cells and as helper T lymphocytes and delayed+ Type hyper sensitivity mediators (DTHS) would explain this immunosuppressive rate of OTP. (ILeri et al, 1996; Karen et al, 2003; Wani, 1996).

4. Oestrogen – Pregesterone Ratio (E:P Ratio)

The role of oxytocin in inducing uterine PGF2 alpha was discussed earlier. However, the release of PGF2 under the action of oxytocin depends on or is controlled by progesterone and oestradiol. It was further indicated that ewes with a high E:P ratio may generate stronger luteolytic signals. It was demonstrated that low progesterone and high oestradiol combination record the largest and sustained increase in PGF2 alpha following oxytocin injection. Trophoblast interferons act locally to suppress the uterine oxytocin receptors in sheep (Karen et al, 2003).

5. Maternal Recognition of Pregnancy

The maternal recognition of pregnancy in sheep and cattle is centered around the production by the trophoblast of type I x interferon (tINF). This tIFN then suppresses uterine oxytocin receptor concentrations (OTr). The oxytocin receptor (OTr) occupancy is associated with oxytocin induced PGF2 alpha release. OTr inhibition may represent the principal antiluteolytic mechanism of tIFN and secretion of the conceptus secretory proteins or bovine recombinant IFN to the uterus reduces OTr. Concentrations in intact and ovarectionized steroid treated ewes . A relationship between the conceptus secretory proteins and the metabolic products and those in the peripheral blood of the dam exists. (Mufti; 1996; Mufti et al, 2000). There are conflicting reports making the action of oestradiol on oxytocin receptor concentration. (Powell et al, 2000). Trophoblastic cells contain interferon on day 14-17 after mating. During maternal recognition of pregnancy goat interferon was detected on day 18 post mating, its absence signifies pregnancy maintenance has been taken over by the corpus luteum. Thus a very thin line exists between maternal recognition of pregnancy and its maintenance or sustenance by the CL. (Gillomot et al, 1998).

6. The Reliability Pregnancy Tests

Various methods used for correctly predicting pregnancy in sheep and goats during gestation have been summarized. The accuracy varies from 70 to100% with different ultrasonic equipment. Different models as well as principles involved have been extensively reviewed (Wani, 1991; Wani et al,1998) and other methods of pregnancy detection during this stage e.g serum progesterone determination, vaginal cytology, laparotomy, estrone sulphate are summarized. Various techniques were also evaluated in assessing mid-gestation. The various pregnancy signs as quoted by various researchers using ultrasonography are summarized. Of late certain anatomical features in the live, developing conceptus in vivo have been reported. This is reviewed and a summary is presented. Various live foetal measurements like Biparietal diameter, Amniotic vesicle diameter, foetal radius and Tibia lengths are reviewed. Various pregnancy related images, histological sections and morphology of endometeruim have recently been published (Wani et al 2007, 2006 abc) where images are presented 6-15.

References

Alacam-E; Dinc-DA; Guler-M; Eroz-S;Sezer-AN, 1988. Veteriner-Fakultesi-Dergisi, Selcuk-Universitesi, 4(1),91-98.

Al-Merestani-MR; Zarkawi-M; Wardeh-M, 1999. Early breeding and pregnancy diagnosis in Syrian Awassi sheep yearlings. *Reproduction -in-Domestic-Animals,* 34(5), 413-416.

Blasco-I; Foklch-J; Echegoyen-E, 1989. *Tecnica-Economica-Agraria,* 20(82), 22-31.

Boscos, C.M; Samartzi, F.C; Lymberopoulos;A.G; Stefanakis, A; Belibasaki, S; 2003. *Reproduction in domestic Animals.* 38(3), 170-174. Assessment of progesterone concentration using enzyme immunoassay for early pregnancy diagnosis in sheep & Goats.

Bozkurt, T; Gundogan, M; Esen, F; Kul.S; 1999. Saglik- Bilimleri- Dergisi- Firat- Universitesi. 13(2),185-180. The use of B mode real time ultrasonography rectally for early pregnancy diagnosis in sheep.

Bretzlaff-KN; Romano-JE, 2001. Veterinary-clinics-of-North-America, *Food-Animal-Practice,* 17(2), 421-434. Advanced reproductive techniques in goats.

Calamari-CV;2001. Avaliacao-de-dois-metodos-de-diagnostico-precoce-de-gestacao-em-ovelhas:Ultrasonografia-transretal-e-detector-de-prenhez-para-pequenos-ruminantes - DPPR-80R.2001, 57pp. Evaluation of two methods for early pregnancy diagnosis inewe: transectal ultrasonography and pregnancy detector for small ruminant (DPPR-80®).

Chalhoub-M; Prestes-NC; Lopes-MD; Trinca-LA; Ribeiro-Filho-AL, 2001. *Ars-Veterinaria,* 17(1), 17-21. Pregnancy diagnosis and embryological/fetal quantification by transrectal ultrasonography in sheep.

Chen-Zao Ying; Lun-SiChun; Cao-Hong; Chen-Zy; Lun-SC; Cao-H, 1996. *Chinese Journal-of-Veterinary-Science- and-Technology*, 26(2), 12-15. Study on early pregnancy diagnosis of ewes by ultrasonic tomographic scanning.

Cuellar-C; Fontanillas-JC; Palacio-M-del; Perez-Fuentes-J, 1990. *Advances-en-Alimentaction-y-Mejora-Animal*, 30(6), 251-257. A comparison of early pregnancy diagnosis in Manchega ewes using endoscopy or ultrasonics.

Chemineau, P; Gauthier, D; Poirier, J.C; Saumande, J. 1982. *Theriogenology*. 13 (3), 313-323. Plasma levels of LH, FSH, Prolactin, Oestradiol 17B and progesterone during natural and induced oestrus in the diary goat.

Cuoghi, F; Castagnetti, G.M; 1989. *Informatore- Agrario*. 47, 27-32. Progesterone in milk- a method of early pregnancy diagnosis

Dickie-MB; Holzmann-A, 1992. *Journal of Veterinary-Medicine- Series*-A,39(7), 525-530. Investigations concerning the use of progesterone tests (Serozyme-Progesterone, Ovucheck) for pregnancy diagnosis of mountain sheep.

Dionysius, D.A; 1991. *Australian. Veterinary Journal*.68(1), 14-16. Pregnancy diagnosis in diary goats using progesterone assay kits.

Flores, J.M; Sanchez, M. A; Nieto, A; Sanchez, B; Gonzalez, M; Garcia, P. 2001. *Theriogenology*. 56(2), 341-55. Delection of estrogen alpha and progesterone receptors and cell proliferation in the uterus during early pregnancy in goat.

Garbayo,J-M; Green,J-A; Manikkam,M; Beckers,J-F; Kiesling,D-O; Ealy,A-D, Roberts, R-M, 2000. *Mol-Reprod-Dev*. 57(4), their cloning, expression and evolutionary relationship to other PAG.

Garcia-A; Neary-MK; Kelly-GR; Pierson-RA, 1993.. Department of Veterinary Clinical Sciences, Purdue University, West Lafayette, IN 47907, USA, 39((4), 847-861. Accuracy of Ultrasonography in early.

Richardson, C 1972; Pregnancy diagnosis in the ewe. Vet Record. 90, 264- 275.

Slosarz, P; Steppa,R;Gadek,A: 1999. *Medychna waterynaryjna* 55(10),686-688. The application of ultrasound for early pregnancy diagnosis in sheep.

Shrief, M; 1997.. M.VSc Thesis, S.K. Univ.Agric.Sc & Technology, Shalimar, J&K India. Investigations on Caprine Ovarian activities during Oestrus cycle.

Tajik-P;Abbas-Veshkini;Sarang Soroori, 2001. Veterinary-Conference-University of Stellenbosch, South-Africa. Measurement of different parts in Chall fetuses to determine the pregnancy age. Proceedings-of-the-5th -International- Sheep.

Taverne, M.A.M; 1991. Tievarztliche. *Monatsschrift*. 78(10),341-345. Application of two dimensional ultrasound in Animal Reproduction Wiener.

Verberckmoes, S; Vandaela, L; Cat,S.de; EL-Ameri, B;Sulon, J; Duchateau, L; Kruif, A.de; Beckers. J.F; Soom, A. Van; de-cat.S; de-Kruif, A;Van-Soom, A;2004. *Vlaams-Diergeneeskundig- Tijdschrift*. 73(2), 119-27. A new test for early pregnancy diagnosis in sheep: determination of ovine pregnancy associate glycoprotein (OVPAG) concentration means of a homologous radio immunoassay.

Wani, G.M, 1981. Ultrasonic pregnancy diagnosis in Sheep and Goats. *World Review of Anim. Prod.* 17, 43-48.

Wani,G.M, 1982;. *World Review of Anim. Prod.* 18,7-13. Laparoscopy in farm Animals.

Wani, G.M; 1991. Proc. Ist. National Seminar on small Ruminant Reproduction, CSWRI, Avikanagar, Rajasthan- India. Use of Laparoscope, ultrasound imaging in small ruminant Reproduction.

Wani-N.A; Shabeena-Mustafa; Misra-AK; Maurya-SN; Mustafa-S; 2003. Pregnancy diagnosis in farm animals - a review. *Indian Journal of Dairy Science* 56 (1),1- 8.

Wani, N.A; Wani, G.M; Mufti, A.M; Khan, M.Z. 1998. *Small Ruminant Research.* 29, 239-240. Ultrasonic pregnancy diagnosis in gaddi goats.

Wani, G.M; Sahni, K.L; 1980. Vet. Research. Journal 2(1), 35-73. An ultrasonic technique for detection of pregnancy in Sheep and Goats.

Wani, G.M; Sahni, K.L; 1981. *Indian. J. Anim. Sci.* 51 (2), 194-197. Ultrasonic pregnancy diagnosis in ewes under tropical field conditions.

Wani, G.M; Sahni, K.L; 1988. *Indian. J. Anim. Sci.* 58(7), 802-804. Ovulation detection by Laparoscopy in Sheep.

Wani, G.M; 1982. *World Review of Anim. Prod.* 18(1), 7-13. Laparoscopy in Farm Animals.

Wani, G.M; 1984a. Dr. Med. Vet.Diss.Tierartzlichen Hochscule, Hannover, Germany. Investigations on Embryo production and culture in goats.

Wani, G.M; 1984b. Ph.D thesis, Indian vet. Res. Inst, Izatnagar- U.P. India. Investigations on ovarian activity, ovulation and frequency diagnosis in sheep and goats.

Wani, G.M; 1989. *Indian Journal Animal Reprod.* 10(1), 44-46. Plasma progesterone concentrations during oestrus cycle in goats.

Wani, G.M, 1996. Embryo Biotechnology in sheep & goats. Published Valley Book House, University Road, Hazratbal, Srinagar.

Wani, G.M; Buchoo,B.A; 1990. *Indian J. Anim. Sci.* 60(5), 564-68. Laparoscopic view of ovine conceptus

Wani, G.M; Buchoo, B.A; 1993. *Indian. J.Anim. Sci.* 63(3), 300. Prediction of Ovarian response in sheep by laparoscopy.

Zarkawi-M; Al-Merestani-MR; Wardeh-MF, 1999. *Small-Ruminant-Research*, 33(1), 99-102. Induction of synchronized oestrous and early pregnancy diagnosis in Syrian Awassi ewes, outside the breeding season.

❑❑❑

Chapter 23

Reproductive Disease Management

Rafiqul Islam & M.Z. Khan

Livestock rearing is one of the oldest professions opted by the rural and landless farmers for their livelihood. Now-a-days, the trend of livestock rearing has been changed from its old practice to a new direction. Use of different reproductive technologies in animals help in improving the production. Reproduction is the important field, where is vast scope to increase the potential of the animals for augmenting production using the assisted reproductive technologies. Reproductive management of animals needs thorough knowledge about the reproductive disorders for their amelioration in a cost effective manner. Reproductive diseases may hamper the production either by infertility in various forms or by developing sterility in animals. Early detection of infertility in animals helps in managing the problems easily and completely. Late detection may deteriorate functional capability of the reproductive organs and thereby timely management may not be possible and the animal may be got sterile. Periodic Screening of animals for detection of different infertility problem is essential for proper sexual health control in herds'. Professor Nils Lagerlof for the first time had recommended for the sexual health control programme in the cattle herds. This programme had helped in detecting infertility problems in different countries of the world and proved noteworthy contributions in reducing the infertility problems in animals (Islam et al., 2006 a). Management of reproductive disorders in animals needs proper data base about the production and reproduction data regarding the reproductive cycles, pregnancy, parturitions and other associated problems that may be encountered on and around the parturition. In an organized cattle farm of Kashmir, the incidence of abortion (0.63 % and 1.32 %) and still births (2.74 % and 2.64 %) has been reported in Cross bred and Jersey cows respectively (Islam and Nadroo, 2006 b). Different Reproductive disorders like dry ewe percentage (23.60 %), abortion (0.25 %), still births (1.60 %) dystocia (1.23 %), Lamb died due to dystocia (0.43 %), Retained placenta (0.03 %), ewes died before lambing (6.87 %), lamb mortality within 24 hours (1.04 %) and lamb died due to poor mothering ability (1.31 %) in Corriedale ewes in an organized farm of Kashmir valley have been reported (Islam et al. 2006 c). Computer applications for making the data base in animal production and reproduction is being applied by the animal scientists, which has shown a new direction to the live stock industry. Industry also demands newer technologies for the detection and amelioration of different reproductive problems for giving maximum out put in the form of livestock products and by products for the upliftment of the mankind. Knowledge level of the progressive farmers, animal scientists must be widened to have an effective reproductive disease management for augmenting production.

The recent trend in animal production is mostly concentrated on the economic production from the livestock. A well managed dairy herd should have 60 % pregnancy rate on first service with an average of 1.3 to 1.7 services preconception. There should be less than 10 percent of cows with reproductive problems. The calving interval should be between 12 to 13 months. The maximum interval of calving to conception should be 100 days. Therefore, the animal with less or no potential for future production should be culled from the farm. Much emphasize should be given for prophylactic measures and advanced managemental practices for reducing infertility in animals.

Causes of Reproductive Diseases

Reproductive diseases may be encountered due to different factors. Most important factors are Managemental faults, Nutritional, Infectious, Hormonal and congenital. The factors may take part either alone or in combinations to create reproductive problems.

1. *Managemental faults* : Managemental procedures maintained in the farms are responsible to a great extent for the incidence of reproductive disorders. Breeding management in animals is one of the important aspects, where the industry may suffer a lot due to the negligence of the workers of

the farm in detecting heat and timely insemination of the animal. Some times due to improper heat detection, the normally cycling animals also may be considered under anoestrum/ acycling animals due to improper heat detection practices. For eradicating this problem, heat detection must be done in farm during the activities of the farm 2 to 3 times daily, at least for 20 to 30 minutes with a close observation to detect the animals in heat. Use of proper teaser animals helps in detecting the animals with silent heat. Now-days, using the satellite devices, the herd manager can detect oestrus in all the cows in the herd very easily. Therefore timely management is possible for the breeding of animals by saving time and precious cycles.

2. *Nutritional causes :* Some of the important reproductive disorders may be occurred due to nutritional deficiencies. Vitamins, minerals particularly trace minerals are important for maintaining the reproductive functions in animals. Some of the minerals like Ca and phosphorus deficiency may lead to pica in animals. This may also lead to inappetence in the animals. The inappetence may develop general weakness and debility in animals and thereby it hampers the reproductive potential in animals. Co is necessary for synthesis of Vitamin B complex by the rumen micro flora. The deficiency of Vitamin A may lead to night blindness in animals. If the deficiency is severe for prolong time it may lead to anoestrum in animals. Winter sterility in animals in temperate area is due to the deficiency of vitamin A as during the period green fodders are not available to the animals. They are provided with low quality roughages, which contain less nutritional levels. The green fodders are rich source of Vitamin a in the form of their precursor carotene. The deficiency of Vitamin A may also lead to abortion in animals. The vitamin a deficiency causes atrophy of the seminiferous tubules and less spermatogenesis in the bulls. Thereby they may be infertile. The deficiency of nutrition may lead to different metabolic diseases in cows like milk fever, Ketosis, Grass tetany etc. Milk fever is mainly due to the deficiency of calcium which may be resulted due to the abnormal absorption of the calcium by the gut due to the deficiency of Vitamin D (cholicalciferol). The other production disease in cow like ketosis is due to the abnormal metabolism for which the cow is deficient in glucose. In sheep the deficiency of glucose may lead to Pregnancy toxaemia which may take the life of the mother. Providing the ewes with good quality molasses during the last months of pregnancy may prevent pregnancy toxaemia. Accordingly reports have shown that injecting vitamin D to the cows 2-3 weeks before parturition may prevent milk fever.

3. *Infectious causes :* Various infectious agents are responsible for producing different reproductive disorders in animals. Some organisms are specific for the specific physiological periods. The site of predilection for Brucella organisms

is pregnant uterus. Different infectious agents like Bacteria, virus, fungus and protozoa, rickettsia and chlamydia may affect the reproductive organs and produces different forms of infertility. Some of the important bacterial agents like Brucella, Leptospira, Listeria and Campylobacter foetus produce abortion in animals. IBR-IPV virus produces abortion in animals. Among the protozooal agents, Tri trichomonas foetus produces abortion and various forms of infertility like early embryonic death leading to endometritis and repeat breeding in animals. Fungal organism is least important in causing abortion in animals. However, Aspergillus species may cause abortion in animals.

4. *Hormonal causes* : Hormonal imbalance is known to cause various kinds of infertility problems. Deficiency of Luteinizing hormone (L.H.) causes cystic ovarian diseases. Due to L.H. insufficiency cows may have ovulatory disturbances leading to repeat breeding. Imbalance in the L.H. level causes improper corpus luteum development leading to early embryonic mortality or fertilization failure due to hostile uterine environment. Anoestrous condition in animals is due to inadequate hypothalamic stimuli for release of gonadotrophic hormones (Shokier, 1959). Estrogen in the pastures or prolong administration of estrogen compound at higher doses may lead to cystic ovarian diseases (Roberts, 1982). Progesterone deficiency in animals may cause early embryonic mortality or habitual abortion at different stages of gestation. The reproductive disorders due to imbalance of hormones are many. Likewise in male reproduction, Prof Nils Lagerlof, an eminent scientist form Sweden has advocated not to use bulls with low endocrine constitution for breeding as the condition is hereditary in nature. The condition will be perpetuated in the offsprings. Thorough knowledge of the animal physiology is must to diagnose the etiology of infertility due to hormonal imbalance. Empirical use of hormones in treating infertility is harmful as they have one or more side effects. Use of hormones in animals for treating reproductive problems should only be carried out after confirmatory diagnosis. Diagnosis may be supported with estimation of hormonal levels in the circulation.

5. *Congenital causes* : Missing or improper development of the reproductive tract or gonads have been reported. The problems caused sterility in most of the cases. However partial or unilateral hypoplasia or aplasia of the organs may reproduce at a lower rate (Infertility). The animals with congenital problems in the genital organs should not be used for breeding as the conditions are hereditary. Hypoplastic conditions of the ovary and testes, segmental aplasia of the reproductive tract both in males and females have been reported. Some of the conditions like hermophroditism, freemartinism in animal causes sterility in animals.

Reproductive Disorders

Different forms of reproductive disorders are reported in animals. The important reproductive disorders/infertility problems in animals are anestrum, repeat breeding, abortion, Still birth, mummified foetus, maceration, pyometra, Endometritis etc. The other types of disorders which are associated with the Pregnancy and parturition of animals are torsion of uterus, dystocia, Uterine inertia, Prolapse of Uterus, retained placenta, Post partum pyometra etc.

Infertility in Animals

1. *Anoestrum* : Anestrum means absence of oestrum in animals. This is the single most infertility condition affecting the dairy animals and thereby causes great economic loss in the form of milk, meat and offsprings. On the basis of presence of functional Corpus luteum in the ovaries anestrum may be classified in to two (Roberts et al., 1982).

a) Class I anestrum: Absence of estrum in the cows due to the presence of functional corpus luteum (CL).

- Anoestrum due to pregnancy: The history or the owners' records can mislead the veterinarians attending the case. Before giving any treatment to the cow in anoestrum, careful rectal examination should be performed to rule out the pregnancy (Physiological anoestrum) and also to ascertain the status of the genital tract and ovary to arrive in a final diagnosis.

  If the cow's genital tract is found normal and healthy then, there is possibility of the cows having been serviced within last 30 days, it is advisable to re examine the cow two to six weeks later.

- Anestrum due to a persistent Corpus luteum.
  - Persistent Corpus luteum with gross uterine pathology: Anestrum is associated with uterine pathology and pregnancy. This is due to the failure of the release of the luteolytic substance from the endometrium or endometrial glands.
  - Pre service or post service pyometra: in most pyometra cases secondary to retained placenta and a post partum metritis, the condition is characterized by a vaginal discharge of pus and a failure of estrum following calving (Post partum anoestrum).
  - Fetal maceration, mummified fetus.
  - Persistent corpus luteum associated with early embryonic or fetal death: if the ovum dies early or before the middle of the cycle, then

oestrus usually occurs at its usual interval; but when embryo dies later, the onset of the next oestrum is delayed. If embryonic death occurs before 90 to 120 days of pregnancy, the embryo or foetus may not be seen when it is aborted due to its small size or it may remain in the uterus and macerate in the uterus and further delay the onset of oestrum until the trophoblast dies and is resorbed in the uterus. This condition has been observed in trichomoniasis, Vibriosis and trauma to the early embryo or rough manipulation of the uterus.

- Subestrum, Weak or 'Silent' Oestrum: Cows that are ovulating and cycling regularly but are not recognized as exhibiting estrum either because it is week or silent oestrum.

- Unobserved Oestrum: This is a managemental fault. If an owner or herdsman does not watch his cows carefully at least 20-30 minutes twice a day or more in a herd, where the cows are free to move about, a number of heat periods will not be observed.

- *Treatment* : Treatment of anoestrum in Class I cows should be based on a careful differential diagnosis, often requiring careful recordings of repeated rectal examinations. Treatment of anoestrum due to weak or sub oestrum, Silent oestrum or unobserved oestrum, requires careful frequent and accurately recorded examinations of the genital tract of the affected cows, careful assessment of the herd management practices and diplomatic education and instructions for the herds man. Cows with a pre-service or post partum anoestrum should be examined at 50 to 60 days post partum or at 15-18 months of age in heifers, if they have not exhibited estrus by that time.

Treatment Procedures

1. Improved Management Practices

- Cows should be observed for heat symptoms carefully 20 to 60 minutes or more, twice or more times per day especially during periods of activity (before and after milking).

- Careful records of the observations.

- Heat Expectancy Charts.

- Use of teaser animal. One or more vesectomised young bulls may be allowed to run with the cattle herd. Bulls selected as 'teaser' animals should have strong, vigorous sex drive.

2. Frequent, Rectal and if Necessary Vaginal, Examinations of the Genital Tract

- This will help to determine the proper time of insemination of cycling cows with weak, silent or unobserved oestrum.

3. Enucleation of the Persistent Corpus Luteum

- Most of the cows show estrum on the third or fourth day after removal of the Corpus luteum. In cows with pyometra or other uterine infections, manual removal is dangerous due to chances of ovaritis and ovaro-bursal adhesions.

4. Estrous Synchronization

- The use of Progesterones orally or in the form of intra vaginal pessary (Wani et. al., ...) may be used to suppress the estrous cycle. Estrus usually occurs 3 to 5 days after withdrawal of the drugs or removal of the pessary from the vagina.

5. Prostaglandin F2ă for Destruction of the Corpus Luteum

- Injecting Prostaglandin preparations once may destroy the persistent corpus luteum in cows. Foetal mummification in cows has been treated successfully with a single injection of Lutalyse (5 ml). Use of hormones should always be done under supervision of qualified veterinarians.

b) Class II Anoestrum : Absence of Oestrum Due to Smooth Inactive Ovaries with no Functional Corpus Luteum

- Cows in anoestrum due to nutritional deficiencies: Minerals, Vitamins and herbal heat inducers have been tried either alone or in combination by several workers to treat anoestrous condition with variable response in Kashmir (Wani et al., 1999; Hussein et al., 2006) and in other parts of India (Madhavan and Iyer, 1993; Deshpande et al., 2000; Ahmed et al., 2003). In temperate zone like Kashmir, the anoestrum in cows has been seen during winters. The winter sterility has also been reported form the temperate area around the globe. The winter sterility is mainly due to the unavailability of good quality greens. Hence the cows are deficient mostly in vitamin A and also due to other essential minerals.

- Cows with hormonal imbalance: Anoestrous condition in animals is due to inadequate hypothalamic stimuli for release of gonadotrophic hormones (Shokier, 1959). Hormonal treatment with Gonadotrophin releasing hormone of anestrous dairy cows and buffaloes with smooth ovaries are reported from different parts of India (Kodagali et al., 1981;

Pattabiraman et al., 1986; Reddy et al., 1994; Singh et al., 2004; Khasatiya et al. 2004). Hormonal treatment with Gonadotrophin Releasing hormone analogue (Receptal @ 200, 300 and 400 μg) has been tried for treating anoestrus cows with smooth ovaries in Kashmir valley. GnRH (Receptal) @ 300 μg was found to be the optimum dose of treatment in treating local cross bred anestrous cows to obtain higher estrus (100 %) and conception rate (80 %) under temperate conditions (Shah, 1996).

- Cows with chronic debilitating diseases: These diseases cause a decreased appetite in the animals leading to anoestrum due to a marked reduction of body weight. Different diseases are namely severe mange infestation, severe chronic acetonemia, chronic traumatic gastritis, Chronic severe pneumonia, Johne's Disease, Actinomycosis that interferes with eating, advanced tuberculosis, Suppurative arthritis etc.
- Cows with Cystic Ovaries. Cows with follicular cysts mostly show frequent and continuous oestrus (Nymphomaniac). Luteal cyst mostly shows anoestrum.
- Senility or old age with loss of teath and inability to masticate properly.
- Environmental influence: Cattle confined in a dark stables and allowed little or no exercise may lead to anoestrum.
- Miscellaneous Condition: Good nourished older cattle in the first 2 to 4 months after parturition when the level of lactation is high. Retained placenta, dystocia, milk fever, ketosis, mastitis, twins delayed the onset of oestrus cycle. This occurs due to reduced level of nutrition with decreased body weight. Spontaneous recovery usually occurs, otherwise should be treated based on the individual case. Congenital defects like freemartin heifers, bilateral hypoplastic ovaries should be culled.

Treatment

1. Cystic Ovarian Conditions

- Removal of the cyst
- Injection of Luteinizing hormone (hCG hormones are used now-a-days successfully in treating this condition).
- Anoestrum due to debility or lack of energy, protein or other minerals and vitamins
- Careful examination
- Analysis of feed (if required)

- Supplementation of the deficient nutrient in the feed or through medication
- Debilitating diseases must be alleviated to bring the animals in to good health with normal body weight.
- Stopping lactation or decreasing the frequency of lactation may help in hastening the return of estrous cycle.
- Provide good quality Green fodders.

Hormonal Treatment

- Injection of FSH and LH hormones (PMSG & hCG)
- Or Injection of Gonadotrophin Releasing Hormone (GnRH).

2. Regular Return to Estrus (Repeat Breeder)

The ovaries and genital tract should be examined per rectum to determine the presence of gross abnormalities such as severe adhesions, anovulation, delayed ovulation or uterine infection. This condition can occur only if there is failure of fertilization or embryonic death before day 12 of the estrous cycle. There are a number of possible causes. The main causes are ovulation failure (anovulation, delayed ovulation), fertilization failure and early embryonic death.

- Infertile bull: If a number of cows and heifers are involved, the bull should be examined. If A.I is used from an approved centre, this can be excluded as a cause of the problem.
- Incorrect timing of service or A.I: this may lead to fertilization failure: The appropriate time insemination is 12 hours from the onset of oestrus in cows.
- Wrong site of semen deposition: The suitable site for deposition of semen in cow is mid to deep cervix.
- Nutritional deficiency or excess: diet and blood profile of the animal should be checked.
- Occluded uterine tubes: the organ is palpated carefully and PSP or starch grain test may be used to confirm. If severe case, the animal must be culled from herd.
- Anatomical defects: careful palpation of the reproductive tracts and gonads of heifer, to look for segmental aplasia. In parous animal, the genital organs are checked for salpingitis, hydrosalpinx, Ovario-bursal or uterine adhesions.

- Endometritis: If there are clinical signs, diagnosis is simple. But sub clinical disease can be diagnosed only by biopsy. The case is treated with antibiotics after choosing suitable antibiotic with antibiotic sensitivity test.
- Delayed ovulation: Diagnosis is difficult manually. However, repeated palpation/ laparoscopy examination of the ovaries can detect it. Treat with GnRH or hCG at the time of insemination. In case of delayed ovulation double inseminations at an interval of 12-24 hours may be tried.
- Anovulation: Diagnosis depends on ovarian palpation 7-10 days after estrus to demonstrate failure of ovulation by absence of corpus luteum. Treat with GnRH or LH like hormones (hCG).

3. Abortion in Animals

Abortion is the termination of pregnancy after organogenesis is complete but before the expelled fetus can survive. If pregnancy ends before organogenesis, it is called early embryonic death. A dead full-term fetus is a stillbirth (its lungs are not inflated). Many etiologies of abortion also cause stillbirths, mummification, and weak or deformed neonates.

Infectious Causes

Bacterial: Brucella Species, Leptospira, Listeria, Campylobacter

Viral: IBR-IPV (Bovine Herpes virus 1), Bovine viral Disease

Protozooal: Tritrichomonas (Trichomonas) foetus infection causes a venereal disease that usually results in infertility but occasionally causes abortion in the first half of gestation.

Mycotic: Fungal placentitis due to *Aspergillus* sp (septic fungi, 60-80% of cases), or to *Mucor* sp, *Absidia* , or *Rhizopus* sp (nonseptate fungi) is an important cause of bovine sporadic abortion

The best specimen is the complete fetoplacental unit in fresh condition, along with maternal serum. The placenta and fetus should be cleaned with water or saline, packed in clean plastic bags, chilled (but not frozen), and rapidly transported to the diagnostic laboratory. If there are multiple fetuses, 3-5 should be submitted with their placentas. It is best to submit calves and foals whole, but in many cases it is more convenient to perform a necropsy and collect samples for submission. The specimens that are routinely used for testing vary somewhat between diagnostic laboratories, but a basic set of samples that will allow thorough examination includes stomach or abomasal contents; heart blood or fluid from a body cavity; lung, liver, kidney, and spleen (some labs also request tissues such as thyroid glands, thymus, heart, brain, abomasum, and stomach; placenta (if available); and dam's serum. These should be submitted in sterile containers to allow for microbiologic cultures. Because they are always contaminated, placentas should not be mixed with other tissues. Most agents, especially bacteria and fungi, infect the placenta and thus gain entry into the aminiotic fluid, which is swallowed by the fetus. Stomach contents can be obtained aseptically, making it the best specimen for detection of fungi and most bacteria. Isolation from the stomach contents is much easier than from the placenta, which is always heavily contaminated. Lungs, liver, spleen, and kidneys are also good for culture. Representative samples of the lung, liver, heart, kidney, spleen, brain, skeletal muscle, thyroid, adrenal, intestines, and placentashould also be submitted in 10% buffered formalin for histopathologic examination.

Non Infectious Causes

The actual incidence of abortions in cows due to genetic factors is unknown. Some genetically caused abortions may not have phenotypically recognizable lesions. Most lethal genes cause early abortion or early embryonic death. Vitamins A and E, selenium, and iron have been implicated in bovine abortions, but documentation based on experiments is available only for vitamin A. While severe trauma may rarely result in abortion (the bovine fetus is well protected by the amniotic fluid), farmers undoubtedly blame too many abortions on the cow "getting bumped. A number of toxins can cause abortion in cows. Ponderosa pine needles can cause abortion if ingested in the last trimester; the cows may become moribund after delivery and hemorrhage excessively. The toxic factor is believed to be estrogenic. Locoweed (*Oxytropis* or *Astragalus* sp) contains an indolizidine alkaloid that can affect the corpus luteum, chorioallantois, and neurons, resulting in abortion or deformities. Broomweed (*Guttierrezia microcephala*) ingestion can also cause abortion, as can coumarins from rat poison, many grasses, or moldy sweet clover. Sodium iodide, IV, has been contraindicated in pregnant cows, but no abortions or adverse effects occurred in pregnant cows treated with a single high dose in recent studies. Mycotoxins, especially those with estrogenic activity, have been implicated in bovine abortions. Nitrates or nitrites have also been incriminated, but experimental evidence is controversial.

Reproductive Disorders Associated with Parturitions

1. Dystocia

In the domestic species dystocia is most common in cattle. Incidence of dystocia is higher in dairy cattle than beef cattle (Roberts, 1982). Dystocia in animals may be due to varied etiology.

a) Maternal Causes

- Fracturs and exostoses of the pelvis
- Small size of the pevis due to breeding at too young age
- Congenital hypoplasia of the birth canal
- Compression or stenosis of the birth canal due to injuries
- Failure of the cervix to dilate (Ring womb in ewes)
- Uterine inertia.

b) Fetal Causes

- Abnormal Presentation, Position and posture of the fetus
- Excessive size of the fetus

Management of Dystocia case : Immediate attention should be given to relieve the dystocia. The case must be presented to the veterinarians immediately. General and specific examination must be initiated by the veterinarians and based on the condition of the case and etiology, suitable treatment must be undertaken. Delay in the management of the case may lead to death of the mother and fetus.

2. Uterine Prolapse

This is an important post partum complication. Cervico vaginal prolapse may also occur pre partum. The uterus is protruded out though the birth canal immediately after parturition. Suitable treatment for the replacement of the oprolapsed organ must be done immediate by a qualified veterinarian. The delay in management of the case may lead to complications like necrosis and laceration of the endometrial tissues. The owner can wrap the prolapsed part with moist cloth or polythene to protect the prolapsed part form laceration till the veterinarian attends the case. The case of pre partum cervico- vaginal prolapse at mid gestation in ewes has been treated successfully with hydroxyprogesterone injection @ 1 ml i/mly to prevent recurrence (Bhattacharyya et. al., 2006).

3. Retained Placenta

Retained placenta is very frequently occurred in dairy animals following parturition chiefly in the dairy cows. The cow when fails to expel the after birth normally within 8 to 12 hours of parturition is defined as retained placenta (Roberts, 1982). It is one of the major reproductive problems, second to anestrus in buffaloes (Khar, 1980) and has been found to be the single largest post-partum complication (Pattabiraman and Bawa, 1977). A heavy loss due to this condition to the dairy industry is reported (Bhalaru and Tiwana, 1986). Irrespective of the etiology of the problem, it causes heavy loss in the dairy industry by creating different uterine complicacy like endometritis, metritis, pyometra etc, which increases the calving interval in cows. Post parturient retention of fetal membranes causes economic loss due to loss of milk and delay in involution of uterus and subsequent conception, early embryonic mortality and the problems of repeat breeding or even permanent infertility (Narashimhan and Deopurkar, 1994).

Causes : Retained placenta is quite common in premature calving (Noakes et al., 2003; Islam and Bhattacharya, 2005). Commonly the organisms, which cause abortion, are responsible for the retention of placenta. The retained fetal membranes are edematous, hemorrhagic, lethery and necrotic and it is a common sequele in aborted cases due to Brucella abortus, Vibrio fetus, Tuberculosis and various mold infections (Roberts, 1982). Hormonal imbalance is also responsible for the retention. If the calf is not allowed to suckle the dam's colostrum, the chance of placental retention is more. In the earlier studies it was seen that in case the calf is not allowed to suckle dams' milk could not stimulate the release of pitocin from the posterior pituitary, resulted in 22.7 % retention of after births as compared with 4.9 % retention on the same farm when suckling was allowed (Roberts, 1982). The hygiene of the shed floor also related with the incidence of retention. It has been established that when many cows calved in the same accommodation in a quick succession, retention of placenta is more likely to occur due to a build up of increased bacterial load in the shed. Old cows are more affected than young ones. Incidence of retained placenta is very high in twin birth and also higher in animals suffering from dystocia (difficult birth) than normal birth. Sometimes the hypocalcaemia in the cow may be undetected and this may aggravate during parturition leads to dystocia and / or ratained placenta. Balanced nutrient is beneficial during the gestation. Low calcium diets along with sufficient supply of Vitamin D will help in increasing the absorption and thereby the mother will obtain sufficient supply of calcium required for the growth of the fetus. Vitamin A and Selenium also have profound role on reproduction and optimum supply of these nutrients minimizes the incidence of retained fetal membrane.

Management and Treatment

The ecbolic drugs (drugs which increase the contraction of the uterus and expel the uterine contents) are recommended for the treatment of retention of the placenta. Nowadays a good numbers of ecbolic preparations are available in the market with very good efficacy and the advantage is that the farmers can use this product with ease. Giving the information about the few leading liquid ecbolic preparations available widely in the market may be helpful for the farmers to manage this problem easily without any technical expertise.

Drugs	Manufacturer	Dose rate
1. Uterotone liquid	Cattle Remedies India LTD. New Delhi	• @ 300 ml orally at an interval of 3 hr upto 12 hours for treatment • 100-125 ml twice daily for five days as prophylaxis and involution
2. Exapar liquid	Dabur Ayurvet Ltd., Laxminagar	• @ 100 ml orally initially followed by 50 ml orally twice daily for five days for treatment • 50 ml orally twice daily for prevention
3. Utrifit	Indian Herbs Research and Supply Co. Ltd., Saharanpur, UP	• @100 ml orally twice daily on the 1st day soon after parturition followed by once daily for 3 days

Number of ecbolic preparations in powder form is also available but in respect of ease in administration and seeing the greater efficacy of this liquid ecbolic, the use of ecbolic in powder form is not discussed. The advantage of the use of this ecbolic preparations is that uterus get cleaned quitely and helps in early involution of uterus and thereby ensures post parturient reproductive health of the animal. Therefore, it is recommended by the veterinarians to use this ecbolic preparation in normal cows also after parturition for enhancing the involution in a faster rate. (Sharma et al. 1972) conducted a clinical trail with Uterotone liquid in buffaloes with retained fetal membrane and reported the expulsion of placenta in 78.5 buffaloes within 7.28 hours. Uterotone liquid has been used with greater success in expulsion of the placenta as well as early involution of uterus in buffaloes (Gautam et al., 2002). Exapar liquid orally @ 250 ml divided in daily dosage of 100 ml on the first two days and 50 ml on the third day in cows with partially retained placenta and found the shedding of placenta in 4 to 8 hr (average 5 hr) and early involution of uterus (26 to 45 days) was also found (Singal, 1996).

Though the age old method of removal of the placenta has been used widely with greater success, some times there is a chance of ascending infection to the uterus, if hygienic practice is not followed to remove placenta which may aggravate the condition. In present day therefore it is advised to expel the placenta with oral medications. The authors have treated so many cases by manual removal of the placenta with the insertion of 3 to 4 antibiotic/ antibacterial boli in the uterus and supported with or without oral ecbolic preparations for five days with greater success without any later complicacy, once the oral ecbolic therapy alone failed to remove the placenta. However, placenta has been expelled in a good number of cows with retained placenta by administering Uterotone liquid @ 300 ml at 3 hour interval upto 12 hr or till the expulsion of placenta from the start of the treatment and on and average after 2 doses the expulsion of the placenta has been observed.

Hormonal preparations like Oxytocin and prostaglandin may be used by increasing the uterine contraction for expulsion of the placenta but due to the one or several side effects use of this hormonal preparations should be discouraged (Balasubramainan *et al* 2003). Moreover, various treatments tried including oxytocin and estrogen were not satisfactory (Arthur, 1979)

If the retained placenta is left untreated, some times it may lead to the prolapse of the uterus for the increased straining. It may lead to the uterine infection and cause severe uterine problems like pyometra, metritis, endometritis which may make the animal infertile for number of days by manifesting different disorders like anestrum, repeat breeding etc., therefore the owner suffers heavy economic loss. Heavy losses due to retention of placenta to the dairy industry have been reported in India (Bhalaru and Tiwana, 1986) and in United Kingdom it costs the dairy Industry more than 25 million Pound/ year (Laven and Peters 1996). Manual removal may not be used but it is must to treat the animals orally with ecbolic preparations or antibiotic/ antibacterial - Urea combinations boli @ 3-4 boli must be inserted for liquefying the placenta and to guard the infection, which will safeguard the uterus of the animal for future reproduction.

References

Ahmed, F.A., Tamuli, M.K. and Akhtar, F. (2003). Effect of Placentrex and Janova in induction of Oestrus in anoestrus crossbred cattle. *Indian Vet. J.*, 80:1078-1079.

Arthur, G.H. (1979). Retention of after births in cattle: A review and commentary. *Veterinary Ann.*, 19 :26.

Balaru, S.S. and Tiwana, M.S. (1986). Consequences of calving disorders in buffaloes. *Indian Journal of Animal Science*. 56 : 274-276.

Balasubramanian, S., Sridevi, P., Prathaban, S. and Pattabiraman, S.R. (2003). *Indian Vet. J.*, 80: 689-690.

Bhattacharyya, H.K, Islam, R., Peer, F.U., Buchoo, B.A. and Choudhury, A.R. (2006). Management of cervico vaginal prolapse in ewe. *Indian Vet. J.*, 83: 881-882.

Deshpande, R.S., Dhoble, R.L and Sawale, A.G. (2000). Efficacy of indigenous drugs in the treatment of post partum anestrum in buffaloes. *Indian J. Anim. Reprod.*, 21(2): 115-116.

Gautam, R.P, Tiwari, R.P. and Koley, K. M. efficacy of Uterotone in induction of uterine contraction in vitro and expulsion of fetal membranes in buffaloes. *Indian Journal of Animal Science*. 72 (9): 754-755.

Hussain, J., Islam, R, Barman, K, Shaheen, M and Ganai, N. A. (2006). Management of post-partum true anestrum in cows with herbal therapy in combination with mineral supplementation. *The Indian J. Field Veterinarians* 2(1): 33-35.

Islam, R. and Bhattacharyya, H. K. (2005). Premature birth in cross bred Jersey Cow-a case report. *The Indian J. of Field Veterinarians* 1(2): 58-59.

Islam, R., Bhattacharyya, H. K. and Khan, M. Z. (2006 a). Sexual Health control in herds. *The North East Veterinarian* 6(3):19-20.

Islam, R. and Nadroo, G. A. (2006 b). Incidence of abortion and still birth in cattle in an organized farm of Kashmir valley. *Indian Vet. J.*, 83 (10): 1127-1128.

Islam, R., Nadroo, G. A., Sarker, T. K. and Bhat, A. S. (2006 c). Reproductive disorders in Corriedale Ewes. *The Indian J. Anim. Reprod.*, 27 (1): 37-41.

Khar, S.K. (1980). Gynaecological problem of buffaloes. In: ICAR sponsored Summer Institute on Buffalo Management System, Haryana Agricultural University, Hissar.

Khasatiya, C.T., Desai, P.M., Dhami, A.J., Dugwekar, Y.G., Panchal, M.T. and Kavani, F.S. (2004). Effect of GnRH and PGF_2a on reproductive performance and progesterone profile in post partum true anestrous and sub oestrus Surti buffaloes. *Indian J. Dairy Sci.*, 57 (5): 324-328.

Kodagali, S.B., Deshpande, B.K. and Sane, C.R. (1981) LH-RH treatment for post - partum anoestrous conditions in Gir cows. *Indian Vet. J.* 58: 487-489.

Laven, R.A. and Peters, A. R. (1996). Bovine retained placenta aetiology, pathogenesis and economic loss. *Veterinary Record*. 15: 131-133.

Madhavan, E. and Iyer, C.P.N. (1993) Treatment of anestrum in crossbred cattle with CoCu-H. *Indian J. Anim. Reprod.*, 14(2):125.

Narasimhan, K.S. and Deopurkar, V.L. (1994). Accidents and diseses incidental to parturition. In: Reproduction in farm animals. 2nd Edition, Edited by C.R. Sane and others., Varghese Publishing House, Bombay.

Noakes, D. E., Parkinson, T. J. and England, G. C. W. (2001) Arthur's Veterinary Reproduction and Obstetrics, 8th edn., W.B. Saunders Company , Philadelphia. pp.

Pattabiraman, S.R. and Bawa, S.J.S. (1977). Incidence of pre and post-partum reproduction disorder in bovines. The Haryana Veterinarian. 16: 99-101.

Pattabiraman, S. R., Veerapandian, E and Quayam, S.A. (1986). Effects of Receptal treatment in anestrous and early post - partum cows and buffaloes. *Indian Vet. J.*, 63 : 409-413.

Reddy, K.R. C., Rao, A. S., Reddy, V. S. C., Yadagiri, B. Sharma, G. P. Rani Reddy, M. and Eswar Reddy, C. (1994). Efficacy of certain non hormonal and hormonal drugs on oestrus induction in post-partum anoestrous buffaloes. *Indian J. Anim. Reprod.*, 15 (2): 127-130.

Roberts, S. J.(1982). Veterinary Obstetrics and Genital diseases. 2nd edition, CBS Publishers and Distributors, New Delhi-110012.

Shah, N.A. (1996). Effect of Gonadotrophin releasing hormone (GnRH) on anoestrous cross bred cows under temperate climatic conditions. MVSc Thesis, Submitted to S. K. University of Agricultural Sciences & Technology (K), Srnagar.

Sharama, S.D., Joshi, B.P. and Rai, P. (1972). A discussion on clinical trails of Uterotone in retained placenta in tropical ruminants. *Indian Journal of Veterinary Research.* 6: 57.

Shokier, A.A . (1959).... Vet. Med. J., 6:149 (cited by Pattabiraman, S. R., Veerapandian, E and Quayam, S.A. (1986). Effects of Receptal treatment in anestrous and early post - partum cows and buffaloes. *Indian Vet. J.*, 63 : 409-413).

Singal, S.P. (1996). Efficacy of Exapar in Post parturient disorders with retained placenta in Bovines. *Indian Journal of Animal Reproduction.* 17 (2): 109-110.

Singh, C.P., Sharad Kumar, Atul Saxena, Ram Sagar. (2004). Induction of Estrus in Post Partum anoestrus cows with GnRH and short term steroid treatment. *Intas Polyvet.* 5(2): 183-185.

Wani, N.A., Buchoo, B.A. and Khan, M. Z. (1999). Treatment of longstanding anestrous cows - a clinical study. *SKUAST J. Res.*, 1: 95-98.

□□□

Chapter 24

Pathological Aspect of Disease Diagnosis

P. Goswami

Diagnosis of a disease depends on proper collection of history, observation of clinical symptoms, lesions and collection of materials for laboratory diagnosis. Most of disease in animals in the field are lacking in proper diagnosis. We shall now discus some of the important diseases occurring in the field and collection of materials for laboratory diagnosis. Accurate diagnosis of diseases depends upon how carefully the suitable morbid material has been collected and dispatched to the diagnostic laboratory. This is an essential requirement in carrying out any laboratory procedure warranted for diagnosis of a particular disease. Similarly we may collect blood, tissue or body fluids and their smears. Before dealing with the procedures it is essential that history and clinical signs are recorded, and postmortem is conducted skillfully and the observed lesions recorded in a lucid manner. The postmortem should be carried out as soon as possible after the death of the animal so that the autolysis and putrefaction does not supervene. Both the changes affect diagnostic work and reduce the chances of recovery of the etiological agent.

Collection of Materials for Histopathological Examination

The material for histopathology should be representative of the lesion and shall include some of the normal tissues also. Small tissue pieces of ½ to 2 cm from organs showing the lesions are collected and put immediately into a proper fixative contained in wide-mouth bottles. The volume of the fixative should be at least 10 times the volume of the tissue. This allows proper fixation of cells and prevents the postmortem changes. A well/ properly fixed tissue is good for cutting sections and staining and provides the typical histopathological picture to the pathologist. The common fixative used are !0% Formol saline solution and Buffered Neutral Formalin Solution. In case of an outbreak of disease material from 5 to 6 or more animals may be collected and dispatched to the laboratory as it may help to give true picture of the disease outbreak.

Specimen Collection for Haematological Examination

a) Collection of Whole Blood from Live Animals

In living conditions, blood is collected from jugular venepuncture of cows, buffaloes, horses, sheep and goats. Blood can also be obtained from ear vein in cattle and buffaloes. In dogs and cats it can be obtained from saphenous vein and in poultry from brachial vein. The hair is clipped and the area sterilized with spirit or 70% alcohol. A 16 gauge needle is suitable for adult horses, cattle and buffaloes whereas in sheep, calves and cats 18 gauge needle is satisfactory.

The blood is collected in anticoagulant vials. Mix the blood and anticoagulant with gentle rocking motion. Vigorous shaking will cause haemolysis. If assay is delayed for over an hour after collection, refrigerate (0-4°C). warm it to room temperature (20°C) and remix gently. Whole blood should not be frozen. Following anticoagulants may be used:

1. *Ethylene Diamine Tetra Acetic Acid (EDTA):* It is the anticoagulant of choice for hematological studies as it does not alter cell morphology. It is available as sodium or potassium salt and prevents clotting by chelating calcium. It is used at the rate of 1.0 to 2.0 mg/ ml of blood. It is not used for plasma to be used for chemical assay.

2. *Heparin:* It is an anticoagulant of choice when plasma is required for analysis as it least interferes with biochemical assays. It is the only anticoagulant which is neutral and does not form any precipitate. Heparin is available as sodium, potassium, lithium or ammonium salts. It is thought to prevent clotting by preventing conversion of prothrombin to thrombin during the clotting process. Heparin is used at the rate of 20 units/ ml of blood. However, it is relatively expensive.

b) Preparation of Blood Films

Both thin and thick blood films are prepared. The smear should be prepared preferably from the fresh blood without using any anticoagulant. Thin blood smears are prepared by placing a small drop of blood on a clean glass microscope slide. The end of the second slide (spreader slide) is placed against the surface of the first slide at 45° angle and drawn back into the drop of blood. When the blood has spread along the most of the spreader slide it is then pushed forwards with a steady, even, rapid motion. A good slide preparation should be thick at one end and thin at the other (feathered edges) and should occupy the central portion of the slide. The smears should be dried quickly by waving it rapidly in the air.

The smears are marked by any glass marking pencil or identification number can also be written with lead pencil on the thick part of the film which is not used for examination.

The smears are fixed with methanol (acetone free) for 2-3 minutes. At least three smears are prepared from one particular case. They are then wrapped in a paper in such a way so as the two smears do not come in contact. Particulars of specimen are written on a paper and dispatched along with the smears.

In case of blood smears to be examined for bacteria, additionally one or two slides should be fixed by repeatedly passing the slide through a flame. Care is taken not to overheat the slide which can be tested on the back of the hand. The slide should feel warm but not hot.

Thick Blood Smears : A thick blood smear allows examination of greater volume of blood than does the thin blood smear. 3 drops of blood are placed on the slide and spread with a wooden applicator stick to make a 2 cm circle. The slide is allowed to dry. It is then placed in beaker filled with distilled water in a slant position with smear side down. It is allowed to remain until the smear gets dehaemoglobinized. The slide is then removed and air dried. It is fixed in methanol for 10 minutes. Thick blood smears are used for detecting parasites.

c) Blood from Dead Animals

Blood is collected for bacteriological examination from the heart. The surface of the heart is sterilized with hot iron spatula. The sealed end of a sterilized pasture pipette is broken and is inserted into the heart and blood sucked. It is then properly sealed, packed and dispatched.

Preparation of Tissue Smears

Several impression smears from cutaneous lesions or tissues collected during surgery or necropsy and the tissue fluids are prepared so that the slides are available for special staining if necessary. Scrapings and swabs are also used for preparation of smears. Smears can be prepared from aspirates directly or following concentration by centrifugation. They are then air dried and fixed with methanol and dispatched.

Collection of Blood Plasma

Blood is collected in a container with the proper anticoagulant. Centrifuge within an hour after collection, for 10 minutes @ 2000-3000 rpm. The plasma is removed with a capillary pipette and transferred to another container labeled with date, time of collection, patients name and clinics number. Refrigerate the sample if centrifugation is not done within an hour. If heparinized plasma has been stored for some time after separation, centrifuge the sample again to remove any fibrin clots.

Collection of Serum

Collect whole blood in a container with no anticoagulant. Keep the bottle at an angle of 20° from the horizontal plane. It is allowed to clot at room temperature for 20-30 minutes. Let it clot thoroughly. It is separated by running a wooden applicator stick or a glass rod around the wall of the container between the clot and wall. It is called "ringing the clot". It is done to prevent haemolysis as the cellular wall is peeled away from the tube wall during centrifugation.

The sample is centrifuged for 10 minutes @ 2000-3000 rpm. The serum is removed with a capillary pipette and placed in a labeled container. Serum can be refrigerated or frozen.

Serum is used for serological and biochemical tests. Pipette out 1cc of the serum in a serum vial with cork stopper. 0.001% concentration of either merthiolate or sodium azide is used as a preservative for serum used for serological tests but not for serum used for neutralization test. Use antibiotics when serum neutralization test is desired.

Prolonged contact of cells with the serum causes the release of the red cell constituents into the serum and this will alter the true serum concentrations e.g. potassium and glucose levels will fall due to glycolysis if the cells are in contact with serum. Refrigerated specimens can be analysed within 24 hours.

When serodiagnosis is required always paired sera samples are collected, one at the start of the disease and another after recovery (3-4 weeks) from the disease.

Collection of Material for Bacteriological Examination

The material is collected with aseptic precautions and dispatched in sterile container. A bad specimen cannot yield good results. Specimen contamination is the most common cause of diagnostic failure. The multiple specimens are kept separately to avoid cross contamination. In case of dead animals, the specimen is taken soon after death, otherwise putrefactive bacteria invade tissues and render them unsuitable for examination. All containers used for holding microbiological specimens must be sterilized. The specimens are collected prior to therapy. If the patient is on antibiotics inform the same. The laboratory can take appropriate counter measures (e.g. diluting the specimen or adding appropriate reagents like penicillinase against penicillin. Avoid contamination of specimens during collection and transportation (e.g. disinfect the skin or collect urine specimen in the midstream). The specimen is collected during acute phase of the disease (e.g. febrile stage of the infection or diarrhoeal stage of the intestinal infections). If anticoagulant is needed (e.g. in case of pleural fluid) this is added before sterilization. Blood is usually collected without anticoagulant and diluted tenfold with nutrient medium.

Deliver the specimens to the diagnostic laboratory by the fastest means within 30 minutes of collection or refrigerate (4°C) immediately. Refrigerated specimens are not left unattended for an extended period. They must be cultured within 2-4 hours.

a. Abortion Specimens

A small piece of placenta with 1 or 2 cotyledons, fetal stomach contents, amniotic fluid and a piece of fetal liver, lung and spleen. In large animal if fetus has aborted before 7 months of gestation send the entire fetus in a leak proof container with several cold packs.

b. Milk Sample

Milk samples are collected aseptically otherwise the microorganisms from the skin of udder and from the air may contaminate the specimen. Use 70% alcohol or 70-90% isopropyl alcohol for wiping first the 2 teats farther from you and then the two near teats. Milk is collected in screw caped bottles after discarding the first few squirts of milk and then milk from each quarter is send in a separate bottle. Sample the 2 teats nearer to you first and then those farther from you so that you do not accidentally brush against the disinfected teats before sampling. Refrigerate the milk samples as soon after collection as possible until they can be cultured. Do not freeze the sample. If mastitis due to *Mycoplasma* is suspected, Ampicillin can be added @ 1-10 mg/ ml of milk to control contaminant bacteria. These samples are left at room temperature so that the *Mycoplasma* organisms can multiply in the sample. Duplicate samples are taken without adding antibiotics for other mastitogenic bacteria.

c. Urine Specimens

Midstream urine samples are collected aseptically by catheter or by cystocentesis. About 5 ml samples are collected in a sterile tube. In this case presence of bacterial count ? 10,000/ ìl suggests an active infection.

d. Fecal Specimens

Freshly voided or collected from the rectum. In large animals disposable rubber gloves can be used. They are inverted, tied off and submitted. In small animals sterilized swabs are used. A sample of 1 gm is needed. Serial samples are submitted as infectious agent like *Salmonella* is excreted intermittently.

e. Specimens from Abscesses and Wounds

Pus is collected with the help of a sterile syringe and wide bore needle. Scrapings from the wall of the abscess have viable organisms. From wounds, the surrounding area is disinfected and superficial debris is lifted. Swabs are taken near the edge of the wound as the centre may contain necrotic debris only. Direct smears can be taken from the wounds. Note the colour, odor and presence of "sulphur granules".

f. Specimens from Eye

Corneal scrapings, a swab rolled into the conjunctival sac, or lacrimal secretions collected with a sterile swab and applying the specimen directly into a blood agar plate. If it is not possible, deposit the end of the swab into about 2 ml of sterile saline and culture within 2 hours.

g. Genital Specimens

Cervical mucous, collected in a sterile artificial insemination pipette introduced through a speculum. About 3 ml of mucous is collected and expressed into vials containing Blair medium. Swabs from the vulvar mucosa, particularly around the 'clitoral fossa' are taken and stored at 4°C and shipped at 4°C with cold packs. If swabs are submitted they should be submitted in Stewarts Agar medium.

h. Specimens from Ear

They are usually collected from dogs. Both ears are individually swabbed which help in differentiating normal from abnormal flora.

i. Culture of Anaerobes

As anaerobes can survive exposure to air for less than 20 minutes, collection of swabs for the purpose of anaerobic culture are useless unless extra care is taken. Anaerobic specimens include blocks of tissues (2 inch cube, minimum) in a closed sterile container, material placed in Becton-Dickinson's Anaerobic Specimen Collector, and pus and exudates collected in a sterile syringe with the air expelled and the needle plugged with a rubber stopper and bent backwards on itself.

Collection of Materials for Isolation of Virus

Virus material for isolation should be forwarded in 50% glycerin saline or preferably in 50% glycerin phosphate buffer (pH 7.3-7.4) and Hank's balanced salt solution. If transport medium is not available tissue is collected in sterile containers, sealed and transported on sufficient ice to the nearest laboratory. Collect the materials that are likely to contain the largest number of virus particles. Collect material from live or recently died animals. The specimens are kept at 4°C and transported to laboratory in the shortest possible time. 2 inch tissue cubes containing both diseased and normal are sampled for virus isolation. Mucosal scrapings are preferred to swabs. The material is collected in sterile screw caped containers using separate container for each sample. Specimens are refrigerated (4°C) as virus content decreases with increase in temperature. If the specimen is delivered within 24 hours they are stored at 4°C and packed with ice in polystyrene insulated cotton for shipment. If more than 24 hours delay will occur snap free at -70°C and ship on dry ice. Influenza virus is best preserved at 20°C. The specimen should be in air tight container otherwise entry of CO_2 gas in the container from the dry ice lower the pH of the fluid killing any pH-labile viruses. When viral disease is suspected antibiotics (penicillin 1000 Units and streptomycin 10 mg/ ml) may be used in transport media and in serum samples dispatched for diagnosis. This inhibits contamination. About 20 gm each of spleen/ lymph node tissues are collected for virus isolation. Material is collected from 5 to 6 or more animals for disease diagnosis and investigations.

Transport Media/ Preservatives

1. 50% Phosphate Buffered Glycerin (pH 7.4)

First prepare M/25 phosphate saline (pH 7.4-7.6) as follows:

Sodium Chloride	8.500 gm
Di-sodium hydrogen phosphate	6.110 gm
Potassium dihydrogen phosohate	0.777 gm
Distilled water	1000 ml

Add equal volume of sterile Neutral Glycerin to sterile M/25 Phosphate Buffer Saline (pH 7.4-7.6) to prepare 50% Phosphate Buffered Glycerin (pH 7.4). add 0.1 ml of 1% Phenol Red solution to 100 ml of that solution so as to give final concentration of phenol red to 0.001%. The solution when sterile should have reddish tinge. Yellow colour indicates contamination. Each vial must contain 10 ml for collection of tissues.

Collection of Material for Fungal Culture

The location of the infection site considerably helps in the diagnosis of a mycotic infection. Materials for superficial mycosis include skin scrapings. These are often contaminated with fast growing organisms on the surface. To minimize contamination the area is first cleaned with 70% alcohol and then with sterile water. Skin scrapings are taken with a blunt scalpel from the lesions edges. Continue scrapping until blood is just drawn. The scrapings are collected in paper envelops. For deep mycosis frozen tissue can be submitted for fungal culture. The container should be marked "CAUTION" as fungi causing mycosis in animals are potentially zoonotic agents. Fluids like blood, cerebrospinal fluid (CSF) or aspirates are submitted in screw-cap tubes.

Collection of Material for Parasitological Examination

Faecal samples are collected for routine diagnosis of alimentary parasitism. They are collected either directly from animal's rectum or immediately when the animal has defecated, in plastic bottles with a light screw cap or zip lip plastic packs. Faeces can be send fresh if examination within a few hours is possible, otherwise mix at a ratio of 1:3 with 10% formalin. Parasites recovered from the digestive tract can be preserved in 70% alcohol or 10% neutral buffered formalin and placed in a leak proof container. Hot 70% alcohol or 5% formalin solution heated to 71-82°C fixes the parasites in an extended state. Before fixing give 2 or 3 washings in physiological saline.

Skin scrapings are preserved in 10% KOH solution to identify mange or mite infection.

External parasites are collected by brushing, combing and shaking the fur, hair or feathers over a large sheet of white paper and then preserved in a vial containing 70% alcohol.

Thin and thick blood smears are collected for identification of haemoprotozoan diseases.

Collection of Material for Toxicological Examination

The stomach and its contents is collected in a wide mouthed bottle of 1 liter capacity. A piece of upper part of the stomach can be taken with its contents. A portion of liver about 500 gm or whole of it along with large portion of spleen and a kidney should be taken. These are preserved in rectified spirit (unmethylated spirit) except in case of suspected poisoning by alcohol, phosphorous, paraldehyde, acetic acid or carbolic acid and other drugs of phenol group, when saturated solution of common salt is used. The viscera are cut in small pieces to ensure penetration of the preservative. The preservative should not fill the bottle completely but only up to two third in height to avoid the risk of the bottle bursting in case of any gas production due to decomposition. The bottles are properly sealed, labeled boldly indicating *"Suspected for Poisoning"* and forwarded for analysis.

Urine is preserved with equal volume of rectified spirit or with a few grains of thymol if rectified spirit is contraindicated. Similarly faeces are preserved in rectified spirit.

Despatch of Morbid Material to Laboratory

1. All materials should be dispatched in properly sealed, leak proof containers on the day of collection, immediately after collection, through a messenger or through air transport or speed post so that it is received by the concerned laboratory within 2-3 days.
2. All bottles/ containers should be properly labeled contained in a packeting material containing sufficient quantity of absorbent material like saw dust or cotton wool. All parcels should be conspicuously marked *"Fragile, with Care"* and bear *"Pathological Specimen"*.
3. All materials should be accompanied with letter giving full details of the specimen like - species, breed, age, sex of the animal; morbidity, mortality, clinical signs, lesions, disease suspected.
4. Specimens packed for rabies, glanders, anthrax should be marked *"Suspected for– – – – – – – –'*.

References

Alexander, D. P., Nixon, D. A., Widdas, W. F. and Wohlzogen, F. X. 1955. Gestational variation in the composition of foetal fluids and foetal urine in sheep. *Journal of Physiology*, 140: 1.

Alexander, D. P., Nixon, D. A., Widdas, W. F. and Wohlzogen, F. X. 1958. Renal function in sheep foetus. *Journal of Physiology*, 140: 14-22.

Aminoor-Rahman, A. N. M., Snibson, K. J., Lee-Chee-Seong and Meeusen, E. N. T. 2004. Effects of implantation and early pregnancy on the expression of cytokines and vascular surface molecules in the Sheep endometrium. *Journal of Reproductive Immunology*, 64 (1/2): 45-58.

Amle, M. B., Chinckar, S. R., Hukari, V. B. and Deopurkar,V. L. 1992. Studies on Foetal Fluids of Buffaloes. *Indian Journal of Animal Reproduction*, 13 (2): 165- 167.

Antunovic, Z., Speranda and M., Steiner, Z. 2004. The influence of age and reproductive status on the blood indicators of ewes. *Archiv-fur-Tierzucht*, 47(3): 265-273.

Arthur, G. H. (1956).In: Vety. Reproduction and Obstetrics, 5th edition, p. 43.

Arthur, G.H. (1957). Some notes on quantities of foetal fluids in ruminants. *British Vety. Journal*, 113: 17.

Bacha J. William, and Jr.Linda M. Wood, 1990. In: Colour Atlas of Vety. Histology. Lea & Febiger, Philadelphia, London. pp 207.

Baetz, A. L., Hubbert, W. T. and Graham, C. K. 1976. Changes of biochemical constituents in bovine foetal fluids with gestational age. *American Journal of Veterinary Research*, 37(9): 1047- 1052.

Banks, W.J. 1986. Applied Veterinary Histology, Hosy Year book, Philadelphia, p. 503.

Bazer, F. W. and Roberts, R. M. 1983. Biochemical aspects of conceptus - endometrial interections. *J. Exp. Zool.*, 228: 373.

Bazer, F.W. and Sharp, D.C. 1991. In: Reproduction in farm animals by Hafez, 6th edition, p. 83.

Bhavsar, B. K., Pattabiraman, S. R. and Venkataswami, V. 1974. Endometrial changes during early pregnancy of Bubalus bubalis. *Indian veterinary journal*, 51(3): 175-176.

Bongso, T. A., Edirisinghe, R. and Athuraliya, D. 1979. Studies on foetal fluids in goat. *Indian veterinary journal*, 56(7): 562- 569.

Bongso, T.A and Basrur, P.K.1976. Foetal fluids in cattle. *Canadian Journal*, 17(2): 38-41.

Cengiz, F., Yaman, K. and Galip, N. 2000. Studies on progesterone, estrogen and calcium values in cows before and after calving. *Veteriner-Fakultesi-Dergisi-Uludag-Universitesi*, 19(3): 37-43.

Chauhan, R. A. S. and Tiwari, S. 1996. Variation in the transaminase and alkaline phosphatase activity of caprine foetal fluids with gestational age. *Indian Veterinary Journal*, 73(10): 1045- 1047.

Chauhan, R.A.S., Tiwari, S. and Datta, I. C. 1992. Changes in the volume and biochemical constituents of caprine foetal fluids with gestational age. *Indian Journal of Animal Reproduction*, 13(1): 72-74.

Cloete, J. H. L. 1939. Onderstepoort. *J.Vet.Sci.Anim.Indust.*, 13: 418.

Dellmann, H. Dieter and Eurell Joann. 1998. In: Textbook of Veterinary Histology,Fifth Edition, Lippincott Williams & Wilkins

Dey, S. K. 2004. Focus on implantation reproduction, 128: 655-656.

Dhami, A. J., Patel, P. M., Lakum, P. D., Ramani, V. P. and Pande, M. B. 2003. Micronutrient profile of blood plasma in relation to age reproductive status of Holstein Friesian cattle. *Indian-Journal-of-Animal-Nutrition*, 20(2): 206-211.

Dukes, H. H. 1955. Protein Metabolism. Physiology of Domestic Animals. 7th Edition, pp 562.

Eulenberger, K. H., Neubauer, K. and Elze, K. 1977. Histological studies of the endometrium and chorioepithelium of sows during early pregnancy. *Monatshefte- fur-Veterinarmedizin*, 32(9):333-336.

Fakowska - Podstawka, M., Wrona, Z. and Krakowski, L. 2002. Biological role of interferontau. *Medycyna-Weterynaryjna*, 58(8): 568-571.

Fergusona, S.E., Robert, A.S. Alison, A. Richard, B. 1995. Diagnosis of uterine cancer bulletin - Memorial Sloan Kettering cancer institute 87, 1995. PP. 645-6451. *Am. J. Obst. Gynaecol.* 1995. 172: 227-235.

Garland, H. O. 1983. Cited in "The effects of pregnancy and sodium intake on water and electrolyte balance in sheep. *British Veterinary Journal*, 144: 147.

Goldstein, M. H., Bazer, F. W. and Barron, D. H. 1980. Characterisation of changes in volume, osmolarity and electrolyte composition of porcine foetal fluids during gestation. *Biology of Reproduction*, 22(5): 1168 – 1180.

Gordon Ian, 2004. In: Reproductive technologies in farm animals. CABI Publishing.

Gould, W. A. 2009. Total Quality Management for Food Industries. Wood Head Publishing Limited. Abington Hall, Abington, Cambridge, CB 216AH, U.K. Ferguson, E.L. and Roberts, H.E. 1995. Morphologenesis of epithelia during development, www.pubmedcentral.nih, www.endogournas.org, www.liebertonline.com

Gray, G. A., Barghardt, R. C., Johnson, G. A., Bazer, F. W. and Spencer, T. E. 2002. Evidence that absence of endometrial gland secretions in uterine gland knockout ewes compromises conceptus survival and elongation. *Reproduction*, 124: 289-300.

Green, W.W. and Winters, L. M. 1945. Tech. Bull. Minn. Agric. Exp., Ta No. 169.

Guillomot, M., Flechon, J. E. and Wintenberger Torres, S. 1982. Cytochemical studies of uterine and trophoblastic surface coats during blastocyst attachment in the ewe. *Journal of Reproduction and Fertility*. 65(1): 1-8.

Hafez, E. S. E. 1987. In: Reproduction in farm animals, 5th edition, Lea & Febiger, Verghese Company

Hafez, E. S. E. and Hafez, B. 2000. Reproduction in farm Animals. Seventh Edition, Lippincott Williams & Wilkins, A Wolters Kluwer Company, pp. 126-139.

Hafez, E. S. E. and Jainudeen, M. R. 1993. Gestation, Prenatal physiology and parturition. In: Reproduction in farm animals. 6th edition, pp. 225-226.

Hafez, E. S. E. and Kamal, M. A. M. 1955. Physiochemical investigation on placental fluids of buffaloes. *Indian Journal of Vety. Science*, 25(3): 39-45.

Hafez, E. S. E., Magharabi, M. M. and Anwar, A. 1958. Some factors effecting blood cholesterol level in buffaloes. *Indian Journal of Vety. Science*, 28(3): 133-138.

Henricks, D. M and Harris, R. B. Jr. 1978. Cytoplasmic estrogen receptors and estrogen concentration in bovine uterine endometrium. *Endocrinology*, 103(1):176-185.

Hewitt, S. C. and Korach, K. S. 2000. Progesterone action and responses in the alpha ERKO mouse. *Steroids*, 65(10/ 11): 551- 557.

Hewitt, S. C., Goulding, E. H., Eddy, E. M. and Korach, K. S. 2002. Studies using the estrogen receptor alpha knockout uterus demonstrate that implantation but not decidualisation associated signaling is estrogen dependent. *Biology of Reproduction*, 67(4): 1268-1277.

Huct-Hudson, M. Y., Andres, G. K. and Dey, S. K. 1989. Cell type specific localization of C-MYC-Protein in mouse uterus-modupition by steroid hormones and analysis of the pre implantation period. *Endocrionology*, 125: 1683-1690.

Hugget, A. St. G. and Widdas, W. F. 1951. The relationship between mammalian foetal weight and conception age. *Journal of Physiology*, 114: 306- 317.

Hugh Y. Yee, Edward S. Jenest, and Fred R. Bowels, 1971. Modified manual or automated o- toluidin system for determining glucose in serum, with an improved aquous reagent. *Clin Chem.*, 17(2):103-107.

Ismail, Z. B., Qureshi, T., Levy, M., Khamas, W. and Nour, A. 2003. Preliminary report on some physiological parameters in pregnant female camels (Camelus dromedarius) in North America. *Journal-of-Camel-Practice-and-Research*, 10(2): 125-129.

Joshi, C. L., Nanda, B. S. and Saigal, R. P. 1976. Histomorphological and histochemical studies on the female genitalia of ageing goat, glands in cervix uteri. *Anatomischer-Anzeiger*, 139(1/2): 193-199.

Joubert, D. M. 1956. A study of prenatal growth and development of components of caprine conceptus. *Journal of Agricultural Science*, 47: 382- 428.

Kancko, J. J. and Corneleous, C. E. 1970. Clinical biochemistry of domestic animals. Volume I, 2nd edition, pp. 313-394.

Khatoon, A. 2007. MVSc, Thesis, Fac. Vet Science and A.H; SKUAST-K, Shalimar Srinagar, Kashmir, J&K India.

Khatri Pershotam, Hamayun Khan and Rind, M. M. 2004. Biometrical studies of reproductive organs of non pregnant buffaloes slaughtered in Hyderabad. *Journal of Animal and Veterinary Advances,* 3(8): 500-502.

Kind, P. R. .M. and Kings, E. J. 1954. *J. Clin. Pathol.,* 7: 322.

King, G. J., Atkinson, B. A and Robertson, H. A. 1981. Development of the intercaruncular areas during early gestation and establishment of the placenta. *Journal of Reproduction and Fertility,* 61(2): 469-474.

Kleflin, A., Zobundzija, M., Martincic, T. and Sladic, Z. 1988. Concentration changes of potassium and sodium ions in the pigs foetal fluids during the first half of gravidity. *Veterinarski - Archiv,* 58 (5): 225- 232.

Kolmer, J. A., Spanbling, E. H. and Robinson, H. W. 1951. Approved Laboratory Techniques.Appleton Century Crofts, New york.

Koul, NA. 2003 (Guide Wani, G.M.) MVSc Thesis, FAC. Vet. Sci A.H; Department of Animal Reproduction Gynecology, SKUAST-K, Srinagar Kashmir.

Kurita, T., Lee - Kijun, Saunders, P. T. K., Cooke, P. S., Taylor, J. A., Dahiya, R. and Cunha, G. R. 2001. Regulation of progesterone receptors and decidualisation in uterine stroma of the estrogen receptor- alpha knockout mouse. *Biology of Reproduction,* 64(1): 272-283.

Lathura, R. A., Gupta, R. C., Khar, S. K. and Baru, P. 1987. Variation in metabolic constituents of caprine foetal fluids during different stages of gestation. *Indian Journal of Animal Reproduction,* 8(2): 133-137.

Lee, K. Y. and De Mayo, F. J. 2004. Animal models of implantation. *Reproduction.* 128: 679-695.

Luna, L. G. 1968. Manual of Histological Staining methods of the Armed Forces Institute of pathology, 3rd Edn. The Blakistan Division, McGraw-Hill Book company, New Work

Lupke, H., Mulling, M. and Sohn, E. 1967. Maternal foetal relationship in cattle, content to total nitrogen, urea, uric acid and creatinine in foetal fluids. *Vety. Bulletin,* 38: 2047.

Malan, A. P. and Curson, H. H. 1937. Onderstepoort. *J. Vet. Sci. Anim. Indust.,* 8: 417.

Manish - Mahawar, Jain, A. K. and Puneet - Kumar, 2004. Clinically important enzymes in Jamunapari and Marwari goats. *Indian-Journal-of-Animal-Sciences,* 74(9): 924-925.

McCrabb, G. J., Egan, A. R. and Hosking, B. J. 1991. Maternal under nutrition during mid pregnancy in sheep. *British Journal of Nutrition*, 65(2): 157-168.

McDonald, L.E. 1980. *Veterinary Endocrinology and Reproduction*, 3rd edition, p. 61, pp. 388-389.

Mellor, D. J. and Slatter, J. S. 1971. Daily changes in foetal fluids during last three months of pregnancy in ewes. *Journal of Physiology*, 217: 573-604.

Mellor, D. J. and Slatter, J. S. 1972. Daily changes in foetal urine and relationship with foetal fluids and maternal plasma during last three months of pregnancy in ewes. *Journal of Physiology*, 227: 503-525.

Michel, G. and Stief, E. 1985. Placentation and foetal development in sheep. *Monatshefte-fur-Veterinarmedizin*, 40(10): 344- 347.

Motta et al. 2001. Yuan et al. 1999 cited references from Reproduction Obility / Infertility Biology, www.abcinfertility.com

Mufti, A. M. 1995. Thesis on Morphological characters and some metabolic constituents at different stages of development of ovine foetus. SKUAST-K, Shalimar.

Mufti, A. M., Wani, G. M., Wani, N. A., Buchoo, B. A. and Khan, M. Z. 2000. Prenatal development of ovine fetus. *Small-Ruminant-Research*, 38(1): 87-89.

Natelson, Scatt, M. L. and Beffa, C. 1951. *Americ. J. Chem, Path.* 21: 275.

Noden M. Drew and Lahunta De Alexander, 1985. In: Embryology of Domestic Animals. Williams & Wilkins.

Owen, J. A., Iggo, B., Scand rett F. J., Stewart, C. P. 1954. *Biochem. J*, 58: 426.

Pareek, P. K., Joshi, C. K. and Vyas, K. K. 1981. Note on the biometry of genital organs of female camel. *Indian Journal of Animal Sciences*. 51(10): 1010.

Paul, S., Ghosh, R. K., Das, P., Mahata, T. K. and Thakur, A. M. 2003. Studies on gross anatomy and histology of the non pregnant and pregnant (upto 90 days) uterus in the non descript cow. *Indian Journal of Animal Health*, 42(2): 178- 181.

Phil, B. 2007. Cited Reproductive health Management. www.streetdirectoruy.com

Phil, B. 2007. Food Quality Management. www.otago.ac.ng.

Pol Kowski, L. and A. Skowran, Eds. Rough sets in knowledge Discovery Management Support System. *European Journal of Operational Research* 84, 1995 www.ca.vu.nl

Pourouchottamane, R., Chatterjee, A., Sheikh, I. U., Sarkar, M. and Gogoi, D. 2005. Blood biochemical constituents of female yaks in different physiological status. *Indian-Veterinary-Journal*, 82(10): 1108-1109.

Raj Sundrum, R. C., Kathireson, D. and Pattabiraman, S. R. 1990. Changes in biochemical constituents of foetal fluids with increase in gestation period in buffaloes. *Indian Journal of Animal Reproduction*, 11(1): 24-28.

Ramadhar Singh, 1988. Ageing and age. Essential of Animal Production and Management. First Edition. Kalyani publisher, pp. 210- 227.

Ramana Rao, K.V. 2006. In: General studies for Agricultural Research services and other Competitive Examinations. 4th Revised and Enlarged Edition, Jain Brothers, New Delhi, pp. 35-36.

Reinhold, J.G. 1953. In: standard methods of clinical chemistry (Reiner M.ed), New York and London, Academic Press, 1: 88.

Reitman, S. and Frankel, S. 1957. A colorimetric method for the determination of serum glutamic oxaloacetic and glutamic pyruvic transaminases. *Am. J. Clin.Pathol.*, 28: 56- 63.

Richardson, C. 1980. In: Vety. Reproduction and Obstetrics (Editor, Noakes David E, Parkinson Timothy J, England Gary C W) Eighth Edition, p. 68.

Roberts J. Stephen, 1971. Veterinary Obstetrics and Genital Diseases, Second Edition (Indian Edition), CBS Publishers and Distributiors, pp. 36-47.

Rodrigues, P. A., Vicenta, W. R. R., Cipolla-Neto, J. and Oliveira, C. A. 1998. Evaluation of the progesterone hormone profile in ewes at early pregnancy, pre-term and early lactation. *Arquivo-Brasileiro-de-Medicina-Veterinaria-e-Zootecnia,* 50(2): 133-139.

Rodriguez-Marquez, J. M., Hernandez-Vasquez, A., Hindalgo-Delgado, G. 2005. Ovine endometrial lining modification during the implantation in natural and induced multiple pregnancies. *Revista-cientifica, Facultad –de- Ciencias-Veterinarians, Universidad-del-Zulia,* 15(4): 326-333.

Roth, T. L., Bateman, H. L., Kroll, J. L., Steinetz, B. G. and Reinhart, P. R. 2004. Endocrine and ultrasonographic characterization of a successful pregnancy in a sumatran rhinoceros (*Dicerorhinus sumatrensis*) supplemented with a synthetic progestin. *Zoo-Biology,* 23(3): 219-238.

Rowe, A. J., Wulff, C. and Fraser, H. M. 2003. Localization of MRNA for vascular endothelial growth factor (VEGF), angiopoitins and their receptors during peri-implantation period and early pregnancy in marmosets (*Callitrix jacchus*). *Reproduction,* 126: 227-238.

Sastry, N. S. R., Thomas, C. K. and Singh, R. A. 1982. Dentition and aging of animals. Farm Management and Poultry Production. Second revised Edition. Vikas Publishing House Pvt. Ltd., pp. 43-50.

Schular, 2002. In: Reproductive Technology in Farm Animals. CABI Publishing, pp. 237-254.

Shalini and Sharma, D. N. 2004. Biometrical study of internal genital organs of gaddi goats. *Indian Journal of Animal Reproduction,* 25(2): 148-150.

Shameha-Intan, A. R., Zuki, A. B., Nornajiban, K. and Zamri-Saad, M. 2005. Microscopic evaluation of uterine mucosa in pregnant and non pregnant rats. *Journal of Animal and Veterinary Advances,* 4(6): 590- 594.

She, BA, Wani, SA, Khan, M.Z. and Wani, G.M. 1999. Bacterial Studies on the uterine secretions of repeat breeding and normal breeding cows under temperate conditions of Kashmir. *Applied Biological Research* 1(2): 153-154.

Singh Yashwant, Sharma, D. N. and Dhingral, L. D. 1979. Morphogensis of the testis in goat. *Indian J. Anim. Sci.*, 11: 925-931.

Singh, S. K, Bhattacharya, A. R and Luktuke, S. N. 1974. Studies on biometry of genital organs of female goat. *Indian Veterinary Journal*, 51(2): 81- 85.

Skidmore, J. A., Wooding, F. B. P. and Allen, W. R. 1997. Placentation during the first 60 days of gestation in the dromedary camel (*camelus dromedarious*). *Journal of Camel Practice and Research*, 4(2): 199-202.

Snedecor, G. W. and Cochran, W. G. 1980.Statistical Methods. 6th Indian Edition. Oxford and IBH Publishing Company, New Delhi

Soliman, M. K. 1975. Studies on physiological chemistry of allantoic and amniotic fluids of buffaloes at various stages of gestation. *Indian Vety. Journal*, 52: 106-112.

Stainer, M. W. 1965. Transfer of radioactive water, urea and glycine between maternal and foetal body fluids in rabbit. *Journal of Physiology*, 178: 127-149.

Surgeev, L., Kleeman and Walker, 1990. Prediction of ovine foetal age using real time ultrasound imaging. Australian Society, *Animal Production*, 18: 547.

Thailender, G., Eriksson, H., Edquist, L. E., Rodriguez-Martinez, H. 1990. Variations in oestradiol and progesterone receptor levels in the porcine myometrium and cervix during pregnancy and parturition. *Journal of Veterinary Medicine*, Series- A, 37(5): 329 -338.

Thomas E. Spencer and Fuller W. Bazer, 1995. Temporal and spatial alterations in uterine estrogen receptor and progesterone receptor gene expression during the estrous cycle and early pregnancy in ewe. *Biology of Reproduction*. 53: 1527-1543.

Wales, R. G. and Murdoch, R. N. 1973. Changes in the composition of sheep foetal fluids during early pregnancy. *Journal of Reproduction and Fertility*. 33(2): 197- 205.

Wallace, L. R. 1948. *Journal of Agricultural Science*, 38: 93, 243.

Wani, G. M and Buchoo, B. A. 1990. Laparoscopic view of ovine conceptus. *Indian Journal of Animal Science*, 60(5): 564 – 568.

Wani, G. M. 1995. Early diagnosis of pregnancy in surrogate mothers (Sheep and Goat). In National Symposium on Embryo Transfer and Genetic Engineering Technologies in Small Ruminants, May 7-8, CIRG, Makhdoom, Farah, India.

Wani, G. M. 1996. Embryo Biotechnology in Sheep and Goats. Publisher Valley Book House, University Road, Hazratbal, Srinagar, Kashmir.

Wani, G. M. 2000. Effect of ocyte harvesting techniques on invitro maturation and invitro fertilization in sheep. *Small Ruminant Research*, 36(1): 63-67.

Wani, G. M. 2002. Multipurpose Management/ Goat Husbandry. Encyclopedia, Dairy sciences, Editor Roginski, H, Academic Press, An imprint of Elsevier Science, London, Boston, Amsterdam, New York, Oxford paris, San Diego, San Francisco, Singapore, Sydney, Tokyo, II: 1259-1270.

Wani, G. M. 2003. Biotechnological advances in goats production. In *National Work-Shope on "Recent Advances in Goat Rearing"*. MPKKV, Agric.Univ. Rahuri, 29-30 December.

Wani, G. M. 2005. Advances in ovine and Caprine Reproduction, www.word-mart.com. Research Paper, 9: 1-25.

Wani, G. M. 2006. Biotechnological Advances in Goat Production. www.word-mart.com. Research paper, p. 10.

Wani, G. M., Khatun, A., Wani, G., Goswami, P. and Khan, M.Z. 2006. Preimplatation endometrial and foetal structural changes in sheep. Paper presented, In *International Symposium on "Emerging Trends in Genomic and proteomic sciences"* Mumbai, India, 15-18 october.

Wani, G.M, Goswani, P, Khantun, A Wani, P and Wani G. 2006a. Mammalian foetomaternal attachment a new invention www.buzzle.com

Wani, G.M, Khantun, A, Goswami, P. Mir, S, Darzi, M.M. and Khan, M.Z. 2006. Early embryonic development in sheep and goats www.buzzle.com.

Wani, G.M. 2008. Ovarian Manipulation and Ovulation www.articlebase.com She, BA, Beg, MA, Khan, M.Z and Wani, F.M. 1998, Studies on the efficiency of RPL-1 for preventing retension of placenta in freshly calved crossbreed cows-a field trail. *Indian J. Indigenous Med* 19(2): 120-123.

Wani, G.M.1984. Investigation on ovarian activity, ovulation and pregnancy diagnosis in sheep and goats. Ph.D.Dissertation. Rohilakhand University, Bareily, Izatnagar.

Xu-HuiZhong, Liu-Dan, Zhong-YuanQiang and Yang-FengXing, 2004. Mineral element contents in the blood of cattle and their variations. *Journal-of-Southwest-Agricultural-University*, 26(3): 236-239.

Zamjanis, R. 1983. Treatment and control of cystic ovarian diseases in dairy cattle a review. Br. Vet. J. 14: 226-237.

Zamjanis, R. 1987. Bovine theriogenology, yesterday, today and tomorrow. Proc. of the society of theriogenology meeting Sept. 16-18, 1987 U.K.

Zietzschmann, O., Krolling, Oh. and Vitmus, 1955. In: Vety. Obstetrics and Genital Diseases (S.J. Roberts eds), 2nd Indian Edition, p. 31.

Zlatkis, A., Zak, B. and Boyle, G. J. 1953. *J. Lab. Clin. Med.*, 41: 486.

Zogno, M. A., Miglino, M. A. and Oliveira, M. F. 2004. Biochemical analysis of fetal fluids and cytology of the amniotic fluid of female moco (*Kerodon upestris*). *Brazilian-Journal-of-Veterinary-Research-and-Animal-Science*, 41(4): 228-235.

□□□

Chapter 25

Reproductive-Abnormalities in Small Ruminants

A. Khatun, G.M. Wani, A.R. Choudhury, R. Islam & M.Z. Khan

ABSTRACT

A total of 86 female genitalia were collected from sheep and goats slaughtered in the abattoir of Srinagar city. The organs were studied for pathological changes in the different parts of the genital organs. Various abnormalities were- parasitic cyst (*Taenia hydatigena*) (8.13%), cystic ovary {follicular cyst (2.33%) or luteal cyst (1.16%)}, endometritis (1.16%), mummified fetus (2%), macerated fetus (3.49%), pyometra (1.16%), ovaro-bursal adhesion (1.16%), double os cervix (1.16%), bent cervix (29% in sheep and 2% in goat). Photographs of the important pathological conditions are incorporated in the paper

Introduction

The abnormalities in the reproductive tract lead to various infertility problems in animals and thereby cause great loss to the livestock sector. The extent of the pathological conditions is important for management of the reproductive disorders to augment the production in scientific lines. With the advancement of science the seasonal natural reproductive cycle is changed to an intensive year round reproductive cycle that ultimately declines the reproductive efficiency. These factors may result in partial or complete reproductive failure and leads to various abnormalities (Hafez and Hafez, 2000). Pathological conditions like follicular cyst in goat (Reddy *et.al.*, 1997), Nilli-Ravi buffaloes (Khan *et.al.*, 1989) and cows (Chaudhari *et.al.*, 2000); luteal cyst in cow (El-Wishy 1976); ewes (Ansari 1978) and buffaloes (Khan *et.al.*, 1989); Endometritis in ewes (Smith *et.al.*,1999), crossbred cows (Remirez-Iglesia *et.al.*,1996) and Nilli Ravi buffaloes (Khan *et.al.*,1989); Fetal mummification in goats (Reddy *et.al.*, 1997) and ewes (Ansari, 1978); Macerated fetus in ewes (Ansari,1978), buffaloes (Khan *et.al.*,1989) and goat (Reddy *et.al.*,1997); Pyometra in cows (Chaudhuri *et.al.*,2000), buffaloes (Khan *et.al.*, 1989) and ewes (Ansari ,1978); Ovaro-bursal adhesion in cows (Chaudhari *et.al.*, 2000), goats (Reddy *et.al.*, 1997) and buffaloes (Khan *et.al.*, 1989); Double os cervix in goat (Reddy *et.al.*, 1997), Bent Cervix in goat (Reddy *et.al.*, 1997) have been reported in the earlier studies. The presence of parasitic cyst in the reproductive tract of animals has been reported in few cases (Cassali and Nasuimento, 1994). The documentation of pathological condition of the female genitalia in small ruminants is scanty. The present paper reports some of the unique pathological changes observed in the reproductive tracts which may highlight some important research area and provide a source of information for management of reproductive disorders in small ruminants.

Materials and Methods

86 female genital organs of sheep and goat were collected from the local abattoirs located in an around Srinagar city. Immediately after collection the samples were brought to the laboratory in the thermos flask for examination. The genital organs were examined from external to internal morphology and various physiological and pathological changes had been recorded. The age of foeti was determined as per the method described by Richardson *et.al.*, (1980). Abnormalities found in the different parts of the reproductive tracts were documented.

Results and Discussions

Parasitic cyst (Fig. 1) : Seven (7) numbers of cysts were collected from six samples. The locations of cysts were either in between rectal wall and vaginal wall or on the external wall of the vagina. Out of six only one animal was pregnant. All the cysts were identified as *Taenia hydatigena* on the basis of scolex (Soulsby, 2005). Cassali and Nasuimento (1994) also reported the presence of cysts in different parts of uterus and in ligamentum latum in ewes, while many authors reported the locations of the cysts in other vital organs like liver in ewes (Abu Samra *et.al.*, 1984); in liver and peritoneal cavity of pig (Blazek *et.al.*,1985); in peritoneal and thoracic cavity, liver, lung and pleura of goat (Pathak *et.al.*, 1982); in pericardium of lamb (Pellegrini *et.al.*, 1982); in pericardium and mesentery in black Bengal goats (Alim *et.al.*, 2002) and in the intestine and duodenum of dog (Buslaeva, 1976).

Cystic Ovary/Follicular Cyst (Fig. 2) : Two follicular cysts out of 86 samples were recorded. The incidence was 2.33%.The incidence of follicular cyst in other reports are 0.27% in goat (Reddy *et.al.*, 1997), 43% in Nilli-Ravi buffaloes (Khan *et.al.*, 1989) 3.35% in cows (Chaudhari *et.al.*,2000). In one case uterine horns were bulging giving an appearance of a gravid uterine horn. A depression was observed nearer to the inter-cornuate ligament in both of the horns and the caruncles were cystic in appearance. Cyst was present in the right ovary and having diameter of 1.1 cm, while the left one contained small to medium follicles with two regressing corpus luteum. In another sample brownish deposition on the caruncles was observed. Cyst was on the left ovary. Right ovary contained small to medium follicles.

Cystic Corpus Luteum (Fig. 3) : The incidence of cystic Corpus Luteum was 1.16%,while in other reports the incidences were 0.6% in cow (EI-Wishy, 1976); 15% in ewes (Ansari, 1978), 1.02% in buffaloes (Khan *et.al.*, 1989). The genitalia looked like a non gravid uterus. The caruncles of the left were more developed compare to the right. The cystic corpus luteum was present in the left ovary while the right ovary contained large, medium and seven numbers of small follicles with one corpus luteum haemorrhegicum and one regressing corpus luteum.

Endometritis (Fig. 4) : The incidence of endometritis was 1.16%, while the incidence in other reports were 9.92% in ewes and 13.51% in nulliparous sheep (Smith *et.al.*, 1999); 12.5% in crossbred cows (Remirez-Iglesia *et.al.*, 1996); 15.31% in Nilli-Ravi buffaloes (Khan *et.al.*, 1989). We have observed the symmetrical uterine horn with flakes of pus internally. Caruncles were hemorrhagic. Left ovary was having small to medium follicles with PCL, whereas right ovary possessed corpus albicans with small to large follicles.

Mummified Fetus (Fig. 5) : The left uterine horn was bulging. The right ovary contained PCL and the left ovary had five follicles. After opening the uterus, the mummified fetus was observed. Only one case of mummification was recorded in sheep (1.16 %). The incidence of mummification was 0.06% in goats (Reddy *et.al.*, 1997) and 7 % in ewes (Ansari, 1978) in the previous reports.

Macerated Fetus (Fig. 6) : Three extra-uterine macerated cases were recorded. The incidence was 3.49%. The reported incidence was 0.3% in ewes

(Ansari, 1978); 2.99% in buffaloes (Khan *et.al.*, 1989) and 0.03% in goats (Reddy *et.al.*, 1997). In all the previous reports, intrauterine macerated cases were observed. The genital organs gave foul smell during examination. Externally both the uterine horns were symmetrical with curved cervix and enlarged vagina. On pressure the vagina gave a typical crepitating sound. After opening the whole tract, the macerated bones were observed mostly in the vagina with slight involvement of cervix. The caruncles were pale in colour with cavitations. The ovary contained CLH with crown and neck with small to large follicles.

Pyometra (Fig. 7) : The incidence was accounted as 1.16% (1 out of 86). The uterus was of irregular shaped with enlarged horns and on opening pus was observed. Left ovary had PCL. The case was suspected for Brucellosis. The other recorded incidences are 0.48% in cows (Chaudhuri *et.al.*, 2000); 5.71% in buffaloes (Khan *et.al.*, 1989) and 0.50% in ewes (Ansari, 1978).

Ovaro-bursal adhesion (Fig. 8) : Externally both the uterine horns were slightly enlarged with slight depression after the normal curve of the horns. The cervix was curved. Both of the ovaries remained attached with their bursa. The left ovary was slightly larger than the right. The caruncles were developed and possessed gelatinous material on the centre. The incidence of this abnormality was accounted as 1.16%. The incidence was also reported in cows (2.9%) by Chaudhari *et. al.*, (2000), in goats (0.31%) by Reddy *et.al.*, (1997) and in buffaloes (13.67%) by Khan *et.al.*, (1989).

Double Os Cervix (Fig. 9) : Cervix was slightly curved with double os. Out of two cases recorded, one possessed developed opening and another had blind end. Left horn of the uterus was more enlarged than the right one. Internally an unimplanted elongated blastocyst was present. Cotyledones were developing and increased circulation of blood to the caruncular and intercaruncular areas of the both horns. Left ovary contained small to large follicles with CLH whereas the right ovary contained medium to large follicles. The incidence of double os cervix was 1.16% in the present study. Reddy *et.al.*, (1997) recorded 0.03% incidence in goat.

Bent Cervix (Fig. 10) : The incidence of bent cervix was high 29% in sheep and 2% in goat which was much higher than the rest of abnormalities. This abnormality was more common along with other abnormalities in sheep. This abnormality is recoded both in pregnant and non pregnant samples. The report of Reddy *et.al.*, (1997) in goats (1.70 %) was in close agreement with the present study.

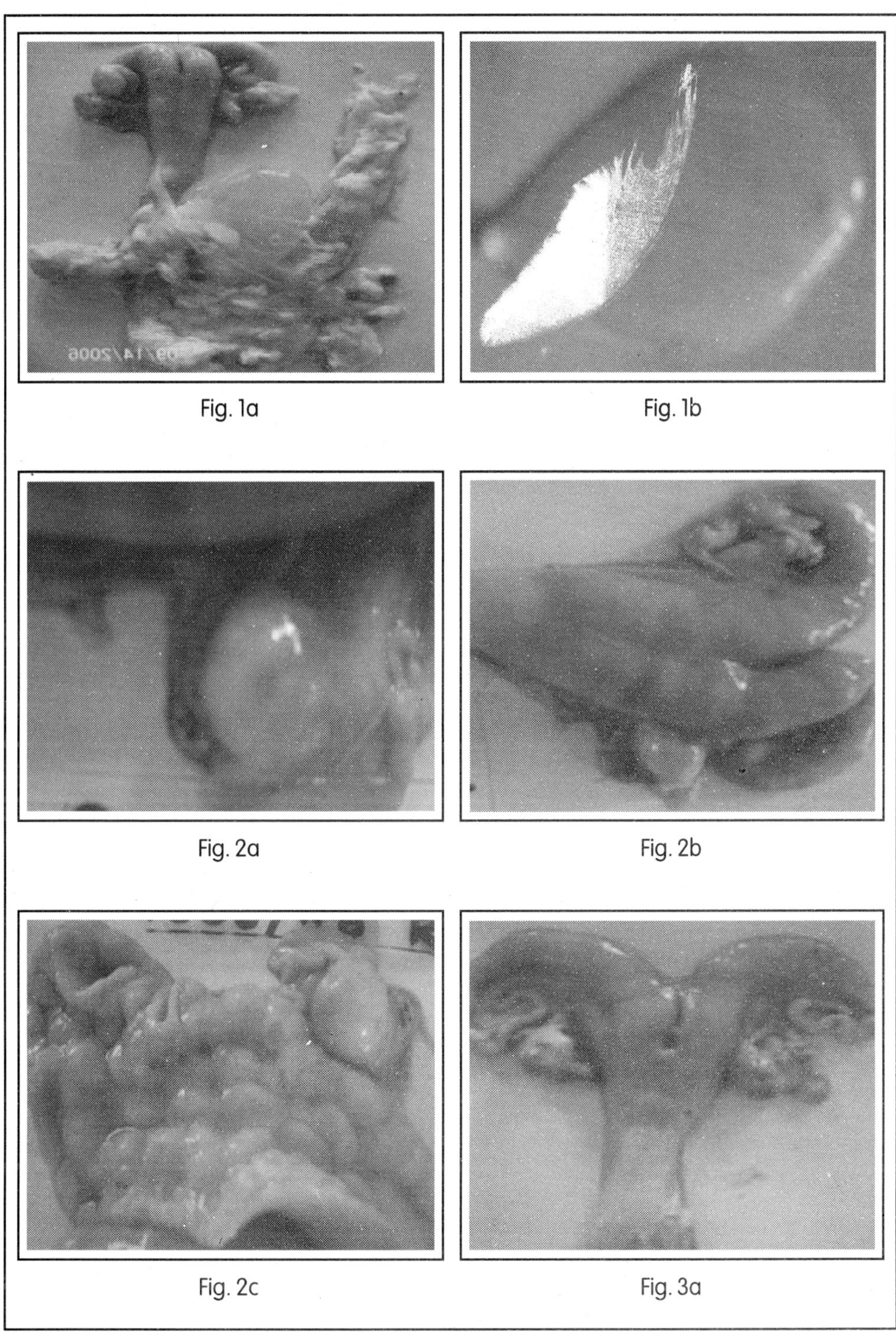

Fig. 1a

Fig. 1b

Fig. 2a

Fig. 2b

Fig. 2c

Fig. 3a

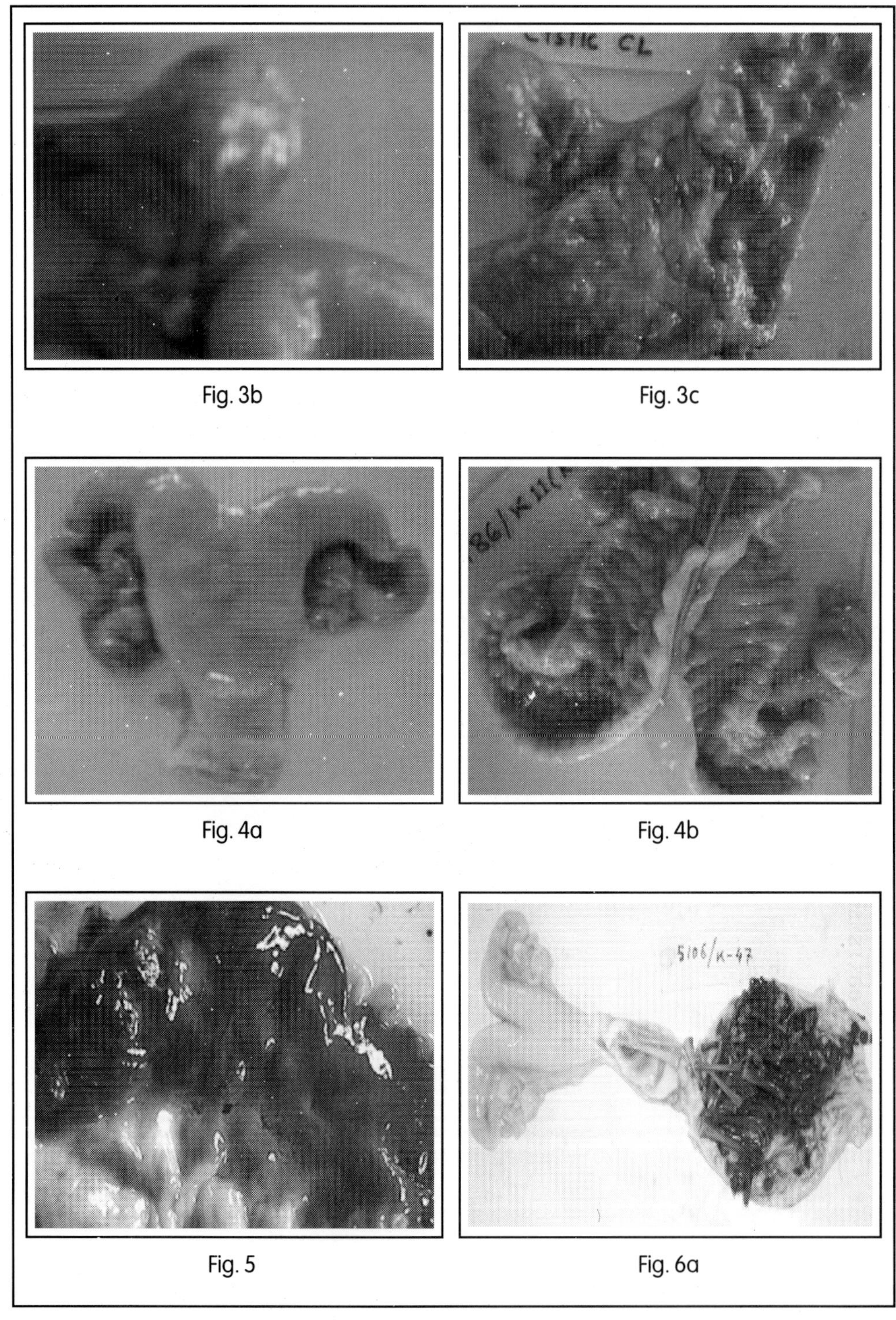

Fig. 3b

Fig. 3c

Fig. 4a

Fig. 4b

Fig. 5

Fig. 6a

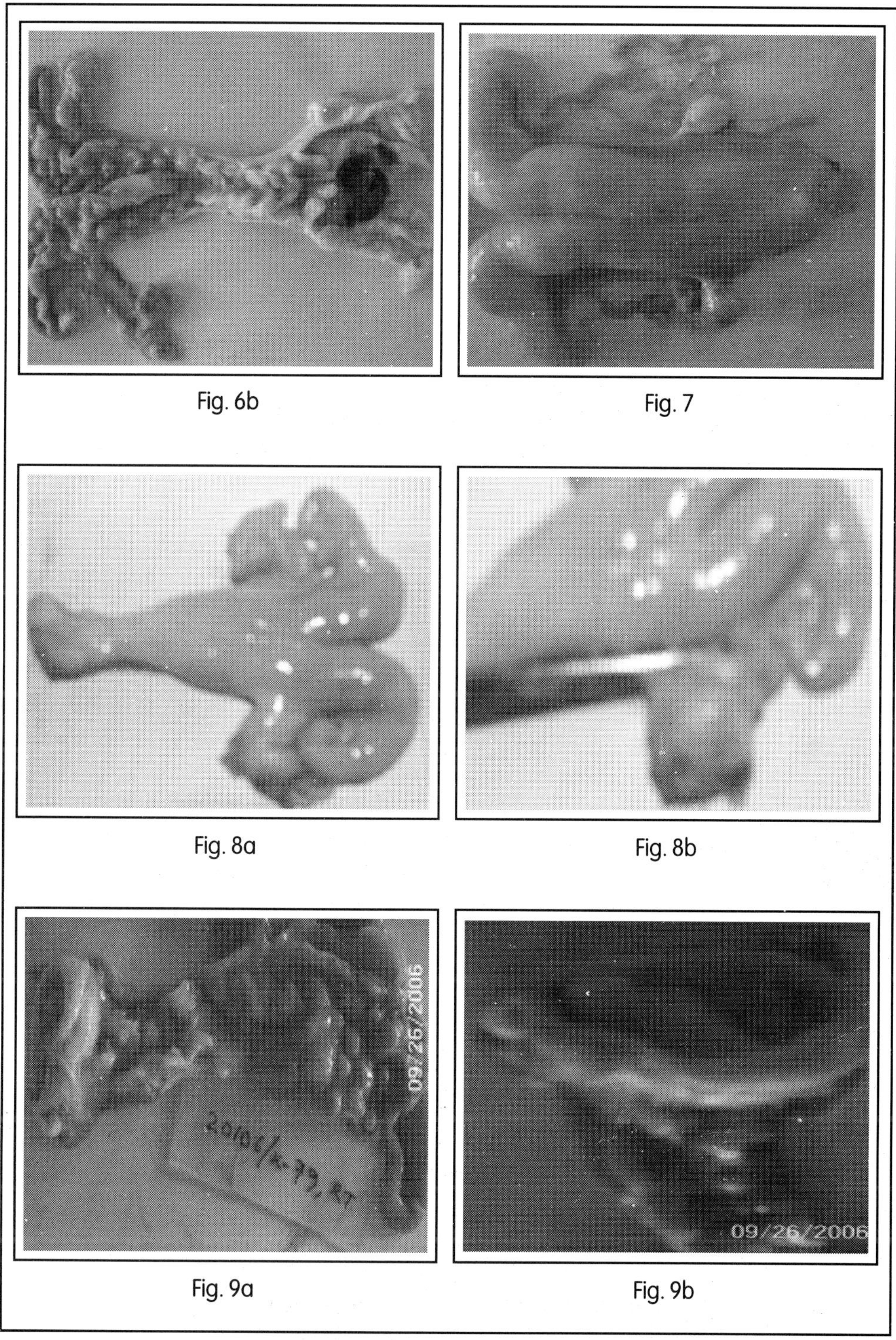

Fig. 6b

Fig. 7

Fig. 8a

Fig. 8b

Fig. 9a

Fig. 9b

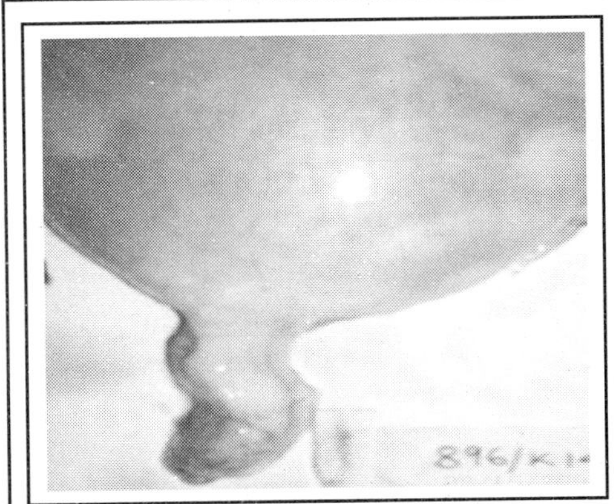

Fig. 10a

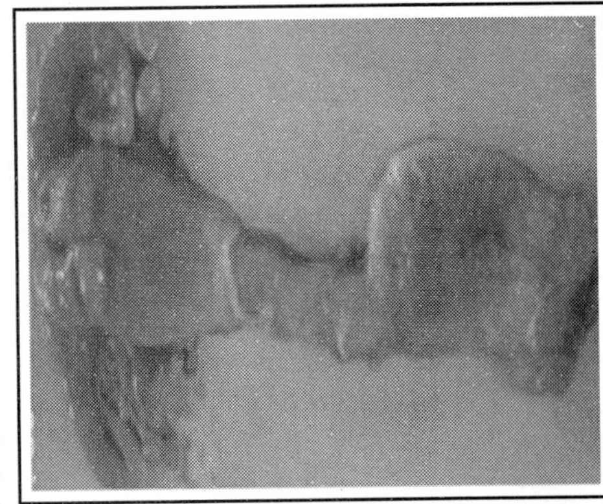

Fig. 10b

References

Abu-Samra, M.T.; Sanousi, S.M.; Idris, S.O.; Bagadi, H.O.; Salam, I.S.A.; Ali, B.H.; Musa, B.E. (1984). Infectious necrotic hepatitis (Black Disease) among saunders sheep. *Revue-d'-Elevage-et-de-Medicine-Veterinaire-des-Pays-Tropicaux*. 37 (4) : 422-429

Agarwal,-S-K;Singh,S-K;Rajkumar,R (2005).Reproductive disorders and their management in cattle and buffalo: a review. *Indian Journal of Animal Sciences*.75 (7): 858-873

Alim-M-A; Islam-M-K; Rahman-M-M; Khan-M-A-H-N-A; Mondal-M-M-H (2002). Further observations on metacestdiosis in Black Bengal goat in Bangladesh. *Pakistan Journal of Scientific and Industrial Research*. 45 (5) : 330-333

Ansari, H. (1978). Functional and pathological conditions of the genital organs of ewes in Tehran. *Revue-de-Medicine-Veterinaire*. 129 (2) : 285-286, 289-296

Blazek, K.; Suhramlova, J.; Hulinska, D. (1985). Pathology of the migration phase of Taenia hydatigena (Pallar, 1766) larvae. *Folia-Parasitologica*. 32 (2) : 127-137

Buslaeva, T.P. (1976). Some data on the biology of Taenia Hydatigena. Sbornik-Nauchnykh-Trudov-Moskouskoi-Veterinarinoi-not-more-than-Akademii-Vazhneishie-Issledovaniya-po-Izuchenyu-Zabolevanii-not-more-than-Sel-Skokhozyaistvennykh-Zhivotnykh. 86 : 99-101

Cassali, G.D.; Nasuimento, E.F. (1994). Three cases of genital cysticercosis in ewes. *Arquivo-Brasileiro-de-Medicine-Veterinaris-e-Zootecnia*. 46 (4) : 422-424

Chaudhari,-S-U-R; Paul -Bokko,-B (2000).Reproductive status, pregnancy wastage and incidence of gross genital abnormalities in cows slaughtered at Maiduguri abattoir,Nigeria. *Pakistan-Veterinary-Journal*. 20 (4): 203-205

El-Wishy, A.B. (1976). A preliminary study of the genital organs of indigenous cows in Uganda. *Beitrage-Zur-Tropichen-Landwirtschaft-Und-Veterinarmedizin*. 14 (1) : 95-107

Hafez, E.S.E. and Hafez, B. (2000). Reproduction in Farm Animals. Seventh Edition. Lippincott Williams & Wilkins A Wolters Kluwer Company. Philadelphia, pp. 261-277

Khan, A.; Khan, M.Z.; Ahmad, M.; Ahmad, K.M. (1989). Pathological studies on reproductive organs of the Nili-Ravi buffalo. *Buffalo Journal.* 5 (2): 197-203

Mufti, A.M.; Wani, G.M.; Wani, N.A., Mir, M.M. and Khan, M.Z. (2000). Prenatal development of ovine foetus. *Small Ruminant Research.* 38 (1) : 87-89

Pathak,- K-M-L; Gaur, -S-N-S; Sharma,-S-N (1982). The pathology of cysticercus tennicollis (Taenia hydatigena) infection in goats. *Veterinary Parasitology.* 11 (2/3) : 131-139

Pellegrini, N.; Renzoni, G.; Taccini, E.; Pambianca, L.A.; Rindi, S. (1981; Publ. 1982). Unusual pericardial localization of Cysticercus tennicollis in a lamb. Anatomo-histopathological findings and pathogenic consideration. *Annali-della-Facolta-di-Medicina-Veterinaria-di-pisa.* 34 : 129-138.

Ramirez-Iglesia, L-N; Soto-Belloso, E.; Gonzalez-Stagnaro, C.; Soto-Castillo, G.; Rincon-Undancta, E. (1996). Postpartum ovarian activity in crossbred primiparous cows with or without periparturient abnormalities. Revista-Cientifica; Faculted-de-Ciencias-Veterinarias,-*Universidad-del-Zulia.* 6 (1): 13-20.

Reddy, K-C-S; Reddy,V-S-C; Rao,A-S; Reddy,J-M; Sharma,G-P (1997).Studies on the incidence of reproductive abnormalities in local non descript female goats. *Indian Journal of Animal Reproduction.* 18 (1) : 51-53.

Richardson, C. (1980). In Vety. Reproduction and Obstetrics (Editor, Noakes David E, Parkinson Timothy J, England Gary C W). 8th Edition, 2001, p.68.

Smith,-K-C; Parkinson,-T-J ; Long,-S-E (1999). Abattoir survey of acquired reproductive abnormalities in ewes. *Veterinary Record.* 144 (18) : 491-496.

❑❑❑

Chapter 26

Involution of Uterus

Basharat Ahmad Ganie

Introduction

Uterine involution refers to restoration of uterus to its normal non pregnant size and function after parturition. It depends on myometrial contractions, elimination of bacterial infection, and regeneration of endometrium.

Lochia,the uterine discharge that normally occurs during puerperium, is composed of mucus, blood, shreds of foetal membranesand maternal tissueand foetal fluids. Lochia ceases by the first week after parturition. The expulsion of lochia and reduction of uterine size are caused by myometrial contractions. This is due to a sustained release of prostaglandins after parturition, which increases the uterine tone and thus promotes involution.

The sterile sterile conditions of the uterus that prevailed during pregnancy are disrupted at parturition. Both pathogenic and non pathogenic bacteria enter the uterus through dialated cervix and rapidly multiply in afavourable uterine environment. The post partum uterus depends on three prosses to prepare it for subsequent pregnancy:

1. A defensive mechanism that includes a massive infiltration of lymphocytes into the uterine lumen to phagocytise the major pathogens present in the early post partum[PP] period.

2. Amassive release of prostaglandins during the first two weeks PP induces myometrial cotractions and the evacuation of fluids and foetal membrane debris from the uterus. The duriation of release of prostaglandins is longer in speacies with a cotyledonary type of placenta than in those with a diffuse type.

3. The time needed to clear the uterus, of bacteria depends on the extent of contamination at parturition, retension of foetal membranes, and production of estrogen. Regeneration of endometrium is completed earlier in speacies with a diffuse placentathan those with a cotyledonary placenta. The endometrium is fully regenerated between 2nd and 3rd week in horses and pigs between 4th and 5th week in ruminants.

Rate & Duration of Uterine Conversion

In cow following the of fetal membrane, the uterine contraction and peristalsis continue as strong rythmical waves that gradually diminish through the 4th day. In the cow, the uterine muscle shorten from 750micron after parturition to 400micronone day latter. From 4th to 8thdays there are only irregular undulation the horn,Benesh. (1)Involution of uterus after paturation was studied using a real-time B-mode ultrasonography in ewes. Ewes that spontaneously delivered showed a physiological regression of the uterus with a transversal diameter of 4. 9 (+_0. 86)cm. on day 1 post partum, which decreased 1. 8(_+0. 140)cm until day 30. the uterine involutionwas delyed in ewes after manual obstetrics and cesarcan section, in addition the incidence for retention of fetal membranes was increased. (2)the rate of uterine involution postpartum was monitored in suckling mule ewes using radio opaque markers radiography. The maximumreduction in the length of uterine body and in the diameters of the horn occurred by 28 days pp,except in one ewe out of 13 ewes,in which the size of utrerus continued for 42days. The presence of intrauterine bacteria did not affect the time for the completion of uterine involution. (3)involution ofb uterus inovines has been investigated following intrauterine adminstreation of oxytertracycline. it was seen that treatmentwith intrauterine oxytertracycline may prolong involution of reproductive tract of ewes after parturition(4)ultrasonography in combination with rectal palpation method has been used to determine the involution period in Sewidish dairy cows with retained foetal membranes {5}. A linear array ultrasound scanner was evaluated for studying uterine involution in Japenese Black cows. Estimates of the croos sectional diameter and area of the uterine horn from ultrasound agreed with with the values measured on removed uterus,therby suggesting that USG was effective for monitoring uterine involution in beef cows. [6]. It has been demonstrated that during involution the muscle layers decrease in mass both by a decrease in cell size and loss of cells. Involution was more rapid for suckled animals than for non suckled animals from day 1 to day 30PP. [7]

Based on rectal palpation during PP period, it has been seen that size of bovine uterus decreased slowly between 4th and 9th days PP. By 10th day the involuting uterus could be completely defined by rectal examination. A marked decrese in uterine size and increase in uterine tone occurred from day 10_14, coinciding with the onset of first estrus in normal cows and a voluminous discharge of uterine lochia. [8]

The following PP rate of regression in weight of the cows uterus has been determined ;10kg at parturition, 5kg at 6 days, 2kg at 12 days, 1kg at 25 days, and 0. 7kg at 50 days. [9] Within 24-36 hrs after a normal parturition the hand could not be passed through cervix and by 4th day PPonly two fingers could be introduced. If the retained placenta was present the rate of closure of cervix was delayed and the hand could still be introduced at 48hrs and often at 72hrs PP. In general the major gross palpable or physical involution of uterus occurred by 25-30 days PP. Changes after 30days s were less marked and more general. Morrow summarized nine reports on utrine involution as determined palpation in which the time for uterine involution varied between 42 and 47 days. [10]

Using rectal palpation, from 4-45 days PP,diameter of the uterus, cervix and length of cervix were significantly affected by "sex of calf", parity {smallest in dams of female calves and in primiparous cows}, genotype and condition of vulva. Length of cervix was significantly affected by sex of calf, parity and genotype, and cervical diameter by parity and vaginal condition. Cows with a discharge showed a delayed genital involution. [11]. The rate of decline in size of the uterus has been determined by palpation in many cows on days 10-14 PP,the size of the post gravid uterine horn declined from 12cm to 7cm in diameter. The rate of uterine regression was quite rapid from 14-25 days PP. There was a decreased rate of involution between days 26 and 39. Between days 40 and 50 there was a little change noted. Prior to day 20 PP, The diameter of cervix was smaller than horn and on 22-25 days it was larger than the horn. [12]

Then a decline to basal levels, and a second small elevation at the time of final elimination of bacteria, from the uterus, or elevated levels during 1st 7 days PP,hten a decline to the basal levels, and a 2nd small elevation before the final elimination of bacteria. [15] It has been seen that when flunixin meglumine {FM}, a cyclo oxygnase {prostaglandin synthesis} inhibitor was given to cows, and blood samples were secreened for conc. Of PGF2á metabolite {PGFM}, there was an inhibition of endogenous synthesis of PGF2á by greater than 80%, which decreased the spontaneous uterine motility and myometrial response to oxytocin {5 IU I/v} and PGF2α {15MG I/V}. But this treatment with FM did not interfere with uterine involution, the return to cyclicity and first PP cycle length. [16] The effect of FM on the utrine involution has also been studied by other researchers in primiparous cows. It was seen after treatment of cows with FM, the plasma conc. Of PGF2α__metabolite and progesterone were determined. The involution of uterus was asscesed by rectal palpation and USG. It was seen that FM did not affect the time of involution, but was not determined for the process of involution as well. However uterine involution as an inflammatory process,flunixin could be, as an anti infilamatory drug beneficial under some conditions. [17]

Intra uterine infusion of 20μg 16,16- dimethyl PG E2. In physiological were given to Holstein cows. Blood samples were taken before; during and after the treatment and rectal palpation was preformed twice weakly. Blood samples were assayed for metabolites of PGF2α; PGF2α, and progesterone conc. Treatment with Dpge2 affected many macroscopic characteristic relating to the uterine involution, such as tone [p<0. 001], length [p<0. 02] and weight [p<0. 05] of the uterus. Follicular development was also affected [p<0. 05] by the treatment. The ratio of plasma PGFM to PGEM was also affected. These results suggest that PGE2 impedes uterine involution in cow. [18]. Saturnine etal showed that when cows were injected with 25mg cloprestenol per day from 1_18pp, treated cows showed more rapid involution of uterus, but the interval from parturition to Ist estrum, no. of inseminations per conception remained unaffected. [19]; Efficacy of prostaglandin-analogue in the prevention of delayed uterine involution and endometritis in dairy cows with placental retention has been studied. It has been seen that the combined with prostaglandins & intrauterine-ox tetracycline helped in normal uterine involution in 30 days with no case of endometritis {20} the treatmentr5 with flunixine meglumine accelerates uterine involution & shorten the calving to first estrous interval in cows with peripheral metrities.

Effect of Estrogen & Progesterone

Marion et al. showed that overectomy & estradiol injections had no effect on uterine involution, while progesterone injections slightly delayed the uterine involution (22), Studies were made of blood progesterone (p-4) levels in cows given different amounts of beta-carotene in the perinatal period. Examination were made to follow the process of involution of uterus, occurrence of first estrous, postpartum stabilization, open period, fertilizations rate. The beta-carotenes dose was shown to increase blood levels which advantageously affect blood p-4 level, particularly at the time of its increase above 3. 2mol/lt of serum, duration of first estrous cycle, involution of estrous etc. It was shown that high beta-carotene levels increase p-4 concentrations in blood of cows in prenatal period (32). Tran rectal examination were carried out on pluparious cows from day 9 to 10 after parturition to the second ovulation. Milk sample were taken from them & accessed for progesterone by RIA. The period from parturition to the time of first PP ovulation & the length of first cycle were positively correlated with the no. of follicular waves on ovaries during this period. During the first 7 2nd estrous cycles the mean progesterone values were 5. 9 ± 4. 78 & 6. 3 ± 5. 16ng/ml. resp. At days 5 to 7 & 5. 56±5. 83 & 11. 4±7. 74mg/ml at day 10 to 12 (24).

A field study of PP period of dairy cows was done by rectal palpation& essaying progesterone profiles in milk by RIA. The cows were submitted weekly to a rectal palpation of genital tract and uterine involution and ovarian activety were considered. The 1st PP estrous was observed AT28. 5day and the CL was detected on 34. 5th day. The period from oestrus to first high progesterone level PP was 5. 5days. The uterine involution period continued to the 24. 1th day. The prosgesterone conc. From the 1st CL of PP was low. [25].

In cattle, the 1st PP dominant follicle has a predliction for the ovary cotralateral to the previously gravid horn. However the presence of an estradiol secreting dominant follicle in the ipselateral ovary is a marker of subsequent fertility, possibly due to alocalised effect of ovarian oestradiol on uterine involution. Dairy cows were treated with estradiol administerd via cervix into previously gravid horn on days 7-10 PP. Uterine involution was monitored by daily trans rectal USG and plasma conc. of prostaglandin metabolite PGFM. It wads seen that there was no effect of estradiol treatment on the diameter of previously gravid horn, or non gravid utrine horn, nor the plasma conc. of PGFM. However in cows in which the 1st PP dominant follicle ovulated during the study, had a smaller diameter of both previously gravid [P<0. 01]and non gravid [p<0. 001] horns compared to those in which the follicle regressed. [26] . The effect of estradiol on PP involution of uterus in sheep has been studied. Ewes were implanted with estradiol implants, sutured with the bursa of ovary ipsilateral to the previously gravid horn. Blood samples were collected for measurement of PGFM, until day 17th PP, when ewes were slaughtered and their genital tracts collected. It was seen that there was no evidence of localized effect estradiol on involution as there was no significant difference between previously gravid and non - gravid uterine horns. [27].

Effect of Equine Chorionic Gonadotropin

This study was done to test the hypothesis that increased estradiol secretion by large follicles in the ovary ipselateral to the previously gravid horn, has a local effect to increase the rate of involution. It was concluded that administration of eCG to increase follicular growth & estradiol production overcome the inhibition of follicular growth in the ovary ipsilateral to previously gravid horn, but did not affect the uterine involution. [28].

Effect of Uterine Infection and Pueriperal Metritis

In female domestic animals duration of uterine involution is mainly determined by the intensity of associatewd infilamation reaction in regard to the type of placentation. In females with diffuse epitheliochorial placentation [mare & sow], uterine involution is fast [2-3]weeks and involves early and intense Uterine contraction and tissue reorganization. In female with cotyledonary epitheliochorical flacentation (ewe goat, cow) uterine involution has and average duration of 4-5 weeks and involves in addition intense vasocougtrition and rediotrisaiton of blood to udder. In female with tonary type uterine involution in proloyed 910-12) weeks and characterized by a late tissue regeneration. (29).

Perusal metrito occurs after 10-20% of calvifa or even non frequently. The effect of Pm treatment with the use of antibiotics, lysozyme, PGF2α et has been studied it was concluded that intrauterine antibiotic introduction together with the I/M injection of lysozyme dinner give the best result in regard to the time of metritis regression and the time of uterine involution. (30). The cow is notally average the domestic species for the almost usiguitious bacterial contamination of the uterine lumen after parturition and a high incidence of the uterine diseases. Such disease not only discomfit uterine but also overian follicle growth and p[unction by perturbation of hypothalamian, (intvitancy and invate immense system resolusy the bacterial contamination and prompt uterine involution. (31) Uterine flushing analyzing showed, the evidence of an inflammatory response in cows with a palphase uterine human at first pp. Mean scorem from microscopical examination of the biopsy specimen use greater in these cows, indicating increased pathologic changes in this group. (32) sequential GnRH & PGF2α application in cows witgh preperal endometriosis positively affected the ovarian function and uterine involution, resulting in improved fertility performance. (33) it has been seen that subclinical bacterial infection of the post partum uterus delays the cercrical and uterine involution in post partum buffaloes, which can interns delays the occurrence buffaloes. (34)

Experimental metritis was induced by infusion of E. coli presioly isolated from ewes with acute metritis. It was seen that puerperal intrauterine infections and subsequent metritis did not affect the rate of involution in ewes. (35) The effect of puerperal uterine infection on involution of uterus was also studied in cows. Infection significantly retarded uterine involution accessed by uterine body diameter and a score of intrauterine fluid. In cows with a severe endometritis although the diameter of uterus return to normal pergravid size, but the uterine fluid remained significantly higher. Ovaran activity as measured by USG scanning of ovaries and plasma progesterone P_4, was also abnormal (prolonged anoestrus, prolonged luteal phase and ovarian cyst) in cows with a severe endometritis. (36). Uterine infections during Pueriperum in cows where associated with elevating levels of PGFM. (37) Bacterial contamination of the bovine uterus and its effects on subsequent fertilization of cattle has been investigated. It was seen that the rate of uterine contamination declined over the involution period from 20-60% the mean time from calving to ist insemination and to caception in each herd group where closely related with number of cows infected at 38-48 days PP. (38)

Use of Intrauterine Antibiotics

Effects of intrauterine administration of oxytetoracyclin on uterine involution in ewes has been studied. It was seen that in oxytetracyclin administered ewes the uterine involution was non significantly prolonged compared to control group. (39)

Gentamicin - sulphate, an amino glyconside antibiotic being a drug of choice in intrauterine antibiotic treatment of bovine chronic or sub clinical infections, has been investigated to inhibit spontaneous as well as oxytocin as well as PGF2α induced contractions of myometrium in non pregnant cows, thereby indicating that gentamicin sulphate may have and adverse effect on involution in cows. (40)

Effect of Suckling

Involution of uterus occurred earlier in the group of suckler cows, and within this group was faster in youngest sub group aged less than 30 months. During the 6 weeks after parturition, placental retention, endometritis and ovarian cysts were recorded in 23. 6% of the sucker cows and 47. 7 % of the control group. The conception rate was also higher in experimental group. (41) In case of beef cows increased parity and milk yield lengthens the duration of uterine involution, while as increased suckling frequency shortens it. (42) the resumption of PP ovarian activity was significantly longer in suckled than non suckled zebu cows, where as there was no effect in Bos taurus taurus.

(43) Cows were subjected to restricted suckling twice daily form 3 days daily from 3 days after calving.

Uterine involution was slower in restricted than in control group. It was concluded that morphological involution of uterus was inhibited by restricted suckling, but the surface epithelium height was influenced by ovarian function regardless of involution of uterus. (44)

Variation in Blood Biochemical Constituents During Involution of Uterus

The blood samples were collected from the day of parturition up to 17th week of PP period in order to estimate blood glucose, serum, cholesterol, serum proteins and triglycerides. The values of these parameters were higher in pregnant cows during early gestation up till the 17th week PP. (45)

Jersey and Friesian cows were studied for evidence of any correlation between puriperal plasma 3 methyl - histidine concentration and uterine involution as myofibril size is reduced uterine and cervical involutions were determined by sequential measurements of the diameters of previously gravid and non gravid horns and cervix using transrectal B-mode real time ultrasound imaging, there was no obvious correlation between 3 methyl histidine conc and uterine involution. (46)

PP involution of bovine uterus was accessed by clinical examination and monitored biochemicaly by measuring the total urinary excretion of hydroxy-proline and collagen cross link, pyridinoline. The pyridinoline : creatinine ratio was a more reliable indicator of uterine resorption than the hydroxyproline: cratinine ratio. The amount of collagen cross line excreted above baseline was equivalent to amount of cross link estimated to have been accumulated during pregnancy. An electron microscope study during the period of peak resorption showed macrophage like cells apparently engulfing or containing collagenous material, indicating phagocytic removal of connective tissue. (47)

Histology of Involuting Uterus

During involution there is an ordered series of events that includes apoptosis of resident cells, neutrophil invasion, resease of degradative enzymes and phagocytosis of cellular débris. Local signals are believed to regulate the progression of involuation in each tissue. It has been shown that mammary glands and uterus express a high levels of "Uterocalin", a protein that specifically induces apoptosis in neutrophils and other leukocytes. In the mammary gland, uterocalin expression is induced by weaning. In both tissues uterocalin is expressed extremely at high levels such that it constitutes an average of 0. 2-0. 5% of total extractable protein at its peak. Epithelial cells in uterus and mammary glands produce utercalin. It has been hypothesized that one of the physiological roles of uterocalin is to induced apoptosis of infiltrating neutrophils and thereby delay the entry of neutrophils into the tissues. It is proposed that the role of uterocalin during involution is to provide a window of time during which the resident cells are protected form the degratative enzymes, free radicals and other products of activated phagocytes to prepare to survive the process of involution. (48)

Biopsy specimens were obtained from cows between 1 and 30 days after calving. Superficial and glandular epithelium regained their non pregnant structure within 3 weeks. The caruncles become coated with epithelium by a month after calving. The presence of phagocytic ability in cells of lamina-propria of the endometrium indicated that tissue belonged to reticulo-histocytic sytem. Resorption took 4-19 days and was important factor in uterine involution. (49)

Changes in Endometrial Gene Expression

In sheep, the uterus undergoes marked remodeling during involution. The return of ovine uterus to a non pregnant state involves termination of a hormonal servomechanism that regulates that endometrial gland morphogenesis and function during pregnancy. Suffolk ewes were ovarohysterectomised on PP days 1,7,14 or 28. serum conc. of oestradiol were high at parturition, declined to PP day 4, peaked on PP day 6. Progesterone was undetectable in ews PP. Immunofloriscence analysis identified both T and B cells within glandular epithelium on PP day one. On day one, expression estrogen receptor alpha was not detected in lunminal epithelium and was low in glandular epithelium, but estrogen receptor (ER) was present in epithelium there after. Progesterone receptor (PR) protein was not detected in endomaterial epithelium on PP day one, but was detected in glandular epithelium thereafter. Between PP days 1 and 7, Er alpha and PR protein increased substantially in the endometrial glandular epithelium. On days 1-28 abundant expression of oxytocin receptor mRNA was detected in glandular epithelium on all PP days. (50)

Immune Parameters

An indirect immunofloresence method for staining and flow cytometric analysis was employed to determine the cell subpopulation of lymphocets. A significant decrease in absolute no. of lymphocytes and subpopulation of

T(CD^{2+}, CD^{4+} CD^{8+}) and B (I_gM^+) cells in dams with PP complications and prolonged uterine involution time was observed. (51)

Resumption of Estrous Cycles /Ovarian Activity

Estrus and ovulation are usually suspended during lactation (lactational anestrus) in several mammalian species, but the inhibitory effects of lactation has been partially or completely overcome in farm animals through selection, improvement in nutrition and weaning. By day 50 PP, about 95% of daily cows will resume ovarian cycles, as compared with about 40% of beef cows. The first ovulation is often not preceded by overt estrus. Sucking and increasing the frequency of milking (4v/s 2 milkings per day) prolong this interval; whereas removal of the calf from the mother shorten it. Development of dominant follicles (10mm) was observed early in the period after calving which caused the resumption of estrous activity, but the conception rate was poor, when the animals were inseminated less than 40 days PP. (52)

References

Amridis, G-S; Leontides, L-L; tassos,-E; Kostonlas, P; fthenalio, G-C. 2001. Funixin meglumin accelerates uterine involution and shorten the calving to first oestrno internal. In cows with frierperal metritis. *Journal of Veterinary Pharmacology and Therapeutics.*; 24(5); 365-367.

Babaci, H;Ayen, E. 2005. Effects of intrauterine administration of oxytetracyclin on uterine involution in ewes. *Iranian Journal of Veterinary Research;* 6 (2 ser. 120): 16-21, 94.

Babaei, H; Ayen, E. 2005. Effects of intrauterine administration of oxytetacyclin on uterine involution in the ewe. *Iranian journal of veterinary Research.* 6:2(ser. 12): 16-21, 94.

Bakana, M; Ekman, T; Kindahl, H. 1994. ultrasonography of bovine postpartum uterus with retained membranes. *Journal of Veterinary Medicine Series* –A. ; 41(9): 653-662.

Bencharif, -D; tainturier,-D; slama, H; Drayas, J-F; Battut, -I; fleni, F. 2000. *Prostglandino and PP period in cow. Revue –de-Medicine –Veterinarie* -2000; 15(5); 401-408.

Benesch, F. 1952. Involution of uterus in cows. Lehrbuch der Tieartzlichen Geburtshilfe and Gynakologie, urban & sechwartenberg, wein – Innsbruck Austria.

Bonlet, M. 1989. efficacy of prostaglandin analogical in the prevention of delayed uterine involution and of endouetritis in dairy cows with placental retention. *Bulletin des-GTV;* (5); 5-12.

Chicoteau, P; Mambone, E;Cloe C;Bassing, A. 1989. uterine involution and PPresumption of ovarian cyclicity in Bos taurus taurus and zebu cows. *Zuthchgiena* 24(6) : 259-264.

Deninshi, Z; Bronichi, M. 1994. concentration of progesterone in blood of cows obtaining different amounts of B-cerotene in food. *Proceeding of 18 world -buiatries congress; 26th congress of the Italian Association of Buatrics, Bologna, Italy, August 29 September,* 2 vol 1; 325-328.

Dhoble, R. L ; Mane, S. D ; Sawle A. G ; Ingawal, M. V. 2004. Variation in blood biochemical constituents during PP period in cross bred cows. *Indian Journal of Animal Reproduction.* 25(1): 14-16.

Dolexl, -R, Karlbeng, -K; Roprtad, -E. 1993. Ovarian activity in ultrasound image and milk progesterone can in PP dairy cows. *Acta -Veterinaria -Brno;* 62 (3/4): 173-178.

Elsheihka,-A-S; Ahmed, -F-O. 2005. Baching up PP cows with PGF_2á *Journal of Med. and Veterinary -Advance.* 2005; 5(5): 506-509.

Erices, J; Ealenberger, K. 1986. Histomorphological changes in endometrium of cows after calving. *Archiv-fur-experimetale -veterinary medizin;* 40 (6) : 826-839.

Gray, C. A; stewart, M. D ; Jhonson, G. A; spencer, T. E. 2003. PP uterine involution in sheep-histoarchitercture and changes in endometrial gene expression. *Reproduction -Cambridge* ; 125 (2) : 185-198.

Hanuser, B; Bostedt, H. 2002. Ulrasonographic observations of the uterine regression in the ewe under different obstetrical conditions. *Journal of veterinary Medicine Series - A,* 49(10) : 511-515.

Izaike -Y, Suzuki, O Shimda, K. 1989. ultrsonographic observation of post partum uterine involution in beef cows. Japanese *Journal of Animal Reproduction;* 35(1), 54-59.

Izaike,y;Suzuki,o;Okans,A;Shimada,K;Oishi,T;Kosugiyam,M. 1989. Influence of parity, milk yield and suckling on uterine involution in beef cows. *Japanese Journal of Animal Reproduction.* 24(6):259-264

Izaike,Y;Suzuki,O;Okans,A;Shimada,K;Oishi,T;Kosugiyam,M. 1989. Influence of parity, milk yield and suckling on uterine involution in beef cows. *Japanese Journal of Animal Reproduction.* 35 (1) : 38-44.

Izak, Y; okano,A ; Suzuki, O ; shimada, K ; kosugyama, M; Oishi, T. 1989. Influence of restricted suckling on uterine involution in beef cows. *Japanese Journal of Animal Reproduction.* 35(1): 45-49.

Janonski, T; Zdunczyk, S ; Mwaenga,E. S. 2001. *Reproduction in Domestic Animals.* 36(5) : 244-246.

Kaezmarowshi; M; Malinowshi, E. 2005. efficacy of some therapy methods for puerperal metritis in cows. *Medycyna Vateriany.* 61(15). 532-535.

Kaidi, R ; Brown, P. J ; David, J. S. E; Robin, S. P. 1991. Uterine collagen during involution in cattle. *Matrix* ; 11 (2): 101 -107.

Kashi,-K; Kurokin,-J; Iindjaru,R; Kash-A vindhl, H. 2003. *Agraarteaduo.* 2003; 14(3); 159-167.

Kozicki, L-E, wollner, L-C, shibatta, F-K; silva, M. A 1998. A field study of PP period of dairy cows by rectal palpation & P_4 conc in milk by radio immenoarsory. *Braxian Journal of Veterinary Research and Animal Science.*; 35 (1/6) 225-228.

Levkut, M ; Pistel, J ; Rejavova, V ; choma a, J; David, V. 2002. comparison of immune parameters in cows with normal and prolonged involution time of uterus. *Veterinari -Medicina;* 47 (10/11) : 277-282.

Lohan, I. S; Malik, R. K; Saini, M. S ; Dhanda O. P ; Baljit Singh. 2000. uterine and ovarian changes during early PP period in Murah buffalo. *Buffalo Bulletin*; 19 (1) : 20-23.

Marion, G-B; Norwood, J-S; Gier, H-T. 1968. uteruo of the cow after parturition. Factero affecting regression. *Amer. J. Vet. Res.* 29(1) -71.

Mateus, L; costa, L; Bernado, F; Silva, J. R. 2002. influence of peuriperal uterine infection on the uterine involution and PP ovarian activity in dairy cows. *Reproduction in Domestic Animals* 37(1) :31-35.

Morion, G. B., Norwood. J. S and Gier, H. T. 1968 uterus of the cow after parturition and post partum regression of bovine carunlces, *J. Dairy Scien.* 42,5, 941.

Morrow, D. A, Robert, S. J and Mc Entoee, K. (1969). Post Partum ovarian activity and involution of uterus and cervix in daily cattle. *Involution of Uterus and Cervix Cor. Vet.* 59,2,190.

Morrow, D. A, Roberts. S. J and Mc Entee, K. 1969. *A review of PP Ovarian Activity and Involution of the cervix & uterus, cer, vet* 59, 2 190.

Morrow, D. A, Roberts. S. J and Mc Entee, K. 1969. *A review of PP Ovarian Activity and Involution of the Cervix & Uterus, cer, vet* 59, 2 190.

Ocal, H ; Yuksel, M ; Ayar, A. 2004. Effects of gentamicin sulfate on the contractility of Myometrium isolated from non pregnant cows. *Animal Reproduction Science.* 84 (3/4) : 269-277.

Odensirk,-K; Tredriusion, G. 1993. The effect of intrusive flexdin treatment during PP period in bovines. *Journal of Veterinary Medicine series A.*; 40(8); 561-568.

Regassa, F; Sheldon, I. M ; Noakes, D. E. 2002. effect of experimentally induced metritis on uterine involution acute phase protein response and PGFM secretion in the PP ewe. *Veterinary Record* 150 (19): 605-607.

Regasse, F, Noakes, D. E. 1999. Acute phase proteins response of ews and release of PGFM in relation to uterine involution and presence of uterine bacteria. *Veterinary Record* -1999; 144 (18) : 502- 506.

Riesen, J. W, saiduddin,s. Tyler, W. J. and casida. 1968. Studies on postpartum cow, *Res Bull, 270, University of Wincosin*, 27.

Ryon, J; Bendichson, L; Nilsen -Hamilton, M. 2002. High expression involuting uterocalin/ $24P_3$, a lipocalin, and acute phase protein. *Biochemical Journal* ; 367 (1) : 271-277.

Saturnino,-H-M; carcagal,-F-D; Margules,-A. P Jr. 1991. effects PGF_2á treatment in holstein ewes. *Julho* vol 2 ;383.

Sheldon, I-M Noahes, D-E, Baylirs,-M, Dobsen, H. 2003. the effect of estradist on PP uterine involution in sheep. *Animal Reproduction Science* ; 78(1/2): 57-70.

Sheldon, I-M. 2004. The postpartum uterus. *Veterinary Clinics of North America, Food animal Practice.* 20(3) : 569-591.

Sheldon, I-M; Noakes, D-e, Rysoft, A-N; Dobson-H. 2003. the effect of intrauterine administration of estradiol on PP uterine involution in cattle. *Theriogenolgy-* 59(5/6); 1357-1371.

Slama, -H; Vaillancourt, D; Goff, -A 1991. Path physiology of peruriferal period; relationship between prostaglandin E_2 7 uterine involution in the cow. *Theriogenoology* ; 36(16): 1071-1090.

Stama, H; Taiuturier, D; Chendi, J; Zaien, -I bencharif J. 1999. Uterine involution in domestic females: a comparable study. *Revue-de-Medicove –Heterie.*; 150 (10): 779-790.

Steveno, R-D; Dinomose, R-P; Ball-L povoro –B. E. : 1995. PP pathologic changes associated with a palpase uterine human in dairy cattle. *Bovine Fractitioner.* ; (29) ; 93 -97.

Takaes, T; Gathy, I; Bagmoey, E; Ramosa, M; Magyari, L; Turi, G. 1989. bacaterial contamination of the uterus and its effects on subsequent fertilization of cattle. *Magyar –Allatorosok – Lasja.* 44(6): 335-341.

Thompson, F. N; Page, R. D ; cook, C. B ; candle, A. B. 1987. PGF_2á metabolite level in normal and uterine infected pp cows. *Veterinary Research Communications.* 11(6): 503-507.

Thuen,-R; Kundig, -H; zerobim,K; Kindahl, - H; Gustaffson, -B-K; Ziegler,-W. 1993. Uterine motility in cow during late pregnancy, parturition & periperum III. Application of fleunixin maglumine & endocrine changes. *Schwerizer –Archiv-fur-Tierheilhunde,;* 135(11/12); 333-344

Tian, W: Noakes, D. E. 1991. plasma 3-methyl histidine conc and uterine involution in the PP cows. *Veterinary Record* ; 128(5): 109-110.

Twardon, J. 1988. Effects of suckling and absence of suckling uterine involution and fertility in cows. Zeszyty. Nankowe –Ahademii –Rolniczej –We-Wrocawin –*Veterynaria.* 44(172)

Usmani, R. H; Ahmad,N; shafiq, P;Mirza, M. A. 2001. Effect of sub clinical uterine infection on cervical and uterine involution, oestrus activity and fertility in PP buffaloes. *Theriogenology*-55(2) : 563-571.

Vasconcelos, -J-L-M; silva, H. M; Reis; Rashan, K-G. 1993 post-portum genital involution in crossbred and grade holstein cows. *Arquivo brasileiro –de- Medicina –Veterinaria –e –Zootemia.* 45 (4); 405-418.

□□□

Chapter 27

Cervical Alterations During Pregnancy in Small Ruminants

P. Goswami & G. M. Wani

The cervix uteri is a thick walled fibromascular tube connecting the body of the uterus and vagina. It a muscular organ composed of connective tissues predominantly elastic fibres. In the non pregnant ewes, the cervical canal is impassable except during oestrous. Five or six hard prominences within the canal assist the sphincter effect of the cervix (Nickel, Schummer & Seiferle, 1973). During the course of gestration, the length of cervix increases and in late pregnancy the wall becomes thicker, with an overall increase in the compliances of the tissues (Cloete, 1939; Abusineina, 1969). This may be related to disaggregation of densely packed collagen fibre in the cervix of pre-partum ewe. This paper will focus some of the important physical and microscopical changes occurred in the cervix of small ruminants during pregnancy.

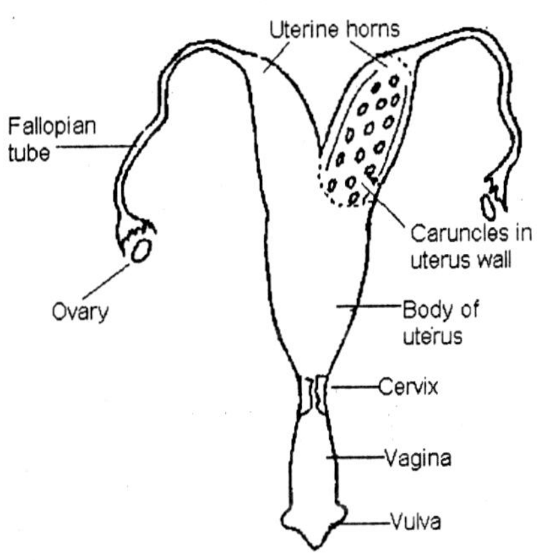

Anatomy of the Cervix

The cervix separates the uterus from the vagina. During pregnancy, it seals and protects the embryo and fetus from the external environment. The gross and microscopic anatomy of the cervix has been studied by various workers. The casts of the inside of the cervical lumen shows its convulated structure consisting five to six circular folds and the second fold being eccentric to the other concentric folds and acting as physiological barrier. The cervical fold in small ruminants varies from five to six folds. In cow four large circular and 15-25 longitudinal primary folds each with many secondary and tertiary folds are present. Cervical mucosa is generally characterized by longitudinal primary fold and most of which maintained continuity throughout the cervix. Superimposed on these secondary folds which is varied in length and depth. Abundant shallow uniformity and parallel longitudinal grooves covers all surface.

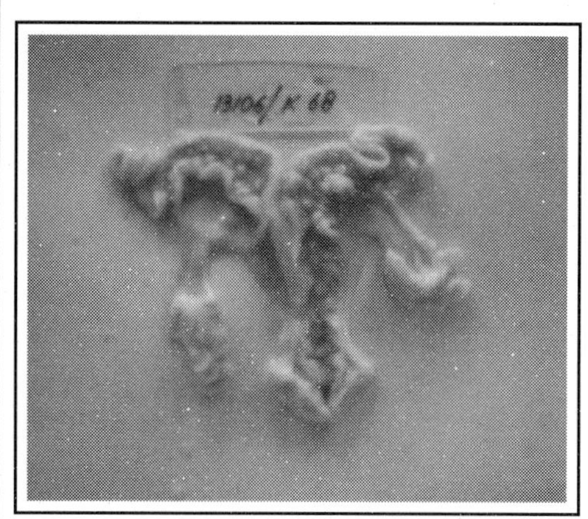

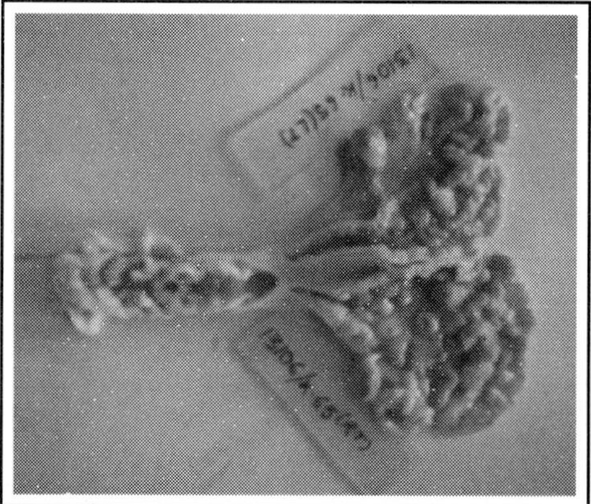

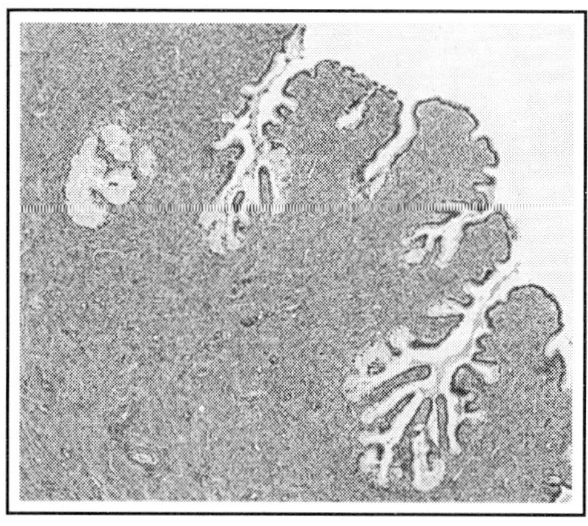

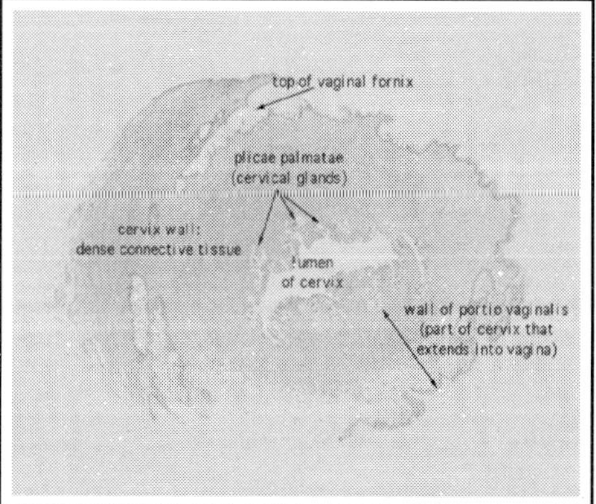

Morphological Changes in the Cervix

Three major changes generally observe in cervix during pregnancy. These are described as growth (physical increase in length and breadth). Softening (changes in tensile properties) and dilation to allow passage of the foetus. The study carried out by different workers showed that ovine cervix shows an increase in width and length in the later stages of pregnancy. The analysis of the constituency of cervices shows increase softening from the mid pregnancy and there after firmness of cervix losses.

A small increase in the degree of hydration of the cervix or dry weight at different gestrational stages has been reported by Fosang et. al. (1994), ward 1968. This may be due to increased tissue mass rather than increase in size of water content. However some author reported no significant changes/ differences in water content of the cervices from non pregnant to pregnant animals. The physical chemical and histological properties of cervix are constant throughout the length of cervix. However, Basset (1958) reported morphological changes in the fibroblast of the broad and sacro-iliac ligament by the 60th day of pregnancy but this information is not supported on ultrastructural studies.

Light Microscopical Changes

Morphologically the most prominent feature of non pregnant cervix is heavy, densely packed collagen fibre interspersed with fibroblast. Small blood vessels are present throughout the depth of the tissue, but most numerous in deepest layer. Smooth muscle bundle are running both longitudinally and transversely in the middle and deeper layer. The figure represents a wall of non pregnant cervix. The lining epithelium is low columnar and secrets neutral mucin. The sub epithelial connective tissue is vascular and contains variety of cells including eosinophil, macrophages, mast cells and plasma cells. The greater proportion of cervical wall is composed of dense fibrous connective tissue consisting of compactly arranged collagen fibre with some fibrocytes and occasional fibroblast embeds in sparse ground substance. The individually arranged smooth muscle fibre forms an incomplete muscularies of which the outer fibre is longer and more prominent than inner fibre. The electron microscopically the collagen fibre shows very compact in arrangement and the scarcity of the ground substance and the presence of fibrocytes. Fosang et. al . opined that there is no significant changes observe between proximal, middle and distal portion of cervix irrespective of stain used. The best stain normally use for differentiation between collagen fibre and the smooth muscle bundle are Massons' Trichrome stain, where the alignment of the collagen fibre along with villi shows projecting towards lumen. In general collagen fibre are large and closely spaced and are organized either longitudinally or obliquely. Section stained with Toludine Blue stain revels metachromatic staining along the collagen fibrils with strong staining of epithelial cells associated mucus. The morphological changes donot become apparent until quite late in the gestration period. The description of non pregnant cervix applied equally to the connective tissues observed in the early stages of pregnancy even to 100 days.

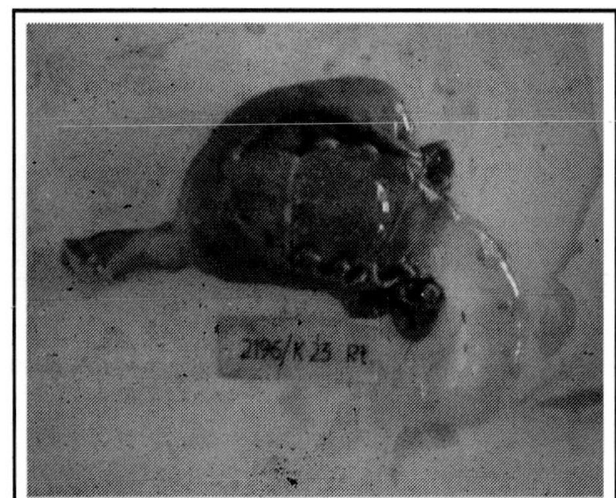

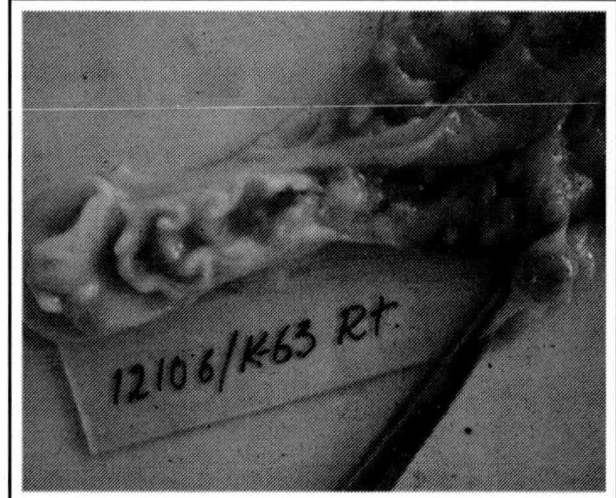

Pregnant cervix showing

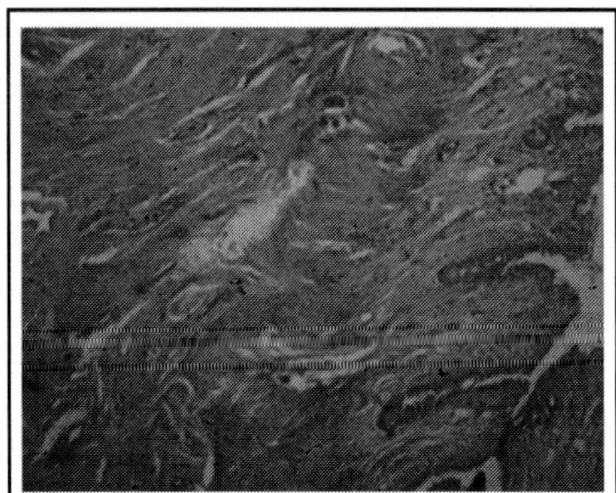

Dense Collagen fibre, inner circular & longitudinal muscular layer with epithelium H&E 4X

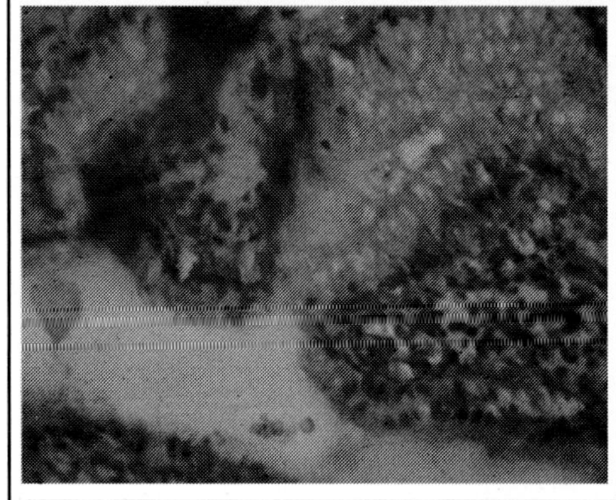

Central cervix Transverse section Loosening of epithelium and collagens layer H&E (pregnant)

The histological section at 100 days of pregnancy revels no virtually distinguish alteration from that of non pregnant cervix (Calder et al). The tall columnar cervical epitheliums are the only changes represents in pregnancy and the secretions are a mixture of acid and neutral mucin. Acidity increases with the pregnancy age. Tissue breakdown and destruction of collagen networks is evident at 140 days of pregnancy. The cells are more widely spaced (empty area) and the collagen fibre losing their organization exposing smooth muscle cells. This can be best seen with Massons' Trichrome. The infiltrating cell at this stage are lymphocytes and monocytes and few eosinophils. In late gestration increased fibroblast activity, smooth muscle hypertrophy, vascular edema and dissolution of collagen fibre bundle are reported by various worker. These findings contrasted with the rigid fibromuscular tissue observe in the non pregnant animals. The appearance of thinner fibre and empty areas between fibres in late pregnancy is lead to decrease concentrations of hydroxyproline in tissue. Collagen fibre dissolution in pregnant cervix has been extensively reported in several species and many authors have reported that active collagenolysis occurring during pregnancy may be the underlying mechanism of cervical softening. Ellowed et al (1981) have shown that ovine cervical explants produce both latent and active collagenase activity, with greater yields of activity in parturient tissue compared with the late pregnancy after 3-5 days in culture. Inflammatory cells invading cervix towards late gestration provide a potential source of collagenase and neutral protinease activity. Eosinophils also have been described as potential bearer of specific collagenase which may be responsible for collagen catabolism (Basset, 1972). At the term the disruption of collagen fibre are more even pronounced with virtually no large fibre remaining. In Haematoxyline & Eosin stain sectioned it sometimes appears very little or no collagen at all. But very little and small fibrils arranging random pattern are seen in Massons' Trichrome stains. In this stage there is heavy infiltration of inflammatory cells among which eosinophils predominant. An area of haemorrhage is also a constant finding along with infiltrating cells. In late pregnancy there is complete network of subepithelial capillaries with a marked increase in the size of the vessels in the outer part of the cervical wall.

Ultrastructure Feature

Ultrastructuraly, non pregnant cervix reveals the typical dense connective tissue with collagen aggravated in closely packed fascicles and fibrocytes embedded in sparse ground substance. The ultrastructural characteristic in late pregnancy are presence of rough endoplasmic reticulum, mitochondria,plasmalemmal vesicle and extensive branching of individual fibres in contrast to the absence of these feature in muscle fibres of the non-pregnant cervix. This description is also similar to early pregnancy stage. The ultrastructural analyses of the cervical connective tissue reflects active changes in tissues, with a reorganization of the cervix prior to the functional changes at parturition.

Changes in Collagen Concentration

The biochemical analysis of hydroxyproline in tissue can be used for collagen concentrations. Study carried out by Regassa et al. (1983) shows the total collagen content of cervix at all stages of pregnancy is significantly greater than that of caruncular mean and the intercaruncular areas. The concentration of hydroxyproline is not changed in cervix during Ist trimester of pregnancy. However the concentration of hydroxyproline progressively decreases at days 100, 140 days and in post partum tissues as compared to the non pregnant tissue(Fosang et. al 1984). The concentration is same between proximal, middle or distal region of the pregnant and non pregnant cervix.

In conclusion it is summarized that uterine cervix of small ruminants became softer during the pregnancy and that some associated changes first appear in early gestartion. There is no significant changes in water content through pregnancy although light increases is associated with cervical size and softening of the tissue. Physical and histological properties are identical in all section along the length of cervix. The changes associated with increasing length of gestration are absolute increase in width and length, relative increases in fibroblasts, smooth muscle and softening; relative decreases in collagen and fibrocytes. But increased vascularisation without any white cell infiltration of the tissue is specifically associated with late gestration.

References

Abusineina M.E. (1969) Effect of pregnancy on the dimendions and weight of the cervix uteri of sheep. *British Vet. J* 125, 21-24.

Amanda J. Fosang, Christopher J. H. Vivien S., Dennis A. L. and Geoffery D. T. (1984) pregnancy related changes in connective tissue of ovine cervix. *Biology of Reproduction* 30, 1223-1225.

Aughey, E, Munro, C. D., Calder, A. A., Coutts, J R. T. & Fleming, R (1981). The histology and ultrastructure of the pregnant sheep cervix uteri. *J. of Anatomy* 132, 448.

Basset, E. G. (1958) Gestational changes in connective tissue. *Nature* 181, 196-197.

Cloete, J.H.L. (1939) prenatal growth in the merino sheep *Journal of Veterinary Science & Animal Husbandry* 13, 417-543.

Calder A.A., Aughey E. Coutts J. Fleming R and Munors C.(1983) Changes pattern of cervix on pregnancy *J. Anat* (1983) 136, 2 389-399.

Ellwood D.A., Anderson, ABM, Mitchell and Turnbill A.C. (1981) Prostanoids, collagenase and cervical softening in sheep. *Am. J. Obst. Gyneol.* 10:281-287.

Hollingsworth, M. (1981) Softening of rat cervix during pregnancy. In the cervix in pregnancy and labour- clinical and biochemical investigations (ed. D. A. Ellwood & A.B.M. Anderson) pp.13-33 Edinburg.

K. June Mullins, R. G. Saacke (1988) Study of the functional anatomy of bovine cervical mucosa with special reference to mucus secretion and sperm transport *Journal of Reproduction and Fertility* (1979) 57 261-266.

Karen Sohan, Rebecca Wiggins and Peter Soothill (1999), Cervical Physiology in pregnancy and labour. *Foetal and Maternal Medicine Review* 11: 135-141 Cambridge.

More J (1984) Anatomy and Histology of the cervix uteri of ewe: A new insight Acta. Anat (basal) 120 (3). 156-9.

Nickel, R., Schummer A. & Seiferle E., (1981) *The viscera of domestic animals* pp.358 and 361 Berlin, Verlag Paul Parry.

Regassa F. and Noakes D. E.(1983) Changes in the weight, collagen concentration and content of the uterus and cervix of ewe during pregnancy. *J Biology* 73, 221-25.

□□□

Chapter 28

Livestock Development Policy and Vision 2050

G.M. Wani

A. Thrust Areas

I) Use of Biotechnology for

a) Enhancing animal productivity

b) Development of sensitive disease diagnostic tests.

c) Production of vaccines and biological products.

d) Conservation of germplasm.

   i) Rare species

   ii) Dieing species

   iii) Local germplasm

e) Reduction in livestock farming costs.

f) Production of low cost life saving drugs.

g) Bioconversion of agriculture, biological and aquatic weeds and wastes into livestock feeds.

h) Efficient progeny testing.

i) Quick multiplication of superior, disease resistant livestock germplasm.

II. Identification of Animal Genetic Resources Well Adapted to a Particular Environment Through

DNA finger printing

Gene maping.

III) Development and Refinement of New Techniques

a) For specific locations

b) For rural-based-low-input oriented livestock production systems

c) For decreasing poverty and improving nutritional standards of poor.

d) Employment oriented livestock entrepreneurship especially to benefit marginal and small farmers.

B. Policy Formulations

i) Rural based simple productive livestock or mixed farming technology innovation. It may involve interdisciplinary and co-operative research ventures like those of:

a) Livestock- crop farming system.

b) Cattle-crop-agro forestry system.

c) Sheep-goat-horticulture-postural system.

d) Duck-fish-mixed farming.

e) Duck-rice-fish-Buffalo farming.

f) Backyard poultry-fish subsistence farming culture.

g) Environment friendly small ruminant (Goat/sheep) backyard farming

h) Rabbit production/Duck weeds cleansing.

This shall aim at improvement in existing system for increased productivity. Sheep farming system is interdependent and closely linked with soil-crop-forest-ecology system and situations. The livestock farming system is more interlinked with agro-climatic and socioeconomic situations. Thus it is a fit venture for co-operative inter departmental research, which should involve central and state government research and development agencies like ICAR institutions, SAU, financial institutions, co-operative, industries and NGO;s.

ii) Identification Preservation and Conservation of Endangered Animals/ Breeds/species/ strains through; Use of Modern Techniques Like

DNA and Biochemical polymorphisms, finger printing,Gene maping Invitro fertilization Ooctye harvesting and fertilization from dead animal Gene cloning.

DNA injection/In vitro foetal and embryonic development.

iii) Livestock productivity recording and progeny testing of sires under varied agroclimatic conditions. The recommendations of Indo-US sub-commission on agriculture.

iv) Production of progeny tested germplasm from improved Ram, goat and other animal sires and is propagation through AI under varied agro-climatic and ecological regions of India. This shall involve farmers herds/ flocks besides organized government and non-government institutional farms.

v) Establishment of sire lamb/kid testing stations by re-organizing and re-orienting existing livestock farms under central, state, ICAR, SAU institutions This shall need a co-operative, inter and intra disciplinary, a institutional and organizational approach. It will be centered around job-oriented missions with target-specific objectives.

vi) Development of a strong farmers livestock productivity data base. The main emphasis shall be centered around economics of animal production system, its marketing and utilization surveys.

vii) Research on low-input-village-level methodology for animal products processing, preservation and marketing provision.

viii) Organization of large livestock co-operatives and societies for breeding of specific species and strain. A herd/flock book concept has to be introduced with strong analytic and interpolative mechanism. This will involve farmers herds/flock under supervision.

ix) Studies on various management and production systems including soil-plant-animal fish mixed farming system.

x) Advanced research on fundamental aspects of pre and post-implantation embryos.

xi) Surveillance of disease incidence and monitoring of new emerging disease of spread across contents/countries.

xii) Study of infertility and sub-fertility problems of improved genotypes obtained through cross-breeding.

xiii) Survey of fodder crop production and pasture evaluation under different agro-climatic conditions. Use of non-conventional feed and fodder resources like agriculture residues and wastes.

xiv) Study of economics of migratory managemental systems of sheep and goat production. Their effect on pastures and environment. Innovation of alternative low-cost managemental systems of producing lean and young lamb kids.

xv) Studies on composition, carrying capacity nutritive value of water weeds and other indigenous plants and their utilization as ruminant fodder.

C) Research Prioritization

i) Identification of superior animals of different livestock species under a given agro-climatic condition and their propagation through technology use. This will help to reduce animal number and increase per animal productivity.

ii) Economic of animal productivity/survey/analysis and module development.

iii) Marketing statistics and consumption trends of livestock products and presentable, competitive packaging.

iv) Herd book/ flock book concept innovations, analytic procedure and package development.

v) Establishment of superior sire evaluation centres and reorganization of livestock breeding farms/zoo-sanitation and biosafety centers..

vi) Establishment of frozen semen banks at national-state SAU-ICAR level to produce superior germplasm for propagation under varied agro climatic conditions. They will incorporate or include a strong research and development team to formulate and innovate new and job oriented missionary technical protocols, within disease-free zones.

vii) Establishment of multiple Embryo Transfer Centres for utilization of superior dams and their use to price 100-200 high ranking sire mothers at national and state level. Introduction of ET programmes under farmers farms whose production capability is much higher than national average and establishment ET centres.

viii) Disease diagnostics in reproduction. Use of hormone assay and recent techniques for quick reproductive assessment and evaluation and development referral clinics at SAU/ICAR levels. Use of laparoscopy, ultrasonography for assessment of reproductive status and problems of different animal species.

ix) Establishment of quality sire producing farms under private sector or by NGO/Govt./ICAR SAU co-operative ventures.

x) Use and management of waste, marshy and fallow lands for fodder Production.

xi) The recommendations of various discipline wise-societies shall form a base for identification new and emerging areas of research and thrust areas in all disciplines.

xii) Research-priority setting-University, ICAR, State Dean-Directors interaction and common drafting of priority in animal sciences.

xiii) Commodity oriented technological development, location specific and species specific technology development funds and missions.

xiv) Reorganization of Research and human resource base, its advance training and remandating research objectives and achievements.

xv) Bridging gaps in requirements, production, packaging, marketing and export orientation.

xvi) Economic animal and poultry production systems, their global comparisons and evaluations.

xvii) Global/Asian Livestock Training Institutional base.

Mandate

1. To formulate and supervise councils policies and work relating to the field of Animal Sciences.
2. To co-ordinate and stimulate operational research programmes and demonstration in different animal systems.
3. To assist the Director General and other officers of the council in implementing approved programmes of research.
4. To perform any other duty that may be assigned.

Brief Review of Work Done

Starting with a cattle plague investigations in 1989.Veterinary Science has to its credit a number of inventions for prevention and diagnosis of diseases like Rindirpest, H.S, Anthrax, Tatanus, BQ, Equine abortions, Buffaloe malaria, Johne's disease, Piroplasmosis, Theileriasis, Equine glanders, F&D, Ranikhet disease etc. The research on Bovine - Immunodeficiency virus could be a handy tool for prevention of HIV, if proper comparisons are made in human disease prevention. In spite of distinguished work for animal disease diagnosis and prevention - critical gaps exists in their utilization for protecting public health. Thus, a new role of this Division is envisioned combining human, animal health concerns with food security and safety. The other wide gaps in its overall impact on production, processing, distribution, marketing of livestock products could be filled by linking it to global - food - chain dynamics through genomic and proteomic research on gene propagation and utilization for improving productivity, reproduction ability, disease diagnosis, therapeutics and animal modeling systems. It shall work in close co-operation with WHO, FAO and other national and international organization.

The gaps could be addressed as new research projects/thrust areas, some of them could be presented as:

- Protecting public health by avoiding emerging diseases threats like Bird flue, Mad-cow disease etc.
- Protecting animals by more refined disease diagnostics using genomic and proteomic research tools.
- Encouraging biomedical - veterinary therapeutic research through drug - disease - targeting. This may involve joint collaboration of medical, biological and industry organizations.
- Improving production systems in tune with Global milk and meat production cost scenario. This will need intensive measures for adopting international Zoo-Sanitation regulations and product safety measures to fit us in Global - food - chain - operatives systems as future exporters of milk and meat products across the globe.

New Project Proposals/Infrastructure Linkages

To address the above goals as thrust areas, new technology use is anticipated as new project proposals. They are aimed to refine our investigative technological cover for unquestionable acceptance in biomedical industry and institutions. An export oriented shift in product technology - transfer shall be the outcome. The emphasis shall be on the following technology interventions;

i. Targeted gene delivery and next generation DNA sequencing, expression, proteomics in health, disease and therapeutics.

ii. Microtubule dynamics to understand neuroregulatory mechanisms.

iii. Molecular and cellular armaments of the innate immune system, using foetal models shown in Annexure-A. This would mean antigen recognitions and immune response studies.

iv. Protein interactions - transcriptional conformational mechanism for disease diagnosis, disease prevention (antisera preparations/biologicals) and physiological regulations.

v. Gene - drug - designing. Molecule targeting under Genome - wide - functional linkage analysis of infective agents. This would include famous German buzzword "Protein repertoire' i.e. antibody based proteomics for tissue profiling animal to create disease spectrum atlas.

vi. Natural receptivity associated proteins for immune protection imaging and mapping. Its use in disease prevention, Organ transplants, Fertility augmentation and control. The new concepts shall fortify human - animal bonds. More emphasis and applications in pet - animal clinics could further accelerate the process and attainment of goals.

Outcome

The outcome of these new technology uses shall bridge the gap in research between biological, medical, agricultural vis-à-vis veterinary and animal sciences. Besides, it would stress on pet, wild, laboratory and comparative veterinary medical - research. New innovative models have been tested by us and shown in (Annexed Figs).

Multidisciplinary Linkages/Collaborations/Referral Center

Veterinary Research in general and at ICAR , in particular has now to cross the institutional barriers. It has to assume a cutting - edge - role across the borders of medical,, biological and agricultural research laboratories. This would need more meetings, stand up packaging and out reach programming. The overarching focus has to be on competitive, innovative and collaborative research. The veterinary and animal science research set up has to be geared up to imbibe new responsibilities. This would prepare profession and professionals for addressing challenges across the food systems, safety, security, consumer lifestyles and preferences.

The close arching and communication links with bio-medical institution, veterinary and agricultural universities, industries, departments, NGO, progressive farmers and agencies in remote local localities shall make it possible to make "animal sciences Institutes of ICAR" a global and national referral center as envisaged.

1. Educational Change

Global prospective – curricula, infrastructure, practical trainings, capacity and skill up-gradation has to be of international standards, if ICAR has to change from National/Asian to International standards. It has the future of being premier international Organization provided a change is made in;

Quality assurance and capacity building Teacher's training and empowerment, generation and reciprocal flow of knowledge and information, collaboration and networking of species/discipline oriented institutes and other Agri/Vet./Bussiness University, colleges/schools.

2. Holistic Education

Classroom teaching has to be combined with farm, farmer, and hospital practices. Diversification of disciplines to meet the requirements of job-market in Animal production, processing, distribution and consumption is the need to be incorporated in our curriculum design research and transfer of technology in the next few years.

3. Educational Up Date

New degree-programmes in processing marketing and management may be initiated. New methods of education using effective communication aids and appliances have to be introduced. A new industry – research participation in animal science and veterinary education too will be effectuated.

4. Developing Faculty and Scientist Leadership

Training faculty and scientific manpower in education and research methods, philosophy, curriculum design etc. should be of prime concern. For on the job training and skill orientation a wide conceptual base is needed. The animal health and production educational scenario is on cross-roads to ensure food safety, security besides animal and human health guarantees. Thus more emphasis has to be on comparative medicine, Animal Modeling for research, emerging new diseases, ecosystem health, exotic, caged animals health and food borne illness. This all demand a refinement in faculty thinking and mind set, which would be possible through training and moral boosting. A policy frame work on these lines shall be my endeavor if selected.

Transfer of Technology Goals

a) Future-Food-Revolution

The future-food-security lies in animal-agricultural protocol. It needs interactions among people to find economic opportunities, biological and geographic resources, and socio-cultural practices to boost our exports or sales. We have a small holder animal-farming sector. Are future production trends favouring this sector? If yes we need to tune our policies to encourage it. A review of Global Milk Pricing is necessary to fix our targets in this direction.

b) Possible Global Milk Pricing

i. Dairy Farm Structure is either small 2-5 animal farms as ours or large more than 50 or 100. The countries like *USA, EU, Canada, Isreal, Hungry, USA, South America and Oceania* have large farm structures. Except Oceania none produce milk at world market price of 15-22 dollar a quintal. Thereby signifying large farm structures are not always economical. Our small holding size of dairy structure is more competitive than West. This advantage needs to be harvested under WTO-Liberalization Global System.

ii. Higher Production Costs.

The costs of milk production are high in Scandinavian Countries and Spain. The yesterday giants are todays loosers. The highest production price is also in counties like *Hungary, Czech Republic, Isreal, Bangladesh, Thailand and parts of China*. Their cost of production is more than global average cost of production viz 15-22 dollars/quintal. Those who produce at world market prices of 15-22 dollars include Estonia, Poland, Chile, Brazil, India, Vietnam and Oceania. Thus, we have tough competition with these nations to gain export markets. The Switzerland, Norway and Canada produce milk at much higher price than EU and USA, their markets in middle east are our export pastures.

iii. Lower Milk Prices

The low milk prices than world market averages are in Argentina and Pakistan. However, they have distribution advantage rather than production axis, as farmers demorcating forums are non-existant. As such they may not pose any competition with us. We have export advantage in Asia as our other neighbour China too produces milk at higher costs. We must prepare ourselves as "*Future milk Exporters in Asia*". This needs a preparation for TOT.

North-West-Export-Zone

The range of states having self sufficiency in milk is shown in Fig.1. We have to make this North-West-Zone comprising of J&K, H.P., Haryana, Punjab, Rajasthan and Gujarat as our potential milk export-zone. We will have to induce mission modes for its animal disease-free-status, so that we meet international-zoo-Sanitary standards. They are our major block in exports of milk. Outreach to farmers on these lines is suggested.

Global Meat Industry

The next thrust area under transfer of technology has to make India "*major meat exporting*" country. The global meat industry is slowly moving in the hands of Brasallians and Chinese farmers. Thanks to our vegetarian majority diets we too are on run. Against, daily needs of 25g/person/day we produce only half of it on per capita basis, but our advantage with goat meat abundance can differentiate our export potential from the rest. Even global super powers USA have more demand for Cheron than what is produced. A policy change and frame work has to be drafted. Our product technology packaging and consumer preferences have to be webbed through transfer of technology to export concerns in particular and in general to home users. The focus has to be centered around hygienic meat production, ecological preservation, processing and product technology. This in principle means "reshaping livestock range and forage management" through effective communication and transfer of technology across its terminal and potential users and beneficiaries.

Generated Technology Transfer

We have to lead in Embryo transfer technology in Asia as has been done by us in dissemination of AI technology, in collaboration with Swedish expertise. Many international experiences are needed in this and other TOT missions. We need bio-technological links to be developed with industry and user departments. Licensing in service mode is desired in technology transfer. We can do it on Cornell University pattern. Strengthening pre-clinical, clinical and basic diagnostic skills to glow animal science Institutes as a word referral clinical centers. This would need emphasis on technician training facilities for physiology, Reproduction, AI, Biochemistry neurosciences, biology and like. Linking, research and extension, faculty, students and farmers under KVK pattern are needed for quick animal science technology transfer. The referral TOT in animal sciences is ready to be switched on in the following area in next few years.

- Rapid diagnostic tools and techniques
- Tannin degradation technology and low- cost feed packaging
- High security disease prevalence checks. May be its sharing of emerging disease pattern with public health experts give us better dividends
- Biotechniques for germplasm propagation and improvements.

Conclusion

Emphasis on

- Global-food- system-chain-factors (GFSCF).
- Research on competitive comparative health and disease.
- Internationalize Veterinary and animal science education.
- Out reach for global farming through TOT.

❑❑❑

Section - II

AGRICULTURE SCENARIO AND ENVIRONMENT

Chapter 29

Agricultural Productivity Gains Through Capacity Building, Eco-Friendly Farming and Extension Education Reforms

G.M. Wani

ABSTRACT

Agricultural Production has entered new era of competitive-low-cost globally oriented enterprise. The cost-benefit ratio decide the mode of Farming. Global marketing needs some changes at base level. Farmers have to be cost conscious and need new strategies of mixing various farming option like Agriculture, Horticulture, Livestock husbandry and Floriculture activities. A new model for incorporating all these fields in an eco-friendly mode is discussed. Appropriate mixed technology in tune with natural resources needed at the moment in India. An agricultural mission for efficient Resource management should aim at bridging gap between current farm yield and potential Farm yield. Use of bio-technology and bio-informatics for identification of economically viable cropping options and combinations can not be ruled out ,however soil and environmental preservation must be kept in mind while using biotech farming options. We should talk of Eco-conducive-germplasm-base instead of high productive - germplasm, un-competable to our natural resources. All these and other aspects of future agricultural scenario in India is discussed in these pages.

India with 1.09 billion people and 60% of them dependent on agriculture can not be leveled with USA who has merely 339 million people and only 0.7 of them dependent or agriculture. Therefore the past trend of aping USA OR EU models of agricultural development won't help. We have to innovate new ways and means conducive to our agrarian conditions and available resource to increase our food production for growing population in future. Some of the original thoughts of the author published as e-papers are reproduced here to suggest new innovative models.

a) Change of Mind Set

We will need a change in the mechanism of technology dissemination thus a reformation model for extension education in India is proposed in these pages. More than 50% of our practicing farmers are women who are not exposed to technology new or old, thus a system of their capacity building and skill up-gradation shall be focused Agricultural and allied technologies tested and available in India have been the replications of commercial farming ventures of the west. I would dare to blame our policy planners for this vulnerable state de-affairs. They professed chemical farming when we were world leaders of organics In the past century when whole world faced the ill effects of chemical farming we are busy in advocating our forgotten options which luckily are still existing in deep and uncommunicable pockects of our country called backward pockects. Our research need a change from individual crop or commodity farming research to mixed farming approaches. My mohy model of 1992 is being quoted with out reference quite often by those whose hollowness of innovative ideas is responsible for our defective policies.

b) Unfinished Agenda Needs, Aims and Goals

Needs

1. Appropriate mixed technology in tune with natural resources.
2. An agricultural mission for efficient Resource management, bridging gap between current farm yield and potential Farm yield.
3. Use of bio-technology and bio-informatics for identification of economically viable cropping options and combinations.

Emphasis

We should talk of Eco-conducive-germplasm-base instead of high productive-germplasm, un-competable with our natural resources.

Economic Farming which means cost

effective use of : Soil
Water resources
No air/water pollution

c) Nutrient Recycling System Options

The chemical fertilizers are to be replaced by use of agricultural waste materials, plant residues, aquatic weeds, left out materials, post harvest wastes, green manure, rainy season over grown vegetation, or abundant vegetation, green weeds, city drains, human excreta, animal manure, forest wastes, agriculture left out substances, bio-mass, bio-fertilizers, stall wastes & even pigeon beats etc.

Nutrient cycling system has to be evolved first by analysing soil status of each agro-climatic zone . The integrated nutrient management system, has to be evolved.

A new, refined, Govt-agri policy is needed. New Institutional support system, excellent marketing networks are to be proposed. For efficient land use we must wed traditional wisdom with improved technology, farmers training and profitable end products. We should advocate:

a) Use of information technology to increase profits from overall farming; cropping, animal aquaculture etc.

b) Analysis of available data on processing, production and its documentation for use in project preparation, policy formulation is a basic necessity. A new sensitive laboratory facility is envisioned for input and output quality assessment.

c) Agriculture of today does not mean more production at environment cost but efficient and economic production with Environmental safeguards.

Introduction

Our 65% employment base is agriculture & its allied sectors like livestock We. Need in future :

- Food and environment security.
- Employment generation through agrobased industry.
- Economic, balanced and hygienic food with original taste and flavours.

We are in a phase of cultural and economic transit. This transformation Process brings,

New livelihood options.

New production & utilization system.

Changed life styles & cultural habits.

I do not know how many of us know about USDA's national Arboretum of Washington DC A 444 acre arboretum now has dropped pesticide use by 75% during 1992-96. They call it new land scope pest Management (NL PM).

It was simple spot spraying i.e spray right where it is needed, i.e on diseased twig or branch or plant and only if it is needed.

This shift from blanket spraying to selected sprays saved 75% of pesticide cost; other saving was through reduced labour cost and eco-risk factors. We need safe equipments and labour health monitoring and analysis of data.

Another area which needs attention is use of less toxic pesticides. We call them bio-rational pesticides. They reduce cost by 12-32% and consist of Horticulture oils insecticidal soaps, insect growth regulators and predators.

We talk of trainings for Information analysis, monitoring disaster, Management of floods droughts. Use of remote sensing and Global positioning system is needed. We in developing countries in future will have Industrial expansion, and more economic security. This will mean less people being attracted to agriculture or in other words agriculture becoming less renumerative. We need improved varieties, their rapid propagation through biotechnological use. We may have little options not to use biotech agriculture or mend fences between traditional, optional and modern biotechniques.

This is where we talk now of molecular farming, using biotechnology for more transgenic crops animals, aquaculture and, commercial floriculture.

New issues in Agricultural and other Sciences in India are

1. Develop data base for patent formulations.
2. We need replacement of top down planning with gross root planning.
3. More farmer's participation reduces risks .
4. Research in past, centered on improved seeds, irrigation, chemical use. We witnessed green revolution, freedom from hunger, but harmed environment. We have to look for alternate options for the execution of research.
5. Extension education, dissemination of information to farmer has to be the first concern of the public and private establishments and planners. Birds eye viewers have to be replaced with visionary professors and professionals.
6. Farmer farming and farm business shall be the First ranking scientific persuit investing at least an equivalent amount on it as is spend on health

or nuclear establishments. Our policy should be Independent and self conscious, farmers, with records of annual balances, gains and losses, risks and benefits is what is meant for future.

Bottom Line Challenge

Bring self- reliant proud farmer trained & Well versed in Agri-business. Farming concepts and international zoo-phytosanitation recommendations and regulations. We need to stop WTO restrictions or options till our farming and farmers are not completely knowledge driven and economically resourceful to face the global farming competitions.

Paradigms of Agri-extension in India

Four paradigms of Agricultural extension consists of Technology transfer, prevalent since colonial times & later National Agriculture Extension Project (NAEP), reshaped in 1970-80 as Training and visit system across Asia. It involved top down approach. Its role needs discussion. The second paradigm called as Advisory work is what is present system of extension, where, government agencies, NGO and technical industries respond to farmers enquiries which we call advisory role. The third paradigm consists of Human Resource Development. This innovation helped Europe and North America. The Universities gave training to rural youth. It is these educated, trained youth which became future farmers in developed countries. What is known as outreach system of colleges and universities is still unknown in our Agricultural Research system including State Agriculture Universities. A half hearted attempt as Agri-business Agri-clinical is still to be backed with University, research, banking and economic support. Is it which can help us to increase productivity in India or Asia needs consideration? (Wani, 2008).

Technology Adoption and Impact

The impact of National Agricultural Research Project (NARP) (1986-92) revealed that technologies generated were few and adoption rate was not high, even then an estimated monetary benefit of Rs.2000/ per hectare was observed. Low relationship between technologies developed and adopted demands a mixed farming technology capsule as 65% of our farmer's are small and marginal farmers. This needs a total shift to left in our agricultural extension methodology. (Reddy 2001, wani, 2008 a b c). T & V programme initiated with central assistance in state has aroused all round technological awareness among agricultural experts. Its dialogue and technology transfer from University (Scientists) to experts agricultural extension workers was nearly total revolution. The (ZAREC) Zonal Agricultural advisory Research and Extension counseling was the best plate form to discuss main issues impeding agricultural development. The missionary publication of production recommendations helps a lot in transfer of technology. The presentation of production recommendations in local language is helpful in increasing productivity in many countries. (Wani, 2004) Brinjal production technology adoption in Parbani, Mahrastra was the result of social participation, risk management guarantee, marketing orientation, information websites and communication methods. (Dudhak et al 2003). Various limitation of transport, extension staff were the reasons of poor productivity gains in Malakhand agency of Pakistan. (Mohamad et al, 2003).

The adoption of legume production technologies under Jhodpur arid conditions was low. The problems as indicated by farmers were small farm size. Small holder farmer was less interested to adopt these technologies. Non existence of market and pricing policy, risk of failure, feeding habits, social taboo and lack of Post Harvest Technology (PHT) were identified. (Yadev et al, 2003). Proper guidance or information support was the major constraint in adoption of improved Wheat cultivation in Himachal Pradesh. It contributed to 16.8% non-adoption rate among farmers. Other contributory factors were technological unawareness (13.33%) sloppy land (11.31%), farm input cost (9.41%), non availability of irrigation (8.29%) and lack of inputs (7.95%) (Puran chand et al, 2001).

Social Structure and Technology Adoption

Schedule caste farmers in Maharastra have low level of rice technology adoption. Among 150 such farmers only 12.6% had high adoption rate (Mankar et al,2004). Adoption of production recommendations of Cashew in Maharastra's Sindhubrug district was 58% (Bhairamkar et al, 2004). Motto of extension programmes should be reach to unreached, teach to un-teached, be known to unknown (Ganesan et al, 2003) (Wani, 1998). Agricultural productivity in Czch Republic increased when information and communication in extension system was improved for small scale farmers (Slavik, 2003).A survey conducted among 120 farmers in Karnool district, Andhra Pradesh, India, revealed that education, social participation, scientific orientation, risk management, mass media exposure, economic and market orientation were positively and significantly correlated with the extent of adoption of recommended hybrid jowar (sorghum) seed practices by the farmers. Majority of the farmers (43.33%) had a medium level of the adoption of the recommended practices. (Kumar et al, 2005). It is suggested that five categories of factors determine the adoption of irrigation systems and are the keys to designing relevant, effective and innovative irrigation extension programmes among horticulturalists. The five categories of factors are : Soil type and topography ; crop quality; sowing time; saving water and crop yield (Kaine & Beswell, 2005). We in J&K under SREP Strategic Research extension plan of Districts notice lack of information on sowing time, seed rate and fertilizer application in almost all the crops. Seed treatment awareness helped farmers to combat paddy blast (Wani, 2006, 2007).

Innovative Change

Methodologies for analysis of farming system, options for development & sustainable eco-friendly farm science technology are the needs of the hour. We do have subject-matter-specialists, but the agents for evaluating the composite effect of various discipline oriented technology cover is yet to be structured. This is exactly where we should intervene and restructure the discipline, programmes and divisions of extension education in the country. Infact a total change from existing slow action, low innovative, conventional and disintegrated extension system has to be refined into analytic , managemental , social, economical, marketable system, so that the whole technology transfer could be evaluated in terms of productivity and monetory gains per hectare. Different cropping and livestock patterns and combinations need to be evaluated & monitored. The impact analysis of wide range of Agricultural technology transfer be it training, entrepreneurship or agri business is needed, both before and after technology application.

The available research results have to pierce the farmers farming practices. Non - adoption is the result of our low productivity levels per unit of land. Inspite of tremendous yield gains the per unit profitability has not raised the economic conditions of our farmers. Our farm economy still fluctuates between season's, rainy years and among regions. Now for any widespread gains from technology, it is essential to know its impact and adoptability rate. It is therefore necessary to develop new technologies after considering socio-organizational arrangements, local existing soil, water, human needs and resources. The local skills need to be known and refinement is made to improve profitability. Our experience under IVLP showed a net productivity gains of 2 q/hac in rice production by improving local village driven plough. Similar innovative ideas local technologies with improved or exotic one gave good results in development of profitable farming. (Wani, 2004,2003, 2005,2006, 2007).

Reform Capsule

i. Chinese Model

Chinese experiences needs a try under our socio-economic conditions. The successful improvement of Sheep and Yak herds in china can be a good example for use in India. Under this project they first took stock of rangeland and livestock resources. Then socio-economic conditions of the livestock breeders were analysed. Livestock and pasture management by rodents was studied. Rasngeland revegatation methods were tested and propagated. Seeded fodders and effect of these improved method of pasture development on disease control and profits from livestock were analysed. (Wagenin gem et al, 2001). Thus, the extension education programmes should ensure facilitation of Farmer empowerment and not dependence on subsidy. This means incorporating studies and programmes of Agri-livestock economics, grass land and rangeland management, crop and tree production in Agri-extension curricula.

ii. African Model

A new participating approach of clubbing local and outside knowledge to solve the agricultural problem is needed. Innovative approaches for soil and water conservation in Ethiopia and Tanzania were helpful. A new model of participatory modified Technology development approach is fast being recognized, as the only way of clubbing traditional and modern technologies for agri-development. (Kib Wana et al 2000).

iii. Dutch Model

The total reformation of agricultural technology transfer shall need refinement of elementary methodologies used in extension consisting of PRA, PERT or other specific managemental, communication and modern GIS-Geographic Information System use. This would involve a total change in Extension Education, objectives, methodology & course curriculum. We need agricultural networking for bringing out results of transferable technologies. Experimental learning and knowledge sharing has to be incorporated, as research on farmers field. Farmers have to be part of our learning and experimentation process. Dutch model could be used. (Leeuwis, 2000) in India to ensure quick technology use. Let us evaluate technology use. It has been seen that some progressive farmers have awareness of technology but being absentee farmers, their workers have no knowledge or skills to use them in the field. Is it time to come out of "Technology dissemination Scenario" to "Technology Use Status in the field" and "Productivity gain assessment models".

iv. Mohi-Model (Proposed Model)

Experience learning, market designs, farm economics, sustainable agriculture, Agri-information, communication skills and farm business shall be the knowledge bank available to Agri-Extension & communication experts, whom we should name "FBM" - "Farm Business Management Experts". The Extension Education is outdated and too oftenly used and misunderstood phrase. Extension includes now every attempt to communicate with farmers. Today's extension agencies can be university scientists, Subject matter specialists, Agri experts, farmers or even partially Agri known NGO or their agents. Thus, the whole blame of others has to be shared by University experts. Agri-University, applied knowledge has to be communicated to farmers as "end product" as pharmaceutical companies do. Thus all the current "building block", known knowledge has to be computerized and stored. This has to be reshaped and re-arranged as knowledge packs for innovative and whole farming capsules rather than one commodity knowledge sharing as in the west. We do have commodity but not composite agriculture packages. A booklet on its composite effects has been published as Broad Based Extension Education (Wani, 1992).Training & visits on farm level failed but farmer-training & Farmer-research participation or demonstrations at farmers field did achieve a great success. With constraints of men and resources and crop insurance in vogue, we must induce researchers to have direct participatory research at farmer's doors, on their fields. This venture can be on cost-sharing basis or with participatory involvement of Industry. (SAMETI, Kashmir model refers). Proc. XXI EECM,2008.

New Approaches

Insect-pest Management research could be an example. Some FAO programme results are encouraging (Matteson, 2000). Rice IPM research must incorporate farmer-training of farmers. New pest management procedures are to be experimented under farmers field. Minimum use of pesticide and eco-friendly application is preferred (Matteson, 2000). Problems and constraints of technology transfer in India have been reported, (Bairathi, 1998). Similarly bio-gas technology use or adoption has been full of constraints detailed. (Prasad and Singh, 1999), these range from technological, financial, administrative and innovative ones. Proper mechanization and training in operation of such bio gas technology is necessary.

An information web networking developed helped farmers to improve their income through training and advise thus participatory technology transfer is a possible mechanism for improvement in Agri-productivity. (Warren, 2003).

Netherlands Model

Quest for knowledge led Netherland farmers to have an association for organic farm produce. This association conducted its own farm research and shared its experiences with other farmer members back in 1980. Soon they felt that their experiences lacked experimental details. They approached Govt agricultural Research Organization and started collaborating with them. The station mandate was to assist the development of farming systems in the area called polder in the village Nagale of Netherland. This combination of research and farming practices at farmers field tried various crop combinations and rotations. The use of nitrogen-fixing crops (Legumes) for maintaining soil fertility and strategic management of pests and diseases, extending or delaying planting of various crops etc. Some experiences of low-weed-crop production by adopting agronomic practices like regular harrowing of lands and use of crop covers were tried. The farmer pressure and self sustaining groups need to be involved in research experiments of SAU/ICAR and all other KVK farms and institutes (Reinders, 2007). This would have both a demonstration and high adoption value. Besides it would be a close to practical field research model (PFRM).

Horizontal, farmer to farmer contacts help dissemination of agricultural information. This horizontal expansion of information from one vegetable grower to another in Suden. (Badri, 2001) helped technology adoption.

Mass media support for rice production has been helpful in Bihar. Out of 5 TV talks, 75% farmers benefited and registered higher production and low disease damages. (1taque, 2000) similar observations are known in J&K. (Wani 2006). Electronic technological diffusion of knowledge at Farmer's fields was helpful, local language and local persons can diffuse ideas and disseminate information and communication faster. It is proposed to have innovative knowledge defusion to women farmers through local women (Rani and Reddy, 2002).

Farm Women Awareness Model

Majority of rice producing farmers in Kangra district of Himachal Pradesh had no extension personnel contact or electronic or print media information. Continuous information on mass media have significant impact on use of seed rate, transplantation time and application of nitrogenous fertilizers in Rice (Thusoo and Sohi, 2000). A gap existed in knowledge of farmers so far as dose of fertilizer application is concerned (Yadav and Singh, 2000).Farm women adoption rate for use of farm implements was good when uses of such implements were demonstrated to them. (Sharma et al 2003).Farm women were given farm implements to use for a fortnight. Majority found them 78% profitable & 77% compatible. (Sharma and Sharma, 2002). Information from research to farmers via extension officers was examined.

The inter system communication helps in technology dissemination of dairy farmers in Iran (Rezvanfer, 2001). Rice cultivation awareness improved when women farmers were trained (Rama Krishnan and Mahendram, 2003). Agricultural Knowledge and information system increased economic efficiency of crop production, out put costs, social upgradation and income generation (Carrasco et al, 2003).

Extension Briefs – Self Help Options (SHO)

Small farmers are neglected by extension services and need proper awareness in Kurram (Haq et al 2004). Agricultural extension specialists must specialize in community structure, development , education learning tools, technology and policy issues. Thus these disciplines or topics need incorporation in Farm business Management curriculum. Partnership process, management and communication should be the main thrust. Co-operative mechanism, farm business management and evaluation and monitoring tools should be used.Seed production programme was successfully implemented by farm women as a self-help group in Dindigul district of Tamil Nadu (Balasubramanian and Thamizoli, 2003). Efficiency of extension lies in bottom up approach in hilly areas. Farmers interest group awareness was most successful (Muhammad-Zafarullah 2003). Small scale farming operations using green manure cover improved soil fertility. Participatory extension system helped more adoption of technology in Uganda. (Miiro et al, 2002).Farmers training and their own information networks helps small holder farmers. Topics including community involvement, international multicultural projects, leadership, learning strategies, new tools and technology, volunteer programme , programme development and evaluation, public policy issues, youth education and partnerships needs introduction in Agri-Extension curricula (now proposed as Farm Business Management). Agriculture, natural resources and health/nutrition shall form the basic of new agri-farm-business extension system (Logan, 2004). The word extension be replaced with "Farm Business Management".

Administrative Bottlenecks

While the Nusa Tenggara Timur Provience is supposedly following the training and visit model, agricultural extension Workers receive very little training. Other major constraints identified include constant restructuring of the Agricultural Department, poor remuneration, oppressive authority, inadequate infrastructure, unclear job direction, geographical conditions and bureaucracy. (Leta et al, 2005). This is the reason of low technology revolution in rural India and Asia too can be overcomed by transparent, NAR System, which need renaming as National Farm utility Research and delivery system. By combining biophysical science with social science, it is possible to increase the impact of biophysical science findings at the farm level (Linehan et al, 2005).

Isreal Model

General model of extension in SAU can be seen from the role of PAU, which needs a total reform. Should we look for Isreal model. Isreal is characterized by topographical variations and a wide range of soils, enabling the growing and raising of a variety of crops and animals. A large increase in population immediately after 1948, required a rapid increase in food production, followed by development of research and extension facilities and services. In recent years Isreal, like many other countries, has faced difficulties in continuing to supply the level of services it previously provided. This difficulty led agricultural organizations to begin assuming the responsibility of partially financing extension for a few years, and later on trying to reflect the cost onto the farmers. Farmers are apparently willing to pay for on-farm extension visits. Under this system, the Flower Board is financing many public extension activities and charging the farmer for partial costs of on-farm visits. This system operates successfully, aided by reports that are filed electronically by agents all over Isreal to the Floriculture Department Head Office, and from the Head Office to the Flower Board. Advantages of the new system are : a tremendous increase in agent-to-farmer commitment; the professional level of agents remains at least the same as before; and the financial situation of the majority of agents is better than before. The disadvantage is that some farmers are consuming less extension than before (Wolpson, 2005).

Global Successful Extension Experiences

California Model

California's land grant extension progamme has been successful in assisting agricultural Industries develop as major world producers. Extension has been a leader in facilitating quality-of-life and economic improvements for rural communities throughout the state. However, population explosion has transformed California into an urban state, with changing societal issues, values and priorities. Agricultural extension programmes continue to directly serve a declining number of clientale that represent less than 1.5% of the population. Financial support for extension is decreasing at the country and state levels, thus such models used by us in SAU could be catastrophic and need change.

Iran Model

Iran is one of the world's top ten horticultural producers. However, post-harvest losses in Iran are six times greater than the world average. Even after establishing an agricultural extension Service, Iran still experiences great problems in reducing post-harvest losses (Jowkar, 2005). Results revealed that education, material possession, mass media exposure, extension contact, scientific orientation and risk preference were significantly and positively correlated with the knowledge level of farmers pre- and post training. (Malite et al., 2005.

Issues

Agricultural development programmes often produce unexpected results. This can be attributed to the fact that the target farmers already have their own knowledge and competencies that determine their practices. In order to be adopted, an innovation has first to be discussed, and then appropriated by a local group of farmers in their system of knowledge and in their system of meaning; subsequently the innovation may be rejected or adopted by all the farmers concerned (Mathieu, 2004).

Problems

The findings also revealed that the attitudes of the respondents were significantly and positively related to education, organizational membership, land holding, contact with extension agent, mass media exposure, socio-economic status, awareness, income and material possession. Lack of proper technical knowledge, lack of irrigation facility and high cost of fertilizers/ chemical were indicated as the three most important problems of the tribal farm women run agriculture farms. (Bharali et al., 2004).

New Vistas

Agricultural extension is evolving world wide, and there is much emphasis today on community-based mechanisms of dissemination in order to bring sustainable change. The factors that make farmer groups successful in disseminating information and technologies need to be analyzed. A mixed methodology, multiple-stage approach was used to obtain data. Dairy-goat farmer groups (n=46) and individual farmers (n=88) were interviewed. Factors that were associated with group success in dissemination included member participation, degree of jealousy within the group, homogeneity of members, group capacity, number of linkages and type of group (Project-supported versus non-supported). Some interventions that may increase the success of groups in dissemination include capacity-building , increasing linkages with other extension stakeholders, providing an environment for groups to form, using established groups (as compared to forming new groups), and encouraging groups to form around common interests rather than for other reasons.

Technology Based Farming

We have to use more technology based cropping system to increase productivity per unit land. Horizontal expansion is not possible. Embargo on indiscriminate use of chemical fertilizers, pesticides and other farm-use-agents is another constraint to increase productivity of food grain. Unfortunately our food grain-production pace has declined. Growth rate of 1-2% has put pressures on our economy. Imports of food grains, an anomaly over past decade, needs problem-cause-analysis. The whole system of National Agricultural Research, extension and field functionaries have registered a fatigue. Similarly, the land degradation, mineral depletion and environmental pollution demands new mechanism to boost productivity. Perhaps use of electronic media, e-extension and agricultural reforms in the shape of revised KVK system, ATMA, village level rural development, self-help groups and agri-information Kioses needs introduction. State Agriculture Management and Extension Training Institute (SAMETI) may help to make adjustments in our farming system so as to integrate agriculture, aquaculture, water conservation and livestock rearing with new technology driven profit earning enterprises. This needs a continued and farmer friendly policies of sustainable agriculture. In these pages we shall discuss role of communication in agriculture technology application for higher productivity and profits per unit of land. Technology awareness and application is must to produce more per unit of land. We do have a chance as our yields are lower than many counties and even our neighboring countries (Samra Samra & sastry 2002).

Organic Farming

Organic farming is advocated as modern technology. We left our traditional organic farming for adopting chemical farming, which landed us in trouble. This rotation of modernity and traditionality taxed us heavily. We perhaps jumped in adopting or testing technologies without comparing them with our own practices. Now reverting back to our own traditional ways is fraught with problems too. The questions often asked are, can we sustain or even maintain our productivity levels by resorting to the organic farming? The Blending of modern technology of today and traditional technology of yester years, may answer this question. Can information and communication skills and technology bridge a new union?

Information and Communication Technology

The present day information and communication technology has trespassed all barriers of race, religion, culture and countries. A Comprehensive study of 23 review papers and a dozen book and journals was presented by (Wani 2005). A detailed description of how communication and information can help production and sustain yields has been discussed (Wani 2006). Strong warning systems for climatic risks, floods and cyclones, pests and mites could help to raise more crops. Farming informatics and awareness packages through print, mass and now E-mails is possible. The role of competitive farming, economic survey and evaluation of farming and women's integration needs attention. Women, the half of agricultural work force is still unawares of the technological skills. The barriers of customs, veil, religion and social bondages could be overcomed by educating them through TV, cassettes, e-mails or other modern communication appliances.

Phyto Sanitation

Technology transfer is easy. We can announce technology practices or even demonstrate them. The key issue is its adoption. India with 25% of its GDP from agriculture spends some 12% of GDP on its subsidies rather than on transfer of technology. Blending subsidies with agricultural exports will need a drastic cut under new WTO agreements. The global market access opportunity limit of 3% import shall further complicate the issues. The international standards of sanitation shall need more awareness at farmer's doors. Our Agricultural exports from 12 agricultural items has been up and now we export around 18.45% agricultural good in the shape of apiculture, floriculture, fresh fruits, mushroom, spices, sugar, molasses, rice, tropical fruit juices, pulp, concentrates and even agro-chemicals. Fruits, nuts and vegetables have increased our export earnings. Our limitations in expanding our exports are infrastructure to provide international biosafe packaging, phyto sanitation & quarantine measures. Our yields too are low to compete with others. Thus, transfer of technology has not to be limited to man methods, publication, leaflets, folders, bulletins, newsletters, journals, magazine, news paper publication, rural farm broadcasts or television interviews but has to be supplemented with video conferencing, massive awareness campaigns through video cassettes, cable net works and other local farm telecasts. The propaganda, publicity and persuasion has to be supplied with communication skills like rural journalism, popular participation, motivation and more so through management of information systems. The farm visits, farmers calls, letters have to be intensified. Farmers need information on markets, bio standards and marketing research and networking mechanisms.

Small Farmer the Small Holder

Another vulnerable class in India is small (holder). Over 65% of our farmers are small holders. The technologies generated are mostly for commercial farmers. The small holders have not only limited hold on land but on information too. Their case is further complicated as they do other work additionally as the small holdings is not sufficient to sustain them. They have limited access to knowledge. Even the word Bauern or Landliche Gebiete (German), peasant or country side (English) campagne, brousse/ paysan(French) and Campo/Campesino Spanish denote that extension work is a mission. One with love for country side and farmers alone could execute this task properly. He has to be well versed with objectives, problem, targets and implementation process. He has to make situation analysis and avoid group clashes in the country side. Thus, the knowledge and experiences of farmers, psychological values, expectations, needs and attitudes are to be organized. This organization and evaluation is not possible without use of modern and applicable communication methods. Welding communication with Agri -technology is the need of the hour.

Communication Methods

Passing on information to farmers is basic fundamental of any extension activity. The basic need for learning process is communication.Factors influencing communication range from person, his personality, his social relation, knowledge, the social and economic parameters of farmers. Their knowledge expectation, experiences and perception needs monitoring and evaluation before technology transfer.

Review of Communication Vs Agriculture Production

Critical Gap yield Reduction

In Kenya, use of advanced information and communication technologies reduced gaps in yields of Agricultural crops between research and farmer's fields. (Oguya and Bellamy, 2001). A country where 70% population is connected with agriculture for livelihood directly or indirectly and 80% of its export is agricultural oriented. Reduction in yield gaps through effective use of information and communication technology will have a significant impact.

Climatic Risks and Communication

The low productivity in Soyabean was found to be due to partial adoption of Production recommendations by farmers in Mahrastra, India. The low yield factors were analyzed. Economic constraints, situational factors and communication gaps on crop production, protection, seed treatment and fertilizer application were found responsible for it (Jaiswal et al, 2002).The modernization of the material and technical information base helped Cuba to increase agricultural production and rural development on a pilot basis (Albelo. et al. 2002).

Precision Agriculture

Precision Agriculture till date has focused on site-specific data collection for soil and crop management. The technologies for the site-specific field operations and automated data recording are available, but precision agriculture rarely involves them for improvement. The application of precision agriculture has to be clubbed with information and communication networking to harvest the gains and to improve productivity. This network may consist of an open software platform, which can be operated by the farmer himself. For efficient communication internet and mobile telecommunication have been identified as important components. The development of an information and communication network integrating modern software (Java, GIS) and hardware (GPS, internet) technologies in a new user friendly manner is necessary to achieve better acceptance of technologies and improved productivity (Lutticken, et al. 2000).

Satellite Data as Source of Communication

Use of satellite data- (Star and Spot - lite) helped time - critical dependant applications in Australia. The Australian centre for remote sensing (ACRES) has introduced a new service to provide satellite data for near real time applications. The STAR (Speedy Transmission after reception) service provides access to digital satellite data products in full resolution or compressed format within 12 hrs of a Satellite overpass. The data obtained from ground stations is processed at a facility via a high speed communication link and high priority processing. This system provides Satellite data on critical applications, like crop yield modeling, pre-harvest crop production forecasting, detecting crop diseases, monitoring crop stress, pest infestation, floods, fires and oil-spills. SPOT - LITE is a low cost, off the shelf satellite data product from ACRES that is ideal for use in Geographical information system (GIS). SPOT-LITE can be accessed at any time via the internet and is available in the form of tiles covering most of Australia (Thankappanm 2001). Advance studies with high applications for increasing agricultural production needs quick dissemination. The effect of rooting zone restriction (RZR) on vegetative and reproductive growth of fruit trees viz grapes, peach and citrus has been investigated. It is known that it improved crop productivity under low availability of water (Wang et al. 2002).

Communication Networking

The available communication facilities for agricultural information in 15 states of India were studied. (Ghosh 2002). The results suggest that while communication networking opens up agricultural economy, it is not cost effective. The communication networking has to become cost effective. An attempt to have better communication between various forest research divisions and other organization interested in sustainable forestry have shown encouraging results (Barbour and Wong 2001). An attempt was made to have quick information flow among and between researchers, extension officers and dairy farmers in East Azerbaijan, Iran. The information input, output and intersystem communication were studied. The communication linkage improved the productivity (Rezvanfar,2001). The basic tools of marketing premotion of fertilizers in India was studied. (Yadav, 2002). The information like advertising, public relation and personal selling was found to be best promoters. In the "Unique Selling" approach of the communicator effects target audience. The impact of integrated approach utilizing computers in agricultural information & dissemination in Greece and Poland was studied. (Tzortzios et al 2001). The gap in technology known and applied at farmers field was found. Researchers lack training in using new information technologies. Thus improvement in Agricultural productivity has to keep pace with advanced communication and information technology using computers. Joint problem solving sessions and communications networking of information on agriculture has improved sugarcane production in Mauritius (Jhoty et al; 2001).

Technological Prospectives

The technological gaps between research yields and Farmers yields are wide. The technological know how generated over years is still on shelf. Much of it has remained frozen if not dead within the fore walls of institutions. The location specific, crop and soil specific innovations to farmer's practices are few. Our integrated & mixed farming practiced by our farmer's needs improvements which is not forth coming. Our research institutes mostly aped the experimentations on commercial farming of west with few modifications and even additions at times. Our traditional agriculture existing still in remote areas encompass the use of farm yard manure, vormiculture & use of agricultural waste materials. The present problems of low soil fertility, reduction in yields and soil salinity have arisen due to excessive use of chemical inputs in farming with little care of the soil. This scenario resulted due to incomplete innovative approaches of research. Our researchers blindly advocated more and more use of fertilizers, pesticides and fungicides, which helped to gain grain revolutions, but left legacy of polluted water, air & environment with degraded soils. These revolutions debarred future sustenance. This was due to poor perception & vision to fore see the net results on long time frame.The immediate need is to make researchers akin with information technology and advanced communication. The rapid evolution of information science demands quick and speedy transfer of technologies, awareness & even subject reviews to farmer's for speedy application. The productivity would be better if technological advances are adopted and their impact is known. The knowledge of computer hardware, software as well interlinking the information dissemination channels and outlets is essential. A new concept of Farmer school with communication system can resolve 70% problems of farmers on spot. Kissan call centre (1551) and the Zraye Paigam shall widen the messages.

Decision Support system

The globalization of agriculture marketing has put forth new challenges. It is not only the yield which matters, but the benefit / cost ratio. The economic viability of an agricultural product has now more importance. The product even if economic and globally competitive has to pass quality tests. The international phyto sanitation standards are going to be tougher. The emergence of diseases like Mad cow disease or even SAARS being linked to animal cultures and other diseases from Agricultural & Livestock products warrants biosafe agricultural produce. The decision have to be supported with information not only on marketing, international safety stands but also on a law regulating agencies and regulations too. Unawareness of these support system could lead us to bankruptcy. Thus a strong case for tools and information management comes into force. The international computer networking and communication systems alone could help in decision making for appropriate and economic viable agricultural production. (Tzortzios et al 2001).

Intelligent Agri-Management

Now we do talk of ecologically based pest management instead of integrated pest Management (IPM) . The buzz word for future in this regard is intelligent pest Management (IPM). As the system now advocated is not the blanket sprays of chemical pesticides but intelligent and well computed programmes of pest management, which incorporates their safe and long term application, economic viability and biosafety of products. Perhaps a well documented interaction between farmers, extension education workers, researchers and policy makers is necessary. This would take years if communication methodology of video conferencing, internet and related satellite technological advancement in communication is not used. This is what would be future intelligent Agricultural Management.

Participatory Management

The whole system of Agricultural research, teaching and extension in NARS needs renewal. A composite ARS system has to be introduced with strong basic of communication and IT. All programmes in the field should be participatory. All agencies be it ICAR, CSIR, RRL, Universities, SAU or other institutions and industry have to be webbed to getter with extension delivery systems in the field. More participation of farmers in planning and execution of project like mega seed or mission horticulture is needed. A true transparent and open system of selection both for Management positions and scientific positions should be advocated. Changing standards for individuals is a crime. Those at top have to exercise truth, justice and fairplay. The role of politics and politicians should be minimal in national extension delivery system. An outreach through communication and IT is the only achievable solution.

Our Vision

Mixed Farm University Culture

Higher productivity gains can be achieved through application of technology and production recommendations at farmer's fields. We have 65% small and marginal farmers whose awareness potential is low. The production system prevailing with these farmers is a mixed farming or composite farming. In contrast to USA and European agriculture our necessity is to increase "Crop -livestock-fish-plant integrated production system with multiple livelihood opportunities". Therefore, we need our own innovative educational and training policies. A mixed agriculture University and Education set ups is our necessity. We are at present going astray to our need. Quick and fast measures are needed to unify our educational system, involving all agriculture and allied disciplines, industries, corporate sectors and farmers institution.

Higher Productivity Concerns

Indian Agricultural pride years of green revolution post 1968 saw reduction in food gain imports and subsequently white, blue and other revolutions sustained our population pressures and agriculture growth. Our agricultural growth rate (AGR) need to be equal if not more to population growth rate (PGR). Our AGR target ought to be double the PGR. This is important as consumption rates, purchasing power and employment prospects increase. An estimated food grain of 210 million tons at present may need to be doubled in next 10 years. We have to achieve high targets of productivity by vertical expansion as horizontal land expansion is just not possible. Dr. M.S. Swaminathan has quoted figures as of 160 million tones of rice from 40m hac of land, thereby setting the productivity target of 4 t/h. Like wise production of 100 million tones of wheat from 25 million tones of wheat from 25 million hectors needs a productivity of 4t/hac. Our aim to double our per hac productivity needs more technical manpower in extension, industry and at gross root level.

The climatic disasters, earthquakes, Titanic tsunami, floods; have effected our agricultural production in the past and additional requirements needs to be kept in mind while planning food security. We need to increase per capita consumption expenditures of Rs.600 per month. We need to bridge the gaps between potential and actual yields at farmers level. The chemical farming hazards of poor soil fertility, low water availability, pollution and environmental concerns impede our agricultural development. Thus refined technology, participatory research and educational modules are needed. The new pressures of Global marketing. World trade and tariff regulations have to be accommodated. This all will need incorporation of new themes like post-harvest management, value addition, packaging, communication, credit and market information services in our course curricula. Thus a new multidimensional change in academic curriculum is envisaged.

Quality Assurance

We feel pride in calling ourselves as the 2nd largest Agricultural Research system (ARS) in World. When we review our performance we are no where in top ten of most cited agriculture publications in the World. USA tops the world list with 3,62,79,842 cited publication/annum, with small country like Switzerland at No.10. The scientific out put in agriculture is highest in USA with 27 lac publication/year followed by Japan, Germany, U.K, France, Canada, Italy, Russia, China and Australia. Our contribution to Agriculture publication is 5.48% only with our share of citation at 2.32%. This demands more focus on Quality Assurance. Our prime agenda should be quality Agricultural Education. Our emphasis has to be on: Academic quality, Accreditation; Desired knowledge, Assessment, Skill and competence building and academic audit. Quality assurance, means strengthening resources, information and maintenance of educational infrastructure. Thus we need to regulate grants and centre-state relations rationally.

Employment Opportunities

We have 45-50 state/deemed or central agricultural Universities and 200 general universalities with 48 agricultural faculties. The total disciplines needing grants may be strengthened in 5 yrs by 1 core grants to each discipline for quality assurance. We produce 10,000 under graduate 5500 post graduates and 1600 Ph.D in agriculture every years . They add to our unemployed pool. For making them self employees in new ventures and for increased employment , they need competence building in Global economics and trade policies. Biotechnology, Bioinformation, Biofertilizers, pesticides and fungicides. New faculty development in all the Universities and colleges is to be executed in coming 5 years. Such as: Pest information and survey; Risk Management Analysis; Decision support system; Geographic information system. A new trust is to be given to course curriculum integrating field practices in a partnership mode with farmer. A teacher-student-farmer-industry, interaction and co-operation is to be developed. A new model of mechanics in Agricultural and allied curricula is to be integrated, unified and fine tuned to end results. This will demand inter and intra faculty harmony and synchronized course curricula at UG, PG and Ph.D level. This has to be fine tuned to our field requirements and location orientations.

Asia Specific Agriculture Association

Indian Economy is a agri-centre economy which supports 70% of our population, as direct rural employment . Forty five percent of the income generated by industries comes from Agri-based (Agro) Industries. Therefore , a vast potential and resource is hidden in it. If we think of Asia specific Agriculture, we have to play a significant role in the region which has 60% of the world population. The region is rich of energy and oil resources and millennium buzz word is open boarders and common market with first priority on peace, confidence and trust. For up-liftmen of region we need training education and human resource utilization. Therefore, Agriculture educational reforms are on our door steps to harvest the gains of common economy in the SARC and total Asian region. These educational reforms should involve schools, colleges and Universities. The re-modeling of curriculum will need incorporation of new emerging era, like competitive global marketing, the climatic, disaster, technology use, restrictions, sustainability, environment, water resource conservation, remediation factors etc. Agri-educational reforms are needed so that ICAR parallels USDA, in governing grant-in aid to whole agriculture sector. An omnibuss act of agriculture in the shape of USA farm bill of 1996 is envisioned. New educational policies so drafted shall be non-discriminative, comprehensive, transparent and accountable.

Access to Education and Training

Access to education and training to people below the poverty line, rural youth and women is to be ensured. This will need a total restructured education infrastructure. A three tyre model is envisioned which consists of: 1. On the job, training opportunities on farm mechanization and agriculture. 2. Training skills, up-gradation and rural orientation at University level, refinement and more innovative participatory mode at farmers field. Roaming teaching taught system on holidays and Sundays. 3. Teacher-student-farmer-industry-interaction-work plans-self learning by living with farmers. It will ensure quality training and job improvement of skilled manpower for use in Asian Agri development Market. 7. Informal-flexible Agriculture Information Services: A flexible curriculum models which should have many options at B.Sc level like:

1. Natural science
2. Agri Science
   - Production system
   - Agri-business
   - Social science
3. International Agriculture
4. Natural Resources
5. Agri-business management
6. Biological engineering
7. Dietetrics
8. Landscape Architecture

Export Orientation in Agri-education

In the present era of bio-safety, phyto and zoo sanitation have assumed tremendous importance. Education and knowledge was safe and secure treasure in the past. It is no more true. The export needs knowledge and new inventions need patenting. Web and web designing have made invisible teachers to unknown students. The students-teacher relationship has raised to spiritual horizons. New targets for future educational planning and policies need to have more information and communication technology. Therefore courses on I&C with computer applications is must. These have been strengthened in the SAU and ICAR institutions in the last few years. However a total connectivity is needed with farmers, farm organizations and utilization departments to harvest the gains of technological reforms to increase our exports. Inspite of ranking I in milk production our exports are meager. This is because of poor -zoo-sanitation and Global lobiest are critical of our disease free status. Therefore, ICAR has to formulate policies and programmes in Agri-export orientation.

Summary

Technology akin to farming existing in an ecosystem need to be generated. It needs location specific perfection and adoption. This could be possible by supporting KVK & Extension system with technology testing and refinement Infrastructure and Resouces. Thus the whole discipline of Agricultural Extension needs reforms. On the basis of a broad review presented in this paper, following recommendations could be deduced to reform and fine tune technology generation, dissemination and its delivery as an agri-pack end product. (APEP)

1. The base of discipline of Agriculture extension may be broaded to incorporate all disciplines, management tools and techniques. It is proposed to name it **"Farm Business Management"**. It should be open to all agricultural & allied undergraduates. The course curriculum needs training and skill up-gradation of graduates and post-graduates in techniques like analysis of data, survey procedures, marketing, economics, rural sociology and farming systems,etc.

2. Various reformation capsules like Chinese model, African experiences, Dutch model, Netherland model, Isreal model, California model and Iranian models have been reviewed. On the basis of these models a new model for technology delivery suitable to indian agro-eco-conditions has been postulated & named as **Mohi-Model.** Once we receive a feed back for the model, logistics and plans to execute it could be formulated.

Suggestion on Agri-eco-Reforms or Policy Change

The reformation towards agriculture which is economic, profitable & enterprise oriented needs:

- Training
- Education and Self
- Employment Opportunities
- This can be possible by adequate knowledge of mordern and traditional agri-practices and there is need to use the following key factors of agrocultural challanges for managing agriculture as an economic

business, rather than as it is inventment which keeps raising land cost as security than its commercial in come.

- Competitive global marketing
- Climatic disaster management
- Technology use and perfection
- Environmental restriction
- Water resource conservation
- Remediation of all effects through excessive use of chemical fertilizers, pesticides & fungicides
- ICAR-USDA Combine is proposed to make ICAR a thinck tank rather than a conglumeration of non-farm actors or spectetors, who never venture to raise any crop or animal agriculture concern is needed. Omnibus model of reforming agri-eco-gains in US, invested on reforms in agricultural education, training, extension & research. What is executed in EU or US 2 decades ahead is aped by us now. Why should not our 2nd biggest NS thinck for itself and bring to some bottom up research modules rather than coping down USEU top down gear in research.

❑ Omnibus agri reform like Model for ICAR-NARS.

I. Research Management and Innovative Composite Packages

i. ICAR research results and end products are directly applicable to farmers. Most of research is commodity based research. Our farms are mixed or composite farming units. They need a wholesome package for their crops, vegetable, cows, poultry and other related trees and plants. None of the ICAR or SAU institutes have come up with such research packages. I would suggest to formulate a group of scientists, extension education personnel and farmers to propose new research projects on the theme proposed.

ii. The attempt to formulate transferable technologies and strategies, the whole national agricultural research system be named as "Farm Business Management and Research" with more emphasis on end results and delivery system. We must involve farmers' field and known practices to evolve result applicable to specific, regions location and ecosystems.

II. NARS (National Agri Research System)

Top priority in ICAR/SAU is given to research for teaching. Most of the funds are spent on tuning research. A review of scientific achievement of last six decades could not fetch Indian Agriculture a single noble-price. This speaks our weakness in planning and execution of research at ICAR level. Is it that the ARS System with emphasis on merit, innovation and ability has changed to some other criterions? I came to know that ASRB standards in assessment have been changed quite frequently. More emphasis is given to oral tuning (Viva) than service, publication or research achievements. We except from a system next only to USA to produce scientific inventions even half as that of USA. On internet, I have checked our contribution (Agri. Research) is less that 2% in citations. This is a serious concern.

III. Leadership in Research Institutes and SAU

The pattern of appointment of Directors, Vice-Chancellors of ICAR institutes, SAU has been quite surprising. An individual is appointed many times after retirement in ICAR and in SAU. This leads to giving such assignment to a favoured few, depriving merit and efficiency. Many of the SAU directors and scientists with equal merit do not find leadership assignment in ICAR institutes. Conversely, when he or she deserved the leadership of SAU, a retired person is either appointed or his tenure is extended. Is this the cause for our negative growth in ICAR.

IV. Reforms Needed

a. Top most priority is to be given to extension education. A new role model needs to be evolved where products of research reach directly to the farming community. I would recommend renaming of the extension education system as Farm Business Management, so that research, management of farms, farmers fields, industry set ups, NGO, extension functionary are directly linked to deliver results. This would mean research on specific locations as farmers' fields. Now that farm insurance is in vogue, we may venture to experiment at farmers' door.

b. SAU/ICAR location-specific research units need to be linked to KVK's to deliver research results.

c. ARS assessment system, emphasizing merit, work and efficiency needs to be revived. The fresh system of career advancement needs further improvement.

d. The direct selection process in ICAR/SAU needs to be improved. There shall be more marks for work done achiements and no oral viva. We must introduce an open presentation for selection of leaders and vicechancellors.

e. The farmers should write the APR's of scientists, directors and Vice-chancellors. Those popular with farming community need to rewarded.

f. Reward and award system in the ICAR and SAU should be transparent.

Area Under Cultivation

S.No	District	Area Sq/Km	No.of villages	Towns	Gross cropped area
1	Anantnag	3984	645	8	103
2	Pulwama	1398	554	4	79
3	Srinagar	2228	175	3	29
4	Budgam	1371	496	1	58
5	Baramulla	4588	660	60	90
6	Kupwara	2379	369	2	46
7	Keh	82668	113	1	9
8	Kargil	14036	129	1	10
9	Ganderbal				
10	Bandipora				
11	Kulgam				
12	Shopian				

000hac

District	Total cultivated area	Net Shown 000 hac	Irrigated 000hac
Anantnag	120	80	49
Pulwama	98	54	33
Srinagar	51	24	17
Budgam	78	53	
Baramulla	144	84	40
Kupwara	67	45	22
Leh	45	9	9
Kargil	19	9	78

Fruit Productivity

District	Area 000hac	Average production MT/hac	Apple T/hac
Anantnag	36	5.7	10.2
Pulwama	29	5.1	8.1
Srinagar	10	4.8	8.4
Budgam	22	2.7	5.9
Baramulla	30	16	18.8
Kupwara	27	2.5	4.2
Leh	1.1	5.4	6.9
Kargil	1.4	1.4	3.3

Area Under Crops

Crop	000hac /area	Productivity
Rice	275	2.1 t/hac
Wheat	247	1.5 t/hac
Maize	385	2.9 t/hac
Veg/fruits	957	
Apple		10.09
Pear		3.68
Cherry		3.00
Walnut		1.39
Almond		0.57
Total		**4.95**

Productivity Fruits t/hac

Fruit	T/hac
Apple	10-12
Pear	2-3
Cherry	3-4
Walnuts	1-1.5
Almonds	0.25-0.5
Total	**4-5**

District Statistics

District	Land holding size in hac
Srinagar	0.36
Baramulla	0.64
Anantnag	0.47
Pulwama	0.69
Budgam	0.57
Kupwara	0.64
Leh	1.41
Kargil	0.77
Jammu	1.09
Udhempur	1.30
Doda	0.93
Kathuwa	1.35
Rajouri	1.29
Poonch	1.22
Total	**0.83**

Area under Different Crops

District	Rice	Maize	Wheat	Veg/Fruit
		000 hac tones		
Srinagar	13	5	0.02	24
Baramulla	32	25	0.48	81
Anantnag	42	22	0.03	75
Pulwama	30	12	0.41	57
Budgam	28	14	0.64	51
Kupwara	17	22	-	46
Leh	-	-	3.36	8
Kargil	-	-	1.37	7
Jammu	58	20	99	116
Udhempur	10	49	40	103
Doda	04	51	07	74
Kathuwa	31	17	47	103
Rajouri	06	44	38	90
Poonch	04	24	11	34
Total	**275**	**385**	**247**	**957**

District	Area 000 hac			Falling	Land fruit	Barren conc.
	Total	Pastured	Net show			
Srinagar	51	3	23	2	8	13
Baramulla	144	13	83	1	15	24
Anantnag	120	08	79	1	13	22
Pulwama	98	09	55	1	12	16
Budgam	79	07	53	1	8	9
Kupwara	67	07	46	-	5	9
Leh	45	01	110	-	3	28
Kargil	19	-	9	-	1	5
Jammu	326	11	107	40	31	70
Udhempur	431	11	28	192	79	108
Doda	411	07	61	219	26	68
Kathuwa	265	11	**58**	73	45	88
Rajouri	233	20	**51**	94	34	72
Poonch	114	18	**28**	37	11	27
Total	**2416**	**126**	**733**		**291**	**582**

Live Stock Statistics in Lacs

District	Cattle	Buffalo	Sheep	Goat	Total
Anantnag	3.7	0.07	3.3	0.3	7.6
Pulwama	2.7	0.06	1.0	0.25	4.9
Srinagar	1.2	0.02	1.4	0.15	3.8
Budgam	2.0	0.01	2.5	0.40	6.3
Baramulla	3.8	0.01	1.6	0.40	4.8
Kupwara	2.4	0.06	0.8	0.60	2.5
Leh	0.3	0.17	1.8	1.20	3.6
Kargil	0.4	0.2	0.8	0.9	1.0
Jammu	4.4	2.8	3.7	2.5	10.4
Udhempur	2.9	1.3	3.9	2.4	9.2
Doda	3.3	0.5	2.8	1.4	9.6
Kathuwa	2.1	0.7	2.0	4.0	6.8
Rajouri	1.3	0.9	1.9	2.5	4.8
Poonch	0.9	0.7	-	2.0	3.6
Total	**38.6**	**7.3**	**30.5**	**19.0**	**91.4**

KVK	Fields	hac	Trainings	Farm beneficiaries
Budgam	951	177	179	3531
Leh	253	42	308	8047
Pulwama	392	93	418	8459
Srinagar	372	78	156	3679
Bandipora	694	151	86	1541
Kupwara	326	52	63	1517
Kargil	356	22	179	4885
Pombay	572	104	73	1976

Yield/Kg/Kanal (Paddy)

Year	Kashmir	Jammu
1970	110	65
1980	125	75
1990	130	80
2006	250	85
Maize		
1970	65	85
1980	50	65
1990	50	100
2006	60	100
Oil seeds		
1970	40	
1980	40	
1990	45	
2006	50	

Paddy/Kanal/in Kgs

2003	260	1976	140
2004	240	1980	205
2005	250	2000	-
2006	250	2005	300
		2007	450

Wheat Fields Kg/Kanal

Srinagar	140	100
Baramulla	156	107
Anantnag	150	107
Kupwara	158	107
Pulwama	145	102
Overall	146	104

Accumulation of Metabolic Products

Molybdenum deficiency – nitrate accumulation.

Citrus- magnese deficiency causes xylose accumulation

Magnese absence – differences in quality sugars (clentoses)

Harmony and balanced fertilizers- soil maintenance crop/plant growth

Vegetable Production q/kanal

Vegetable	Experimental q/ha	Actual
Tomato	47	25
Brinjal	47	23
Capsicum	63	28
Okra	13	04
Chilli	10	02
Potato	15	10
Carrot	12	11

Gross Domestic Production

Dairy Contribution

1950-51	:	15.5%
1980-87	:	25.0%
1997-2007	:	27%

New Approaches

a. Plants deficiency syndromes are latent until critical symptoms occure. We must thus detect the plant /crop hidden hunger and not wait for disease symptoms to have remedies. Thus a new approach of diagnostic, agricultural and horticulture chemistry is needed. We have been using foliàr testing diagnostics on payment basis and other AMC realized Rs 80,000 as fee during first year.

b. Nutrient indexing
   Active nutrient
   Critical limits
   Total elemental analysis
   Atomic absorption spectrophotometer

c. Biochemical indicators
   Micro nutrients- catalyze
   Metabolism
   Enzyme system
   Rib nuclease activity index of zinc hunger.

References

Badri-OA; 2001. School of Rural Extension Education and Development, Ahfad University for Women, Omdurman, Sudan. 18:1, 35-50; 20.

Balasubramanian and Thamizoli, 2003, Social differentiation in the horizontal transfer of knowledge: a case study from south India. *J. of Agri.Edu.and Extn.*2003, 9:2, 51-60; 3.

Bharali et al, 2004. Influences of extracellular calcium and iron on membrane sensitivity to bisulphite in the mosses.*Jr. of Bryology.* Vol.26No.1 of March 2004, pp 53-59

Carrasco et al, 2003, Absorbing the shocks: the case of food security, extension and the agricultural knowledge and information system in Haryana, Cuba.*Journal of Agri. Edu.Extn.* 2003, 9:3,93-102;9.

Davis et al, 2004, Extending technologies among small-scale farmers in Meru, Kenya: ingredients for success in farmer groups.J.ol.of Agri.Edu.and Extn.2004; 10 (2):53-62.

Ganesan et al, 2003,Reaching the unreached: basics of extension education, Tamil Nadu Agri.University, Coimbatore -641 003, India. Reaching the unreached: basics of Extn.Edu. 2003:xii+387.

Kaine and Beswell, 2005, An innovative approach to irrigation extension in horticulture. *Acta -Horticulture.*2005; (672): 177-183.

Khalache and Khaire, 2007, Social-ersonal and socio-economic characteristics of the fig growers and information sources used by them in fig cultivation technology. International J. of Agri.Sci.2007;3(1):258-20.

Kibwana- OT, 2000. Bringing together local and outside knowledge for innovation in land husbandry in Tanzania and Ethopia - a comparative case study. Jr. of Agri. Edu. & Extn. 7:3, 133-142.

Kumar et al, 2006, Adoption of cultivation and post harvest technology of tomato by farmers., Karnataka. *J. of Agri Sci.*2006;19(1); 76-79.

Linehan et al, 2005, Combining biophysical and social science to achieve change through targeted extension. *Acta-Hort.2005L672* :211-216.

Logan, 2004, Galaxy II Conference, Exceeding expectations through eamwork, Salt Lake City, Utah, USA, 21-25 September 2003.

Leta et al, 2005, Human Resources and scientific output in Brazilian science. Jr. Aslib proceeding, 0001 - 253X. Vol.57

Leeuwis,2000. Reconceptualizing Participation for Sustainable Rural Development. *Agri. J.* 2000.

Malite et al, 2005, Relationship of farmers' traits with knowledge of rapeseed-mustard production technology. *Jol.of Oilseeds Res*, 2005: 22(1): 159-161.

Mathieu, 2004, The meaning of practices: farmers; conceptions in agricultural development strategies,*Jol.of Agri.Edu. and Extn*, 2004; 10(3):101-109.

Matteson-PC, 2000. Insect pest management in tropical Asian irrigated rice. *Annual Review of Entomology*, 45: 549-574; 164.

Muhammad Akram, 2003. Main factors affecting extension activities. *Sarhad Jr. of Agri.* 19:1, 163-168; 7.

Muhammad-Zufarullah 2003, Improving extension efficiency through bottom-up approach: an evidence from remote areas of Northern Pakistan. *Sarhad J.of Agri.* 2003, 19:4,591-594;5.

M.M.Jowkar, 2005. The Importance of Horticultural Extension in Reducing post Harvest losses. IV International Symposium on Horticulture Education, in Iran.

Prasad-RB; Singh-S, 1999. Constraints in adoption of bio-gas technology. *Jr. Applied Biol.*, 9:1, 100-102.

Puran Chand et al 2001, Constraints of the farmers in the use of improved wheat technology. *Himachal-Journal of Agri.Res.*2001, 27:1-2,95-102;7.

Rama Krishnan and Mahendran, 2003, Impact of TANWA training on knowledge gain of trainees about paddy cultivation practices.*Journal of Res*-ANGRAU, 2003, 31:3,68-71:3.

Rani and Reddy, 2002, Knowledge and adoption of farm and home technology of farm women. *Journal of Research*-ANGRAU,2002,30:2,101-104;6

Reddy, M.V; 2001, Relationship between number of technologies developed at NARP stations of ANGRAU and their adoption. *Journal of Res.* ANGRAU, 29:1, 42-46.

Reinders, Hans Peter, 2007 Learning together for organic farming India. March 2007, 9(1): 9-11.

Sharma and Sharma, 2002. An overview of the field of family business study, current status and directions for future. A handbook of research on Extension.

Sharma et al 2003, Gain in knowledge and attitudinal change in hill farm women regarding drudgery reducing farm implements through action research, CCS Haryana Agri.Univ.Hissar . *Annals of Agri.Bio-Res.*2003, 8:1,81-82:3.

Slavik, 2003., Information systems in Czech Agriculture:implications for research and the study of agricultural extension. *Jol.of Agri.Extn.Edu.*,2003:94(4):177-184..

Souvik et al, 2005 Appropriateness of recommended agricultural water-management technologies as perceived by the personnel of research extension system: a study in the eastern region of India. *J.of-Agri.Education and Extn* 2005; 11(1/4): 17-26

Thusoo and Sohi, 2000, Extension contacts and mass media exposure of the farmers in relation to the use of improved rice technology. *Himachal-Journal of Agri.Res*, 2000, Publ.2001, 26:1-2, 137-142.

Vasantha and Buchareddy, 2006, Perceptions and attributes of IPM technologies by cotton farmers of Guntur District. *J. of Res.ANGRAU*, 2006;34(2): 53-61.

Wani, G.M.2007. Success Stories of Agribusiness / Agri-Clinics - a tool of Agri Entrepreneurship held at Department of Extension Education, Institute of Agri.Sci. Banaras Hindu University, Varanasi.

Wani, G.M.2007. Recent Trends in Policy initiatives and technological intervention. Goat Production as Livelihood resource. Lead paper presented at Sri Venkateshwar Veterinary University, Tirupati, 20-22 June, 2007 pp.

Wani, G.M.2007. Communication for Sustainable Agriculture Production. Proc. Communication conference held at University of Hyderabad, Nov. 2007.

Wani, G.M.et al 2000.Higher Education and Human Resource development : Knowledge perspective in current millennium. Proc. Int. conf. on Higher Education for human development. Published AIU, New Delhi. Feb. 22-24, 2000. PP239-247.

Wani, G.M et al 1999. Integration of NSS in Agricultural Universities. Invited paper, Proc. Symposium onm NSS, University of Agric. Sci. Dharward, Karnataka 27-30 July, 1999.

Wani G.M; 2006. Self sufficiency in Rice Production in Jammu and Kashmir. Presented for publication in magazine on world Food Day,2007 at Directorate of Extension, Krishi Vistar Bhawan, New Delhi,

Wani, G.M. 2003. Appraisal of institutionalized Farmer's Training Activities. Lead Paper: National Seminar on Extension Strategy for Promoting Development initiatives among farming community, held at G.B.Pant.Univ of Agric. Sci & Technology, Pantnagar, Uttranchal, June 18-20, 2003.

Warren, 2003, SWARD: participatory knowledge transfer at work. *Journal Agri.Edu.Extn*, 2003 9:1,43-50; 21.

Yadev et al 2003, Socio-economic constraints in technology transfer of arid legumes, Hisar, India, 15-16 May 2002. Advances-in-arid-legumes-research,2003 497-500.

□□□

Chapter 30

Success Stories of Agribusiness/ Agri-Clinics-A Tool of Agri Entrepreneurship

G.M. Wani

Introduction

Agricultural development over years has been the result of continuous agri skill generation and its popularization. The earliest agriculture was animal domestication over thousands of years ahead, man domesticated wild fowl, dog, goat and smaller animals, whom he could overpower easily and subjugate to his sub-ordination. Agriculture thus since beginning has been the results of trails, experiments and experiences over years, learned first though behavioral changes, psychic reoccurrences, memories passed through parents to children and later on through doing and learning and now through sharing experiences and writing them or dotting them as an Entrepreneurship concern.

Entrepreneurship Concept

The Entrepreneurship adds economic profits and cost-benefit ratios to Agricultural Output. Entrepreneurship is dominated by four factors like:

a. Social systemic changes
b. Support system availability and use
c. Resource base and its utilization
d. Self confidence, exploration work capacity and intellectual potency.

An entrepreneur has to have a thinking of his own, a capacity building interest in acquiring needed technique. An explorative and analytic faculties to judge the way of procuring cheap raw material. He must be equipped with "knowledge" and mindset to use and benefit out of it.

Farm Business

A potential entrepreneurship must strive from getting maximum output. Decades back agricultural development and industrial setups was a public sponsored and heavily subsidized but over time "knowledge" explosion in India Agriculture, have brought us on threshold of a system, where wide distances exist between industry and farm business. Where huge subsidies are benefiting Agro-Industrialists. The Farmers who use fertilizers or agro-chemical are crushed under economic pressures. The gaps between technology generated and technology use at farmers door is increasing day after day. The farm technology adoption rates are not more than 20-30% by any higher prospective. The use of information and communication technology (I&CT) for reducing the gaps and increasing productivity is the need of the hour (Wani, 2005). The modern technology and knowledge flow is fast expanding and bringing change. It demands more educated and trained farmers. Our education system has produced more literates but not educationally trained youth to earn their own bread. They after attaining graduation in agriculture and allied sectors, beg for job. The system has to be corrected to make these graduates as employers and not employees. I wrote a treatise as back as 1992, emphasizing a system. Germans are smart to have Farmers school, Farmer business training institutes, practical agri-farmers training centres and like, where way farmer or animal husbandry man is essentially a trained fellow. The banking system is so organized that they are on the door of convocation hall to sell their agri-business and agri-clinics to graduates, without any personal investments. Banks are so smart, that they have surveyed the villages who need vets or agri-graduates or have attained land and all facilitation, so that agricultural or veterinary or even other medico-biological graduates are used as bank investment. This is what is envisaged in India under agriclinic, Agribusiness venture. We have trainings not in the hands of banks but universities.

Success of Agri-business

A systematic liaison and support system between Govt. banking and University culture has made this otherwise an remunerative and lucerative programmes into a failure in spite of its personal monitoring of PMO. The success of Agri-business and Agri-clinics success rates are shown in table 1. The universities involve and their success stories are shown in table 2. Both these details are distressing inspite of huge moral, financial support from Govt. This is inspite of subsidiary support table 3.

Various ventures are listed in table 4-7. Agricultural professionals are getting converted into Agribusiness and agri-clinical experts. More than 14,000 applicants and 615 agriclinics came to existence in Indian 12 states. The agri clinic trained persons in J&K many number in hundreds. Among them 34 have registered agri clinics earning a handsome profit annually.

We visited Bandipora district and unregistered Agriclinics were earning a handsome salary, more than the Rahbar-e-zerat or Agriculture Asstt. A visit documentary is enclosed and shall be shown. It consisted of Agribusiness viz sale of pesticides, cattle feed, poultry feed and agri-extension services. At a small village in Papchan, one agri graduate Mr. Iqbal Shah earns Rs. 10,000/ - per month by selling the services and input. At a distance of few kms. In same district one Mr. Khyatlani owns a big poultry farm and earns around Rs. 20,000/- per month. Both these entrepreneurs employ 2-3 persons at present. Similarly, the success shown by one Mr. Shah at Malangam in Agri products and pesticide sale and one Mr. Bhat in Dairy production and milk product sale earn a handsome income besides generating employment for poor.

Farmer as Entrepreneur

Indian Farming and farmer has to change if proper WTO recommendation and GATT agreements are to be followed.

The present day poultry scenario has emerging high profile agri-business prospects in India.

The conversion of poultry farmer's into poultry entrepreneurs shall make the present day 6% contribution of poultry products to 25% share of Global market from India and China. This when translated into action shall increase employment generation by manifolds. The introduction of rural based Vanraja, Gramapriya, Giriraja, Cari Gold and vast other locally grown varieties of poultry have adopted well to our agri-rural base. The market acceptability is higher than exotic poultry concerns. Research to farmers doors in generating free-rang-poultry is like BT cotton hybrid spreading through villages of India and assuring high returns and exports (wani, 2007).

Poultry as Agri-Business

Dr. Gordon Butland, president of Global poultry strategies presents "Backyard poultry production" as a tool of alleviating poverty and malnutrition. We have tried to distribute "birds" under free-rang system in all our KVK's our results were excellent and income generation was totally in favour of the Agri-business and agri-clinics as will be shown in case histories and success stories.

A grand show of using poultry, rabbit meat processing introduction at SKUAST-K have innovated white meat usage. Our own preparation could be seen in Figs 1-3.

This all will need the involvement of Agri-Veterinary and food processing technocrats to develop rural-based establishments so as to faster export and fast returns.

Holistic Vision for Livestock Enterprise

Improving income, employment and self-reliance are among educated graduates and un-employed youth especially women needs fostering community development, women empowerment, environmental protection. Rural-based backyard poultry subscribes to all these norms and could be a rich resource for developing agri-entrepreneurship. Govt. of India is liberally financing such agri-business ventures and a proposed infrastructure cost set-up can be seen in table 8. A vast and finance assured schemes are available for agri-graduates for establishing poultry ventures (table 9). An initial allocation of 107 crores for initiating nucleus breeding farms. Further more provision of hatcheries to provide chicks to more than 2 lac farmers and farm women will need many agri-business centres for providing basic germplasm, medicine and above all training.

Some of the success stories in animal husbandry section can be reproduced as follows:

a. Backyard Poultry and Incubation

Though the Vanraja are the most suitable for back yard poultry, they do not have habit of broodiness. There is a problem among the farmer to get a broody hen in all season. KVK solve this problem of hatching by installing small unit of hatchery. Every month 15-20 farmers are benefited by purchasing chicks for backyard poultry. There 200 back yard poultry units of Vanraja. Each farmer is rearing 10 to 25 in the backyard. There is a good demand and response for the chicks and eggs of Vanraja. KVKs are now planning to expand this unit.

b. Semi-stall-fed Goat Rearing

KVK's made an intervention to improve this enterprise by conducting short durational training programmes for rural youth. Similarly exposure visit were organized on goat feed, breed and health management. More emphasis was given on O)smanabadi goat and up-gradation in selected non-descript goat breed by osmanabadi pure buck and given the knowledge about semi stallfed goat rearing concept.

c. Broiler Production

KVK has conducted many durational training programmes for 165 trainees. Due to training and demonstrations awareness was increased about contract farming in broiler production with private sector which provide chicks, feed and medicine and after 40 days purchases Rs. 3 to 3.50 per kg on live weight and FCR basis and changed their attitude. They acquired skills through learning by doing at KVK demonstration unit.

The technology has been adopted by 10 percent of youths now in the radius of 20 km there are 27 poultry units having capacity of 5000-10000 poultry birds on contract farming basis. These self employed rural youth earning Rs 10000-15000 per lot.

Recently a seminar-cum-farmer's meet was arranged at SKUAST-K on 26-27th of Oct.2007. The knowledge -sharing and use for making agricultural graduate and scientists was emphasized by our worthy Chancellor. A vision of poverty alleviation through backyard poultry intervention was the theme of the seminar. Many belts in Gurez, Tangdar, Telail and Zanskar are rearing native livestock species. Who are better suited and need improvement and identification. The cooking methods will need more expansion and scientific intervention for export. More emphasis has to be made on:

- Safe feed and food.
- Organic fodder and food.
- Operational excellence and modern mechanization to improve quality of indigenous enterprises.
- Local family management to farm business management and seller-buyer medol adoption.

High Value Agri-business

Rapid growth rate in high value commodities in Indian agriculture promises 40% total output. The sectors assuming importance for export earnings are Fruits, milk vegetable and poultry. Thus Agriclinic training centres should focus on these commodity oriented training. The sector may need more than 1 lac young entrepreneurs to achieve national goals and not the more 14000 applicants.

What is Needed

- Openness and transparency
- Simple banking
- Credit facility
- Mission and Training

Seikh Kabab

Rista

Rabbit meet curry

Rabbit meet reganjesh

Fig. 1

Fig. 2

Fig. 3

Table 1 : Agribusiness

S.No	Name of the state	No. of Success Stories
1	Andhra Pradesh	261
2	Assam	15
3	Bihar	519
4	Chattisgarh	63
5	Goa	1
6	Gujarat	158
7	Haryana	20
8	Himachal Pradesh	39
9	Jammu & Kashmir	28
10	Jharkand	15
11	Karnataka	629
12	Kerala	15
13	Madhya Pradesh	134
14	Manipur	37
15	Nagaland	2
16	Orissa	74
17	Pondicherry	1
18	Rajashtan	557
19	Punjab	23
20	Tamil Nadu	231
21	Uttar Pradesh	990
22	Uttarancahl	49
23	Weat Bengal	51
24	Maharashtra	842
	Total	**4754**

Table 2

S.No.	Name of the Organisation & sucees in Agri business	No. of Sucess
1	Shree Maa Guru gramodhyog Sansthan (SMGGS), UP	626
2	M/s. Terra-Firma Bio Technologies Ltd.(TFBTL), Banglore, Karnataka	489
3	Krishna Valley Advanced Agriculture foundation (KVAAF), Sangli, Maharashtra	251
4	Mitcon Consultancy Services Ltd .(MCSL), Pune, Maharashtra	224
5	Mahatama Phule Krishi Vidapeeth (MPKV), Pune, Maharashtra	203
6	Jaipur School of Business & Finance Management (JSBM), Rajasthan	221
7	SRISTI Foundation, Patna, Bihar	197
8	Rajasthan Institute of Co-operative Education & Management (RICEM)	180
9	Indra ganghi institute of Co-operative Management (IGICM), UP	175
10	Voluntary Association for people Service, Virudhanagar, Tamil Nadu	146
11	Participatory rural development Initiatives (PRDIS), Hyderabad, Andhra Pradesh	124
12	Institute of Entrepreneurship Development, Patna, Bihar	118
13	University of Agricultural Science (UAs-BLR), Bangalore, Karnataka	107
14	SCADA Computre Centre Patna, Bihar	107
15	State institute of Managemeny of Agriculture (SIME) UP	89
16	R.C.V.P Naronha Academic of administration (RCVP), Bhopal, MP	80
17	M.R.Rmoraraka GDC Rural Research foundation (MRMGDC), Rajasthan	69
18	Krishi Vigyan Kendra (PIRENS) Mahrashtra	69
19	SKC Educational Trust (SKC), Basaith, Bihar	68

contd...

20	Vasant Prakash Vasakh Pratistan (VPVP) Sangli, Mahrashtra	67
21	Confederation of aquaculture fishers and Welfare Organisations (CAFWO), AP	64
22	Indra Gandgi Krishi Vishwa Vidyalaya, Rajpur, Chhattisgrah	63
23	JAI Research Foundation (JRF), Vapi, Gujarat	62
24	Centre for entrepreneurship development (CED), Hyderabad, Adhra Pradesh	57
25	Centre for Human Development (CHD), Bhudaneshwar, Orissa	53
26	Vivekananda Research and Traning Institute (VRTI), Kutch, Gujarat	51
27	Maharana Pratap university of Agriculture & Technology (MPUAT), Rajasthan	50
28	G.B.Pant university of Agriculture & Technology (GBPUAT) Uttaranchal	49
29	Bidhan Chandra Krishi Viswavidyalaya (BCKV), West Bengal	43
30	Institute if Co-operative Management (ICM), Manipur	39
31	Allahabad Agriculture Institute -Deemed University (ADU), UP	38
32	Nationlal Research Centre for Agro-forestry (NARCF), UP	37
33	State institute of Agriculture of Management (SIAM), Rajashthan	37
34	Indian Society of Agri business Professional, Himachal Pradesh	36
35	M.P.water Land management Institute (WALMI), Bhopal, MP	35
36	International Institute of bio-Technology and Toxicolgy (IIBAT), Tamil Nadu	35
37	Entrepreneurship Development Institute of Indian (EDII), Bhat, Gujarat	32
38	Nagarjuna Agricultural Research and Development Institute (NARDI), Hyderabad, Andhra Pradesh	32

contd...

39	University of Agricultural Sciences (UAS_DHR), Dharwad, Karnataka	31
40	Baramati Agricultural Development Trusts Krishi Vigyan Kendra, Maharahtra	29
41	Centre of Agriculture and Rural Development Studies (CARDS), Tamil Nadu	21
42	Sher-e- Kashmir University of Agricultural Science and Technology, Srinagar	20
43	Dip Narayan Singh Regional Instiute of Co-operative Management (DNSRICM), Bihar	19
44	Gujarat Agricuture University (GAU), Anand, Gujarat	15
45	Agro Clinical Development Trust (ACDT), Tamil Nadu	15
46	Shanmuga Arts, Science Technology & Research Academy (SASTRA), Tamil Nadu	15
47	Assam Agriculture university (AAU), Jorhat, Assam	14
48	C.S.Azad University of Agriculture and Technology (CSAUAT), UP	14
49	The Agriculture and Promotion and Investiment Corporation of Orissa Ltd. (APICOl0)	14
50	CCS Haryana Agriculture University (CCSHAU) Hisar, Haryana	14
51	Indian Society of Agri business Professionals, Amritsar, Punjab	11
52	Centre Institute of Agricultural Engineering (ICAR), Bhopal, MP	9
53	Indian Society of Agribusiness Professionals, Haryana	9
54	Indian Society of Agribusiness Professionals, Jharkand	9
55	Kerala Agriculture university (KAU), Tissur, Kerala	8
56	Netaji Subhash regional Institute of Co- operative Management (NSRICM), West Bengal	8
57	Society for rural Industrialization (SRI), Ranchi, Jharkand	8
58	Central Institute of Post Harvest Engineering & Technology (CIPHET), Punjab	6
59	Agroweb Online (Pvt) Ltd. (AOPL), Indore, MP	5

contd...

60	Indian society of Agri business Professionals, Srinagar, Jammu and Kashmir	4
61	Indian veterinary research Institute (IVRI), UP	4
62	Acharya NG Ranga Agriculture University (EEI), Hyderabad, Andhra Pradesh	4
63	Dr. Y.S. Parmar University of Horticulture and Forestry (YSPUHF)	3
64	Punjab Agriculture Management & Extension Training Institute (PAMETI), Punjab	3
65	Rajendra Agriculture University (RAU), PUSA, Bihar	3
66	Narendra Deva University of Agriculture & Technology (NDUAT), UP	3
67	Agriculture Co-operative Staff Traning Institute (ACSTI), Jalandhar, PunJab	3
68	Sher-e Kashmir university of Agricultural Science and Technology (SKUAST), Jammu	2
69	Indian Society of Agribusiness Professionals, Jammu, Jammu and Kashmir	2
70	Indian society of Forest Management (IIFM), MP	1
71	Sardar Vallabh Bhai Patel University of Agriculture and Technology	1
	Total	**4754**

Table 3 : Under Agri-clinics/ Agri-business norms thereof

S.No	Means of Finance	General Category	Weaker Category
1	Promoter's contribution	10%	5%
2	Soft loan from NABARD	0%	5%
3	Total promotion contribution	10%	10%
4	Subsidy	25%	33.33%
5	Eq..—-	355	43.33%
6	Term loans	65%	56.67%
	Total	100%	100%

Table 4 : List of Ventures

1. Soil and water quality cum inputs testing labotatories (with Atomic Absorption Spectrophotometers)
2. Pest surveillance, diagnostic and control services
3. Maintenance, repairs and custom hiring of Agricultural implements and machinery including micro irrigation systems (sprinkler and drip);
4. Agri Service Centres including the three activities mentioned above (Group activity);
5. Seed Processing units
6. Micro-Propagation through Plant tissue Cultural Labs and hardening Units.

Table 5 : List of Ventures

1 Setting up of Vermiculture units, Production of bio fertilizers,bio-pesticides, bio-control agets;

2 Setting up of apiaries (bee-Keeping) and honey &bee products' processing units;

3 Provision of Extension Consultancy Services;

4 Facilitation and agency of agricultural insurance services;

5 Hatcheries and production of fish finger-lings for aquaculture;

Table 6 : List of Ventures

1. Provision of livestock health cover, setting up of veterinary dispensaries & services including frozen semen banks and liquid nitrogen supply;
2. Setting up of Information Technology Kiosks in rural areas for access to various agriculture related portals;
3. Feed Processing and testing.
4. Value Addition Centres
5. Setting up of Cool Chain from the farm Level onwards (Group Activity)

Table 7 : List of Ventures

1. Post Harvest Management Centres for sorting, grading, standardization, storage and packaging;
2. Setting up of Metallic/ Non-Metallic Storage structures (Group activity)
3. Retail marketing outlets for processed agri-products.
4. Rural marketing dealerships of farm inputs and outputs.
5. Projects in any other service oriented activities in agriculture and other allied areas can also be considered.

Table 8 : Poultry Agri business

S.No.	Component/ Cost	I	II	III	IV	V	Total	Remarks
1.	Satellite Hatcheries	100	50	-	-	50	200	Assuming cost of each Satellite unit around Rs.10.00 lakhs and Proposing 10 and 5 units during Ist and 2nd year respectively and another 5 units during 5th year.
2.	Mother Units	420	60	300	-	-	780	Assuming Single Mother Unit cost to be around rs 1.2 lkhs during Ist and 2nd year and rs 1.5 lahks during 3rd year and Proposing for Establishment of 350,50 200 unts during Ist, 2nd and 3rd year respectively
3.	Poultry traning cum-extension Centres for traning of trainers, poultry link workers etc.	50	30	30	-	-	110	Assuming cost of each Centre to be around Rs. 2.00 lakhs and proposing 25.15,15 Centres during Ist, 2nd and 3rd year respectively
4.	Poultry Proceesing Units	-	-	500	500	500	1500	Assuming cost of each Mini Plants with dressing capacity of 500 birds per hour and proposing 5 units to be set up each year during the 3rd, 4th and 5th year
5.	Skill up gradation and training of beneficiaries	420	400	450	500	500	2270	Assuming 70,000 beneficiaries are trained during the Ist year and 50,000 each during the subsequent years, Further only honorarium is proposed @ Rs. 600, 800, 900, 1000, 1000 during Ist, 2nd, 3rd, 4th & 5th year respectively

Table 9 : Poultry business Targets.

Sl. No.	Component/costs	Year I	Year II	Year III	Year IV	Year V
1.	Cumulative no. of nucleus breeding farms	35	40	40	40	40
2.	Cumulative no. of mother units	350	400	600	600	600
3.	No. of beneficiary families ('000)	70	120	180	180	180
4.	Financial assistance towards fixed cost of bene ficiary families (at the rate of Rs.1300) Rs in lacs	910	650	780		
5.	Subsidy towards input costs of beneficiary families (assuming averagebatch size of 30 birds) - Rs. In lacs	1,607	1,683	1,760	459	
6.	Subsidy towards fixed cost of mother units (at the rate of Rs. 18,000 i.e 20% of Rs. 90,000) Rs. In lacs.	63	9	36		
7.	Interest free loan (net of recovery) to mother units (at the rate of Rs. 27000)- Rs. In lacs	95	14-19	54-23= 31	Rec. of 32	Rec. of 32
8.	Poultry training cum extension centres for training of trainers, poultry link workers etc. (25,15 and 15 centres to be set up in the Ist, 2nd and 3rd years at the cost of Rs. 2 lacs each) Rs. In lacs.	50	30	30		
9.	Skill upgradation and training of beneficiaries (70,000 to be trained in Ist year and 50,000 in each subsequent year) Rs. In lacs	420	300	350	350	400
10.	Administrative expenses, monitoring & evaluation - Rs, in lacs	120	130	140	160	200
	Total cost	**3265**	**2797**	**3127**	**937**	**568**

□□□

Chapter 31

Eco-Friendly Agricultural Productivity with Indigenous Techniques in Kashmir

G.M. Wani

ABSTRACT

Agricultural Production has to be in tune with ecological preservation. The age old agricultural productivity methods and practices need to be clubbed in today's farming system. It is not true that all what came to be known in past few centuries is only science. The use of cultivation methods, selection of seeds . plant and animal germplasm have been improving year after year. These indigenous practices have come to existence through centuries, experience, tests and notes of farmers around the globe. The faster communications in this modern era have now made a dent on technology dissemination. The evils of past unidirectional more production oriented policies have given enough food but has robbed us of our ecological resources. Now we face soil degradation, erosions, denudation of forests and great dangers to our biodiversity and eco-environment.

An attempt has been made in these pages to present our own experience in Kashmir ,firstly to define some of the indigenous techniques used in the past which were eco-friendly. Secondly, a combination of the modern and indigenous technology use in agriculture has been compared in term of individual crop and fruit production. Lastly policy and innovative models of crop and livestock production has been briefly reviewed.

Competitive global marketing needs some changes at base level of our farming systems. Farmers have to be cost conscious and need new strategies of mixed farming option like combined agriculture, horticulture, livestock husbandry and floriculture activities. A new model for incorporating all these fields in an eco-friendly mode is needed.

An agricultural mission for efficient resource management should aim at bridging gap between current farm yield and potential Farm yield. Use of bio-technology and bio-informatics for identification of economically viable cropping options and combinations can not be ruled out; however soil and environmental preservation must be kept in mind while using biotech farming options. We should talk of eco-conducive-germplasm-base instead of high productive germplasm base. Minimum use of chemical fertilizers has been tried with increased crop production. Such organic farming options resulted into 19% enhancement of production per hectare in capsicum, cabbage, rice, apple, Onions and other crops. The use fertilizers were replaced from 25-75% with organic manures like agricultural waste material, dalweeds, cow-dung, forest waste and other non chemical substitutes including microorganisms. All our results are presented in this paper.

The micro-nutrient availability in mud, green waste and other cheep byproducts have been analyzed. The soil composition and remediation factors to restore ecological balance in degraded lands is discussed. New options for combined farming of crops, animals, plant, fish, poultry and other biological partners need to be projected for a more safe farming with ecological resource management. The blend of indigenous and more recent technology for resource management of eco-systems should be considered as future models.

Historic view of Indigenous Agriculture in Kashmir

The famous Chinese traveler, Huen Tsiang around 631 AD stayed in Kashmir for 2 years. His travel itinerary commonly known as Huen Tsiang report mentions about pears, wild plum, peach, apricot and grape of high quality being cultivated in Kashmir. All of them were organic and propagated through indigenous techniques. Most of them have been lost in wilderness of new cultivars introduced by invaders, be them Afgans, British, Mogul, Dogras and like till date. A comprehensive agricultural policy was initiated by Sher-e-Kashmir, Sheikh Mohammad Abdullah soon after revolution of 1947. The policy was centered round improvement through organic and indigenous means. Very little attention was paid to execute this policy. The policy itself has now been reduced to archives. Fruits too have a mention in Rajtarangni around 1000 BC or about 3000 years back. Indigenous fruits and nuts were of special mention in these historic epics.

The indigenous fruits and its nut valleys are remembered by village names like Koel (habitat of quil), Chaeri-van (village or woods of apricot), Chaeri-kot (village of apricot); Badamwari (valley of almonds), Tangdar (pear orchard); Chunt-var (apple orchards); Aloocha bagh(cherry garden), like it many places are known from traditional agricultural and allied products as described (Wanchoo, 2009).

The Kashmir practices of burring hair, nail forest or house hold waste and other agricultural waste in good olden days of primitive agricultural development could be patented as mother of organic farming in modern world. Kashmir vegetable floating garden, hosieries, perfumes, waternuts and variety of special varieties of fruits and animal products were the best inventions of its time. Production potential and qualitative traits of indigenous chicken of Kashmir have been documented. The study comprises of 51 villages covering 461 households. Maturity age, live weights and other production traits are described. The quality of eggs and meat under free range system is economic, Meat taste is preferred over modern poultry rearing. (Iqbal and Pampori, 2008). A chronological details of various equipments, rearing and cultivation methods has been studied and shown in fig 1-10.

Under the aegis of State Agricultural Management and Extension Training Institute, (SAMETI), the author as Director of the autonomous Institute, got a survey-cum-study of many villages, blocks, Tehsils and districts of Kashmir. The staff of SKUAST, KVK's, ATMAs and NGO's were webbed together to prepare detailed scientific Research & Extension plans for these places. Some of the results shall be presented in these pages (Tantray et al, 2007, 2008).

Agricultural Resource base of Kashmir

Kashmir - a geographic area of 138942 sq kms. It has an altitude ranging from 305 masl to 9610 masl. The annual rainfall is 92.6 mm (Ladakh), 650 - 1000 m (Kashmir valley districts) and 1115 mm (Jammu). The dependence on agriculture was 80% and in the past it contributed 36.3% to net domestic product. From 16% of total area under agriculture only 5% is net cultivable area. There has been eight times increase in net domestic product from 1980-81 to 2001-02 in agriculture, inspite of war-like situation prevailing in the area. Such a great enhancement denotes efficiency of our indigenous technologies. Even on date very few imported technologies on various agricultural crops are used. (Table 1-IV). The Rice productivity using low level of fertilizers in valley districts are higher than the Indian averages. (Wani, 2007, 2008). Our fertilizer consumption is 38.2 kg/ hac as against 170 kg/ hac in Punjab. We have already reached a mark of 3-4 t/hac in Rice with indigenous technology. Our agricultural resources under local techniques and prospects of economic growth have been reported (Siraj, 2006).

Not only fertilizer, but other insecticides and pesticides are used on very low levels in Kashmir, in comparison to advance countries or other states in India.

Our horticulture sector has shown a tremendous increase over years. The area and production has increased. A review has recently been published by the author (Wani, 2008) and some of the features are shown under indigenous techniques and Eco-Economic Agricultural features of Kashmir (Table IV a - i). Various commodities like vegetable have increased by 28%, however an increase in area has been upto a magnitude of 73%. Livestock has shown self sufficiency in milk but other products like meat is still deficient. A detailed study on these commodities have been reported by this author (wani, 2001, 2002, 03, 04,05,06,07,08) Fish capture was 4227 tonnes in 1956, which has increased to 19010 tonnes in 2008.

Kashmir Soils

The classification of Kashmir soils has been made much earlier than the modern day soil texture nomenclature. Indigenous local vernacular classification of Kashmir soils and their English translated names are as:

Local name → English
Ghurut Meyech → Clay soils
Sekil meych → Sandy soils
Nambelh zameen → Water logged
Lemb zamin → Flood affected soils
Surzamin → Good soils
Wooder zamin → Karehawa land
Dal zamin → Floating gardens

Kashmir soils are predominantly loamy soils. They are poor in lime, high in Magnese content. The rich in humus soils were called Ghurut Meych and red or yellow soils poor in humus, are called wooder meych or Doonae meych. Agricultural productivity under local indigenous technology had been preserving eco resources. The per capita income of Kashmir in rural areas was higher than Indian rural averages. J&K rural monthly incomes were 15.40 dollars against national (Indian) averages of 9 dollars. Similarly, the urban 21.99 dollar monthly income averages in J&k are higher than 12 dollar averages of India per capita monthly income. Per annum income in 1994 - 2008 has increased from 241-324 dollars in J&K, thereby signifying that our indigenous technology use have been increasing our productivity base with resources maintenance.

Horticulture Produce

The increased area under horticulture has resulted into upward swing of both per capita income and use of chemicals as pesticides or fungicides in apple. However, the bulk of our nut, stone fruit production is organic, devoid of any intensive use of chemicals. The use of local medicaments like nicotinized water (Huqa water - Jajeer Aab), ashes and other local medicines and herbal sprays did reduce inseet-pest incidences. The use of mud with salt on cracks or cankers in plants had been used. The use of organic manures with reduced "N" fertilizers and yield increases have been demonstrated in our experiments (Graphics I-III). Use of hot iron pinning, local herbs as livestock and human medical cure have been in use since ages. Saffron, Khazaban and many herbs and shrubs like iron rich. Handh and willow leaves to reduce fever are still in use under far away pockets of Gurez, Tangdar etc. Many such indigenous technologies have been described. (Wani, 2001, 2009, 2006,2007).

Land Holdings

Marginal farmers holding less than one hac of land are predominant (80%) the ranges are 0.05-1.0 hac. The district wise average land holding is as follows :

Anatnag	0.49 hac	Pulwama	0.64
Srinagar	0.37	Badgam	0.54
Baramullah	0.58	Kupwara	0.55
Leh	1.38	Kargil	0.78
Jammu	0.96	Udhampur	1.11
Doda	0.84	Kathua	1.16
Rajouri	1.19	Poonch	0.94

Small land holding is a limitation to use industrialized produce, chemical pesticides, fungicides or even heavy machines and implements. Thus production of 80% food grains, more than 100% milk and apple (12t/ha), which is higher than Chinese average under indigenous to semi-modern technology use is an achievement of local talent and technology use. Various productive figures of Kashmir and organic manure and option analysis has been presented in graphic i-iii & tables (I-V). The economic prospects, of Jammu and Kashmir and its agricultural productivity and ecological resources have been reviewed (Siraj, 2006).

Global Review of Indigenous Technologies

A detailed account of rice, livestock and other resources have been described (Wani,2007,2008). Frontline demonstrations on maize have indicated excellent local germplasm with better yields under local cultural practices (Nahvi et al, 2008).

Mountain ecosystems are conserved by simple means locally. These methods used in China, India, Pakistan consist of planting trees on river beds after floods recede or on canal bank. The easily availability of moisture during summer and shade helps to conserve resources and are thus efficient natural resource management. These simple methods have been tested as valuable resource management methods indigenously in Asia since last many generations (Shi et al, 1999). Hetzel, 1977 has described nomadic life ways in Australia. Juming cultivation (Singh & Singh, 1980) is a traditional agriculture. Capitalist model of food production is damaging environment (Fleury and Mollard, 1976). However, one can say use of chemical harmful to environment can be hazardous in long run for holistic agriculture and nature.

Bees as Polinizers (Japanese Experiment)

Stingless bee (Hymenopetra: Apidae) were natural polinizers in Uganda (Byarugaba, 2004)· Chemical use as pesticides and fungicides have reduced their numbers. These little known organisms are efficient pollinators and are facing risk of disappearce. (Byarugaba, 2004). Thus use of chemical pesticides are harmful to ecosystems. Bee keeping in Kashmir is a traditional culture, local indigenous profession, consisted of fixing small wooden or earthen boxes in the mud and stone walls of the cowsheds or houses. This is done to keep them warm during winter and rains. They were structured in cow shed walls or in grain/grass bin stores in traditional Kashmiri houses and cow or food storage sheds. Synonymous reports are available from other Asian Countries. (Petersen, 2005). Japan too have now used the stingless bees as polinizers (Amano, 2004).

Milk Production in India

Milk production in district Karnal under usual local upkeep of animals was around 12.26 Lakh liters of milk per day. With advanced technology it may increase to 15.5 lakhs liters. However, the cost of production and chemical and inorganic feed supplement use may have residual harmful effects (Sing and Kalra 1995). A detailed discussions on global use of advancement and organized dairy farming has been discussed (Wani, 2008). The organized dairy production systems produce milk at higher global price than the small holder mixed livestock farmer of India. Indian goats under tethering. (zero grazing), range system (extensive and semi-intensive) or under stall fed intensive feeding systems have been more economic and efficient. A shift of use for indigenous technology is gaining more popularity. It promises more livelihood options and sustained eco-preservation and resource management. (Wani 2006, 2005 Wani 2008; Karthikeyan et al, 2006).

Ethnovet Practices

Ethnoveterinary medicine i.e. clubbing of information of traditional medicinal plants and herbs for use in livestock disease control has been successful (Int. Instt. Of Rural reconstruction 1994). We in Kashmir have been using it since long.

Trypanotolerent cattle and their propogation in Ethiopia, Gambia, Kenya, Zaire, Tago and cote d'/ voive is reported. The higher capital cost of eridication of tsetse flies or the trypanosoma control drugs could be reduced. (Illay, 1993).

Many modern dairy technologies have low adoption among Indian farmers. Hardly 1/3rd use improved balanced feeds, 1/10th adopt vaccination cover. The reason is high cost of production in comparison to use of local indigenous technologies (Dixit and Sinha 1993)·Dahi (curd) prepared locally by indigenous techniques in Pakistan did process lactic acid bacteria (Masud et al, 1991). Even the low temperature pasteurization using high-tech membrane processing does not eliminate all micro-organism, but the costs are too prohibitive (Eckner and Zottola, 1991).

Herbal Cure

An indigenous herbal ointment Himax applied topically were successfully used for wound healing in bovine (Varshney and Verma 1990)

The presence of bacteriophages and bacteria are present in a mixed strain of cheese starter. Thus mitigation of bacteria presence is never of Zero level even with high tech in USA. (Lodics and Steenson 1990).

Many indigenous knowledge methods have been used successfully to control foot & mouth disease in India. (Uma-Sah et al, 2002). The author also treated various food rot and FMD cases using concentrated $KMNO_4$ cauterization of wounds. We also treated post operative, laparotomy/ laparoscopy wounds without sepsis by cauterizing them with higher concentrations of potassium permagnate (Wani, 1984-85, 1995-1996, 2005). Many Indigenes herbs and leaves have been used to treat wounds and boils in Kashmir traditional Hakmat. Even hepatitis is cured using herbal preparation in Hamdard University by Hakim Abdul Jameel.

Deworming helps increased milk production (Sahoo and et al, 2003). Many indegnious herbs are available in Kashmir and were successfully used for deworming. Blood constituents differ in local cattle with those of exotic, thus differentiation of disease resistence could be probed (Sharma et al 1998)Many diseases like swine fever, FMD, diarrhea, helminthes and skin infection are cured using ITK in Nagaland (Rohilla et al, 1999) Lans and Brown 1998, used Ethnoveterinary medicines in Trinided and Tobago. The plants used were Azadirachta indica, and Curcumalonga. Latest SAGE technology was used to confirm trypanatolerent local bovines (Berthier et al, 2005). Leptospira infection were more in crossbred than local cattle, buffaloes or goats (Dwivedi et al, 2006). In Kenya small holder farmers use herbal preparations for cattle diseases. The successful herbs are endangered species and need conservation and resource management (Njoroga & Bussmann, 2006). Goats use herbages at higher heights 1.65m than cattle (1.47m) or sheep 0.87m. Thus, the grazing damages by nibbling at higher heights is more prevalent in goats. The knowledge could help regeneration of denuded lands. (Sanon et al 2007).

Iranian Najde bovine breeds have BOLA-DRB-2 locus gene highly polymorphic and thus may be used for its improvement (Torbate, et al, 2006). Paddy straw and de-oiled ground nut cake (DGNC) were used successfully to grow bullocks (Nanada et al, 2006) Chinese medicines used for toxoplasmosis in cows (Liu-Defu 2004). Many indigenous wisdom in herbal use is described. (Srivastava et al, 2006; Kanwar & Sharma, 2006).

Livestock Indigenous Technology (Methodology & Results)

Retention of placenta syndrome in cows has been successfully prevented using RPL-1 an indigenous drug in Kashmir local and CB cattle. A field trial was conducted to evaluate the efficacy of a new herbal formulation, RPL -1, on freshly calved cows for the prevention of placental retention. Twenty one crossbred jersey x Hoistein cows were divided into two groups. The treatment group (n=15) was given a single dose of 50g of RPL-1, mixed with molasses, immediately after calving and every six hours at a dose of 25 g each, until the placenta was shed completely. Thereafter, RPL -1 was administered twice daily at a dose of 25 g for 5 days. The control group was left untreated. The cows were observed to determine the time it took to expel the placenta, the time of disappearance of the lochial discharge and time it took for the uterus to involute. The results showed that placenta was expellsed 153 ± 81.41min (range 60-240 min.) after parturition in the treatment group compared to 435 ± 102 min (range 180-720min) in the nontreated control group. The lochial discharge in the treatment group disappeared completely in 5.9+or 0.39 days (range 4-9 days), while in the non-treated control group, the lochial discharge disappeared in 9.33 ± 1.08 days (range 6-14 days). In the treatment group, the uterus was palpable completely by 15.53 ± 0.30 days (range 14-18 days) as against 20.63 ± 1.01 days (range 18-20) days in the non-treated control group. Involution of the uterus occurred significantly earlier in treated group. (Seh et al, 1998; Joshi, 2006; Kanwar and Sharma, 2006).

Survey & Study Reports of Uri Tehsil

i. Location Description

Uri is one of the Tehsils of district Baramulla in Kashmir spread over an area of 25192 hectares with a population of 104159 persons which comprises of 54989 males and 49170 females. It lies in the North-Western parts of district Baramulla at 73° 59′ 18.26 E and divided in two administrative blocks viz: Uri and Boniyar having 33 Panchayat halquas with a number of 90 villages. The Tehsil is well demarcated in sub tropical and temperate agro-climatic zones, where summer temperature average goes up to 36° C and occasionally falls below 0° C in temperate areas. It is surrounded by lofty mountains with deep ravines, terraced land and is one of the boarder Tehsil. The river Jehlum is flowing through the area and is the major source of Hydroelectric Power production of the J&K State. The population is mostly settled on mountain terrains and scattered in small hamlets with low population density and poor road connectivity. Altitude of the area varies from 1100 meters to 2200 meters. Farmers are having small size of land holdings on these terraces.

Soils are product of weathered mountainous rocks and are very much productive from agricultural point of view. Clay and organic mater content of these soils is comparatively high due to combined effect of precipitation, temperature and vegetation. The total area consists of 6573 hectares as cultivable and 18619 hectares as uncultivable. Under cultivable area 2805 hectares is irrigated and 3768 hectares as unirrigated. Instead of marginal and small holdings the major occupation of the people is agriculture and allied activities. Due to the after effects of earthquake, most of the fertile lands have got eroded and the human populations prefer to work as construction labourers to make their living in addition to the farming enterprises.

ii. Cropping Patterns

The farming in the area is dominated by growing of crops like maize, paddy, wheat, oats, pulses (mostly rajmash), both as irrigated and rain fed. Among the vegetables both Rabi and kharief vegetables are grown in the small kitchen gardens for the domestic purposes only. However, growing of peas and potato as commercial crops is also seen at some places. In higher altitude belts mono-cropping and at lower belts double-cropping is in practice. High altitude hills are covered with Pine and Deodar trees and at foothills there is a scattered plantations like pear, walnut, peach, apple, olive, lemon, and wild apricot etc. The number of fruit trees (local cultivars/ wild) varied from 8-10 trees per house hold. The old walnut trees and uncared pear trees are seen in good numbers in the area.

iii. Population Statistics

Livestock holding size varies from 4-5 (local/ cross bred) cows, bulls, buffaloes. Sheep & goats. They are reared using indigenous methods. The reported sheep and goat population is 81940 and 27390 respectively. The numbers are 3-4 per household. Indigenous poultry birds few in number for domestic purposes is a common practice with some of the farm families. Traditional method of bee keeping is still in practice in some village hamlets.

Methodology

The reorganization process of extension system has lead to the establishment of District ATMAs (agricultural technology and management agencies) and a SAMETI (State agriculture management extension training institute) in Jammu & Kashmir. The Director Extension Education SKAUST-K is also looking after as Director SAMETI. In order to prepare SREP for the agro-climatic zones of the valley, training programmes and conducting of PRA has now been a regular feature of the SAMETI activities. It is in this connection that Director SAMETI instructed to prepare a Strategic Research and Extension Plan (SREP) through PRA techniques by deputing multidisciplinary team of experts /scientists of SKUAST-K to Uri tehsil of district Baramulla from 16th to 21st of July 2007. The team comprised of scientists/ experts from Extension Education, Agronomy, Pomology, Soil Science, Floriculture, Home Science, Plant protection and Olericulture. The team has been trained prior to these assignments by MANAGE Hyderabad – a Govt. of India concern, which governs the SAMETI too.

i. Rapport Building

In order to have a brief idea about the climatic conditions, natural resources, ecology, needs of farming community and opportunities etc. prior to make a plan of PRA the team first of all visited the office of SDAO, Uri, (SDAO = Sub District Agriculture Officer).

An interaction meeting was arranged with the officers and field functionaries of development departments, NGOs, scientists and farmers of the locality and information was gathered. This interaction meeting was also helpful in deciding the representative villages for PRA according to the agro-ecological situations. In each representative village group of farmers of different ages were gathered in the schools/central places of the villages and they were taken in to confidence through informal discussions. This exchange of ideas and information was the basis of PRA.

ii. Conducting of PRA

For the purpose of SREP, the Tehsil was divided in to three agro -ecological situations (AES) mainly on the basis of altitude because tehsil. Uri comprises of prominent high and low altitude belts with entirely different cropping pattern.

The various AESs as identified are given below:

S.No.	Name	Situation	Representative village	Altitude
1	AES-1	Sub-tropical	Isham	1300 masl
2	AES-2	Temperate	Limber	1800 masl
3	AES-3	Sub-temperate	Noorkhah	1420 masl

iii. Transect Walk

After the selection of three representative villages the team along with the villagers had a transect walk of each village to acquaint themselves with the village resources and its boundaries.

iv. Mapping

Some of the farmers were requested to draw map of the each village using the materials like drawing sheets, pencils, erasers, etc. provided by the resource personnel. While others were encouraged to make correction on the flow charts prepared. This exercise helped to draw social and resource mapping of the village.

v. Wealth Ranking

Wealth ranking was done by direct and indirect questioning of the farmers on the basis of gross income/month/household without hurting their sentiments and accordingly two income groups i.e. resource rich and resource poor were identified in each village. The number of lower income groups/ resource poor with an income up to Rs 2000/month is the basis of consideration.

vi. Existing Farming System

To know the existing farming systems, different farming enterprises with their contribution and combination being followed were critically analyzed and identified. The farming systems differ according to agro-ecological situations.

vii. Gap Analysis

The recommended agricultural practice was compared with the existing practice and extent of gap evaluated. The specific reasons for the gap were enquired and the farmers were encouraged to propose strategies for filling up the gaps. On the basis of above findings SREP has been formulated for the representative village.

Results of the Study

The village Isham is spread over an area of 342.37 hectares with only 80 hectares under cultivation of which only 42 hectares are irrigated. Rest half is rain-fed. Isham comprises of five Pattis (hamlets) namely Takiya, School, Masjid, wahpora and Madah, with three irrigation canals, viz- Satur-Kul, Dardhkote - Isham-Chakra-Kul and Isham-Kul. The number of total households is 180, having population of 1070 people (Males: 683 & Females: 587). According to the revenue records of the region, about 87 persons belonged to Schedule Tribe category.

a. Resource (Income groups)

S.No	Income group (Rupees/Month)	No of Families	Percentage
1	Up to 8000	113	62.78
2	8001-12000	45	25.00
3	More than 12000	22	12.22
	Total	**180**	**100 %**

b. Wealth Ranking

S.No	Category	No of Families	Percentage
1	Resource poor	106	58.89
2	Resource Rich	74	41.11
	Total	**180**	**100.00**

c. Existing Farming Enterprise

S.No	Enterprise	No. of households	Percentage
1	Agriculture	159	88.5
2	Horticulture	125	69.5
3	Animal Husbandry	122	68.0
4	Sheep Husbandry	54	30.0
5	Poultry Keeping	104	58.0

d. Existing Farming System

S.No	Existing Farming system	No. of households	Percentage
1.	Agriculture + Horticulture+ Animal Husbandry	94	52.0
2.	Agriculture + Horticulture+ Poultry Keeping	54	30.0
3.	Agriculture + Horticulture+ Sheep Husbandry	32	18.0
	Total	**180**	**100.0**

e. Contribution of Different Enterprises (%)

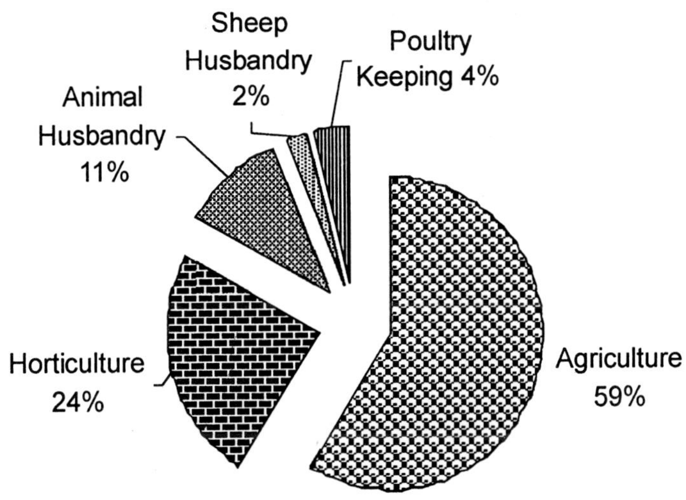

Indigenous Techniques

A graphic representation of animal housing, water driven grain mill, milk processing equipments and appliances used in indigenous postharvest techniques in agriculture can be seen in Fig 1-10. Their description and details shall be presented in the conference.

Conclusion

a. Eco-farming Options

- Use Fertilizer only where deficiency is detected
- Spray Pesticide only when needed
- No flooding Water use efficiency during irrigation
- Use disease resistant High yielding varieties (HYV)
- Use alternate cropping modules
- Establish Field data bank/germplasm banks
- Know Farmers practices and farming packages
- Indigenous technique be used as alternative to high-input-eco-destructive technologies in agriculture
- Use of mud, agriculture waste and bye products to substitute soil repairs is suggested on the basis of analytic maps presented
- Use of indigenous culture practices, with blend of cost-effective eco-friendly material use is recommended. Various comparisons have been presented in various tables.

b. Contribution of Different Enterprises (%)

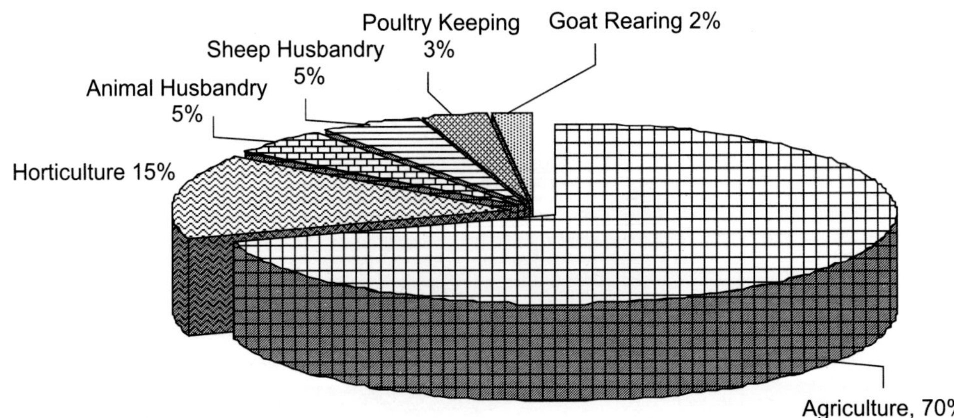

Table I : Productivity under local agricultural practices & suggestions for improvement.

Agriculture						
S. No.	**Item of package**	**Recommended Practices**	**Existing practices**	**Gap in adoption**	**Cultivation Method**	**Farmers proposed extension strategies**
1	**Paddy**					
i.	Seed rate	50 -60kg/ha	100-120kg/ha	partial	Semi Organic	Use of indigenous technology advocated.
ii.	variety	Shalimar rice-1, Jehlum K-39	Chinese variety K-39	Partial	Semi Organic	Use of indigenous technology advocated.
iii.	Transplanting time and method	appropriate time is 25th may-10 June with 3-4 seedlings/ hill	First fortnight of June and planting 8-10 seedlings /hill	Partial	Semi Organic	Use of indigenous technology advocated.
iv	Fertilizer application (Urea,DAP, MOP)	Urea-10kg, DAP6.5kg, MOP1.7kg/ kanal	Urea 5kg and DAP5kg/kanal	Partial	Semi Organic	-do-
v	Method of fertilizer application	Basal + Top doses	Using basal dose of the available fertilizers only	-do-	-do-	-do-
vi	Insect/ pest problems	IPM	Nil	Full	Organic	Organic
vii	Disease problem	IDM	Nil	Full	Local methods used	Needs improvement of local methods
viii	Yield	3-4 quintal/ kanal	1.3-1.6 quintal/ kanal	Partial	Semi Organic	Organic produce should be certified to price more.
2	**Maize**					
i	variety	Mansar, C-6, C -8	Local (Francece)	Full	Unawareness -Showing preferences for growing white maize for their food - local techniques	Research on local cultivars

contd...

ii	Seed rate	1 kg /kanal for line sowing and 1.5kg/kanal for broadcast	4-5 kg/kanal through broadcast only	Partial	-To ensure good plant population -More labour intensive	Demonstration on benefit of line sowing cultivation which is also feasible for intercropping practicing in the area
iii	Fertilizer	7 kg urea, 6.5 kg DAP, 2.5 kg MOP and 1 kg ZnSo4/ kanal	Applying only urea @ 2 kg /kanal at knee high stage	Partial	Advise Organic manure use.	Encourage for organic farming.
iv	Insect/pest /Diseases	Control of cut-worm ,maize stem-borer and sheath blight	Sowing more seed than the recommended seed rate	Full	Organic	Indigenous technology use advocated.
v	Yield	2.5-3 q/kanal	1-1.2 q/kanal	Partial	Growing of local cultivars under indigenous technology	Indigenous technology use advocated.
3.	**Wheat**					
i	Variety	HS 240	Local	Full	Organic	Local Techniques
ii	Seed rate and method	5-6 kg/kanal in line sowing	8-10 kg/kanal through broadcast only	Partial	Semi Organic	Use of indigenous technology advocated
iii	Seed treatment	Carboxin @ 2g/kg of seed	Nil	Full	Organic	Organic
iv	Yield	30-40 q/ha	10-12 q/ha	partial	Semi Organic	Indigenous technology use advocated.

Horticulture Features

Pear, apricot, walnut low chill apple, olive and lemon are grown in the area. Pear and apricot, matures a month earlier than valley and fetch higher returns to the growers. Walnuts of the areas are not up to the mark due to higher temperature at maturity. Olive of best quality is grown in the area. Lemons are found growing in lower areas but are of poor quality.

S. No	Packages	Recommended	Existing	Gap (%)	Cultivation Method	Farmers proposed strategy
4.	**Walnut**					
1.	Variety	Hamdan, Sulaiman or better clonal selections	Seedling origin plants	100	Non availability of planting material and unawareness	Indigenous technology use.
2	Spacing	8 x 8	No systematic planting of orchard	100	unawareness and undulated lands	-do-
3	Manure	20kg /Tree	nil	-	unawareness	-do-
4	Fertilizer	Urea-3915g DAP-1085g MOP-2490g	nil	-	unawareness	-do-
5	Pest and disease management	Spray of insecticide and fungicides	Nil- but walnut weevil is the major problem.	-	unawareness	-do-
6	Irrigation	Required	Nil - though plenty of water is available.		unawareness	-do-
5.	**Pear**					
1.	Variety	Bartlet, Max red Bartlet, Chinese sand pear,	Bartlet, William Bartlet	50	Lack of new introduction	-do-
2	Spacing	6 x 6	No systematic planting of orchard	100	unawareness and undulated lands	-
3	Manure	20 /Tree	nil	-	unawareness	-
4	Fertilizer	Urea-1065g DAP-650g MOP-1500g	nil	-	unawareness	-do-
5	Pest and disease management	Spray of insecticide and fungicides	Nil	-	unawareness	- do-
6	Irrigation	Required	Nil - though plenty of water is available.		unawareness	-do-
6.	**Apricot**					
1.	Variety	Charmagz , Kaisa, Gilgiti sweet,Amba, Quetta, Sukerpara	Unidentified varieties	50	Lack of new introduction	-do-
2	Spacing	6 x6	No systematic planting of orchard	100	unawareness and undulated lands	-
3	Manure	10 /Tree	nil	-	unawareness	-
4	Fertilizer	Urea-850g DAP-325g MOP-11250g	nil	-	unawareness	-do-

contd...

5	Pest and disease management	Spray of insecticide and fungicides	Nil	-	unawareness	-do-
6	Irrigation	Required	Nil - though plenty of water is available.		unawareness	-do-
7.	**Olive**					
1.	Variety	Leccino, Coratina, Pendolino Frontoio Ramya, Nikitshaya	Leccino, Coratina, Pendolino Frontoio	50	Lack of lack of high oil yielding varieties like Ramya, Nikitshaya	-do-
2	Spacing	6 x6	No systematic planting of orchard	100	unawareness and undulated lands	-
3	Manure	20 /Tree	nil	-	unawareness	-
4	Fertilizer	1.2kgN $0.80kgP_2O_5$ $0.40\ k_2O$	Applying only N in form of Urea	80	unawareness	-do-
5	Pest and disease management	Spray of insecticide and fungicides	Nil	-	unawareness	-do-
6	Irrigation	Required	Nil - though plenty of water is available.		unawareness	-do-

Remarks

There is a large wild vegetation of Indian olive (*Olea cuspidata)* which may be rejuvenated through top working of high yielding varieties. Lemons are growing in lower areas are of poor quality and there is a good chance for introduction of kinnow in the area for economic production. Allison and Abbott varieties of kiwifruit and day neutral straw berry can also be introduced for crop diversification.

Vegetables

S.No	Packages	Recommended	Existing	Gap	Cultivation Method	Farmers proposed strategy
Kharief Vegetables						
1.	Kharif vegetables and varieties	Cole crops, cucurbits, Solaneaceous and beans etc. with different high yielding varieties.	growing only few selected vegetables like kale, cucumber, bottle gourd, chillies brinjal tomato and potatoes using local seeds	Partial	Semi Organic	Indigenous Technology use advocated.

contd...

2.	Nursery raising of vegetables	Open, Hot bed and polyhouse technologies	Growing nursery in open beds.	Partial	Semi Organic	-do-
3.	Level of Cultivation	Domestic and commercial	Growing only for Domestic purposes in their backyards	Partial	-do-	-do-
4.	Plant protection	Seed treatment and IPM	Not undertaking any plant protection measure and face problems of cut worms, borers and wilt in chillies.	Full	Organic	Organic
5.	Use of FYM and Fertilizers	Use of sufficient quantity of well decomposed FYM and balanced dose of fertilizers	Using only FYM and a little quantity of urea	partial	Semi Organic	Indigenous Technology use advocated.
Rabi Vegetables						
1.	Rabi vegetables and varieties	Root, leafy, bulbs and tuber vegetables	Growing only selected vegetables like Kale, turnip and radish	Partial	-do-	Indigenous Technology use advocated.
2.	Level of cultivation	Domestic and commercial	Growing only for Domestic purposes in their backyards	Partial	-do-	-do-
3.	Post Harvest	Various post harvest technologies for vegetables	Following only natural sun drying methods in some vegetables	Partial	-do-	-do-

Livestock (Cattle, sheep, goat and poultry)

S.No	Item	Recommended	Existing	Gap	Cultivation Method	Farmers proposed strategy
1.	Breed 1. Cattle	cross-bred	Local and cross-bred	Partial	Semi Organic	Indigenous technology use advocated.
	2. Sheep & Goat	cross-bred	Local and cross-bred	Partial	-do-	-do-
2	Feed	Fodder and concentrate ration	Mostly Fodder	Partial	-do-	-do-
3.	Health	Regular dosing and vaccination	Nil	Full	Organic	Organic

contd...

4.	Milk yield	10-15 litres/day/cow	3-5 litres/day/cow	Partial	Semi Organic	Indigenous Technology use advocated.
5.	Poultry 1. Breed 2. Feed 3. Health 4. General Management	Broilar (COBB), Vanraja Recommended balanced concentrate ration as per age and type Regular dosing and vaccination Recommended scientific practices	Local Mostly feeding on household wastes Nil Practicing indigenous methods (Backyard poultry)	Full Full Full Partial	Organic Organic -do- Semi Organic	Organic -do- -do- Indigenous technology use advocated.

Table II : Health status of women in Islam (percentage of disorders found)

S.No	Health condition	Females (age-group)		
		5-20 years	20-35 years	Above 35
1	Normal Health	50	35	40
2	Malnutrition and anemia	45	40	24
3	Orthopedic Problem	2	25	65
4	Stomach Problem	-	25	30

Table III : Analysis of the main nutrient intake of the lactating mothers (Adult-moderate worker)

S.No	For lactating mothers	K.cal	Proteins(gms)	Calcium(mg)
1	Recommended allowances/day	2725	70	1000
2	Nutrient intake /day	1945	43	685
3	Gap found/day	-780	-27	-315

Table IV : Existing Farming Enterprise

S. No	Enterprise	No. of households	Percentage
1.	Agriculture	412	85.83
2.	Horticulture	91	18.96
3.	Animal Husbandry	137	28.54
4.	Sheep Husbandry	63	13.13
5.	Goat Rearing	91	18.96
6.	Poultry Keeping	274	57.08
7.	Bee Keeping	46	9.58
8.	Mushroom	48	10.00

Table V : Existing Farming System

S. No	Existing Farming system	No. of households	Percentage
1.	Agriculture +Animal Husbandry+ Sheep Husbandry/ Goat Rearing	423	88.1
2.	Agriculture + Horticulture + Poultry Keeping	57	11.9
	Total	**480**	**100%**

Table VI : Eco-Economic agricultural features of Kashmir

a. Land use pattern in J&K
Net geographical area 10.139 Mha

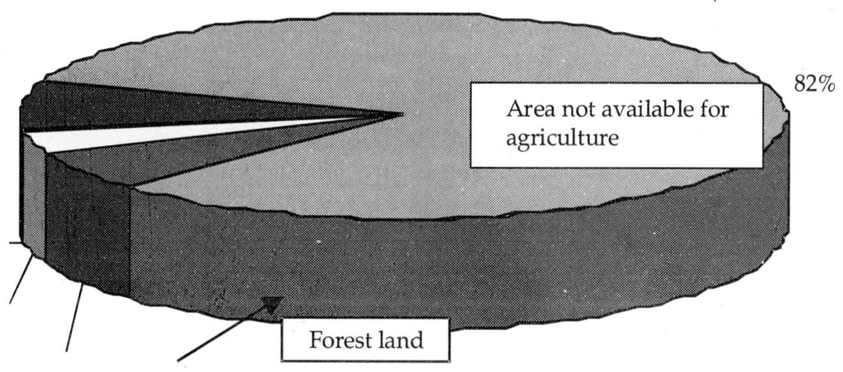

6.50% forest land,
3.30% pastures land
0.90% fallow land
7.40% Net shown land

b. Kashmir Economic Changes

GDP% of Sector	Financial Year	
1993-94		2001-02
Agriculture	32.9	29.9
Industry	21.8	17.4
Services	45.4	53.7

Rural/ Urban/PC/monthly expenditure

	J&K US$	All India US$
Rural	15.43	9.0
Urban	21.9	20
Total	16.9	12

c. Average Productivity (APPLE)

Year	Area (Ha)	Production (M. T)	Productivity
2001-02	90080	909583	10.09
2002-03	94874	953946	10.06
2003-04	100702	1041538	10.34
2004-05	107925	1093275	10.12

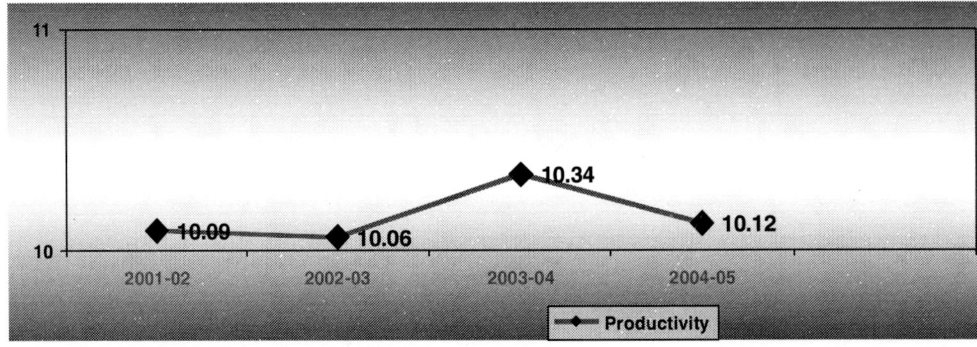

d. Apple Production trends (2002)

Country	Area (Hectares) in Lakhs	Production (M.T.)	Productivity (MT/HA)
China	24	2,40,07,000	10.00
U.S.A.	1.86	48,50,000	26.08
India	2.31	13,80,000	5.97
J&K	0.9	9,09,583	10.10

e. Horticulture Product statistics of J&K

Product	2004-05 (Lac Tons)	2005-06
1.	Apples	10.93 11.52
2.	Cherry	0.074 0.08
3.	Pears	0.40 0.42
4.	Walnut	1.005 1.083
5.	Almond	0.135 0.143
6.	Others	0.002 0.002

f. Soil fertility status of kashmir Kg/hac

Altitu	N	P	K	S	Z	Mn	Cu
Lo	195	30	298	17	0.8	36	2.6
M	190	35	327	20	0.8	64	2.6
H	207	37	346	22	0.88	39	2.0
CR	350	30	200	20	1.0	-	2.0

g. Soil improvement through organics

Materials %	N	P	Micronutri
Mudwaste	0.8	1-4	55
FYM	1-2	1-2	40
Decioled cakes	1-2	1-4	45
Municipality waste	1-2	1-3	45
Green waste	0.5-2	1-5	50

h. Indian energy resources-2012

Coal	1.15 lac MW
Gas /LPG	0.31
Nuclear	0.12
Hydro	0.58
Total	2.15

Kashmir alone has a potential of generation more than 1 mw of hydroenegy if all cannals, nallahs and water resources are retaped, redirected and chanalised both for energy and irrigation.

i. Production Cost Per Unit Of Energy

• Coal/Lignite		6-20 cents
• *Natural gases*		4-9
• Nuclear		10-14
• Biomass		8-12
• Hydropower		2-8
• Photovolters		25-50
• Wind		4-5
• Crop residues in India	2001	70 ML/T
• Estimates	2021	200

Graphic-I : Organic Farming Options Tested

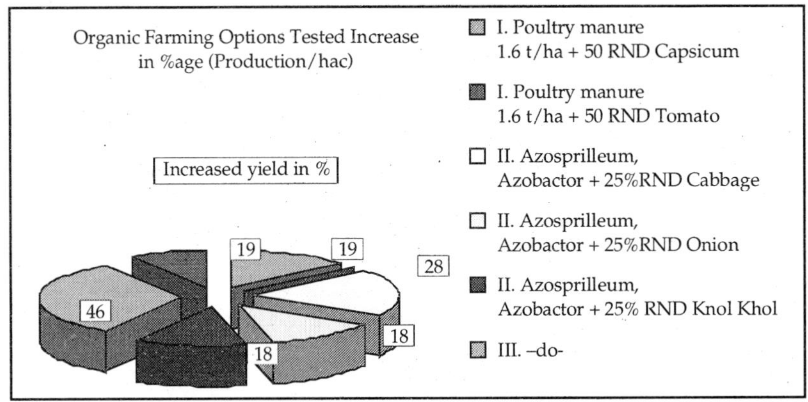

Graphics-II : Organic Apple Production N % age

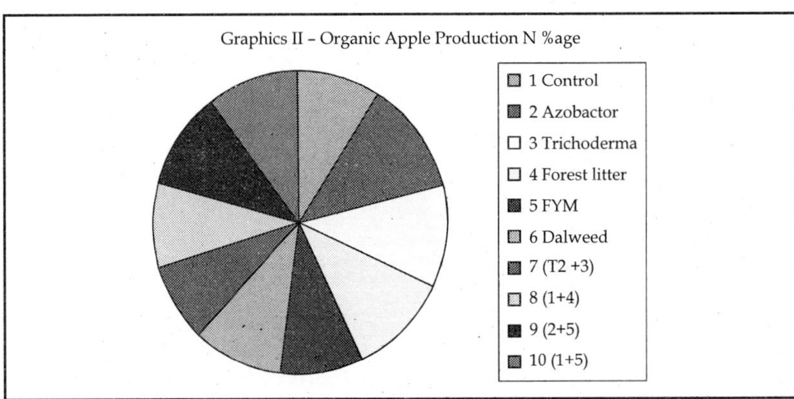

Graphics-III : Organic Rice Production Yield/ha Yield of Rice Tones/per hac.

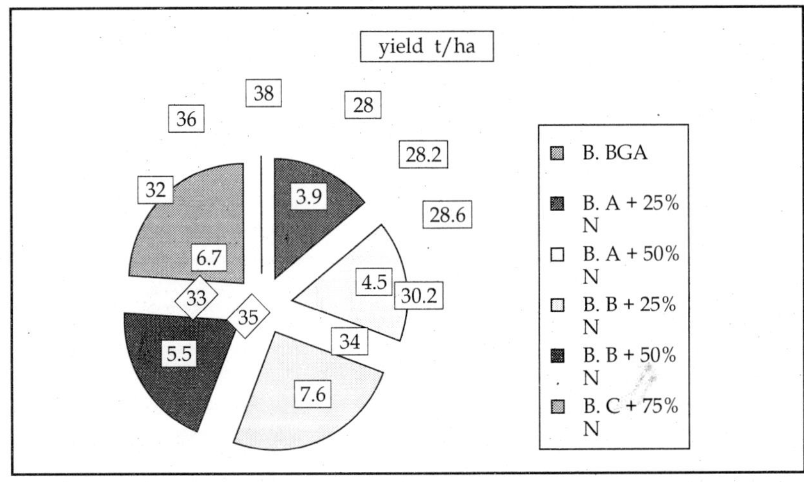

Fig. 1 : Water Mill

Fig. 2 : Summer Livestock Shelter

Fig. 3 : Summer Livestock Shelter

Fig. 4 : Post Harvest Equipments

Fig. 5 : Summer Livestock Shelter

Fig. 6 : Post Harvest Equipments

Fig. 7 : Plough

Fig. 8 : Goat Winter Shelter

Fig. 9 : Farm House

Fig. 10 : Indigenous Wooden Implement

Fig. 1-10 : Indigenous Technology Mapping

References

Amano K, 2004. Attempts to introduce stingless bees for the pollination of crops.*Bulletin Feed and Fert.Tech.(167):1-9.*

Berthier, D; Chantal,I; Thevenon,S; Marti,J; Piquemal D and Maillard J.C, 2005. Bovine transcriptome analysis by SAGE technology during an experimental. Academy of Sci. 1081:286-299.

Byarungaba, D.2004. Stingless bees (Hymenopteva:Apidas) of Bwindi impenetrable forests, Uganda *Int.J. of Tropical Insect.Sci.24(1):117-121.*

Dixit, P.K and Sinha, M.A 1993. Dairy farming technologies in rural Bangalore. *Indian J. of Dairy and Biosciences. 4:44-52.*

Dwivedi,-H-P; Mahesh-Kumar; Srivastava,-S-K, 2006. Risk factors analysis for prevalence of different Leptospira sp. in ruminants of sub-Himalayan Kumaon region. *Indian-Journal-of-Veterinary-Medicine*. 2006; 26(2): 100-103

Eckner,-K-F; Zottola,-E-A, 1991. Potential for the low temperature pasteurization of dairy fluids. *J. of Food Prot. 54(10):793-797*

Fleury, A and Mollard, A. 1976. Agriculture, the social system and the environment. Published IREP- CNEEJA, Grenoble, France, INRA, Cards, Feb, Marach, 1977.

Hetzel, B.S, 1977. The nutrition of Aborigines in the eco-system of central Australia. *Proc. Nut. Soc. of Australia* 2: 28-31.

Itty,-P, 1993. Economics of trypanosomiasis control: research implications. Future of livestock industries in east and southern Africa: Proceedings of the Workshop held at Kadoma Ranch Hotel, Zimbabwe, 20-23 July 1992

Joshi, B.K, 2006. Traditional knowledge and land use Management in Indian Central Himalayas. *Indian J. Traditional Knowledge* 5(1): 34-40.

Kanwar, P and Sharma, N. 2006. Traditional storage structures prevalent in Himachali homes. *Indian J. Traditional Knowledge* 5(1): 98-103.

Karthikeynan, C; Veeraragavathathan; D, Karpagam; D & S, Ayissa Firdous, 2006. Cow based indigenous technologies in dry farming. *Indian J. of Traditional Knowledge*. 5(1): 47-50.

Lans, C and Brown,G , 1998. Ethnoveterinary medicines used for ruminants in Trinidad and Tobago. *Pre. Vety. Med.* 35 (3):149-143.

Liu-Defu, 2004. Diagnosis and treatment of toxoplasmosis in cows uring traditional Chinese. *J. of Vety.Sci. & Tech.* 34(6):73-74.

Lodics,-T-A; Steenson,-L-R, 1990. Characterization of bacteriophages and bacteria indigenous . *J. of Dairy Sci. 73 (10):2685-2696.*

Masud,-T; Sultana,-K; Shah,-M-A, 1991. Incidence of lactic acid bacteria isolated from indigenous dahi. Asian-Australasian-*Journal-of-Animal-Sciences*. 1991; 4(4): 329-331.

Nanda,-J; Swain,-R-K; Sahu,-B-K; Samantaray,-D-Pl, 2006. Effects of strategic supplementation on work efficiency of bullocks. *Ani. Nut.Feed Tech.* 6(2):167-176.

Njoroge, G.N & Bussmann, R.W. 2006. Herbal usage and informant consensus in Ethnoveterinary management of cattle. *J. of Ethno.*108(3):332-339.

Singh.A and Kalra,-K-K, 1995. Milk Production potential or Rohtak district.*J. of Dairy Sci. 48(9): 554-558.*

Srivastava, S.K, Attri, B.L and Pandey, H. 2006. Indigenous wisdom for the use of gaint weed in disease and pest management. *Indian J. Traditional Knowledge* 5(1) : 83-86.

Iqbal.S and Pampori, Z.A (2008). Production potential and qualitative traits of indigenous chicken of Kashmir. Livestock Research for Rural Guide for preparation of development (LRRD News) 20 (11): 1-10.

Sanon,-H-O; Kabore-Zoungrana,-C; Ledin,-I, 2007. Behaviour of goats, sheep and cattle and their selection of browse species on natural pasture in a Sahelian area. *Small-Ruminant-Research.* 2007; 67(1): 64-74

Singh, G.B and Singh,A, 1980. Shifting cultivation in the N.E region. *Indian Farming* 29(10): 17-20.

Sen,-B-A; Beg,-M-A; Khan,-M-Z; Wani,-G-M, 1998. Studies on the efficacy of RPL-1 for preventing retention of placenta in freshly calved crossbred cows - a field trial. *Indian-Journal-of-Indigenous-Medicines.* 1998; 19(2): 120-123.

Sharma,-M; Bisoi,-P-C; Mohanty,-B-P,1998. Serum constituents of indigenous and crossbred cattle. *Indian J. of Ani.Sci.* 68 (5):474-475.

Shi-PeiLi; Li-WenHua; Shi-PL; Li-WH; Kessler-E, 1999. Rehabilitation of degraded mountain ecosystems in southwestern China *Ambio, .28:5, 390-397.*

Uma-Sah; Shantanu-Kumar; Fulzele,-R-M; Alok-Pandey, 2002. Indigenous knowledge system for combating foot and mouth disease in cattle in India. *Indian-Journal-of-Dairy-and-Biosciences.* 2002; 13(2): 145-148

Wani, G.M.2008. Food Security under new global marketing system. Lead paper Proc. 4th JK Science Congress, University of Kashmir, 2008.

Wani, G.M.2008. Eco-friendly food production through technology outreach. Guest lecture series I delivered at : Hamdard University, Faculty of Medicine (Unani) New Delhi, Dec. 2008.

Wani, G.M, 2008. Global status of Biotech/G M Crops/ Products, regulations thereof. S&T Dept. sponsored seminar at Kashmir University, Biotechnology dept. Srinagar, Kashmir, Lead paper I.

Wani G.M., (2007) Recent Trends in Policy Initiatives & Technological intervention Goat production as livelihood resource, presented at Sri Venkateswara Veterinary University, *Tirupati w.e.from 20-22, June,2007*

Wani, G.M. 2001. Enhancement of chevon production through reproductive management. In: Proceedings of IV National Seminar of Indian Society for Sheep and Goat Production & Utilization (ISSGPU), held at Bharathiar University, Coimbatore from November 9-11, 2001.pp.60-68.

Wani, G.M. 2002. Goat Husbandry/multipurpose management. Encyclopedia of Dairy Sciences. Edited by Roginishi, Hubert, Enquay, Jhon W. and Fox patrict, Academic Press, London, NW 17B4MS192.

Wani, G.M. 1984a. Augumentation of fertility in Goats. Seminar-cum-Training programme on goat rearing and production sponsored by D.O.Extension, GOI at CIRG, Makhdoom, Mathura.

Wani, G.M. 1984b. Embryo Transfer and its application in goats. Seminar cum Training programme on goat rearing and production sponsored by D.O.Extension, GOI at CIRG, Makhdoom, Mathura.

Wani, G.M. 1996. Embryo Transfer in sheep and goats. (Lead paper). Proceedings XIII National Convention of Indian Society for the study of Animal Reproduction, GBPUA&T, Pantnagar.

□□□

Chapter 32

Employment Opportunities for Agricultural Graduates

G.M. Wani

Introduction

In the past few years the growth of Agricultural Universities have increased from few in 1940 to more than 40 by 2004. The enrolled students number increased from mere half a thousand to nearly 20 thousands between this period. As on today we have 260 colleges in India imparting educational degrees in Agriculture alone. They consist of various disciplines including veterinary, horticulture, sericulture, forestry, floriculture and a number of other sub-disciplines as plantation crops, medicinal herbs, mushroom, seed production, dairy production, agri-management and like. A number of Government department in India employ agricultural and allied graduates. The mode of recruitment is through respective State & Central Govt. public service commissions.

Beside professional employment, the agricultural graduates can compete in the national or State administrative or financial and allied services. Similarly many banks, insurance companies, private industries, private concerns and development companies. The NGO do also employ agricultural and allied graduates. The graduates have to go through employment new and other professional Journals through Radio, TV and both central and State Govts.

also give advertisements. A vast number of professional, managerial and industry related advanced courses after agri or allied graduation are also advertised. They include business management, financial management or related disciplines in corporate sector. A diploma in mass media, computers or even commerce shall fetch a graduate additional employment chances in Govt. and private sectors. Opportunities in foreign countries for agri-graduates are also there, however this would need diploma in international agriculture and a driving licence besides a passport.

For graduates a number of courses are available for Masters. Even on date we have 1/3rd positions vacant in SAU and ICAR Institutes, for which annually ARS examination is being held by Agriculture Scientists Recruitment Board. The recruitment to scientists and Assistant professors is open to all M.Sc /M.V.Sc or allied post graduates and Ph.D holders. Ministry of Agriculture in collaboration with NABARD and other banks have launched Agri-business Agri clinic courses. They are 3 months courses on the job training to Agri-graduates free of cost. In fact all the expenses are borne by the Agri-business school. A network of such training schools have been established. We have also a dozen in J&K. There is a strong future for agri-graduates in private sector. They can set up crop, fruit nurseries or open seed sale and production centres. We have a number of successful agri-graduates earning more than Rs. 20,000/- a month in various parts of the valley. The whole agricultural and allied sector has a strong employment generation potential. We have identified 15 such agri business ventures, which can be profitable employment ventures. A great potential of employment is hidden in Nursery plants for fruits and vegetables. An investment of Rs. 30,000/- per year may give a net profit of Rs. 2 lakh per annum.

Australia also provides many jobs like in Europe and USA for Indian Agri-graduates. The respective embassies or the internet employment advertise many help Agri graduates. For Australia information can be had on www.dookie. Uniemb.ed.am/. More personal information can be had on phone 0427355721 or email saparker@uniemeb.edu.an. Likewise many websites are available around the globe and in India too. New ventures financed by banks are Agri.biotechnologies and precision farming for organic Agriculture.

Agricultural development over years has been the result of continuous agri skill generation and its popularization. The earliest agriculture was animal domestication. Over thousands of years man domesticated wild fowl, dog, goat and smaller animals, whom he could overpower easily and subjugate to his sub-ordination. Agriculture thus since beginning has been the results of trails, experiments and experiences over years, learned first though behavioral changes, psychic reoccurrences, memories passed through parents to children and later on through doing and learning and now through sharing experiences and writing them or dotting them as an Entrepreneurship concern.

Entrepreneurship Concept

The Entrepreneurship adds economic profits and cost-benefit ratios to Agricultural Output. Entrepreneurship is dominated by four factors like:

a. Social changes
b. Support system availability
c. Resource base and its utilization
d. Self confidence, exploration work capacity and intellectual potency.

An entrepreneur has to have a thinking of his own, a capacity building interest in acquiring needed technique. An explorative and analytic faculties to judge the way of procuring cheap raw material. He must be equipped with "knowledge" and mindset to use and benefit out of it.

Farm Business

A potential entrepreneurship must strive from getting maximum output. Decades back agricultural development and industrial setups was a public sponsored and heavily subsidized but over time "knowledge" explosion in Indian Agriculture, have brought us on threshold of a system, where wide distances exist between industry and farm business. Where huge subsidies are benefiting Agro-Industrialists. The Farmers who use fertilizers or agro-chemical are crushed under economic pressures. The gaps between technology generated and technology use at farmers door is increasing day after day. The farm technology adoption rates are not more than 20-30% by any higher prospective.

The use of information and communication technology (I&CT) for reducing the gaps and increasing productivity is the need of the hour (Wani, 2005). The modern technology and knowledge flow is fast expanding and bringing change. It demands more educated and trained farmers. Our education system has produced more literates but not educationally trained youth to earn their own bread. They after attaining graduation in agriculture and allied sectors, beg for job. The system has to be corrected to make these graduates as employers and not employees. I wrote a treatise in 1992, emphasizing a system where our agriculture graduates should be employers not employees. Germans are smart to have Farmers school, Farmer business training institutes, practical agri-farmers training centres and like, where every farmer or animal husbandry man is essentially a trained fellow. The banking system is so organized that they are on the door of convocation hall to sell their agri-business and agri-clinics packages to graduates, without any investments. Banks are so smart, that they have surveyed the villages who need vets or agri-graduates or have attained land and all facilitation, so that agricultural or veterinary or even other medico-biological graduates are used as bank investment. This is what is envisaged in India under agri-clinic, Agribusiness venture. We have trainings not in the hands of banks but with universities. There is practically little interaction between universities or banks or employers.

Success of Agri-business

A poor liaison and support system between Govt. banking and University culture has made this otherwise an remunerative and lucerative programmes into a failure inspite of its personal monitoring by PMO. We visited Bandipora district and unregistered Agriclinic graduates were earning a handsome salary, more than the Rahbar-e-zerat or Agriculture Asstt. Agribusiness viz sale of pesticides, cattle feed, poultry feed and agri-extension services were common. At a small village in Papchan, one agri graduate Mr. Iqbal Shah earns Rs. 10,000/- per month by selling the services and input. At a distance of few kms in the same district one Mr. Khyatlani owns a big poultry farm and earns around Rs. 20,000/- per month. Both these entrepreneurs employ 2-3 persons at present. Similarly, the success shown by one Mr. Shah at Malangam in Agri products and pesticide sale and one Mr. Bhat in Dairy production and milk product sale earn a handsome income besides generating employment for poor. All these agricultural graduates have started their own business concerns without any bank or university help.

Farmer as Entrepreneur

Indian Farming and farmer has to change if proper WTO recommendation and GATT agreements are to be followed. The present day poultry scenario is emerging as high profile agri-business in India. The conversion of poultry farmer's into poultry entrepreneurs shall make the present day 6% contribution of poultry products from India and China to 25% share of Global market. This when translated into action shall increase employment generation by manifolds. The introduction of rural based Vanraja, Gramapriya, Giriraja, Cari Gold and vast other locally grown varieties of poultry have adopted well to our agri-rural base. The market acceptability is higher than exotic poultry concerns. Free-rang-poultry is like BT cotton hybrid spreading through villages of India and assuring high returns and exports (wani, 2007).

Poultry as Agri-business

Dr. Gordon Butland, president of Global poultry strategies presents "Backyard poultry production" as a tool of alleviating poverty and malnutrition. We have tried to distribute "birds" under free-range system in all our KVK's, our results were excellent and income generation was totally in favour of the Agri-business and agri-clinics.

This all will need the involvement of Agri-Veterinary and food processing technocrats to develop rural-based establishments so as to foster export and fast returns.

Holistic Vision for Livestock Enterprise

Improving income, employment and self-reliance among educated graduates and un-employed youth especially women needs fostering community development, women empowerment, environmental protection. Rural-based backyard poultry subscribes to all these norms and could be a rich resource for developing agri-entrepreneurship. Govt. of India is liberally financing such agri-business ventures. Some of the success stories in animal husbandry section can be reproduced as follows:

Backyard Poultry and Incubation

Though the Vanraja are the most suitable for back yard poultry, they do not have habit of broodiness. There is a problem among the farmer to get a broody hen in all season. KVK solve this problem of hatching by installing small unit of hatchery. Every month 15-20 farmers are benefited by purchasing chicks for backyard poultry. There are 200 back yard poultry units of Vanraja. Each farmer is rearing 10 to 25 in the backyard. There is a good demand and response for the chicks and eggs of Vanraja. KVKs are now planning to expand this entrepreneurship by Agri-graduates.

Semi-stall-fed Goat Rearing

KVK's made an intervention to improve this enterprise by conducting short duration training programmes for rural youth. Similarly exposure visit were organized on goat feed, breed and health management. More emphasis was given on Osmanabadi goat and up-gradation in selected non-descript goat breed by osmanabadi pure buck and given the knowledge about semi stallfed goat rearing concept. This experience in Andra Pradesh shows a great potential for agri-graduates own business.

Broiler Production

A KVK has conducted training programmes for 165 trainees. Due to training and demonstrations awareness was increased about contract farming in broiler production with private sector which provide chicks, feed and medicine and after 40 days purchases them back. They acquired skills through learning by doing at KVK demonstration unit and earned money. The technology has been adopted by 10 percent of youths now in the radius of 20 km. There are 27 poultry units having capacity of 5000-10000 poultry birds on contract farming basis. These self employed rural youth earning Rs 10000-15000 per lot.

A seminar-cum-farmer's meet was arranged at SKUAST-K on 26-27th of Oct.2007. The knowledge -sharing with farmers by agricultural graduates and scientists was emphasized by His Excellency the Chancellor and Governor of J&K. A vision of poverty alleviation through backyard poultry intervention was the theme of the seminar. Many belts in Gurez, Tangdar, Tillail and Zanskar are rearing native livestock species. They are better suited and need improvement and identification. The cooking methods will need more expansion and scientific intervention for export. More emphasis has to be made on:

- Safe feed and food.
- Organic fodder and food.
- Operational excellence and modern mechanization to improve quality of indigenous enterprises.
- Local family management to farm business management and seller-buyer mode adoption.

This will need graduate farmers who are trained. Thus future farmer has to be a farmer graduate as in USA, Germany etc.

High Value Agri-business

Rapid growth rate in high value commodities in Indian agriculture promises 40% total output. The sectors assuming importance for export earnings are Fruits, milk, vegetable and poultry. Thus Agriclinic training centres should focus on these commodity oriented trainings. The sector may need more than 1 lac young agri-graduate entrepreneurs to achieve national goals and not the mere 14000 un-employed agriculture graduates.

Technological Prospective

The technological gaps between research yields and Farmers yields are wide. The technological know how over years is on shelf. Much of it has remained frozen if not dead within the fore walls of institutions. The location specific, crop and soil specific innovations to farmer's practices are few. Our integrated & mixed farming practiced by our farmer's, needs packages recommending mixed farming improvements which is not forth coming. Our research institutes mostly aped the experimentations on commercial farming of west with few modifications and even additions at times. Our traditional agriculture existing still in remote areas encompass the use of farm yard manure, vormiculture & use of agricultural waste materials. The present problems of low soil fertility, reduction in yields and soil salinity have arisen due to excessive use of chemical inputs in farming with little care of the soil.

This scenario resulted due to incomplete innovative approaches of research. Our researchers blindly advocated more and more use of fertilizers, pesticides and fungicides, which helped to gain grain revolutions, but left legacy of polluted water, air & environment with degraded soils. These revolutions debarred future sustenance. This was due to poor perception.

Thus immediate need is to make researchers akin with information technology and advanced communication. The rapid evolution of information science demands quick and speedy transfer of technologies, awareness & even subject reviews to farmer's for speedy application. The productivity would be better if technological advances are adopted and their impact is known. The knowledge of computer hardware, software as well interlinking the information dissemination channels and outlets is essential. This will consume hundreds of Agri-graduates in mass media, information and communication.

Our Vision

Mixed Farm University Culture

Higher productivity gains can be achieved through application of technology and production recommendations at farmer's fields. We have 65% small and marginal farmers whose awareness potential is low. The production system prevailing with these farmers is a mixed farming or composite farming. In contrast to USA and European agriculture our necessity is to increase "Crop -livestock-fish-plant integrated production system with multiple livelihood opportunities". Therefore, we need our own innovative educational and training policies. A mixed agriculture University and Education set ups is our necessity. We are at present going astray to our need. Quick and fast measures and needed to unify our educational system, involving all agriculture and allied disciplines, industries, corporate sectors and farmers institution.

Higher Productivity Concerns

Indian Agricultural pride years of green revolution post 1968 saw reduction in food gain imports and subsequently white, blue and other revolutions sustained our population pressures and agriculture growth. Our agricultural growth rate (AGR) need to be equal if not more to population growth rate (PGR). Our AGR target ought to be double the PGR.

This is important as consumption rates, purchasing power and employment prospects increase. An estimated food grain of 210 million tons at present may need to be doubled in next 10 years. We have to achieve high targets of productivity by vertical expansion as horizontal land expansion is just not possible. Dr. M.S. Swaminathan has quoted figures as of 160 million tone's of rice from 40m hac of land, thereby setting the productivity target of 4 t/h. Like wise production of 100 million tones of wheat from 25 million tones of wheat from 25 million hectors needs a productivity of 4t/hac. Our aim to double our per hac productivity needs more technical manpower in extension, industry and at gross root level.

The climatic disasters, earthquakes, Titanic tsunami, floods; have effected our agricultural production in the past and additional requirements needs to be kept in mind while planning food security. We need to increase per capita consumption expenditures of Rs.600 per month. We need to bridge the gaps between potential and actual yields at farmers level. The chemical farming hazards of poor soil fertility, low water availability, pollution and environmental concerns impede our agricultural development. Thus refined technology, participatory research and educational modules are needed. The new pressures of Global marketing. World trade and tariff regulations have to be accommodated. This all will need incorporation of new themes like post-harvest management, value addition, packaging, communication, credit and market information services in our course curricula. Thus a new multidimensional change in academic curriculum is envisaged, which may open new opportunities for Agricultural and allied graduates.

Quality Assurance

We feel pride in calling ourselves as the 2nd largest Agricultural Research system (ARS) in World. When we review our performance we are no where in top ten of most cited agriculture publications countries in the World. USA tops the world list with 3,62,79,842 cited publication/annum, with small country like Switzerland at No.10. The scientific out put in agriculture is highest in USA with 27 lac publication/year followed by Japan, Germany, U.K, France, Canada, Italy, Russia, China and Australia. Our contribution to Agriculture publication is 5.48% only with our share of citation at 2.32%. This demands more focus on Quality Assurance. Our prime agenda should be quality Agriultural Education. Our emphasis has to be on:

Academic quality, Accreditation; Desired knowledge, Assessment, Skill and competence building and academic audit. Quality assurance, means strengthening resources, information and maintenance of educational infrastructure. Thus we need to regulate grants and centre-state relations rationally. This will open golden opportunities to our graduate across the borders.

Employment Opportunities

We have 36 state/deemed or central agricultural Universities and 20 general universalities with 48 agricultural faculties. The total disciplines needing grants may be strengthened in 5 yrs by 1 core grants to each discipline for quality assurance. We produce 10,000 under graduate 5500 post graduates and 1600 Ph.D in agriculture every years. They add to our unemployed pool. For making them self employees in new ventures and for increased employment , their competence building in Global economics and trade policies. Biotechnology, Bioinformation, Biofertilizers, pesticides and fungicides etc are to be enhanced. New faculty development in all the Universities and colleges is to be executed in coming 5 years. Such as :

Pest information and survey; Risk Management Analysis; Decision support system; Geographic information system. A new trust is to be given to course curriculum integrating field practices in a partnership mode with farmer. A teacher-student-farmer-industry, interaction and co-operation is to be developed. A new model of mechanics in Agricultural and allied curricula is to be integrated, unified and fine tuned to end results. This will demand inter and intra faculty harmony and synchronized course curricula at UG, PG and Ph.D level. This has to be fine tuned to our field requirements and location orientations and Employment opportunities will be the end result.

Asia Specific Agricultural Education

Indian Economy is a agri-centre economy which supports 70% of our population, as direct rural employment . Forty five percent of the income generated by industries comes from Agri-based (Agro) Industries. Therefore , a vast potential and resource is hidden in it. If we think of Asia specific

Agriculture, we have to play a significant role in the region which has 60% of the world population. The region is rich of energy and oil resources and millennium buzz word is open boarders and common market with first priority on peace, confidence and trust. For up-liftmen of region we need training education and human resource utilization. Therefore Agriculture educational reforms are on our door steps to harvest the gains of common economy in the SARC and total Asian region. These educational reforms should involve schools, colleges and Universities. The re-modeling of curriculum will need incorporation of new emerging era, like competitive global marketing, the climatic, disaster, technology use, restrictions, sustainability, environment, water resource conservation, remediation factors etc.

Agri-educational reforms are needed so that ICAR parallels USDA, in governing grant-in aid to whole agriculture sector. An omnibuss act of agriculture in the shape of USA farm bill of 1996 is envisioned. New educational policies so drafted shall be non-discriminative, comprehensive , transparent and accountable. They will pave a way to total employment in Asia for Agri and allied graduates.

Access to Education and Training

Access to education and training to people below the poverty line, rural youth and women is to be ensured. This will need a total restructured education infrastructure. A three tyre model is envisioned which consists of:

1. On the job, training opportunities on farm mechanization and agriculture.
2. Training skills, up-gradation and rural orientation at University level, refinement and more innovative participatory mode at farmers field. Roaming teaching taught system on holidays and Sundays.
3. Teacher-student-farmer-industry-interaction-work and planed-self-learning by living with farmers. It will ensure quality training and job improvement of skilled manpower for use in Asian Agri development Market. The utilization of Rehbar-e-Zerat in this paid Agri-services to farmers shall provide them job and money.

Informal-flexible Agriculture Information Services

A flexible curriculum models which should have many options at B.Sc level like:

1. Natural science
2. Agri Science

- Production system
- Agri-business
- Social science

3. International Agriculture
4. Natural Resources
5. Agri-business management
6. Biological engineering
7. Dietetrics
8. Landscape Architecture

These inclusions at B.Sc or Post graduate level shall help in graduate employment.

Export Orientation in Agri-education & Employment

In the present era of bio-safety, phyto and zoo sanitation have assumed tremendous importance. Education and knowledge was safe and secure treasure in the past. It is no more true. The export needs knowledge and new inventions need patenting. Web and web designing have made invisible teachers to unknown students. The students-teacher relationship has raised to spiritual horizons. New targets for future educational planning and policies need to have more information and communication technology. Therefore courses on I&C with computer applications is must. These have been strengthened in the SAU and ICAR institutions in the last few years. However a total connectivity is needed with farmers, farm organizations and utilization departments to harvest the gains of technological reforms to increase our exports. Inspite of ranking I in milk production our exports are meager. This is because of poor -zoo-sanitation and Global lobiest are critical of our disease free status. Therefore, policies and programmes in Agri-export orientation and the training to our agri-vet. And allied graduates is a must. Training of these graduates in management, mass media, export-import ventures, industry and related agrobased ventures like sales of fertilizers, pesticides, seeds etc is needed.

References

Bishnapada, Sehthi, 2007. Backyard Poultry in Orissa. Orissa Review, January, 2007. pp 48-52.

Dipak De, 2004. Workshop Background and recommendations In communication support for sustaining extension services edited Dipak De, Jirli,B and Ghadi,K. Deptt. of Extension Education, Institute of Agric. Sci, BHU, Varanasi.

Dobrowski, Z.T.2000.The necessity of changes in the methodology of development and implementation of integrated pest management.*Pre-conference Symposium – Progress in Plant Protection, held at Nowoursynowska, Poland, 40(1): 334 -342.*

Frewer, L.J; Howard, C and Shepherd R. 1998. The influence of initial attitudes on responses to communication about genetic engineering in food production. Agriculture & Human values. 15(1):15-30.

Hardman, PA; Darroch, MAG; Ortmann, GF; Trienekens, JH, and Omta, SWF,2002. Improving co-operation to make South African fresh apple export value chain more competitive.*Proc. 5th Int. Conf. on Chain and network management in agribusiness and food industry. Noordwifk, Netherlands 6-8 June, 2002 PP 434-443.*

Jacques Diouf 1998,Director General, FOA*Report on World Food Day, 26th Oct, 1998 Rome.*

Luttichen, R.E; Robert, P.C; Rust,R.H and Larson, W.E.2000. Development of an internet based communication and information network to progress the implementation of precision agriculture.*Proc. 5th int. conf. on Perecision Agriculture, Bloomington, Minnesota, USA, 16-19 July,2000. Publ. 2001 CD. Page 1-14.*

Report of the Quinquennial, Review team for KVK Zone V, AP & Maharastra, September,,2001, *Edited Wani, G.M, C.M.Singh. Gogoi, DK; Sandhu and Suilac C,9th September,2001, Hyderabad.*

Sher-Muhammad; Garforth, C, and Kausar Almas, 2001. Adoption of recommended agricultural technologies by farmers as an indicator of effective communication by extension field staff.*International Journal of Agriculture and Biology. 3(1):13-15. b*

Sidiqui, M.A.A, 2004. Plant Resource Management for Entrepreneurship Development. Agri-Clinics and Agri-Business, Deptt. of Extension Education, SKUAST-K, Shalimar, Srinagar- J&K.

Sutton, A.L; Ong. H.K; Zulkifli, I; Tee, TP and Liang, JB. 2002.The role of Education and Technology Transfer in Livestock Waste Management. *Proc. 4th Int. Livestock Waste Management symposium 19-23 May,2002, PP 281-288. Published Malaysian Society of Animal Production, Selangor, Malaysia.*

Wani, 2004. Role of communication in enhancing agricultural production – a Vision . In Communication support for sustaining extension services edited. Dipak De, Jirli, B and Ghadi,K. Deptt. of Extension Education, BHU, Varanasi, PP 1-26.

Wani, G. M.2000, World Food Day Bulletin, 16 Oct, 2000, Directorate of Extension Education, SKUASTK, Shalimar Srinagar, Kmr.

Wani, G.M. 2004. Agriculture Production & Entrepreneurship – Preface to the proceedings on Plant Resource Management for entrepreneurship development, Directorate of Extension Education, SKUAST-K Shalimar, PP 1-4.

Wani, G.M.2003 Appraisal of Training needs of Farmers. *Lead paper presented at National Seminar on "Apraisal of Institutionalized Farmers' Training Activities'organized by Indian Society for Agriculture Extension at GBPU of Agriculture & Technology, Pantnagar, U.P.*

Wani,G.M.2007. Message, "Backyard poultry Production". Seminar on Backyard poultry Farming for women empowerment & nutritional security Oct. 26-27, 2007, SKUAST-K, J&K PP.

Yahaya, M.K and Omokhaye,S.B.2001. Cocoa farmer's perceived influence of communication channels on utilization of improved cocoa seed technologies in owan -east local Government area of Edo State. Moor - Journal of Agricultural Research 2(2) : 199-207.

❑❑❑

Chapter 33

The Horticulture Development in Jammu and Kashmir, Past, Present, Future Research and Outreach Initiatives

G.M. Wani

Abstract

Agricultural Production has entered new era of competitive-low-cost production oriented enterprise. The cost-benefit ratio decide the mode of Farming. Global marketing needs some changes at base level. Farmers have to be cost conscious and need new strategies of mixing various farming option like Agriculture, Horticulture, Livestock husbandry and Floriculture activities. A new model for incorporating all these fields in an eco-friendly mode is discussed. Appropriate mixed technology in tune with natural resources, is needed at the moment in India.. An agricultural mission for efficient resource management, bridging gap between current farm yield and potential Farm yield is needed. Use of bio-technology and bio-informatics for identification of economically viable cropping options and combinations is a must. We should talk of eco-conducive-germplasm-base instead of high productive - germplasm, uncompetable to our natural resources(Wani,2008)

India with 1.09 billion people and 60% of them dependent on agriculture can not be leveled with USA who has merely339 million people and only 0.7 of them dependent or agriculture.

The horticulture sector of J&K state is most vibrant sector for economic development. The per capita income of J&K is 324 US dollars which is more than national average. The monthly expenditure of J&K is also higher than national average. The urban /rural differences are lesser due to horticultural earnings of the farmers. The increased income from horticultural crops has resulted in diversification of cereal land to horticultural crops. Various features of horticulture in J&K have been presented in this review (Table 1-10) Area under horticultural crops have increased from mere 12 thousand hectares in 1953-54 to 221 thousand hectars as on date. Most of the denuded forest area, high altitude rice, wheat and maize lands have been diversified to more renumerative horticultural crops. The per unit land profits from horticultural crops is more than the cereals. With increased environmental degradation and low water availability, the diversification to horticultural plantations is on increase. Our overall productivity of 1.2 tons per hectare was very low in 1953-54 It too has increased to around 4.95 t/hac. Thus, our horticultural development has shown both vertical and horizontal expansion. Various factors favouring horticultural productivity have been analysed. A strength and weakness constraint analysis have been applied and shown in graphics, & tables As shown in tables the average productivity of apple in 2004-05 has increased to 10.12 t/h^{-1}, which is marginally higher than China, double than national average and $1/3^{rd}$ of that in USA (Table 8). Thus, we need more advanced technological application for increased productivity of apple in J&K. All quantitative and qualitative traits have to be exploited through second generation biotechnological tools and molecular and nuclear techniques. New germplasm propagation using DNA identification and mapping systems are needed to increase our 20% A grade quality fruits to around 40% superior quality fruits. Perhaps new ladder hedge system of orchard management as in USA needs a try to combine ornamental lawn hedge with horticultural apple hedges. Various developments of Indian horticulture have been summed up in Fig 1-3. Various projects, schemes and options have been discussed and presented A new mechanization and research use is suggested to have biosafe and marketable fruits at international level.

Introduction

The horticulture sector in India has an important role in economic development of rural poor. Livelihood options and cultural development are key factors related to horticultural development. India has a variety of horticultural crops including fruits, vegetable, roots, tubers, flowers, ornamental plants, medicinal, aromatic plant species and other plantation crops. India is the second largest producer of fruits after China and contributes 11% of all vegetables and 15% of all fruit production in the world. Majority 65-70% of fruits and vegetables produced in India are consumed domestically. Two percent of fruits and vegetables are being processed. Out of the total production only 1%is being exported Post harvest losses account to 20-30% of the stored fruits. The major contributor to Gross domestic product of J&K is horticulture (Wani, 2007, Wani, 2008, a,b,c). As per medical recommendations of 120 g fruits and vegetable per person per day, our national demand works out to 63.million tones. Horticultural global market share for India may be around 15%. We dominate markets in Sri Lanka, Maldives, Nepal and Bangladesh. Our share is relatively less in countries like Saudi Arabia, Singapore, Isreal, Kuwait, Pakistan, Oman and Brunei. With enhancing trade in SARC countries, we may have more opportunities of expansion in domestic Asian markets. Thus a thrust towards horticultural export orientation is linked with peace and union among SARC countries. We share cultural, religious and food habits, thus our products could be preferred over others. This needs proper Post-harvest and market-led extension and outreach system. Given proper attention we may capture markets of Japan, Hong Kong, China and other countries too.

Research Review

Various investigations on use of organic fertilizers, orchard management, disease and agronomic practices have been reported (Srivastava et al 2004). Growth responses of caladium cultivars, influence of gibberelic acid and thiourea on growth and flowering, crop duration, planting density etc. have been reported, Nazki et al (2005), Ahmed & Ali, (2004). Evaluation of pomegranate cultivates (Mir et al 2007), Genetic divergence studies (Verma et al 2006), azotobacterization of mango in Jammu region (Kerni & Gupta, 1986) russeting (Chib, 1990. Fruit cracking (Muttoo et al, 1999) & post-harvest measures (Ahmed et al, 1999).have been investigated.

Various experimentation on floricultural crops like gladiolus, dutch iris, (Khan et al, 2002), fruit drop (singh et al, 2002) have been concluded.The annual reports from RRL Jammu, SKUIAST J&K and other ICAR projects give details of research projects, investigation methods and new knowledge developed in the horticulture sector in Jammu and Kashmir state.. On the basis of these research reviews, the major problems and their solutions can be assessed as follows.

A review of Annual Report of SKUAST-J 2007-8 shows that many research options in post harvest technology have been successfully tested.Tomatoes, Guava and Papaya quality parameters under strorage have been standardized to reduce losses. (Liddhoo,2008, Sharma et al, 2008). The extent of parasitism on cole crops,okra hybrids and advanced stidies on Knol khol, Brinjal, Broccoli, Tomatoes and other horticultural crops have been initiated.Among ornamental crops studies on Gladiolus pathogens is the prominent one to reduce losses of cut flowers. New variety of Marigold to replace existing variety, Pusa

Narangi is on cards.The Brunella nava and B.cento varieties of lilium yielded more cut floweres. Many experiments at Sartingal and Gwari farmas of Jammu Division on apple sprays and other investigations on pear, peach, apricort, pomegranate, and walnut cultivars has been described (Annual report, 2007-2008).

The varietal trails and post-harvest tests on peca nut and strawberry are reported. (SKUAST-J Annual Reports 2007-8)The Chandhlar Strawberry cultivar has been rated as best for Jam. A 20% unripe fruit mix does not alter the jam quality. This can serve as a major post -harvest mechanism to reduce the existing losses. A refrigerated van facility for strawberries has helped its marketing and various studies on fumigation of Litchi with sulphur dioxide has helped to increase shelf life. Anar fruit borer was detected on other fruit crops too signifying thrust on pest and insect management.

FICCI Suggestive Measures

For more aggressive development of horticulture in J&K state, FICCI has stressed for development of:

i) Marketing facilities

ii) Employment and income generation options

iii) Setting of export promotion zones

Export thrust on strawberry, mushroom and cumin seed, berries, wild fruits and indeginous vegetables is needed. The local horticultural germplasm needs investigative and annovative projections. As local germplasm is well suited to our environment and needs less water. The new market options for organic fruits produced in J&K need prioritization and certification under WTO regulation.International biosafety measures,Grading and packaging of available fruits, initiatives for development of new varieties of fruits like kiwi fruit, wild apricot, black cherry, broccoli and mushroom are the suggested ways to development of this sector.. The options for air transport, cargo from Leh, Srinagar and Jammu to centres of consumption like Delhi, Chandigrah & Mumbai need exploration. The agro-based industrial set ups using low grade apple for juice, shampoo, biofuels and use of available horticultural and flouriculture wastes for marketable commodities as gift packs, scents, ornamentals, dehydrated flowers, perfumes and bio-fuel are discussed (wani, 2008).

Development of Horticultural Markets

In the recent past department of Horticulture Planning and Marketing J&K has established Fruit and vegetable markets within the state. The department is in the process of strengthening three main operating markets of Nowpora (Sopore),Parimpora (Srinagar) and Narwal (Jammu). So far multipurpose grading, packing and auction sheds have been constructed at Nowpora Sopore. Further 17 satellite markets have been established within the state. Market information service has been introduced to collect and disseminate market intelligence through NET connectivity within and outside state. A regular agricultural marketing news as Zariye Khabernama has been started in Kashmir Door Darshan.

Table 1 : Some of the problems faced by the Horticulture sector

- Lack of storage facilities.
- Lack of Processing and value addition units.
- Low productivity of especially under rainfed and low input conditions.
- Premature fruit fall/ dropping.
- Low cost packages and transport difficulties especially cold chain mechanization.
- Heavy post-harvest losses and low promotional efforts for market led extension services.
- Low germplasm propagation infrastructure.
- Organic premotions.
- The present fertilizer consumption in J&K is 38.3 kg/ha as compared to 170 kg/ha in Punjab. In contrast the estimated quantity of nutrients mined by crops in Kashmir is 48 kg/ha. Thus we have better options for organic fruit production in J&K.

Agro-techniques Developed by SAUS

- Fruit /vegetable nursery under poly-house.
- Spray schedule for apple and other fruits
- Leaf testing for fertilizer evaluation before utilization.
- Dehulling of walnut.
- Walnut rejuvenation and propagation technology
- Apricot solar dryers.
- Orchard management through use of polinizers and bee-farming.

Table 2: New Varieties developed by SAU.

Crop	No. of varieties released	Name of varieties	Salient features of genotypes
Apple	5	Shreen	Yielding 50-60 kg/tree, scab resistant, small medium size jucy fruits.
		Firdous	Yielding 50-60 kg/tree, medium crisp, sweet slightly acidic fruits.
		Akbar	High yielding, medium size, red, colour fruit.
		Lal Ambri	
		Sunheri	
Walnut		Hamdan	Semi-dwarf, 15 kg/ha, 14g nut with 54% shelling.
		Suleiman	Semi-dwarf, 30kg/ha, 21g nut with 52% shelling.
Almond	4	Mukhdom	Medium size nut with 7 q/ha yield
		Parbat	Small kernet with 5 q/ha yield
		Waris	Late bloomer, medium size kernel, 5-7 q/ha yield potential.
		Shalimar	Mid bloomer, smooth kernel with yield potential 5q/ha.
Capsicum	1	Nishat	Darl green, better shelf life, moderately resistant to disease, 50% advantage over California wonder.
Chilies	1	SC-107	Yield advantage of 60% over local varieties.
Turnip	1	Nageen-1	Suited forvalley during rabi season and for hilly area during summer.
Carrot	1	Chaman-1	Early maturing, tender, tastey. Yield potential of 25-26 tonnes/ha.

Table 3 : Strategies for development.

- Most of the fruit crops are rain-fed, need development of irrigation facility in the form of drip/ sprinkler irrigation.
- Site specific nutrient management technique need to be adopted to improve and economise fertilizer use in fruit and vegetable crops.
- Development of cost effective, eco-friendly crop production and plant protection technologies free from all hazardous chemicals with minimal damage to soil, water, air, plants and animals.
- Diversification of fruit cultivation, cultivation of dry fruits in rainfed areas.
- Use of local genotypes of wild fruits like nuts, berries etc.
- Production of fruits, vegetables and flowers in off-season and creation of facilities for making off-season products thereof.
- Rejuvenation of sick and unproductive horticulture units such as orchards, fruit nurseries and processing units etc.
- Promotion and popularization of hybrid technology in vegetables.
- Establishment of tissue culture laboratories and green houses for mass multiplication of disease free planting material, tubers/corms and seeds.
- Degraded soils of the fruit crops where crops are growing for decades due to soil erosion or in soil chemical change, need to be given attention to ameliorate it.
- Establishment of modernized fruits and vegetables processing/preservation units.

Table 4 : Infrastructure facilities needed for post harvest management of horticulture industry in J&K state.

A.Genral for all Fruits and in All Districts

1. Improvement in Road Connectivity
2. Creation of pack House facilities in the production area
3. Provision of precooling and storage facilities
4. Enforcement of Grading act.
5. Efficient Transport facilities.
6. Alternate packaging material.
7. Value addition centre for fruit and vegetable processing.
8. Fruit Insurance scheme.
9. Fruit mandies at district level.
10. Market Information system.
11. Provision of facilities for walnut processing.

12. Provision of Facilities for processing of saffron.
13. Facilities needed for cut flower trade.
14. Medicinal and aromatic plants. Promotion and livelihood options.
15. Creation of necessary cold storage at airport.
16. Establishment of quality control laboratory.
17. Technological awareness of the farmer.

B. For Walnut Trade

Walnut is an important nut crop grown in the state of J&K. the state produces 86263 tones of walnut form an area of 61723 hectares with the productivity of 1.39 tones/ha and had monopoly in the export to many countries of the world. Walnut cultivation is common in Badrawah, Poonch, Kupwara, Baramulla, Bandipora, Ganderbal, Budgam, Srinagar, Anantnag and other hilly areas of J&k.

Walnut are consumed as luxury snacks as well as commercially used for preparation of bakery products, chocolates, ice cream, ornaments, oils, confectionary and salad products. Shells are sued in glue and plastics and for making solutions for cleaning and polishing metal surfaces.

Constraints in Walnut Trade

a) Awareness of maturity indicies
b) Method of harvesting
c) Non-scientific de-hulling

Post Harvest Bottlenecks

a) Improper storage, drying, grading practices at farm level.
b) Lack of awareness about sanitary conditions to handle nuts.
c) Lack of integrated handling system to manage the nuts.
d) Non adoption of international standards of grades.
e) Lack of processing/storage facilities/orchard management.

C. Apple Post Harvest Measures Needed

Apple is an important principal crop of J&K state. The state produced 1041417 MT of apples from 100702 ha of orchard area during year 2005-2006.The state contributes 57% of national production of apples, while as Himachal Pradesh contributes 24%. The state produces early, mid and late season cultivars of apple. The commercial apple varieties are Red Delicious, Golden Delicious, Ambri, White Dotted Red, Royal Delicious kulu Delicious, Hazratbali, Benoni.

Post harvest handling of apple encapsulated many management decisions and proceses that are involved in harvesting, handling, storage, packing and transport of apple fruit necessary to provide consumer with an acceptable end product. The characteristics of apple which determine its acceptance in market place are size, shape, colour, texture, flavour, juiciness, freedom from blemishes and bruises. A proper export oriented packaging, sanitary evaluation and eco-friendly shall improve its consumer acceptability under WTO regime.

D. Cherry Postharvest Options Needed

Cherry is important fruit crop of the valley which is being marketed outside the state. In Kashmir cherry is the first fruit crop which comes to the market and fetches good prices. The quality of fruit is good in cool climate. Cherries after harvest are to be pre-cooled , stored packed and carried under refrigerated conditions to the terminal market. The fruit transport under refrigerated conditions remain healthy for a longer period, compared to the fruits stored under ambient conditions. Storage, packaging, transport and preservation is needed at local level.

For export of horticulture produce we have to integrate production activities with post harvest management. Infrastructural facilities for Pre-cooling, packaging, grading ,processing centers, refrigerated vans, cold storage, cargo handling facility at airport, laboratories for pesticide residue analysis, good road connectivity, market promotion through IT network, credit facility for exporters and awareness programmes are essential pre requisites for export of horticultural produce from the state.

Table 5 : Major Problems in Horticulture

Low Productivity

10. Poor quality planting materials & its inspection mechanization.
11. Low disease diagnostic appliances.
12. Poor orchard management.
13. Low capital- small farmer – holidings.
14. Poor technological applications.
15. Poor post-harvest handling, industrial base.
16. Poor quality control measures and laboratories.
17. Marketing network modernization.
18. Poor data base and electronic use.
19. Poor infrastructure.

New Options Needed

15. Germplasm base, analysis, collection of materials across the state.
16. Standardization of rootstock for all important crops.
17. Development of horticultural cropping systems for all districts.
18. Certification and quality control laboratories.

Table 6 : Current Schemes for promotion of Horticulture

Horticulture Technology Mission

Mini Mission I

In order to provide technology support, this mini mission aims to supply of Nucleus/ Basic Seed and Planting Material of Horticultural crops. Standardization of Production and Protection Technologies, and Technology refinement and training through on-farm trials. The "Indian Council of Agricultural Research" is the nodal agency implementing the programes under this mini mission.

Mini Mission II

Department of Agri. & Cooperation is the nodal department for the mini mission. The mini mission primarily aims at increasing the production and productivity of the horticulture produce in the region through integrated implementation of various governmental programmes based on compact area Development approach. The major components of this mini mission are:

1. Area expansion
2. Creation of water sources.
3. On-farm water management.
4. Production of planting materials
5. Transfer of technology through training, frontline.
6. Demonstration, publicity and training of trainers
7. Popularization of Organic Farming and Use of Bio-fertilizers
8. Promotion of popularization of agricultural equipments
9. Promotion of integrated pest management
10. Establishment of plant health clinics
11. Establishment of tissue/ leaf analysis laboratories.
12. Entrepreneurial development of women farmers.

Base line survey to identify the amount and type of the work being contributed by the women in the field of horticulture and other agricultural activities, their role in the decision making process and problems faced by them in farming.

Table 7 : Average Productivity of Apple.

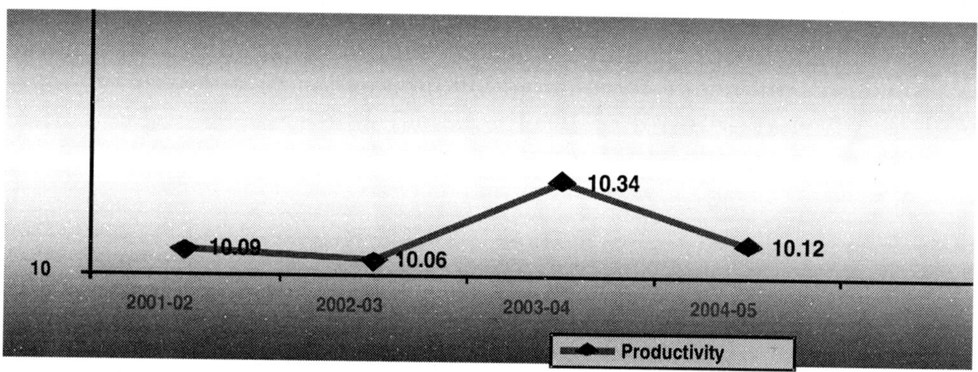

Table 8 : Apple Production trends in leading countries (2002)

Country	Area (Hectares) in Lakhs	Production (M.T.)	Productivity (MT/HA)
China	24	2.40	10.00
U.S.A.	1.86	48.50	26.08
India	2.31	13.80	5.97

Table 9 : Marketing of Major Horticulture Products

S.No.	Product	2004-05 (Lac Tons)	2005-06
1.	Apples	10.93	11.52
2.	Cherry	0.074	0.08
3.	Pears	0.40	0.42
4.	Walnut	1.005	1.083
5.	Almond	0.135	0.143
6.	Others	0.002	0.002

Table 10 : Export of Fresh Fruits out of J&K

Year	Production (Lakh MT)	Exported (Lakh MT)	%age of fruit exported	% increase in export
2000-01	8.372	5.016	59.91	-
2001-02	10.008	6.467	61.78	28.92
2002-03	10.457	7.053	67.45	9.06
2003-04	11.657	7.624	65.40	8.09
2004-05	12.176	8.194	67.29	7.47

Table 11 : Export of Walnut outside the Country

Year	Quantity Exports		Foreign Exchange earned (Lakh INR)
	In-Shell (in MT)	Kernell	
1999-00	508	4485	60.55
2000-01	597	7145	109.94
2001-02	987	6415	117.78
2002-03	656	6976	121.23
2003-04	300	6001	99.30
2005-05	257	5417	92.82

Table 12 : Key Result Area: Enhancing Productivity of fruit crops

Bottlenecks

- Area expansion carried out mechanically without reference to variety/hybrid to be promoted, based on recommended technology.
- Research available with ICAR (CITH) or SKAUST not involved for developing technically sound area expansion.
- No system in place for sourcing and propagating high quality plant material through SKUAST or based on SKUAST's evaluation.
- No mechanism in place to monitor how subsidy on area expansion was utilized by beneficiary.

Table 13 : Fruit Production Trends

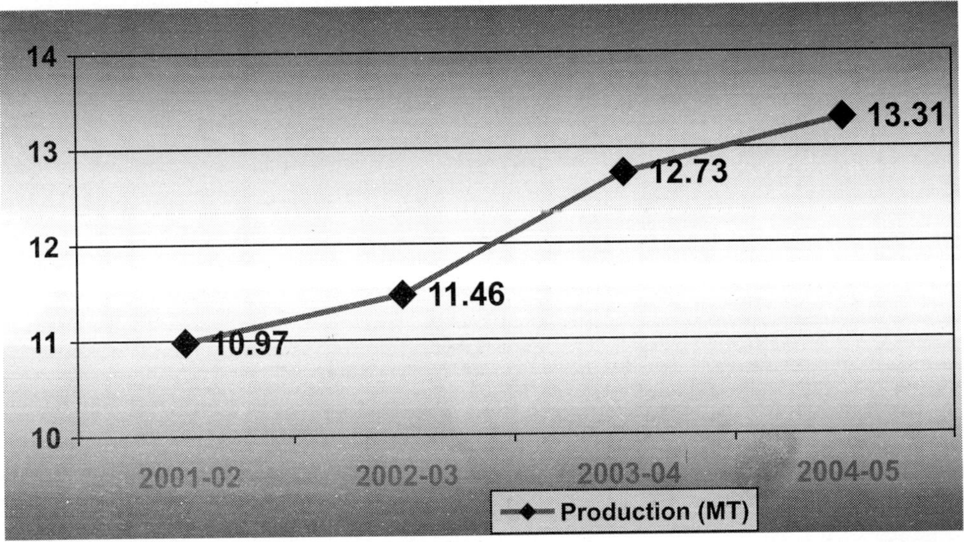

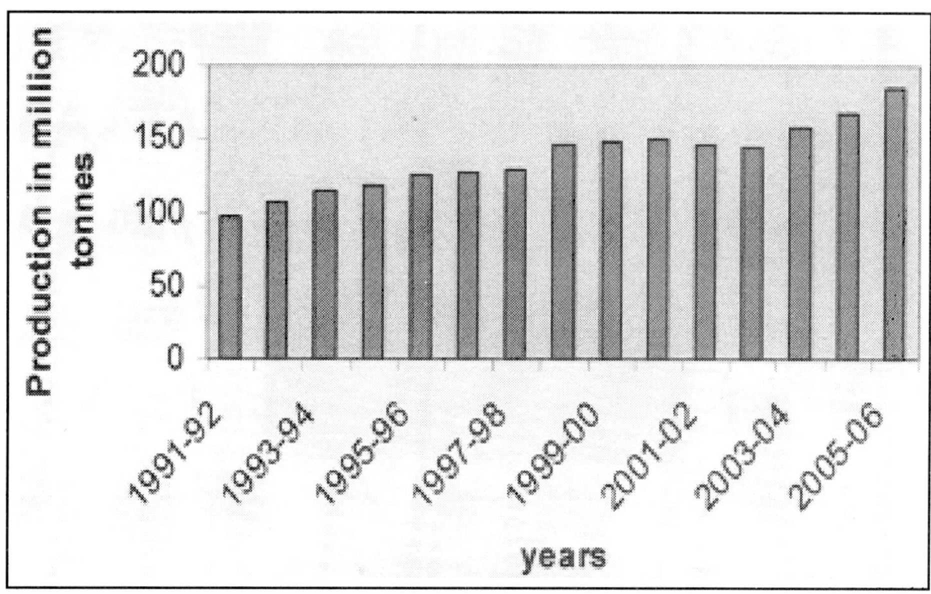

Fig 1 : Indian Horticulture Production at Glance

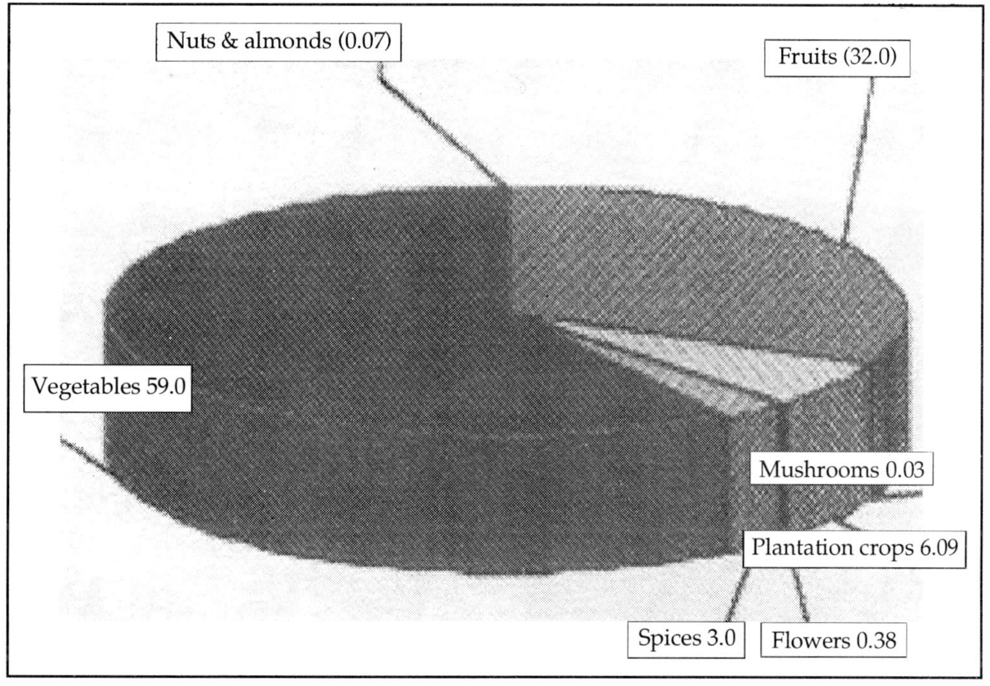

Fig 2 : Share of major horticulture crops (in %)

References

Ahmed, F; Khajwall, M.H and Ahmad F.1999; Effects of harvest dates and post harvest calicium chloride treatments on physico-chemical charteristics of pear ev Bartellet applied Biological Research.1(1):43-46.Annual Report 2005 to2008.

SKUAST J Annual Reports 2007-2008 edited Kher, D, published by SKUAST Jammu www.skuast.org.

Annual Reports SKUAST K2001-2007.

www.skuastkashmir.ac.in.

Ahmed M. J and Liaqat Ali 2004; Growth response of caladium cultivars. *Sarhad Journal of Agriculture* 20(4):497-500.

Chib, H.S.1990 ; Investigatios on ressuting effect of some fungicides on apple in Kashmir. *Research and Development Reporter* 7(1):73-76.

Gupta, O.P. 1986; Determination ofmmaturity standards of Kashmir apple ev Red Delicious. *Research and Development Reports*.3(1):76-79.

Lidhoo, C.K.2008; Hot air drying characteristics of tomatoe slices at optimum drying temperature. *Journal of Research SKUAST*-J 7:62-67.

Khan F.U; Nazki,I. T ; John, A Q and Misra R. L 2002; Effect of bulb grade and spacing on duch iris. Proc.of national symposium on Indian flouriculture in the new millennium,ed s .misra; published by Indian society of ornamental horticulture Banglore,Feb,2002 PP 150-153.

Kerni, P N and Gupta,A 1986; Growth parameters affected by azotobacterization of mango seedlings in comparison to different nitrogen losses. *Resaerch and Development Reporter*. 3(2):77-79.

Muttoo,A.H;Raina,.Bland Nathu,B.L 1999; Fruit cracking studies in some commercial cultivars of Litchi(Litchi chinensis Sonn). *Advances in Plant Sciences* 12(2)543-547.

Mir,M.A, Mir,N.M; Bhat;A.R and Naik, H. R 1999; Effect of harvest date and canopy location on the quality of sweet cherries cv Misri.(Bigareau noir gressa) *Applied Biological Research* 1(1):39-41.

Sharma. I; Koul ,R.K and Bhat A. 2008; Effect of different treatment combinations of Guava and Papaya on quality and storability of ready to serve beverages. *SKUAST J. Research* 7(1):68-78.

Research Highlights SKUAST- K 2001-2007 and SKUAST -J 2004-0. Directorate of Research, SKUAST-K and SKUAST -Jammu and Kashmir.

Wani, G.M.2008, a. Communication for Sustainable agricultural production. WWW.Articlebase.com 19.2.2008. Free publication on web. science.

Wani, G.M.2008 b. Capacity building and E. extension for enhancing agricultural productivity in India. WWW.Articlebase.com 18.2.2008. Free Publication on Web. Science.

Wani, G.M.2008 c. Success stories of Agribusiness/.Agriclinic- a tcol for agricultural entrepreneurship. WWW.Articlebase.com. 19.2.2008.

Wani, G.M.2007. Role of communication in enhancing Agricultural production - A vision. WWW.word-mart.com/html/communication - Agricultural- html.

□□□

Chapter 34

Extension Education Reformation
I - New Reformation Models

G.M. Wani

ABSTRACT

Technology akin to farming existing in an ecosystem need to be generated. It needs location specific perfection and adoption. This could be possible by supporting KVK & Extension system with technology testing and refinement. Thus the whole discipline of Agricultural Extension need reforms. On the basis of a broad review presented in this paper, following recommendations could be deduced to reform and fine tune technology generation, dissemination and its delivery as an agri-pack end product (APEP).

1. The base of discipline of Agriculture extension may be broaded to incorporate all disciplines, management tools and techniques. It is proposed to name it **"Farm Business Management"**. It should be open to all agricultural & allied undergraduates. The course curriculum needs training and skill up-gradation of graduates and post-graduates in techniques like analysis of data, survey procedures, marketing, economics, rural sociology and farming systems etc.
2. Various reformation capsules like Chinese model, African experiences, Dutch model, Netherland model, Isreal model, California model and Iranian models have been reviewed. On the basis of these models a new model for technology delivery suitable to our agro-eco-conditions has been postulated & named as **Mohi-Model**. Once we receive an acceptance for the model, logistics and plans to execute it could be formulated.

Introduction

Four paradigms of Agricultural extension consists of Technology transfer, prevalent since colonial times & later National Agriculture Extension Project (NAEP), reshaped in 1970-80 as Training and visit system across Asia. It involved top down approach. Its role needs discussion. The second paradigm called as Advisory work is what is present system of extension, where, government agencies, NGO and technical industries respond to farmers enquiries which we call advisory role. The third paradigm consists of Human Resource Development. This innovation helped Europe and North America. The Universities gave training to rural youth. It is these educated, trained youth which became future farmers in developed countries. What is known as outreach system of colleges and universities is still unknown in our Agricultural Research system including State Agriculture Universities. A half hearted attempt as Agri-business Agri-clinical is still to be backed with University, research, banking and economic support. Is it which can help us to increase productivity in India or Asia needs consideration ?

Technology Adoption and Impact

The impact of National Agricultural Research Project (NARP) (1986-92) revealed that , technologies generated were few and adoption rate was not high, even then an estimated monetary benefit of Rs. 2000/per hectare was observed.

Low relationship between technologies developed and adopted demands a mixed farming technology capsule as 65% of our farmer's are small and marginal farmers. This needs a total shift to left in our agricultural extension methodology (*Reddy 2001*).

T & V programme initiated with central assistance in state has aroused all round technological awareness among agricultural experts. Its dialogue and technology transfer from University (Scientists) to experts agricultural extension workers was nearly total revolution. The (ZAREC) Zonal Agricultural advisory Research and Extension counseling was the best plate form to discuss main issues impeding agricultural development. The missionary publication of production recommendations helps a lot in transfer of technology. The presentation of production recommendations in local language is helpful in increasing productivity in many countries (*Wani, 2004*).

Brinjal production technology adoption in Parbani, Mahrastra was the result of social participation, risk management guarantee, marketing orientation, information websites and communication methods. (*Dudhak, et al 2003*).

Various limitation of transport, extension staff were the reasons of poor productivity gains in *Malakhand agency of Pakistan. (Mohamad et al, 2003).*

The adoption of legume production technologies under Jhodpur arid conditions was low. The problems as indicated by farmers were small farm size. Small holder farmer was less interested to adopt these technologies. Non existence of market and pricing policy, risk of failure, feeding habits, social taboo and lack of Post Harvest Technology (PHT) were identified. (Yadev et al, 2003).

Proper guidance or information support was the major constraint in adoption of improved Wheat cultivation in Himachal Pradesh. It contributed to 16.8% non-adoption rate among farmers. Other contributory factors were technological unawareness (13.33%) sloppy land (11.31%), farm input cost (9.41%), non availability of irrigation (8.29%) and lack of inputs (7.95%) (Puran chand et al, 2001).

Social Structure & Technology Adoption

Schedule caste farmers in Maharastra have low level of rice technology adoption. Among 150 such farmers only 12.6% had high adoption rate (Mankar et al,2004). Adoption of production recommendations of Cashew in Maharastra's Sindhubrug district was 58% (Bhairamkar et al, 2004). Motto of extension programmes should be reach to unreached, teach to un-teached, be known to unknown (Ganesan et al, 2003) (Wani, 1998). Agricultural productivity in Czch Republic increased when information and communication in extension system was improved for small scale farmers (Slavik, 2003).

A survey conducted among 120 farmers in Karnool district, Andhra Pradesh, India, revealed that education, social participation, scientific orientation, risk management, mass media exposure, economic and market orientation were positively and significantly correlated with the extent of adoption of recommended hybrid jowar (sorghum) seed practices by the farmers. Majority of the farmers (43.33%) had a medium level of the adoption of the recommended practices. (Kumar et al, 2005). It is suggested that five categories of factors determine the adoption of irrigation systems and are the keys to designing relevant, effective and innovative irrigation extension programmes among horticulturalists. The five categories of factors are : Soil type and topography ; crop quality; sowing time; saving water and crop yield (Kaine & Beswell, 2005). We in J&K under SREP Strategic Research extension plan of Districts notice lack of information on sowing time, seed rate and fertilizer application in almost all the crops. Seed treatment awareness helped farmers to combat paddy blast (Wani, 2006, 2007).

Innovative Change

Methodologies for analysis of farming system, options for development & sustainable eco-friendly farm science technology are the needs of the hour. We do have subject-matter-specialists, but the agents for evaluating the composite effect of various discipline oriented technology cover is yet to be structured. This is exactly where we should intervene and restructure the discipline, programmes and divisions of extension education in the country. Infact a total change from existing slow action, low innovative, conventional and disintegrated extension system has to be refined into analytic, managemental, social, economical, marketable system, so that the whole technology transfer could be evaluated in terms of productivity and monetory gains per hectare. Different cropping and livestock patterns and combinations need to be evaluated & monitored. The impact analysis of wide range of Agricultural technology transfer be it training, entrepreneurship or agri business is needed, both before and after technology application.

The available research results have to pierce the farmers farming practices. Non - adoption is the result of our low productivity levels per unit of land. Inspite of tremendous yield gains the per unit profitability has not raised the economic conditions of our farmers. Our farm economy still fluctuates between season's, rainy years and among regions. Now for any widespread gains from technology, it is essential to know its impact and adoptability rate. It is therefore necessary to develop new technologies after considering socio-organizational arrangements, local existing soil, water, human needs and resources. The local skills need to be known and refinement is made to improve profitability. Our experience under IVLP showed a net productivity gains of 2 q/hac in rice production by improving local village driven plough. Similar innovative ideas local technologies with improved or exotic one gave good results in development of profitable farming *(Wani, 2004, 2003, 2005, 2006, 2007)*.

Reform Capsules

(i) Chinese Model

Chinese experiences needs a try under our socio-economic conditions. The successful improvement of Sheep and Yak herds in china can be a good example for use in India. Under this project they first took stock of rangeland and livestock resources. Then socio-economic conditions of the livestock breeders were analysed. Livestock and pasture management by rodents was studied. Rasngeland revegatation methods were tested and propagated. Seeded fodders and effect of these improved method of pasture development on disease control and profits from livestock were analysed. *(Wagenin gem et al, 2001*). Thus, the extension education programmes should ensure facilitation of Farmer empowerment and not dependence on subsidy. This means incorporating studies and programmes of Agri-livestock economics, grass land and rangeland management, crop and tree production in Agri-extension curricula.

(ii) African Model

A new participating approach of clubbing local and outside knowledge to solve the agricultural problem is needed. Innovative approaches for soil and water conservation in Ethiopia and Tanzania were helpful. A new model of participatory modified Technology development approach is fast being recognized, as the only way of clubbing traditional and modern technologies for agri-development (*Kib Wana et al 2000*).

(iii) Dutch Model

The total reformation of agricultural technology transfer shall need refinement of elementary methodologies used in extension consisting of PRA, PERT or other specific managemental, communication and modern GIS-Geographic Information System use. This would involve a total change in Extension Education, objectives, methodology & course curriculum. We need agricultural networking for bringing out results of transferable technologies. Experimental learning and knowledge sharing has to be incorporated, as research on farmers field. Farmers have to be part of our learning and experimentation process. Dutch model could be used. (*Leeuwis, 2000*) in India to ensure quick technology use. Let us evaluate technology use. It has been seen that some progressive farmers have awareness of technology but being absentee farmers, their workers have no knowledge or skills to use them in the field. Is it time to come out of "Technology dissemination Scenario" to "Technology Use Status in the field" and "Productivity gain assessment models".

(iv) Mohi – Model – (Proposed Model)

Experience learning, market designs, farm economics, sustainable agriculture, Agri-information, communication skills and farm business shall be the knowledge bank available to Agri-Extension & communication experts, whom we should name "FBM" – "Farm Business Management Experts". The Extension Education is outdated and too oftenly used and misunderstood phrase. Extension includes now every attempt to communicate with farmers. Today's extension agencies can be university scientists, Subject matter specialists, Agri experts, farmers or even partially Agri known NGO or their agents. Thus, the whole blame of others has to be shared by University experts. Agri-University, applied knowledge has to be communicated to farmers as "end product" as pharmaceutical companies do. Thus all the current "building block", known knowledge has to be computerized and stored. This has to be reshaped and re-arranged as knowledge packs for innovative and whole farming capsules rather than one commodity knowledge sharing as in the west. We do have commodity but not composite agriculture packages. A booklet on its composite effects has been published as Broad Based Extension Education (Wani, 1992).

Training & visits on farm level failed but farmer-training & Farmer-research participation or demonstrations at farmers field did achieve a great success. With constraints of men and resources and crop insurance in vogue, we must induce researchers to have direct participatory research at farmer's doors, on their fields. This venture can be on cost-sharing basis or with participatory involvement of Industry. (SAMETI, Kashmir model refers). Proc. XXI EECM, 2008.

New Approaches

Insect-pest Management research could be an example. Some FAO programme results are encouraging (Matteson, 2000). Rice IPM research must incorporate farmer-training of farmers. New pest management procedures are to be experimented under farmers field. Minimum use of pesticide and eco-friendly application is preferred. (Matteson, 2000).

Problems and constraints of technology transfer in India have been reported, (Bairathi, 1998). Similarly bio-gas technology use or adoption has been full of constraints detailed. (*Prasad and Singh, 1999*), these range from technological, financial, administrative and innovative ones. Proper mechanization and training in operation of such bio gas technology is necessary.

An information web networking developed helped farmers to improve their income through training and advise thus participatory technology transfer is a possible mechanism for improvement in Agri-productivity. (Warren, 2003).

Netherlands Model

Quest for knowledge led Netherland farmers to have an association for organic farm produce. This association conducted its own farm research and shared its experiences with other farmer members back in 1980. Soon they felt that their experiences lacked experimental details. They approached Govt agricultural Research Organization and started collaborating with them. The station mandate was to assist the development of farming systems in the area called polder in the village Nagale of Netherland. This combination of research and farming practices at farmers field tried various crop combinations and rotations. The use of nitrogen-fixing crops (Legumes) for maintaining soil fertility and strategic management of pests and diseases, extending or delaying planting of various crops etc. Some experiences of low-weed-crop production by adopting agronomic practices like regular harrowing of lands and use of crop covers were tried. The farmer pressure and self sustaining groups need to be involved in research experiments of SAU/ICAR and all other KVK farms and institutes (Reinders, 2007). This would have both a demonstration and high adoption value. Besides it would be a close to practical field research model (PFRM).

Horizontal, farmer to farmer contacts help dissemination of agricultural information. This horizontal expansion of information from one vegetable grower to another in Suden (*Badri, 2001*) helped technology adoption.

Mass media support for rice production has been helpful in Bihar. Out of 5 TV talks, 75% farmers benefited and registered higher production and low disease damages. (1taque, 2000) similar observations are known in J&K.*(Wani 2006)*. Electronic technological diffusion of knowledge at Farmer's fields was helpful, local language and local persons can diffuse ideas and disseminate information and communication faster. It is proposed to have innovative knowledge defusion to women farmers through local women (Rani and Reddy, 2002).

Farm – Women Awareness Module

Majority of rice producing farmers in Kangra district of Himachal Pradesh had no extension personnel contact or electronic or print media information. Continuous information on mass media have significant impact on use of seed rate, transplantation time and application of nitrogenous fertilizers in Rice (Thusoo and Sohi, 2000). A gap existed in knowledge of farmers so far as dose of fertilizer application is concerned (Yadav and Singh, 2000).

Farm women adoption rate for use of farm implements was good when uses of such implements were demonstrated to them. (Sharma et al 2003).

Farm women were given farm implements to use for a fortnight. Majority found them 78% profitable & 77% compatible. (Sharma and Sharma, 2002). Information from research to farmers via extension officers was examined. The inter system communication helps in technology dissemination of dairy farmers in Iran (Rezvanfer, 2001). Rice cultivation awareness improved when women farmers were trained (Rama Krishnan and Mahendram, 2003). Agricultural Knowledge and information system increased economic efficiency of crop production, out put costs, social upgradation and income generation (Carrasco et al, 2003).

Extension Briefs – Self Help Options (SHO)

Small farmers are neglected by extension services and need proper awareness in Kurram (Haq et al 2004). Agricultural extension specialists must specialize in community structure, development , education learning tools, technology and policy issues. Thus these disciplines or topics need incorporation in Farm Business Management curriculum.

Partnership process, management and communication should be the main thrust. Co-operative mechanism, farm business management and evaluation and monitoring tools should be used.

Seed production programme was successfully implemented by farm women as a self-help group in Dindigul district of Tamil Nadu (Balasubramanian and Thamizoli, 2003). Efficiency of extension lies in bottom up approach in hilly areas. Farmers interest group awareness was most successful (Muhammad-Zafarullah 2003). Small scale farming operations using green manure cover improved soil fertility. Participatory extension system helped more adoption of technology in Uganda. (Miiro et al, 2002).

Farmers training and their own information networks helps small holder farmers. Topics including community involvement, international multicultural projects, leadership, learning strategies, new tools and technology, volunteer programme , programme development and evaluation, public policy issues, youth education and partnerships needs introduction in Agri-Extension curricula (now proposed as Farm Business Management). Agriculture, natural resources and health/nutrition shall form the basic of new agri-farm-business extension system (Logan, 2004). The word extension be replaced with "Farm Business Management".

Administrative Bottlenecks

While the Nusa Tenggara Timur Provience is supposedly following the training and visit model, agricultural extension Workers receive very little training. Other major constraints identified include constant restructuring of the Agricultural Department, poor remuneration, oppressive authority, inadequate infrastructure, unclear job direction, geographical conditions and bureaucracy. (Leta et al, 2005). This is the reason of low technology revolution in rural India and Asia too can be overcomed by transparent, NAR System, which need renaming as National Farm utility Research and delivery system.

By combining biophysical science with social science, it is possible to increase the impact of biophysical science findings at the farm level (Linehan et al, 2005).

Isreal Model

General model of extension in SAU can be seen from the role of PAU, which needs a total reform. Should we look for Isreal model. Isreal is characterized by topographical variations and a wide range of soils, enabling the growing and raising of a variety of crops and animals. A large increase in population immediately after 1948, required a rapid increase in food production, followed by development of research and extension facilities and services. In recent years Isreal, like many other countries, has faced difficulties in continuing to supply the level of services it previously provided. This difficulty led agricultural organizations to begin assuming the responsibility of partially financing extension for a few years, and later on trying to reflect the cost onto the farmers. Farmers are apparently willing to pay for on-farm extension visits. Under this system, the Flower Board is financing many public extension activities and charging the farmer for partial costs of on-farm visits. This system operates successfully, aided by reports that are filed electronically by agents all over Isreal to the Floriculture Department Head Office, and from the Head Office to the Flower Board. Advantages of the new system are : a tremendous increase in agent-to-farmer commitment; the professional level of agents remains at least the same as before; and the financial situation of the majority of agents is better than before. The disadvantage is that some farmers are consuming less extension than before (Wolpson, 2005).

Global Successful Extension Experiences

California Model

California's land grant extension progamme has been successful in assisting agricultural Industries develop as major world producers. Extension has been a leader in facilitating quality-of-life and economic improvements for rural communities throughout the state. However, population explosion has transformed California into an urban state, with changing societal issues, values and priorities. Agricultural extension programmes continue to directly serve a declining number of clientale that represent less than 1.5% of the population. Financial support for extension is decreasing at the country and state levels, thus such models used by us in SAU could be catastrophic and need change.

Iranian Model

Iran is one of the world's top ten horticultural producers. However, post-harvest losses in Iran are six times greater than the world average. Even after establishing an agricultural extension Service, Iran still experiences great problems in reducing post-harvest losses (Jowkar, 2005). Results revealed that education, material possession, mass media exposure, extension contact, scientific orientation and risk preference were significantly and positively correlated with the knowledge level of farmers pre- and post training. (Malite et al , 2005.

Issues

Agricultural development programmes often produce unexpected results. This can be attributed to the fact that the target farmers already have their own knowledge and competencies that determine their practices. In order to be adopted, an innovation has first to be discussed, and then appropriated by a local group of farmers in their system of knowledge and in their system of meaning; subsequently the innovation may be rejected or adopted by all the farmers concerned (Mathieu, 2004).

Problems

The findings also revealed that the attitudes of the respondents were significantly and positively related to education, organizational membership, land holding, contact with extension agent, mass media exposure, socio-economic status, awareness, income and material possession. Lack of proper technical knowledge, lack of irrigation facility and high cost of fertilizers/ chemical were indicated as the three most important problems of the tribal farm women run agriculture farms. (Bharali et al, 2004).

New Vistas

Agricultural extension is evolving world wide, and there is much emphasis today on community-based mechanisms of dissemination in order to bring sustainable change. The factors that make farmer groups successful in disseminating information and technologies need to be analyzed. A mixed methodology, multiple-stage approach was used to obtain data. Dairy-goat farmer groups (n=46) and individual farmers (n=88) were interviewed. Factors that were associated with group success in dissemination included member participation, degree of jealousy within the group, homogeneity of members, group capacity, number of linkages and type of group (Project-supported versus non-supported). Some interventions that may increase the success of groups in dissemination include capacity- building , increasing linkages with other extension stakeholders, providing an environment for groups to form, using established groups (as compared to forming new groups), and encouraging groups to form around common interests rather than for other reasons. (Davis et al, 2004). Some methods in this direction are documented (Wani, 2008 a).

Reference

Badri-OA; 2001. School of Rural Extension Education and Development, *Ahfad University for Women, Omdurman, Sudan. 18:1, 35-50; 20.*

Balasubramanian and Thamizoli, 2003, Social differentiation in the horizontal transfer of knowledge: a case study from south India, *J.of Agri.Edu.and Extn.2003, 9:2,51-60;3.*

Bharali et al, 2004. Influences of extracellular calcium and iron on membrane sensitivity to bisulphite in the mosses.*Jr. of Bryology,* vol.26No.1 of March 2004, pp 53-59

Carrasco et al, 2003, Absorbing the shocks: the case of food security, extension and the agricultural knowledge and information system in *Haryana, Cuba.Journal of Agri. Edu.Extn. 2003, 9:3,93-102;9.*

Davis et al, 2004, Extending technologies among small-scale farmers in Meru, Kenya: ingredients for success in farmer groups.*Jol.of Agri.Edu.and Extn.2004;10(2):53-62.*

Ganesan et al, 2003,Reaching the unreached: basics of extension education, Tamil Nadu Agri.University, Coimbatore -641 003, India. *Reaching the Unreached: Basics of Extn.Edu. 2003:xii+387.*

Kaine and Beswell, 2005, An innovative approach to irrigation extension in horticulture, *Acta –Horticulture.2005; (672): 177-183.*

Khalache and Khaire, 2007, Social-ersonal and socio-economic characteristics of the fig growers and information sources used by them in fig cultivation technology, International *J. of Agri.Sci.2007;3(1):258-20.*

Kibwana- OT, 2000. Bringing together local and outside knowledge for innovation in land husbandry in Tanzania and Ethopia - a comparative case study. *Jr. of Agri. Edu. & Extn. 7:3, 133-142.*

Kumar et al, 2006, Adoption of cultivation and post harvest technology of tomato by farmers., Karnataka *J. of Agri Sci.2006;19(1); 76-79.*

Linehan et al, 2005, Combining biophysical and social science to achieve change through targeted extension.,*Acta-Hort.2005L672):211-216.*

Logan, 2004, Galaxy II Conference, Exceeding expectations through eamwork, *Salt Lake City, Utah, USA, 21-25 September 2003.*

Leta et al, 2005, Human Resources and scientific output in Brazilian science. *Jr. Aslib Proceeding*, 0001 - 253X, vol.57

Leeuwis, 2000. Reconceptualizing Participation for Sustainable Rural Development. *Agri. J.* 2000.

Malite et al, 2005, Relationship of farmers' traits with knowledge of rapeseed-mustard production technology. *Jol.of Oilseeds Res, 2005: 22(1): 159-161.*

Mathieu, 2004, The meaning of practices: farmers; conceptions in agricultural development strategies,*Jol.of Agri.Edu. and Extn, 2004; 10(3):101-109.*

Matteson-PC, 2000. Insect pest management in tropical Asian irrigated rice. *Annual Review of Entomology, 45: 549-574; 164.*

Muhammad Akram, 2003. Main factors affecting extension activities. *Sarhad Jr. of Agri. 19:1, 163-168; 7.*

Muhammad-Zufarullah 2003, Improving extension efficiency through bottom-up approach: an evidence from remote areas of Northern Pakistan, *Sarhad J.of Agri.2003, 19:4,591-594;5.*

M.M.Jowkar, 2005. The Importance of Horticultural Extension in Reducing post Harvest losses. IV International Symposium on Horticulture Education, in Iran.

Prasad-RB; Singh-S, 1999. Constraints in adoption of bio-gas technology. *Jr. Applied Biol., 9:1, 100-102.*

Puran Chand et al 2001, Constraints of the farmers in the use of improved wheat technology. Himachal-*Journal of Agri.Res.2001, 27:1-2,95-102;7.*

Rama Krishnan and Mahendran, 2003, Impact of TANWA training on knowledge gain of trainees about paddy cultivation practices.*Journal of Res-ANGRAU, 2003, 31:3,68-71:3.*

Rani and Reddy, 2002, Knowledge and adoption of farm and home technology of farm women, *Journal of Research-ANGRAU,2002,30:2,101-104;6*

Reddy, M.V; 2001, Relationship between number of technologies developed at NARP stations of ANGRAU and their adoption. *Journal of Res. ANGRAU, 29:1, 42-46.*

Reinders, Hans Peter, 2007 Learning together for organic farming Leis India, *March 2007, 9(1): 9-11*

Sharma and Sharma, 2002. An overview of the field of family business study, current status and directions for future. A handbook of research on Extension.

Sharma et al 2003, Gain in knowledge and attitudinal change in hill farm women regarding drudgery reducing farm implements through action research, *CCS Haryana Agri.Univ.Hissar . Annals of Agri.Bio-Res.2003, 8:1,81-82:3.*

Slavik, 2003., Information systems in Czech Agriculture:implications for research and the study of agricultural extension, *Jol.of Agri.Extn.Edu.,2003:94(4):177-184.*

Souvik et al, 2005 Appropriateness of recommended agricultural water-management technologies as perceived by the personnel of research extension system: a study in the eastern region of India, *J.of-Agri.Education and Extn 2005; 11(1/4): 17-26*

Thusoo and Sohi, 2000, Extension contacts and mass media exposure of the farmers in relation to the use of improved rice technology, *Himachal-Journal of Agri.Res, 2000, Publ.2001, 26:1-2, 137-142.*

Vasantha and Buchareddy, 2006, Perceptions and attributes of IPM technologies by cotton farmers of Guntur District, *J. of Res.ANGRAU,2006;34(2): 53-61.*

Wani, G.M.2007. Success Stories of Agribusiness / Agri-Clinics - a tool of Agri Entrepreneurship held at Department of Extension Education, Institute of Agri.Sci. Banaras Hindu University, Varanasi.

Wani, G.M.2007. Recent Trends in Policy initiatives and technological intervention. Goat Production as Livelihood resource. Lead paper presented at Sri Venkateshwar Veterinary University, Tirupati, 20-22 June, 2007 pp.

Wani, G.M.2007. Communication for Sustainable Agriculture Production. Proc. Communication conference held at University of Hyderabad, Nov. 2007.

Wani, G.M.et al 2000.Higher Education and Human Resource development : Knowledge perspective in current millennium. Proc. Int. conf. on Higher Education for human development. Published AIU, New Delhi. Feb. 22-24, 2000. PP239-247.

Wani, G.M et al 1999. Integration of NSS in Agricultural Universities. Invited paper, Proc. Symposium onm NSS, University of Agric. Sci. Dharward, Karnataka 27-30 July, 1999.

Wani G.M; 2006. Self sufficiency in Rice Production in Jammu and Kashmir. Presented for publication in magazine on world Food Day,2007 at *Directorate of Extension, Krishi Vistar Bhawan, New Delhi,*

Wani, G.M. 2003. Appraisal of institutionalized Farmer's Training Activities. Lead Paper: National Seminar on Extension Strategy for Promoting Development initiatives among farming community, held at *G.B.Pant.Univ of Agric. Sci & Technology, Pantnagar, Uttranchal, June 18-20, 2003.*

Warren, 2003, SWARD: participatory knowledge transfer at work *Journal Agri.Edu.Extn, 2003 9:1,43-50; 21.*

Yadev et al 2003, Socio-economic constraints in technology transfer of arid legumes, Hisar, India, 15-16 May 2002. *Advances-in-Arid-Legumes-Research,2003 497-500.*

□□□

Chapter 35

Extension Education Reformation
II-Technology Use and Knowledge Upgradation

G.M. Wani

ABSTRACT

The technology delivery system need information on social structure and adoption procedures. An innovative change to induce more adoption has been discussed. New approaches, women empowerment, self help options, administrative bottlenecks, issues, problems and new vistas have been discussed.

Knowledge reasoning for low adoption, crop production constraints, communication links, technology use, assessment and information credibility has been reviewed. New modiums of private extension, eco-friendly agriculture, technology gaps, have been analysed and reforms identified.

New proposals for cast sharing of technology delivery, Farmer – training and knowledge upgradation, Farmers field schools have been postulated to increase use of technology on shelf. The generation of technology itself is in critical stage. The top-down system of research planning is not completely replaced with bottom-up-approaches. Even with initiation of innovative projects by ICAR like NIAP, the total involvement of researchers, extension functionaries, farmers and farming system is invisible. Thus, a total change in technology generation, adoption & refinement at farmers field is proposed. A portion of adoptive & applied research could be undertaken on farmers field by research scientists. It could be participatory research by scientists, extension educators and farmers. The failure risks can be covered under crop insurance. A Netherland model of Farmer-Scientist-participatory research on farmer's field can be applied. The delivery mechanism thus can be simple, transparent and faster. Extension & Extension educators could be trained on field to carry their functions on a large area.

Introduction

We may not forget our challenges. Man in Agricultural sector has always faced them and came out successful. Roughly man as hunter 30,000 years ago feed 6 million people globally. Primitive agriculture 3000 years later helped him to feed 60 million people. Intensive agriculture 300 years ago fed 600 million bellies. Today chemical farming or industrial agriculture feeds 6000 million people. We hope our fourth paradigm shift consisting of modernization blend with sustainability shall feed twice as fed today, provided we move now to precision & intelligent agriculture as viewed in these and other documents (Wani, 2008 a,b,c). This could be possible with introduction and total facilitation of farmer empowerment. This would mean involving farmer in experimental learning, farmer to farmer exchanges, knowledge thrust, search for methods, interaction. This would mean Farmer field schools, participatory technology development and total mass involvement of research and teaching for advancement of farming. This will need a paradian shift in Extension Science reforms. This will mean a community supported agricultural extension system. We need a transparent democratic, economic, knowledge dissemination system. This could possibly be called as "Farm Business Management Venture" or system.

A change towards integrating crop, fruit, livestock, water conservation and climatic changes in our agricultural extension projects, missions and programmes is needed (*Wani, 2007*).

Knowledge Reasoning for Low adoption

Adoption of Rice technology in farmers of Maharastra's Ratngiri district was influenced by family size, occupation, social participation, mass media exposure, income and other extension contacts,. Only 12.6% farmers from SC communities has high technology aware ness (Mankar et al, 2004)

Adoption of a technology needs communication, knowledge decision making power and risk guaranties. Training of Mushroom, helped farm women to earn their livelihood independently in Tamil Nadu respondents from coimbatore, Priyar districts showed women awareness and training was helpful. (Syatha & Palarisway, 2006). Mushroom training helped young women and men to have a gainful employment. (Proc. 21 EECM, SKUAST-K, 2007).

The data were gathered from 100 farmers in Maharashtra, India, through personal interview using an interview schedule. It was found that medium educated farmers with medium socioeconomic status, procured information from progressive farmers. 50% of the farmers who adopted drip irrigation had medium level of adoption of the recommended practices, 25% of the respondents had low level of adoption and 19% had high level of adoption. (Katkar & Ahive, 2006)

The impact analysis revealed that approximately 55% of the cauliflower growers adopted vermi-compost in adopted villages as well as in neighboring areas (Singh et al, 2005).

The study was conducted to identify the constraints encountered by farmers (n=120) in adopting the recommended pearl millet (Pennisetum glaucum) production technology in dry land farming in Sirsa District, Haryana, India. The farmers were asked to rate the identified constraints of production as very serious, serious and not so serious. It was found that the constraints such as high price of herbicides, fungicides and other pesticides, and lack of irrigation facilities, were very serious. (Chander et al, 2006).

Crop Production Awareness Constraints

A study was conducted to assess the constraints encountered by farmers in practicing wheat production technologies, and obtain feed back/ suggestions to minimize or eradicate the constraints. The data were gathered by administering a pre-tested questionnaire to 320 farmers from Uttar Pradesh, India during 2001-02, based on the study. The following were the constraints in implementing wheat production technologies: lack of timely funding to arrange inputs, lack of knowledge about the recent technologies, availability of technologies that are costly and requiring more inputs, and lack of proper marketing facilities. Farmers (84%) expressed that demonstration would minimize the constraints, 79% of the farmers suggested to delay sowing of drought- and blight-resistant varieties of wheat, and 80% of them favoured strengthening of cooperative facilities. (Prasad et al, 2006)

A study in J&K on maize production revealed that 82% farmers had technology awareness through radio and 62% through TV. Majority of 70% of the maize growers had medium level of awareness (Lakshimikant et al 2005).

The use of Farm Yard Manure was restricted 27% farmers among maize growers of J&K state. The recommended practices of fertilizers was adopted by 19% (nitrogen) 24% (Phosphorus) % 2% (Potash). Under these circumstances low levels of rainfed maize production can be increased if wholesome recommended practices are employed. (Lakshamikant and Chandagri, 2005). A low level of FYM use is the reason for low production of Saffron, rice & fruits. (Wani, 2008). Under 3000 front line demonstration in last few year, with application of production recommendation, maize production was doubled (4.3 t/h^{-1} Vs 2.1 t/h^{-1}) (Wani 2008 d), (Nehvi, et al, 2007).

Communication Links with Scientists

A study was conducted to determine the communication channels and methods used by scientists to establish linkages with extension personnel. The data were gathered from 102 scientists in four districts of Karnataka, India. It was found that almost all scientist working in different centres used phone calls to link with extension personnel. The regularity of use depends on their needs. Professors, associate professors and assistant professors had regularly used written communication to provide information to extension personnel. It was also recorded that lesser professors participated in workshops, meetings, trainings, exhibitions, field visits, farmer-scientist interactions and results demonstrations, compared to associate and assistant professor (Gupta and Chandragi, 2005). The low percent of professors, Associate Directors and other research scientists interaction with farmers is responsible for knowledge gap. A proper reward system in SAU/ NARS needs to be incorporated to give due credit to professors/Associate & Asstt. Prof. for Extension innovations (wani, 2004).

A study was conducted in four districts of northern Karnataka (Dharwad, Belgaum, Gulbarga and Bellary), India, to determine the performance of scientists (n=102) in different linkage activities with extension personnel. To measure the performance of the scientists, a scale was developed by following a standardized procedure. Results showed that majority of the professors (85.71%), assistant professors (82.25%) and associate professors (69.23%) belonged to medium performance category in linking with extension personnel. Moreover, associate professors were observed to be in-charge with most of the research projects. Associate and assistant professors often participated in field days, demonstrations and other activities (Gupta and Chandargri, 2005).

We have also observed that extension delivery system in J&K is mostly by Assistant or Associate professor rather than professors. This is very critical that participation of professors in under graduate teaching and extension in minimal. The role of research is also decimal. A policy change is thus needed to stimulate research, extension and teaching among senior agricultural scientists. Perhaps research on Farmers field could be a possibility now that, we have crop insurance in vogue. (Wani, 2007).

Field results of Technology Use

This study was conducted in Belgaum district of North Karnataka, , India , during the year 2004 to assess the adoption of production and post harvest technology for Tomatoes (Lycopersicon esculentum) and to identify the problems encountered by farmers (n=120) in adopting the technology. Majority of the farmers cultivated tomato hybrids, planted during June-July, followed 75-100 g per ha seed rate, adopted 60x40 cm spacing and applied 5-10 tonnes of farmyard manure per ha. Majority (86.66%) of the respondents used 5% neem seed kernel extract for fruit borer and 58.33% farmers used Bordeaux mixture for fruit rot. With respect to postharvest technology, 66.67% of the respondents sorted manually, 46.67% followed manual grading and none of the respondents had taken up processing of Tomato. Lack of technical knowledge and guidance about improved cultivation practices and postharvest technology, high fluctuation in market price and high transportation cost were the major problems, faced by the farmers in adopting the production and postharvest technology (Kumar et al, 2006).

Mass media and visual media helped animal husbandry and related activities through scientist /farmers interaction, North -24 pargnas of West Bengal (Roy et al, 2006). A study was conducted to determine the knowledge level of dairy farmers regarding selected animal husbandry practices for profitable dairy farming. The data were gathered by administering a questionnaire to 100 dairy farmers from Nadia district, West Bengal, India . Results showed that majority of the dairy farmers had high knowledge level on artificial insemination, medium knowledge level on deworming and other animal husbandry practices, and low knowledge level on feeding animals with green fodder. It is suggested that the results of this study can be utilized in formulating, planning and implementing transfer of technology that promotes the use of scientific method in dairy farming (Islam et al, 2006) The farmers adopted less of the recommended practices of fertilizer application, seed rate per hectare and spacing while, none of them used any plant protection measures to control pests and diseases. It is suggested that the extension agencies should take part in the implementation of various extension activities and popularizing the re commended practices of wheat in wheat growing areas so as to increase wheat production and consequently, the total agricultural production (Waman and Ahire, 2006).

The determinants in dairy extension system perceived by Directors of Animal Husbandry and Veterinary Sciences, Chief Veterinary Officers, Managing Directors, General Managers and Managers in dissemination of information were studied. The study was conducted in twenty-five states of India. The total numbers of determinant selected for the study were seven. Out of these the "facilitating adequate availability of credit to support the programme taken up for the rural poor" and resource and income generation of vulnerable section of the rural population were much required. However, "Skill formation and skill upgrading projects to provide self and wage employment among the rural poor' and increasing production and productivity of milk and milk products' were least required in dairy shared learning extension system. (Suman & Ram, 2005).

A study was done on social capital in accelerating the spread of agricultural technologies such as integrated pest management (IPM). The study was done in a village in Munoz, Nueva Ecija, Philippines, in response to the problem of slow diffusion of agricultural technologies. Both quantitative and qualitative methods were used in investigating the process of sharing and learning of IPM. Results show that social capital in terms of kin networks and spatial proximity such as found among farm neighbours are essential in the learning process and hence, diffusion of IPM among Filipino farmers. Kin networks and farm location are important considerations in snapling (Logan, 2004) participants strategically for the long season IPM farmer field schools (FFS). (Palis et al, 2005).

Technology Assessment

The assessment of recommended water-management technologies was performed after their documentation from different organizations working in the field of water management in two eastern India states viz. Orissa and West Bengal. The perception of 30 members of the research system regarding the feasibility of these technologies elucidated that out of 86 documented recommended water -management technologies , 40 were having feasibility scores of 4.0. Eight technologies with score of 3.0 and the rest of the 38 technologies with a score between 3.0 and 4.0 on a feasibility continuum range from 1.0 (not feasible) to 5.0 (highly feasible). Out of 40 recommended technologies(already assessed as highly feasible by the research personnel), extension personnel have perceived 16 and 10 technologies as highly appropriate and feasible, respectively. While six and four technologies were found to be less appropriate and feasible (Souvik et al, 2005). The gap in research extension perception need remedial measures by link research to farmers fields or allowing KVK's to have adoptive research in districts. Remodeling of extension - Research linkage is needed.

This study was conducted to ascertain the effectiveness of different communication media for transfer of cotton production technology within cotton growers (n=120) selected from four village of Hisar district, Haryana, India , during 1999. The farmers were given training on cotton production technology through four selected communication media namely, lecture plus discussion, printed material plus discussion, audio plus discussion and visual plus discussion. The study revealed that farmers gained maximum knowledge (18.67%) when the improved cotton production technology was communicated through visuals plus discussion and gained minimum through lecture plus discussion (9.47%). In the study, visuals plus discussion was found most effective, followed by printed material plus discussion, audio aids plus discussion, and lecture plus discussion teaching methods. (Kumar et al, 2005).

Farm Information Credibility

The study aims to ascertain the relative credibility of sources of farm information as perceived by the televiewer and non-viewer farmers residing under the penetration range of Muzaffarpur Doordarshan Kendra of Bihar state, India. A sample of 120 televiewer and 80 non televiewer farmers were randomly drawn from eight villages under the same viewing jurisdiction and data were collected through a pre-tested interview schedule. The findings suggest that demonstration, progressive farmers, agricultural scientists, television and village level workers were recognized as the most credible sources of farm information among televiwer farmers. Progressive farmers, demonstration, friends, relatives and agricultural scientists were considered as the credible sources of farm information among non-viewer farming groups. The study further revealed that the viewers perceived television as the fourth most credible source whereas non-viewers considered it as the sixth most credible source of farm information. The implications of the results are also discussed in view of an effective communication strategy for the transfer of technology (Ansari and Singh, 2005).

A study was conducted to(a) examine the personal, social, economic and psychological characteristics of fig (Ficus carica) growers; (b) identify the sources of information used by the fig growers; and (iii) obtain the suggestions of the respondents to overcome the problems. The data were gathered by interviewing 200 fig growers in 20 villages in Pune District, Maharashtra, India, during January 2005. Results showed that most of the fig growers are middle-aged and are educated upto the 10th grade. Most of them have medium-sized families with medium level of experience on fig cultivation. All the respondents suggested that they need more information regarding effective and efficient marketing and credit system. They also recognized their need for information on prices during harvesting, subsidy for drip irrigation and use of low cost technology for processing figs and cold storage (Khalache and Khaire, 2007).

Farmers Training Needs

This study was conducted to;(i) examine the characteristics of paddy farmers; (ii) assess the knowledge level of paddy farmers regarding integrated management practices in paddy cultivation; and (iii) explore the relationship between characteristics of paddy farmers and their knowledge level. Data were gathered from 120 paddy farmers in Maharashtra, India. Results of the study indicated that the respondents have medium knowledge level regarding various integrated management practices. Thus, the farmers, should be made aware of the availability and benefits of integrated management practices through television, radio, newspapers and extension literature. It is suggested that training programmes may be organized for farmers regarding scientific use of fertilizer, plant protection through biological control measures and water management (Ahire and Kiran, 2007). A detailed training reforms have been reported (Wani, 2003).

Private Extension

A study was conducted to determine the types of agricultural extension services suitable for privatization as perceived by agricultural scientists (n=15) and officers (n=20) of the department of agriculture and horticulture of Andhra Pradesh, India. Majority of the agricultural scientists preferred the privatization of technological services: Soil analysis, irrigation water analysis, providing information on value addition techniques/ production of products for export, supply of weedicides, transportation facilities, provision of credit facilities from cooperative banks, farmers' association, distribution of farm literature, and dairy services. Moreover, officers of the department of agriculture and horticulture also recognized the need for private services on the following technological services: providing information on growth regulators; plant protection measures on IPM; supply of fertilizers, manures and pesticides; supply of agricultural implements and other equipment; transportation facilities; and the conduct of field visits/tours and dairy services (Kumar and Reddy 2006).

Reference

Ansari and Singh, 2005, Information source credibility among televiewer and non-televiewer farmers. *J. of Interacademicia 2005; 9(4):606-613.*

Gupta and Chandragi, 2005, Communication channels and methods used by the scientists for establishing linkages with extension personnel , Karanataka-*J. of Agri.Sci, 2005 :18(4):1129-1132.*

Gupta and Chandragi, 2005, Development of scale of measure the performance of scientists on different linkage activities with the extension personnel. Karnataka –*J. of Agri.Sci, 2005; 18(4):1013-1017.*

Katkar and Ahire, 2006, A study on adoption of drip irrigation system in Maharashtra State., International *Jol. of Agro.Sci, 2006; 2(2):335-337.*

Khalache and Khaire, 2007, Social-ersonal and socio-economic characteristics of the fig growers and information sources used by them in fig cultivation technology, International *J. of Agri.Sci.2007;3(1):258-20.*

Kumar and Reddy 2006, Type of agricultural extension services suitable for privatization as perceived by agricultural scientists and officers of agriculture and horticulture departments.J. of Research-*ANGRA U,2006;34(2):62-69.*

Kumar et al, 2005, Effectiveness of different communication media for transfer of cotton production technology, *J. of Cotton-Res. And Dev.2005; 19(1):120-123.*

Kumar et al, 2006, Adoption of cultivation and post harvest technology of tomato by farmers., Karnataka *J. of Agri Sci.2006;19(1); 76-79.*

Lakhsimikand et al 2005, A study on knowledge level of rain maize growers in jammu District of Jammu and Kashmir State., Karnataka-*J. of Agri.Sci.2005; 18(4):1138-1139.*

Palis, et al 2005. Lineage analysis of the hemangioblast . Agri. Jr. sept. issue, 2005.

Puran Chand et al 2001, Constraints of the farmers in the use of improved wheat technology. Himachal-*Journal of Agri.Res.2001, 27:1-2,95-102;7.*

Reinders, Hans Peter, 2007 Learning together for organic farming Leis India, *March 2007, 9(1): 9-11.*

Singh et al, 2005, Development and dissemination of horticultural technologies by the Punjab Agricultural University, Ludhiana, India, *Acta-Horticulture 2005;(672):265-269.*

Souvik et al, 2005 Appropriateness of recommended agricultural water-management technologies as perceived by the personnel of research extension system: a study in the eastern region of India, *J.of-Agri.Education and Extn 2005; 11(1/4): 17-26*

Suman and Ram, 2005.Evaluation of biokinitic parameters pharmaceutical waste waters using aerobic oxidation. Vol. 40 issue 1, 2005

Vasantha and Buchareddy, 2006, Perceptions and attributes of IPM technologies by cotton farmers of Guntur District, *J. of Res.ANGRA U,2006;34(2): 53-61.*

Waman and Ahire, Awareness and adoption of wheat production technology by farmers, International-*J.of Agri.Sci.2006;2(1):44-46.*

Wani, G.M.2007. Success Stories of Agribusiness / Agri-Clinics - a tool of Agri Entrepreneurship held at Department of Extension Education, Institute of Agri.Sci. Banaras Hindu University, Varanasi.

Wani, G.M.2007. Recent Trends in Policy initiatives and technological intervention. Goat Production as Livelihood resource. Lead paper presented at Sri Venkateshwar Veterinary University, Tirupati, 20-22 June, 2007 pp.

Wani, G.M.2007. Communication for Sustainable Agriculture Production. Proc. Communication conference held at University of Hyderabad, Nov. 2007.

Wani, G.M.et al 2000.Higher Education and Human Resource development : Knowledge perspective in current millennium. Proc. Int. conf. on Higher Education for human development. Published AIU, New Delhi. Feb. 22-24, 2000. PP239-247.

Wani, G.M et al 1999. Integration of NSS in Agricultural Universities. Invited paper, Proc. Symposium onm NSS, University of Agric. Sci. Dharward, Karnataka 27-30 July, 1999.

□□□

Chapter 36

Extension Education Reformation III-Farmers Perception and Farmers School

G.M. Wani

Abstract

A mixed technology pack consisting of awareness on generation of total farm income is needed. Our farming is a composite farming comprising crop, fruit, livestock, fish and allied production systems. We need a composite technology pack and not a commodity research end product. Thus a total change and reformation in National Agricultural Research System (NARS) is needed to shape it as farmer friendly. We may name it NFR&ES, i. e National Farm Research & Extension System or simply Indian Farm Research and Business Management System.

Introduction

Future of agricultural extension education lies in its advancement through research, methodological refineness and village-linked mechanization. The disciplines of extension education has to be reshaped, incorporating explorative methods, survey techniques, computing skills, economic and social back ground analysis. An attempt to characterize barriers in transmission of technologies to the baseline farming systems needs a reformation in extension education in India and Asia. A detailed reformation model in the form of

Broad Based Agricultural extension has been published (Wani, 1992). The booklet describes a model approach for Kashmir Valley districts. With the establishment of SAMETI in SKUAST-K, as an autonomous Institution, a series of documents for improving productivity in whole agriculture i,e crops, fruits and livestock have been prepared. (Wani, 2008, a & b). On the basis of these reviews and documents, the present document is 4th of the series, reviewing technology dissemination, impact & productivity changes. It concludes with a reformation model for Extension Education - a Vision.

Farmers Perception

A study was conducted to examine the perception of cotton farmers(n= 210) of Guntur District, Andhra Pradesh, India, regarding the attributes of integrated pest management (IPM) technologies. Majority of the cotton farmers of the area perceived that the initial cost of IPM technologies is high, and IPM technologies give meager and irregular net profit, but may be beneficial in the long run. Moreover, farmers stated that IPM technologies are not feasible given the situation of the district, culturally not acceptable, not necessary, not socially recognized, have more cognitive complexity and scarce resource complexity, reversible, and consume more labour. Regarding the practicality of IPM technologies, the respondents felt that the technologies can be easily demonstrated and tried out, and have reliable point of origin. It is suggested that the success of IPM technologies depends on farmers awareness, community approach, timeliness and availability of inputs. (Nasantha & Buchareddy, 2006).

Eco-friendly Technology

This study was conducted to analyse the constraints encountered by farmers (n=245) in employing eco-friendly technologies in cultivating rice in Andhra Pradesh, India. Ineffectiveness of the recommended dosage of spray solutions (51.42%), and difficulty in maintaining recommended depth of water after transplanting (23.26%), were the major problems perceived by the farmers. Based on the results, 55.51% farmers suggested that adequate trainings must be conducted on improved management practices to enhance farmers's skills, and 32.24% farmers said that more demonstrations must be conducted on farmers's fields, for effective adoption of eco-friendly technologies in rice cultivation (Reddy et al, 2006).

The study was conducted to analyse eco-friendly technologies for rice cultivation, which exploit renewable resources to control pollution upto considederable level and to recycle wastes. The data were gathered from 245 farmers in three regions of Andhra Pradesh, India it is suggested that extension agencies should concentrate on major suggestions i.e., conducting required number of training programmes for enhancing the skills of rice farmers in various management practices. Conducting more number of demonstration on the efficacy of eco-friendly technologies in the field, rather than mere publicity. Extension should be strengthened so that it is demand driven, market driven and farmer led. Visits of experts should be increased to farmers' field region wise and block wise for better comprehension and diagnosis of field problems. (Reddy et al 2005).

This study was conducted to analyse the attitudes of rice farmers (n=245) in Andhra Pradesh, India, towards eco-friendly technologies. Results showed favourable attitude of farmers towards eco-friendly technologies. The results indicate the need for the extension agency to train and educate the farmers regarding the advantage of adopting eco-friendly technologies to decrease environmental pollution, obtain high quality produce and sustainable yields,(Reddy et al, 2006).

Technology Gaps

The study was conducted in Junagadh-1 and Junagadh-2 subdivisions of Rajkot division, where National Watershed development programme has already been implemented, to find out the extent of technological gaps in adoption of soil and water conservation technologies and factors responsible for it. The results revealed that the technological gap was higher in adopting the percolation tank-cum-farm ponds, strip cropping and mulching practices in beneficiary farmers (BFs). But in non-beneficiary farmers (NBFs), the technological gap was higher in the practices of contour cultivation, contour bunding along with vegetative barriers, afforestation, recharging well and tubewell, deep ploughing and graded bunding in addition to above three practices mentioned for BFs. The study clearly indicated that overall technological gap in NBFs was significantly higher with a mean difference of 24.57% compared to BFs. It was also observed that variables namely, size of land holding , social participation, employment status, opinion leadership, extension participation, localite-cosmopolite value orientation knowledge and attitude were significantly correlated with the technological gap (Popat et al 2006).

Farmer Decision Process

This study was conducted to analyse the direct and indirect relative influence of the selected independent variables with each other to analyse the relative influence of other dependent variable, i.e., differential innovation decision process of rice eco-friendly technologies. The data were gathered from 245 farmers in Andhra Pradesh, India, and analysed using path coefficient. The path diagram clearly shows that the variables social participation, farm size, and extension participation were crucial variables. This might be due to the fact that many variables were channeling their substantial indirect effects on differential innovation decision process of rice eco-friendly technologies through them. The variables exposure to electronic media and cosmopoliteness showed significant positive direct effects. Moreover, maximum direct negative effects were shown by fatalism and religious beliefs (Reddy et al, 2006).

Farmer Cost-sharing – Nigerian Model

Providing adequate and stable funding for agricultural extension service in Nigeria has been a major problem since the expiration of the World Bank's funding arrangement. In order to solve this problem, cost -sharing of agricultural technology delivery is seen as a tenable privatization policy option. Considering the fact that agricultural technology transfer in Nigeria has mainly been publicly funded, introducing cost-sharing arrangement is expected to meet some obstacles. Hence, this study ascertains the perception of farmers and extension agents on the constraints and strategies towards effective cost -sharing of agricultural technology delivery in Nigeria. This study was carried out in six geopolitical zones of Nigeria. Multistage random sampling technique was applied in the selection of respondents. A sample size of 267 farmers and 272 Agricultural Development Programme (ADP)staff participated in the study. Means, standard deviation, exploratory factor analysis and test statistics were used in realizing the objectives. The results show that the major constraints to effective cost-sharing of agricultural extension service in Nigeria are weak institutional development, extension system lapses, lack of cooperation by farmers, uncertainties experienced in agriculture, conflicts and corruption. The major strategies for effective cost-sharing arrangement include building political support for cost sharing , establishment of farmers' cooperatives to serve as avenues for collection of payments, creating enabling legislation for cost -sharing and increasing the number of extension staff. The study recommends proper dissemination of information on cost-sharing before implementation, creating enabling legislation, decentralizing the extension system and building the capabilities of extension staff (kukwuone et al, 2006). Should we use this model of Nigeria.

Farmers Knowledge Level

The study was conducted to determine farmers' knowledge level regarding the major aspects of pearl millet production technology in dryland farming. The data were gathered by administering a questionnaire to 120 farmers in Sirsa District, Haryana, India. Results showed that 54.17% of farmers possessed medium knowledge level whreas 26.67 and 19.16% of them have high and low knowledge level, respectively. Regarding aspect-wise knowledge level of production technology, 75% of farmers possessed high knowledge in preparatory tillage, 78% had low knoeledge level about improved variety, 90% of farmers had low level of knowledge on seed treatment and 68% possessed medium knowledge level about sowing rate. For sowing date and FYM application, 82% and 70% of farmers had high and medium knowledge level, respectively. Farmers possess low level of knowledge about chemical fertilizers application 48%. In mechanical weed control, 85% had high knowledge level and 82% had low knowledge about chemical weed control. On plant protection measures, 50.83 and 70% farmers had low knowledge level about insect control and disease control, respectively (Chander et al, 2006).

Farmers Field Schools

The Farmer Field School (FFS) approach is gaining prominence as an extension approach but its usefulness in influencing farmers to understand and uptake new technologies is still a question lingering in the mind of many FFS practitioners. This study was conducted to determine the effectiveness of the approach in knowledge acquisition, adoption and dissemination of soil and crop management (S&CM) technologies among small-scale farmers in Kenya. Eight technologies were scaled -out using the approach and the conventional extension methods. A survey with an ex-post facto design was employed with a sample frame consisting of 940 farmers. A sample of 60 FFS graduates and 60 non-FFS (NFFS) farmers was chosen for the study. There was a significant difference in knowledge acquired in S&CM technologies by FFS compared to NFFS farmers. About 50% of FFS farmers had acquired high to very high level of the knowledge of all the eight technologies disseminated while the majority (80%) of the NFFS farmers had acquired less than 50% of the same knowledge. About 45% of the FFS farmers had adopted 50% of the technology components while only 17% of NFFS farmers had adopted the same percentage of the technology components. The FFS participants were significantly (P .05) better disseminators of S&CM technologies than the non-FFS farmers. It was recommended that all stakeholders in Agricultural development in Kenya be sensitized on the potential effectiveness of FFS methodology in scaling-up agricultural technologies and also monitor how the application of knowledge acquired by FFS farmer is impacting on their livelihoods. (Bunyatta et al, 2006).

Conclusion and Work Plan 2008

Community supported agriculture system needs to be evolved. It should encompass food production, marketing, distribution, quality assurances and pricing decisions. A transparent system of research, outreach (extension) industry profiting based on Farmers production prices need to be evolved. Research and extension needs to be clubbed to give end products to farmers and not the research publication which are never transferable to farmers. Communication process within extension system need to be informative, analytic and fact gathering exercises. The total change in Extension delivery system should begin with curriculum changes to suit new name of "Farm Business Management. It needs incorporation of Agricultural sciences awareness with particular emphasis on Agribusiness, appraisals, consultancy, risk management, data analysis, credit management, marketing, public relations, economics and policy issues, farm management and international agriculture - public private partnerships, Farm co-operative ventures, rural banking, Agro-industrial managements, feed/seed services and programmes need to be added to the projects and mission on agricultural technology delivery system.

Farmer schools, women awareness and technology impact analytic projects need to be financed by ICAR. A broad national based project as Agriculture and National resource utilization needs to be financed including all SAU Extension Education Directorates. This should cover areas of farm production, production cost, price structure, service cost, input - output ratio with respect to livestock, fish, agriculture, forestry, sericulture and horticulture products. Quality trainings in food processing and preservation to stimulate new graduate to set up food processing units. A composit packaging of technology, their use and cost benefit ratio analysis should be the areas of operation of the project. Once the basic idea of reforms is conceived by ICAR logistics could be worked out.

Capacity Building and e.Extension for enhancing Agriculture Productivity in India

G.M.Wani

Director Extension Education

Sher-e-Kashmir University of Agricultural Sciences and Technology of Kashmir, Shalimar, Srinagar, 191121

P.O.Box : 461, GPO, Srinagar

To be held at

National Agriculture Science Centre (NASC), New Delhi

w.e.from 12.02.2008 to 16.02.2008

Author with Farmers

References

Badri-OA; 2001. School of Rural Extension Education and Development, *Ahfad University for Women, Omdurman, Sudan. 18:1, 35-50; 20.*

Balasubramanian and Thamizoli, 2003, Social differentiation in the horizontal transfer of knowledge: a case study from south India, *J.of Agri.Edu.and Extn.2003, 9:2,51-60;3.*

Bharali et al, 2004. Influences of extracellular calcium and iron on membrane sensitivity to bisulphite in the mosses.Jr. of Bryology, vol.26No.1 of March 2004, pp 53-59

Bunyatta et al, 2006, Farmer field school effectiveness for soil and crop management technologies in Kenya. *J. of International Agricultural and Exn.Edu.2006; 13(3):47-63.*

Chander et al, 2006, Farmer;s knowledge level on major aspects of pearl millet production technology in dryland farming, *Enviornment-and-Ecology , 2006; 245 (Special 3!): 868-871.*

Kukwuone et al, 2006, Constraints and strategies towards effective cost-sharing of agricultural technology delivery in Nigeria: *Perception of Farmers and Agricultural Extension Personnel.*

Davis et al, 2004, Extending technologies among small-scale farmers in Meru, Kenya: ingredients for success in farmer groups.*Jol.of Agri.Edu.and Extn.2004;10(2):53-62.*

Ganesan et al, 2003,Reaching the unreached: basics of extension education, Tamil Nadu Agri.University, Coimbatore -641 003, India. *Reaching the Unreached: Basics of Extn.Edu. 2003:xii+387.*

Muhammad Akram, 2003. Main factors affecting extension activities. *Sarhad Jr. of Agri. 19:1, 163-168; 7.*

Puran Chand et al 2001, Constraints of the farmers in the use of improved wheat technology. Himachal-*Journal of Agri.Res.2001, 27:1-2,95-102;7.*

Rani and Reddy, 2002, Knowledge and adoption of farm and home technology of farm women. *Journal of Research-ANGRAU,2002,30:2,101-104;6*

Reddy et al 2005,, Eco-friendly technologies in rice cultivation- a constraint analysis, *J. of Res-ANGRAU, Rajendranagar, 2005;33(4):76-78*

Reddy et al, 2005, Path coefficient analysis of independent variables with differential innovation decision process of rice eco-friendly technologies. *J.of Res.ANGRAU,* Rajendranagar, Hyderabad-*30, India.*

Reddy et al, 2006, Content analysis of attitude of rice farmers towards eco-friendly technology, *J. of Res.ANGRAU,2006;34(1):72-77.*

Reddy et al, 2006, Eco-friendly technologies in rice cultivation- a constraint analysis. *J. of Res. ANGRAU, 2006; 34(2):44-47.*

Reddy, M.V; 2001, Relationship between number of technologies developed at NARP stations of ANGRAU and their adoption. *Journal of Res. ANGRAU, 29:1, 42-46.*

Reinders, Hans Peter, 2007 Learning together for organic farming Leis India, *March 2007, 9(1): 9-11.*

Singh et al, 2005, Development and dissemination of horticultural technologies by the Punjab Agricultural University, Ludhiana, India. *Acta-Horticulture 2005;(672):265-269.*

Thusoo and Sohi, 2000, Extension contacts and mass media exposure of the farmers in relation to the use of improved rice technology. *Himachal-Journal of Agri.Res, 2000, Publ.2001, 26:1-2, 137-142.*

Vasantha and Buchareddy, 2006, Perceptions and attributes of IPM technologies by cotton farmers of Guntur District. *J. of Res.ANGRAU,2006;34(2): 53-61.*

Wani, G.M.2007. Success Stories of Agribusiness/Agri-Clinics - a tool of Agri Entrepreneurship held at Department of Extension Education, Institute of Agri.Sci. Banaras Hindu University, Varanasi.

Wani, G.M.2007. Recent Trends in Policy initiatives and technological intervention. Goat Production as Livelihood resource. Lead paper presented at Sri Venkateshwar Veterinary University, Tirupati, 20-22 June, 2007 pp.

Wani, G.M.2007. Communication for Sustainable Agriculture Production. Proc. Communication conference held at University of Hyderabad, Nov. 2007.

Wani, G.M.et al 2000.Higher Education and Human Resource development : Knowledge perspective in current millennium. Proc. Int. conf. on Higher Education for human development. Published AIU, New Delhi. Feb. 22-24, 2000. PP239-247.

Wani, G.M et al 1999. Integration of NSS in Agricultural Universities. Invited paper, Proc. Symposium onm NSS, University of Agric. Sci. Dharward, Karnataka 27-30 July, 1999.

Wani G.M; 2006. Self sufficiency in Rice Production in Jammu and Kashmir. Presented for publication in magazine on world Food Day,2007 at *Directorate of Extension, Krishi Vistar Bhawan, New Delhi,*

Yadev et al 2003, Socio-economic constraints in technology transfer of arid legumes, Hisar, India, 15-16 May 2002. *Advances-in-Arid-Legumes-Research,2003 497-500.*

❑❑❑

Chapter 37

Capacity Building and E-Extension for Enhancing Agriculture Productivity in India

G.M. Wani

Abstract

The green, white and blue revolutions gave us food security. The high yielding varieties and new technologies were webbed with chemical farming. Even today we have critical gaps existing in productivity of food grains between technology using farms and traditional farming practices. The chemical farming resulted in the soil degradation, water pollution, soil erosions and salinity. By now, we face land degradation problems in 173 million hectares which is around 53% of cultivated land. Annually we loose 5000 Million Tones of top soil with NPK losses of 5-8 Million tones per year.

In Mahrastra a survey showed that the depth of black soil was 60 cms in 1910 which has reduced now. About 18% of it has turned as shallow land. Reduced soil depth has resulted into low productivity, increases soil runoffs and drought like conditions. Therefore to avoid these ill effects we have to link strong information and communication methods for soil mapping, annual rainfall data, rain and climatic forecasts with farming operations (Wani, 2005).

Resource conservation & their proper utilization needs adequate knowledge, which could be obtained through advanced satellite system and relied back through communication mechanism. Thus, for enhancing agricultural production, communication tools have to be used.

It is time for sustainable agriculture. We should make use of mud waste, farm yard manure, waste cakes, municipality wastes and green wastes for increasing phosphorus, nitrogen and micro nutrients in the soil. It has been found that these substitutes have a capacity of giving 1-2% of Phosphorus and Nitrogen and provide about 40-45% micro nutrients. Therefore, a new strategy to use crop stubs, waste tree leaves, waste orchard residues and other agriculture waste material for improving soil has to be advocated.

Although we use pesticides worth 3500 crores. In urban India we use 1000 crores of pestices in domestic homes. In 1948 malaria took lives of 2-8 million people in Malaysia. With use of DDT the toll of death reduced to 17 in 1963. Thus a balance between food security and environment purity is to be maintained in developing countries. This denotes maintenance of a balance between food security and environmental protection or purity. The ways and means in this direction shall be discussed in this paper.

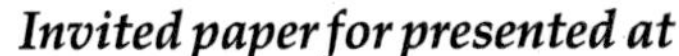

STEERING COMMITTEE

Chairman
Prof. M.S. Swaminathan

Co-Chairmen
Shri M.V. Rajasekharan

International Conference on

Conservation Farming Systems & Watershed Management in Rainfed Areas for Rural Employment & Poverty Eradication (ICON-FARM)

February 12-16, 2008, New Delhi (India)

Sponsored by

Ministry of Agriculture, Govt. of India

Acceptance Vide President SCSI No. SCSI/IC - Abstract/2007-08/1533 dated 26-12-2007

Introduction

The green, white and blue revolutions gave us food security. The high yielding varieties and new technologies were webbed with chemical farming. Even today we have critical gaps existing in productivity of food grains by use of technology and those produced at traditional farmer's field. The chemical farming resulted in the soil degradation, water pollution, soil erosions and soil salinity. By now we face land degradation problems in 173 million hectares which is around 53% of cultivated land. Annually we loose 5000

Million Tones of top soil with NPK losses of 5-8 Million tones per year. In Mahrastra a survey showed that the depth of black soil was 60 cms in 1910 which has reduced now. About 18% of it has turned as shallow land. Reduced soil depth has resulted into low productivity, increases soil runoffs and drought like conditions. Therefore to avoid these ill effects we have to link strong information and communication methods for soil mapping, annual rainfall data, rain and climatic forecasts with farming operations (Wani, 2005). Resource conservation & proper utilization needs adequate knowledge, which could be obtained through advanced satellite system and relied back through communication mechanism. Thus, for enhancing agricultural production communication tools have to be used.

Land Holding

We have to use more technology based cropping system to increase productivity per unit land. Horizontal expansion is not possible. Embargo on indiscriminate use of chemical fertilizers, pesticides and other farm-use-agents is another constraint to increase productivity of food grain. Unfortunately our food grain-production pace has declined. Growth rate of 1-2% has put pressures on our economy. Imports of food grains, an anomaly over past decade, needs problem- cause-analysis.

The whole system of National Agricultural Research, extension and field functionaries have registered a fatigue. Similarly, the land degradation, mineral depletion and environmental pollution demands new mechanism to boost productivity. Perhaps use of electronic media, e-extension and agricultural reforms in the shape of revised KVK system, ATMA, village level rural development, self-help groups and agri-information Kioses needs introduction. State Agriculture Management and Extension Training Institute (SAMETI) may help to make adjustments in our farming system so as to integrate agriculture, aquaculture, water conservation and livestock rearing with new technology driven profit earning enterprises. This needs a continued and farmer friendly policies of sustainable agriculture. In these pages we shall discuss role of communication in agriculture technology application for higher productivity and profits per unit of land. Technology awareness and application is must to produce more per unit of land. We do have a chance as our yields are lower than many counties and even our neighboring countries (Samra & sastry 2002).

Organic Farming

Organic farming is advocated as modern technology. We left our traditional organic farming for adopting chemical farming, which landed us in trouble. This rotation of modernity and traditionality taxed us heavily. We perhaps jumped in adopting or testing technologies without comparing them with our own practices. Now reverting back to our own traditional ways is fraught with problems too. The questions often asked are, can we sustain or even maintain our productivity levels by resorting to the organic farming? The Blending of modern technology of today and traditional technology of yester years, may answer this question. Can information and communication skills and technology bridge a new union?

Information and Communication Technology

The present day information and communication technology has trespassed all barriers of race, religion, culture and countries. A Comprehensive study of 23 review papers and a dozen book and journals was presented by (Wani 2005). A detailed description of how communication and information can help production and sustain yields has been discussed (Wani 2006) . Strong warning systems for climatic risks, floods and cyclones, pests and mites could help to raise more crops. Farming informatics and awareness packages through print, mass and now E-mails is possible. The role of competitive farming, economic survey and evaluation of farming and women's integration needs attention. Women, the half of agricultural work force is still unawares of the technological skills. The barriers of customs, veil, religion and social bondages could be overcomed by educating them through TV, cassettes, e-mails or other modern communication appliances.

Phyto Sanitation

Technology transfer is easy. We can announce technology practices or even demonstrate them. The key issue is its adoption. India with 25% of its GDP from agriculture spends some 12% of GDP on its subsidies rather than on transfer of technology. Blending subsidies with agricultural exports will need a drastic cut under new WTO agreements. The global market access opportunity limit of 3% import shall further complicate the issues. The international standards of sanitation shall need more awareness at farmer's doors. Our Agricultural exports from 12 agricultural items has been up and now we export around 18.45% agricultural good in the shape of apiculture, floriculture, fresh fruits, mushroom, spices, sugar, molasses, rice, tropical fruit juices, pulp, concentrates and even agro-chemicals. Fruits, nuts and vegetables have increased our export earnings. Our limitations in expanding our exports are infrastructure to provide international biosafe packaging, phyto sanitation & quarantine measures. Our yields too are low to compete with others. Thus, transfer of technology has not to be limited to man methods, publication, leaflets, folders, bulletins, newsletters, journals, magazine, news paper publication, rural farm broadcasts or television interviews but has to be supplemented with video conferencing, massive awareness campaigns through video cassettes, cable net works and other local farm telecasts. The propaganda, publicity and persuasion has to be supplied with communication skills like rural journalism, popular participation, motivation and more so through management of information systems.

The farm visits, farmers calls, letters have to be intensified. Farmers need information on markets, bio standards and marketing research and networking mechanisms.

Small Farmer- the Small Holder

Another vulnerable class in India is small (holder). Over 65% of our farmers are small holders. The technologies generated are mostly for commercial farmers. The small holders have not only limited hold on land but on information too. Their case is further complicated as they do other work additionally as the small holdings is not sufficient to sustain them. They have limited access to knowledge. Even the word Bauern or Landliche Gebiete (German), peasant or country side (English) campagne, brousse/ paysan(French) and Campo/Campesino Spanish denote that extension work is a mission. One with love for country side and farmers alone could execute this task properly. He has to be well versed with objectives, problem, targets and implementation process. He has to make situation analysis and avoid group clashes in the country side. Thus, the knowledge and experiences of farmers, psychological values, expectations, needs and attitudes are to be organized. This organization and evaluation is not possible without use of modern and applicable communication methods. Welding communication with Agri -technology is the need of the hour.

Communication Methods

Passing on information to farmers is basic fundamental of any extension activity. The basic need for learning process is communication.

Factors influencing communication range from person, his personality, his social relation, knowledge, the social and economic parameters of farmers. Their knowledge expectation, experiences and perception needs monitoring and evaluation before technology transfer.

Review of Communication Vs Agri Production

i. Critical Yield Gap Reduction

In Kenya, use of advanced information and communication technologies reduced gaps in yields of Agricultural crops between research and farmer's fields. (Oguya and Bellamy, 2001). A country where 70% population is connected with agriculture for livelihood directly or indirectly and 80% of its export is agricultural oriented. Reduction in yield gaps through effective use of information and communication technology will have a significant impact.

ii. Climatic Risks and Communication

The low productivity in Soyabean was found to be due to partial adoption of Production recommendations by farmers in Mahrastra, India. The low yield factors were analyzed. Economic constraints, situational factors and communication gaps on crop production, protection, seed treatment and fertilizer application were found responsible for it (Jaiswal et al, 2002).

The modernization of the material and technical information base helped Cuba to increase agricultural production and rural development on a pilot basis (Albelo. et al. 2002).

Precision Agriculture

Precision Agriculture till date has focused on site-specific data collection for soil and crop management. The technologies for the site-specific field operations and automated data recording are available, but precision agriculture rarely involves them for improvement. The application of precision agriculture has to be clubbed with information and communication networking to harvest the gains and to improve productivity. This network may consist of an open software platform, which can be operated by the farmer himself. For efficient communication internet and mobile telecommunication have been identified as important components. The development of an information and communication network integrating modern software (Java, GIS) and hardware (GPS, internet) technologies in a new user friendly manner is necessary to achieve better acceptance of technologies and improved productivity (Lutticken, et al 2000).

Satellite Data as Source of Communication

Use of satellite data- (Star and Spot – lite) helped time – critical dependant applications in Australia. The Australian centre for remote sensing (ACRES) has introduced a new service to provide satellite data for near real time applications. The STAR (Speedy Transmission after reception) service provides access to digital satellite data products in full resolution or compressed format within 12 hrs of a Satellite overpass. The data obtained from ground stations is processed at a facility via a high speed communication link and high priority processing. This system provides Satellite data on critical applications, like crop yield modeling, pre-harvest crop production forecasting, detecting crop diseases, monitoring crop stress, pest infestation, floods, fires and oil-spills.

SPOT - LITE is a low cost, off the shelf satellite data product from ACRES that is ideal for use in Geographical information system (GIS). SPOT-LITE can be accessed at any time via the internet and is available in the form of tiles covering most of Australia (*Thankappanm 2001*).

Advance studies with high applications for increasing agricultural production needs quick dissemination. The effect of rooting zone restriction (RZR) on vegetative and reproductive growth of fruit trees viz grapes, peach and citrus has been investigated. It is known that it improved crop productivity under low availability of water. (*Wang et al. 2002)*

Communication Networking

The available communication facilities for agricultural information in 15 states of India were studied. (*Ghosh 2002*). The results suggest that while communication networking opens up agricultural economy, it is not cost effective. The communication networking has to become cost effective.

An attempt to have better communication between various forest research divisions and other organization interested in sustainable forestry have shown encouraging results. (*Barbour and Wong 2001*)

An attempt was made to have quick information flow among and between researchers, extension officers and dairy farmers in East Azerbaijan, Iran. The information input, output and intersystem communication were studied. The communication linkage improved the productivity. (*Rezvanfar, 2001*)

The basic tools of marketing premotion of fertilizers in India was studied. (*Yadav, 2002)*. The information like advertising, public relation and personal selling was found to be best promoters.

In the "Unique Selling" approach of the communicator effects target audience. The impact of integrated approach utilizing computers in agricultural information & dissemination in Greece and Poland was studied. (*Tzortzios et al 2001*). The gap in technology known and applied at farmers field was found. Researchers lack training in using new information technologies. Thus improvement in Agricultural productivity has to keep pace with advanced communication and information technology using computers.

Joint problem solving sessions and communications networking of information on agriculture has improved sugarcane production in Mauritius (*Jhoty et al; 2001*).

Technological Prospective

The technological gaps between research yields and Farmers yields are wide. The technological know how generated over years is still on shelf. Much of it has remained frozen if not dead within the fore walls of institutions. The location specific, crop and soil specific innovations to farmer's practices are few. Our integrated & mixed farming practiced by our farmer's needs improvements which is not forth coming. Our research institutes mostly aped the experimentations on commercial farming of west with few modifications and even additions at times. Our traditional agriculture existing still in remote areas encompass the use of farm yard manure, vormiculture & use of agricultural waste materials. The present problems of low soil fertility, reduction in yields and soil salinity have arisen due to excessive use of chemical inputs in farming with little care of the soil.

This scenario resulted due to incomplete innovative approaches of research. Our researchers blindly advocated more and more use of fertilizers, pesticides and fungicides, which helped to gain grain revolutions, but left legacy of polluted water, air & environment with degraded soils. These revolutions debarred future sustenance. This was due to poor perception & vision to fore see the net results on long time frame.

The immediate need is to make researchers akin with information technology and advanced communication. The rapid evolution of information science demands quick and speedy transfer of technologies, awareness & even subject reviews to farmer's for speedy application. The productivity would be better if technological advances are adopted and their impact is known. The knowledge of computer hardware, software as well interlinking the information dissemination channels and outlets is essential. A new concept of Farmer school with communication system can resolve 70% problems of farmers on spot. Kissan call centre (1551) and the Zraye Paigam shall widen the messages.

Decision Support System

The globalization of agriculture marketing has put forth new challenges. It is not only the yield which matters, but the benefit / cost ratio. The economic viability of an agricultural product has now more importance. The product even if economic and globally competitive has to pass quality tests. The international phyto sanitation standards are going to be tougher. The emergence of diseases like Mad cow disease or even SAARS being linked to animal cultures and other diseases from Agricultural & Livestock products warrants biosafe agricultural produce. The decision have to be supported with information not only on marketing, international safety stands but also on a law regulating agencies and regulations too. Unawareness of these support system could lead us to bankruptcy. Thus a strong case for tools and information management comes into force.

The international computer networking and communication systems alone could help in decision making for appropriate and economic viable agricultural production. (*Tzortzios et al 2001*).

Intelligent Agri-Management

Now we do talk of ecologically based pest management instead of integrated pest Management (IPM) . The buzz word for future in this regard is intelligent pest Management (IPM). As the system now advocated is not the blanket sprays of chemical pesticides but intelligent and well computed programmes of pest management, which incorporates their safe and long term application, economic viability and biosafety of products. Perhaps a well documented interaction between farmers, extension education workers, researchers and policy makers is necessary. This would take years if communication methodology of video conferencing, internet and related satellite technological advancement in communication is not used. This is what would be future intelligent Agricultural Management.

Participatory Management

The whole system of Agricultural research, teaching and extension in NARS needs renewal. A composite ARS system has to be introduced with strong basic of communication and IT. All programmes in the field should be participatory. All agencies be it ICAR, CSIR, RRL, Universities, SAU or other institutions and industry have to be webbed to getter with extension delivery systems in the field. More participation of farmers in planning and execution of project like mega seed or mission horticulture is needed.

A true transparent and open system of selection both for Management positions and scientific positions should be advocated. Changing standards for individuals is a crime. Those at top have to exercise truth, justice and fairplay. The role of politics and politicians should be minimal in national extension delivery system. An outreach through communication and IT is the only achievable solution.

Our Vision

1. Mixed Farm University Culture

Higher productivity gains can be achieved through application of technology and production recommendations at farmer's fields. We have 65% small and marginal farmers whose awareness potential is low. The production system prevailing with these farmers is a mixed farming or composite farming. In contrast to USA and European agriculture our necessity is to increase "Crop -livestock-fish-plant integrated production system with multiple livelihood opportunities". Therefore, we need our own innovative educational and training policies. A mixed agriculture University and Education set ups is our necessity. We are at present going astray to our need. Quick and fast measures are needed to unify our educational system, involving all agriculture and allied disciplines, industries, corporate sectors and farmers institution.

2. Higher Productivity Concerns

Indian Agricultural pride years of green revolution post 1968 saw reduction in food gain imports and subsequently white, blue and other revolutions sustained our population pressures and agriculture growth. Our agricultural growth rate (AGR) need to be equal if not more to population growth rate (PGR). Our AGR target ought to be double the PGR.

This is important as consumption rates, purchasing power and employment prospects increase. An estimated food grain of 210 million tons at present may need to be doubled in next 10 years. We have to achieve high targets of productivity by vertical expansion as horizontal land expansion is just not possible. Dr. M.S. Swaminathan has quoted figures as of 160 million tones of rice from 40m hac of land, thereby setting the productivity target of 4 t/h. Like wise production of 100 million tones of wheat from 25 million tones of wheat from 25 million hectors needs a productivity of 4t/hac. Our aim to double our per hac productivity needs more technical manpower in extension, industry and at gross root level.

The climatic disasters, earthquakes, Titanic tsunami, floods; have effected our agricultural production in the past and additional requirements needs to be kept in mind while planning food security. We need to increase per capita consumption expenditures of Rs.600 per month. We need to bridge the gaps between potential and actual yields at farmers level. The chemical farming hazards of poor soil fertility, low water availability, pollution and environmental concerns impede our agricultural development. Thus refined technology, participatory research and educational modules are needed. The new pressures of Global marketing. World trade and tariff regulations have to be accommodated. This all will need incorporation of new themes like post-harvest management, value addition, packaging, communication, credit and market information services in our course curricula. Thus a new multidimensional change in academic curriculum is envisaged.

3. Quality Assurance

We feel pride in calling ourselves as the 2nd largest Agricultural Research system (ARS) in World. When we review our performance we are no where in top ten of most cited agriculture publications in the World. USA tops the world list with 3,62,79,842 cited publication/annum, with small country like Switzerland at No.10. The scientific out put in agriculture is highest in USA with 27 lac publication/year followed by Japan, Germany, U.K, France, Canada, Italy, Russia, China and Australia. Our contribution to Agriculture publication is 5.48% only with our share of citation at 2.32%. This demands more focus on Quality Assurance. Our prime agenda should be quality Agricultural Education. Our emphasis has to be on:

Academic quality, Accreditation; Desired knowledge, Assessment, Skill and competence building and academic audit. Quality assurance, means strengthening resources, information and maintenance of educational infrastructure. Thus we need to regulate grants and centre-state relations rationally.

4. Employment Opportunities

We have 45-50 state/deemed or central agricultural Universities and 200 general universalities with 48 agricultural faculties. The total disciplines needing grants may be strengthened in 5 yrs by 1 core grants to each discipline for quality assurance. We produce 10,000 under graduate 5500 post graduates and 1600 Ph.D in agriculture every years. They add to our unemployed pool. For making them self employees in new ventures and for increased employment, they need competence building in Global economics and trade policies. Biotechnology, Bioinformation, Biofertilizers, pesticides and fungicides. New faculty development in all the Universities and colleges is to be executed in coming 5 years. Such as :

Pest information and survey; Risk Management Analysis; Decision support system; Geographic information system. A new trust is to be given to course curriculum integrating field practices in a partnership mode with farmer. A teacher-student-farmer-industry, interaction and co-operation is to be developed. A new model of mechanics in Agricultural and allied curricula is to be integrated, unified and fine tuned to end results. This will demand inter and intra faculty harmony and synchronized course curricula at UG, PG and Ph.D level. This has to be fine tuned to our field requirements and location orientations.

5. Asia Specific Agricultural Education

Indian Economy is a agri-centre economy which supports 70% of our population, as direct rural employment. Forty five percent of the income generated by industries comes from Agri-based (Agro) Industries. Therefore , a vast potential and resource is hidden in it. If we think of Asia specific Agriculture, we have to play a significant role in the region which has 60% of the world population. The region is rich of energy and oil resources and millennium buzz word is open boarders and common market with first priority on peace, confidence and trust. For up-liftmen of region we need training education and human resource utilization. Therefore, Agriculture educational reforms are on our door steps to harvest the gains of common economy in the SARC and total Asian region. These educational reforms should involve schools, colleges and Universities. The re-modeling of curriculum will need incorporation of new emerging era, like competitive global marketing, the climatic, disaster, technology use, restrictions, sustainability, environment, water resource conservation, remediation factors etc.

Agri-educational reforms are needed so that ICAR parallels USDA, in governing grant-in aid to whole agriculture sector. An omnibuss act of agriculture in the shape of USA farm bill of 1996 is envisioned. New educational policies so drafted shall be non-discriminative, comprehensive , transparent and accountable.

6. Access to Education and Training

Access to education and training to people below the poverty line, rural youth and women is to be ensured. This will need a total restructured education infrastructure. A three tyre model is envisioned which consists of:

1. On the job, training opportunities on farm mechanization and agriculture.

2. Training skills, up-gradation and rural orientation at University level, refinement and more innovative participatory mode at farmers field. Roaming teaching taught system on holidays and Sundays.
3. Teacher-student-farmer-industry-interaction-work plans-self learning by living with farmers. It will ensure quality training and job improvement of skilled manpower for use in Asian Agri development Market.

7. Informal-flexible Agriculture Information Services

A flexible curriculum models which should have many options at B.Sc level like:

1. Natural science
2. Agri Science
   - Production system
   - Agri-business
   - Social science
3. International Agriculture
4. Natural Resources
5. Agri-business management
6. Biological engineering
7. Dietetrics
8. Landscape Architecture

8. Export Orientation in Agri-education

In the present era of bio-safety, phyto and zoo sanitation have assumed tremendous importance. Education and knowledge was safe and secure treasure in the past. It is no more true. The export needs knowledge and new inventions need patenting. Web and web designing have made invisible teachers to unknown students. The students-teacher relationship has raised to spiritual horizons. New targets for future educational planning and policies need to have more information and communication technology. Therefore courses on I&C with computer applications is must. These have been strengthened in the SAU and ICAR institutions in the last few years. However a total connectivity is needed with farmers, farm organizations and utilization departments to harvest the gains of technological reforms to increase our exports. Inspite of ranking I in milk production our exports are meager. This is because of poor -zoo-sanitation and Global lobiest are critical of our disease free status. Therefore, ICAR has to formulate policies and programmes in Agri-export orientation.

References

Albelo, V.M.F; Ramachandran, V.K. and Swaminathan, M. 2002. Cuba, an experience of rural development.*Proc. Int. Conf. on Agrarian relations & Rural Development in less Developed Countries.*January 3-6,2003, Pub. Tulika Print Communication Ltd. New Delhi - India P 506-535.

Barbour, R.J, and Wong, A.H.H. 2001.Sustainable production of forest Products.*Psroc. IUFRO Division Published by Pacific Northwest Res. Station, USDA, Forest Service, Portland, USA. 2001.*

Dobrowski, Z.T.2000.The necessity of changes in the methodology of development and implementation of integrated pest management.*Pre-conference Symposium – Progress in Plant Protection, held at Nowoursynowska, Poland, 40(1): 334 -342.*

Frewer, L.J; Howard, C and Shepherd R. 1998. The influence of initial attitudes on responses to communication about genetic engineering in food production. *Agriculture & Human values.* 15(1):15-30.

Gosh,N.2002 Infrastructure cost and labour income in agriculture.*Indian Journal of Agricultural Economics. 57(2): 153-168.*

Hardman, PA; Darroch, MAG; Ortmann, GF; Trienekens, JH, and Omta, SWF,2002. Improving co-operation to make South African fresh apple export value chain more competitive.*Proc. 5th Int. Conf. on Chain and network management in agribusiness and food industry. Noordwifk, Netherlands 6-8 June, 2002 PP 434-443.*

Sutton, A.L; Ong. H.K; Zulkifli, I; Tee, TP and Liang, JB. 2002.The role of Education and Technology Transfer in Livestock Waste Management. *Proc. 4th Int. Livestock Waste Management symposium 19-23 May,2002, PP 281-288. Published Malaysian Society of Animal Production, Selangor, Malaysia.*

Jacques Diouf 1998,Director General, FOA*Report on World Food Day, 26th Oct, 1998 Rome.*

Jhoty, I; Autrey, L.S.C and Hogarth, D.M, 2001. Precision Agriculture for Sugar Cane in Mauritius.*Proc. XXIV. Congress of Int. society of sugar cane technologists – Brisbane, Australia, 17-22 Sept., 2001 Vol 1 page 47-50.*

Luttichen, R.E; Robert, P.C; Rust,R.H and Larson, W.E.2000. Development of an internet based communication and information network to progress the implementation of precision agriculture.*Proc. 5th int. conf. on Perecision Agriculture, Bloomington, Minnesota, USA, 16-19 July,2000. Publ. 2001 CD. Page 1-14.*

Report of the Quinquennial, Review team for KVK Zone V, AP & Maharastra, September, 2001, *Edited Wani, G.M, C.M.Singh. Gogoi, DK; Sandhu and Suilac C,9th September,2001, Hyderabad.*

Rezvanfar, A.2001 Information flow – a study of dairy husbandry technology in east Azerbaijan – Iran. *Iranian Journal of Agric. Sci. 32(1): 17-19.*

Samra, J.S and Sastry, G.2002, Conservation of natural resources for sustainable productivity. *Final US1F Project report, Bulletin No T-47/D-32, Central Soil & Water conservation Research and Training Institute, Dehradun, PP 1-7.*

Sher-Muhammad; Garforth, C, and Kausar Almas, 2001. Adoption of recommended agricultural technologies by farmers as an indicator of effective communication by extension field staff.*International Journal of Agriculture and Biology. 3(1):13-15.*

Tzortzios, S; Bojar, W; Adm, G; Gitsakis, N and Juliszewski, T.2001. An integrated approach of computers utilization in the field of Agriculture in Greece and Poland., *Proceedings XXIX - CIOSTA GIGR-V. Congress, Krakow, Poland, 25-27 June, 2001. PP229-234.*

Tzortzios,S; Bojar,W; Adam,G; Gitsakis, N; and Juliszewski, T.2001. An integrated approach of computer utilization in the field of agriculture in Greece and Poland.*Proc. XXIX. CIOSTA - GIGR-V. Congress, Krakow, Poland, June 25-27, 2001, Page 229-234.*

Thankappan, M. 2001. Access to Satellite data for time critical applications; Star and Spot -Lite. First Australian Geospatial information and Agriculture conference. Sydney, Australia, 17-19 July 2001 & 497-506.

Wang, S.P; Zhang, C.X; Luo, J.H; Shao, H, GUO, Q.H and Zhu, L.N. 2002. Advances in research of rooting zone restricted fruit growing. *Journal of Fruit Science 19(5): 298-301.*

Wani, G. M.2000, World Food Day Bulletin, 16 Oct, 2000, Directorate of Extension Education, SKUASTK, Shalimar Srinagar, Kmr.

Wani, G.M.2003 Appraisal of Training needs of Farmers. *Lead paper presented at National Seminar on "Apraisal of Institutionalized Farmers' Training Activities' organized by Indian Society for Agriculture Extension at GBPU of Agriculture & Technology, Pantnagar, U.P*

Yadav, D.S. 2002 Reorientation of fertilizer promotion strategy.*Fertilizer Marketing news. 33(1):1-8.*

Yahaya, M.K and Omokhaye,S.B.2001. Cocoa farmer's perceived influence of communication channels on utilization of improved cocoa seed technologies in owan -east local Government area of Edo State. Moor - *Journal of Agricultural Research* 2(2) : 199-207.

□□□

About the Author

The author and editor of this book Prof. Dr. Ghulam Mohy-Ud-Din is gold medalist and first position in BVSC and AH, MVSC, Dr. Med vet from reputed and world class I Universities of Udaipure, Rajasthan, GBPUA & T, Pantnagar and world first Veterinary University of Hannover, Germany. His Ph.D from IVRI as in service ARS scientist and his dedicated and merituous services as first batch ARS scientist at CIRG crowned him Sr. scientist (S-2) grade through assessment and a (DAAD) German Academic Fellowship for post doctorate degree in Germany. His pioneering work on ultrasonography, laparoscopy, ovarian manipulation's, embryo culture, post-partum oestrus, post-partum involution and micro manipulation has been published in leading professional Journal of world like world review of Animal Production, Small Ruminant Research, zuchthygiene, besides leading Indian and Asian Journal of repute like Indian Journal of Animal Sciences, Indian Journal of Animal Reproduction and Indian Journal of Small Ruminants. He is honored as referee and reviewer of all these Journal on Small Ruminant Research and Biotechnology.

He has served as editor of IVRI / CIRG, Makhdoom Annual Reports 1979-85, SKUAST news editor 1992-97 and National Editor of Indian Journal of Animal Reproduction 2006-09. He has published more than 200 original research papers, research reviews, project reports, conferences papers and training manuals, besides 12 books, booklets and has contributed chapters in the world famous Encyclopedia of Dairy Sciences besides book chapters in books, written by Vetrons in India and abroad. His work on ultrasonography, laparoscopy and embryo biotechnology is translated in many languages and name has become a legend in this filed. One needs to press his pen name Wani G.M. Hundreds of bytes citing him appear on the screen of your computer on www.google.com. His bio-data, achievements and recognitions are carried by various international websites, such as www.word-mart.com; www.buzzle.com; www.articlebase.com; www.piwet.pulawy.pi; www.vfu.cz; www.aightgenossen.ch; www.medwid.nic.in; www.blackwell.synegy.com. He is recognized by Winrocks International as a registered professional, academic partner of EU and has dozens international and national awards. His work on DNA finger printing was first of its kind in 1990 and the honorarium bestowed to him by donors was donated by him as Wani-Gall, Gelderman fellowship for his students which has accumulated to Rs. 27 lacs as foundation funds for this Wani-Gall-Geldermann fellowship to be advocated by FAST-Kashmir. His research as student and guide has earned many awards and fellowship to his students, who are now leading project leaders of famous national and international laboratories. His work on ova-pic up through laparoscopy, GnRH as therapy for infertility and earey Luteal phase ovarian manipulation are basis of modern follicular dynamics. His work on foetal- uterine interactions - a priority area of research marked in USA in 2004 is a pioneering innovative work yet to be initiated in the world of science. Prof. Wani has served as ARS scientist at IVRI, CIRG, Veterinary University Hannover, Germany, Agriculture University SKUAST-K and Hohenhein, Germany as research Scholar, Professor and Director, he has been national and international consultant, chairman and expert reviewer of ICAR eminent scientists progress report, DST, ICAR projects proposals and a guest professor/ visiting scientist in various countries. He has visited Europe, Russia, and many other continents for seeking and delivering knowledge. His earlier books on pregnancy diagnosis, embryo biotechnology, have been out saled and the present book brings a new and modern outlook on these topics besides new advancement in science and Technology since 1986-96. He is presently engaged as Chief Consultant Animal Husbandry and Fishries, RFNA, Govt. of India, New Delhi, after his hon'ble retirement as Director, State Agricultural Management & Extension Training Institute, SKUAST-K, Shalimar.